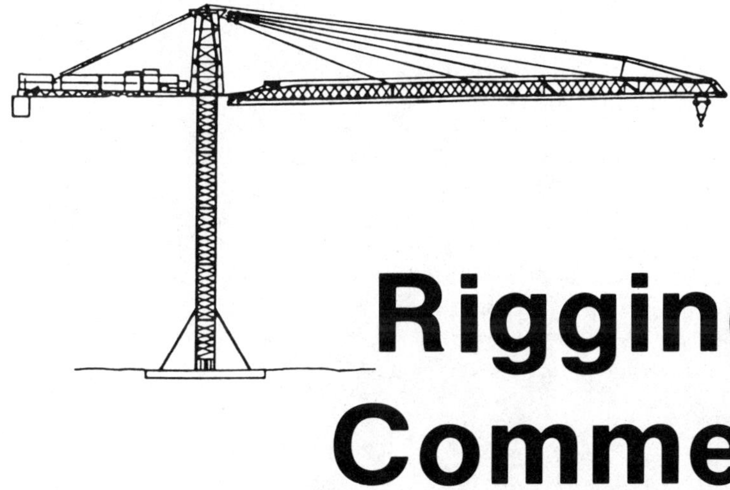

Rigging for Commercial Construction

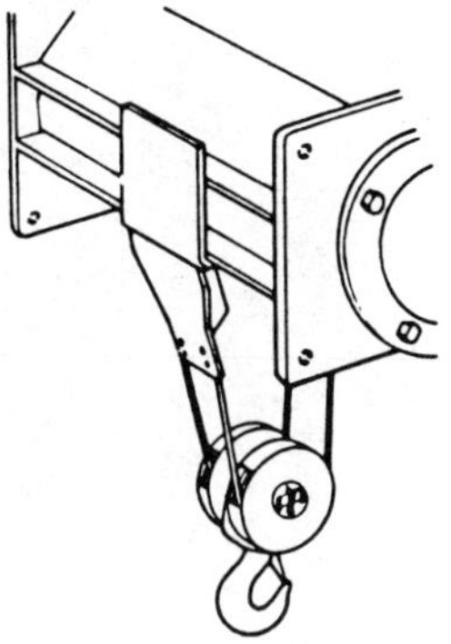

Rigging for Commercial Construction

Reston Publishing, Inc.
A Prentice-Hall Company
Reston, Virginia

ISBN 0-8359-6729-8

10 9 8 7 6 5 4 3 2 1

Printed in the United States of America.

Contents

Acknowledgments

Portions of this book were taken—with permission of the publishers—from the *Rigging Manual* published by the Construction Safety Association of Ontario, *Rigging* published by the Department of the Army, the *Wire Rope Users Manual* published by the American Iron and Steel Institute and the *Boilermaking Manual* published by Apprenticeship Training Programs Branch, Ministry of Labor (Province of British Columbia).

The following companies supplied information contained in this publication:

- American Hoist & Derrick
- Dresser Industries Inc., Crane and Hoist Operations
- Pennsylvania Power and Light Co.
- Shepard Niles Crane & Hoist Corp.
- Silent Crane & Hoist Co., Inc.
- Whiting Corp.

1
Wire Rope

INTRODUCTION

Webster defines a machine as an "assemblage of parts that transmits forces, motion, and energy one to another in some predetermined manner and to some desired end." In and of themselves then, both ropes and chains are machines. The geometry—or configuration—of their cross section and the method and material of their manufacture are precisely designed to perform "in some predetermined manner and to some desired end." Hence, as befits any useful machine, it is imperative that the potential uses of ropes and chains be fully recognized, that their functional characteristics be understood, and that the procedures for proper maintenance be scrupulously adhered to. By giving recognition to these generally accepted concerns, the rigger can be reasonably certain that the maximum service life and safety will be realized for every rope and chain installation and application.

Today, the most common rope used by the rigger is that made from wire. To use it safely and efficiently, the rigger *must* know wire rope thoroughly—how to handle it and how to splice it.

SIX REQUIREMENTS OF SELECTION

Many factors influence the selection of wire rope. The strength of the rope, although of major importance, is only one of these factors. It is advisable, therefore, to use only ropes of the correct size, grade, type, and construction as specified by either the equipment manufacturer or the rope manufacturer who must base their recommendations on actual working conditions.

There are six requirements which must be considered when selecting a wire rope:

1. The rope must possess sufficient strength to take the maximum load that may be applied, with a factor of safety of at least 5 to 1, and 10 to 1 when it is used to carry personnel. The wire ropes that are supplied as rigging on mobile cranes must possess factors of safety as follows:
   - live or running ropes that wind on drums or pass over sheaves
     = 3.5 to 1 under operating conditions
     = 3.0 to 1 when erecting the boom
   - pendants or standing ropes
     = 3.0 to 1 under operating conditions
     = 2.5 to 1 when erecting the boom.
2. The rope must withstand repeated bending without failure of the wires from fatigue.
3. The rope must resist abrasion.
4. The rope must withstand distortion and crushing.
5. The rope must resist rotation.
6. The rope must resist corrosion.

Table 1-1 illustrates how the design and construction of the rope can be varied to give the best characteristics for specific service demands.

BASIC COMPONENTS AND CONSTRUCTION

Wire rope consists of three essential components. These, while few in number, vary in both complexity and configuration so as to produce ropes for specific purposes or characteristics. Basically, these three components of a standard wire rope design are: (1) *wires* that form the strand; (2) a *core*; and (3) the *multi-wire strands* laid helically around the core (Fig. 1-1).

The *wire* for rope is made in several materials and types; these include steel, iron, stainless steel, Monel, and bronze. By far, the most widely used material is high-carbon steel. This is available in a variety of grades, each of which has properties related to the basic curve for steel rope wire. (Wire rope manufacturers select the wire type that is most appropriate for requirements of the finished product.) "Iron" type wire, for example, is actually a low-carbon steel and has fairly limited use except for older elevator installations. However, when iron is used for other than elevator application, it is most frequently galvanized.

Steel wire strengths are appropriate to the particular grade of the wire rope in which they are used. These grades of wire rope are traction steel, mild plow steel, plow steel, improved plow steel, and extra improved plow steel. (While

Table 1-1: Variations of Design and Construction

Wire Rope Service Requirements	Wire Rope Design Characteristics
Strength: The wire rope must develop sufficient strength to support the load plus the necessary factor of safety.	The strength of a wire rope depends on its size, grade of wire, and type of core.
Flexibility or Resistance to Bending Fatigue: The wire rope must have the ability to bend over small sheaves or wind onto relatively small drums without the wires breaking due to bending fatigue.	Strands containing a large number of small wires have greater resistance to bending fatigue than strands containing a few large wires. Lang Lay has greater fatigue resistance than Regular Lay. Preforming increases the wire rope's resistance to bending fatigue.
Resistance to Abrasion: The wire rope is subjected to wear or abrasion as it passes through operating sheaves under high pressure or comes in contact with stationary objects.	Large outer wires are better able to withstand abrasive wear. Lang Lay provides greater resistance to wear than Regular Lay.
Resistance to Crushing: Some wire ropes distort or flatten when they are forced to operate under heavy pressure in grooves that do not provide ample support or on drums where multiple lay winding occurs.	An Independent Wire Rope Core (IWRC) provides greater support for strands under heavy bearing pressures. The coarser wire rope constructions provide greater resistance to flattening on drums.
Resistance to Rotation: A wire rope may rotate as the load is applied. This could be undesirable for load control and might lead to rapid deterioration of the wire rope.	Special nonrotating constructions are available for specific applications. Regular Lay provides greater stability than Lang Lay, and wire ropes with an IWRC twist less than those with fiber cores.
Resistance to Corrosion: Wire rope may corrode if in contact with corrosive elements or it may rust when exposed to atmospheric conditions over a long period of time.	Galvanized or stainless steel wire offers excellent protection against corrosion. Special lubricants can also inhibit the development of rust.

steel grade names originated at the earliest stages of wire rope development, they have been retained and serve as indicators of the strength of a particular size and grade of rope.) The strength of plow steel forms the basis for calculating the strength of all steel rope wires, and the tensile strength of any grade is not constant, but varies with the diameter—being highest for the smallest wires.

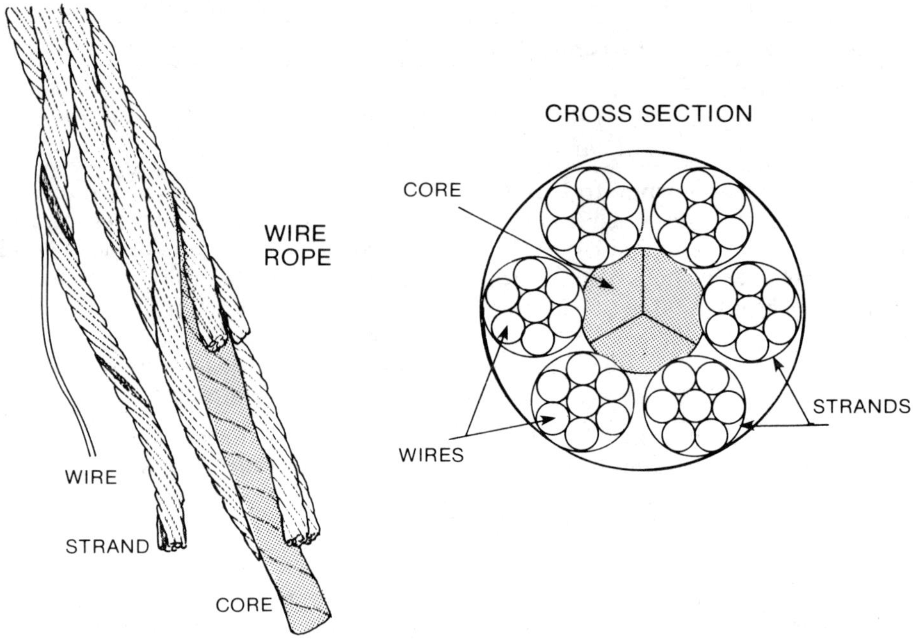

Fig. 1-1: The three basic components of a typical wire rope.

The most common finish for steel wire is "bright" or uncoated. Steel wires may also be galvanized (zinc coated). "Drawn galvanized" wire has the same strength as bright wire, but wire "galvanized at finished size" is usually 10% lower in strength. In some special applications, tinned wire is used, but it should be noted that tin provides no sacrificial (cathodic) protection for the steel as does zinc. Stainless steel, Monel metal, and bronze wire ropes are highly resistant to corrosion and should be used where corrosive situations exist.

The *core* is the intrinsic foundation of wire rope and is made of materials that will provide proper support for the strands under normal bending and loading conditions. Core materials include fibers (hard vegetable or synthetic) or steel. The steel core consists either of stranded wires or of another independent wire rope. The three most commonly used core designations are: *fiber core* (FC), *independent wire rope core* (IWRC), and *wire strand core* (WSC) (Fig. 1-2). Catalog descriptions of the various available ropes include these abbreviations to identify the type of core.

The *strands* are made up of two or more wires, laid in one of many specific geometric arrangements or in a combination of steel wires with some other materials such as natural or synthetic fibers. Although it is conceivable that a strand can be made up of any number of wires, or that a rope can have any number of strands, in the United States the majority of wire ropes are designed with six strands. Major U.S. strand classifications are 7-, 19-, 37-, 61-, 91-, and 127-wire.

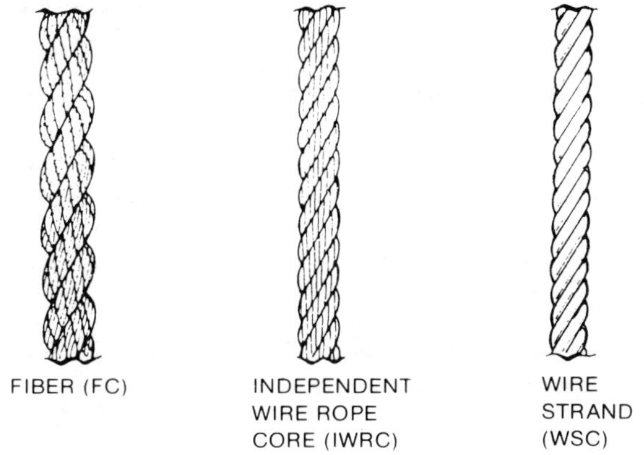

Fig. 1-2: The three basic wire rope cores. In selecting the most appropriate core for a given application, a qualified manufacturer should be called upon for guidance. Fiber cores, for example, are not recommended for applications involving elevated temperatures or high peak loads.

Wire Rope Identification and Construction

Wire rope is identified not only by its component parts, but also by its construction, i.e., by the way the wires have been laid to form strands and by the way the strands have been laid around the core.

Figures 1-3A and C show strands as normally laid into the rope to the right—in a fashion similar to the threading in a right-hand bolt. Conversely, the "left lay" rope strands (Figs. 1-3B and D) are laid in the opposite direction.

The first two drawings (Figs. 1-3A and B) show *regular lay* ropes. Following these are the types known as *lang lay* ropes. Note that the wires in regular lay ropes appear to line up with the axis of the rope; in lang lay rope the wires form an angle with the axis of the rope. This difference in appearance is a result of variations in manufacturing techniques. Regular lay ropes are made so that the direction of the wire lay in the strand is opposite to the direction of the strand lay in the rope. Lang lay ropes (Figs. 1-3C and D) are made with both strand lay and rope lay in the same direction. Finally, Fig. 1-3E shows the type called *alternate lay,* consisting of alternating regular and lang lay strands.

Of all wire rope types in current use, *right regular lay* is found in the widest range of applications. Currently, *left lay* rope is used less frequently. As for *alternate lay* ropes, these are used for special applications.

Circumstances that favor the use of lang lay ropes derive from two unique advantages over regular lay ropes. Lang lay ropes: (1) are more resistant to bending fatigue, and (2) have a greater wearing surface per wire across the crown of the strand. The total wearing surface area of the rope is, for practical purposes, the same for both regular and lang lay ropes—with the same geometric construction and depth of wear. The eventual wear on the equipment

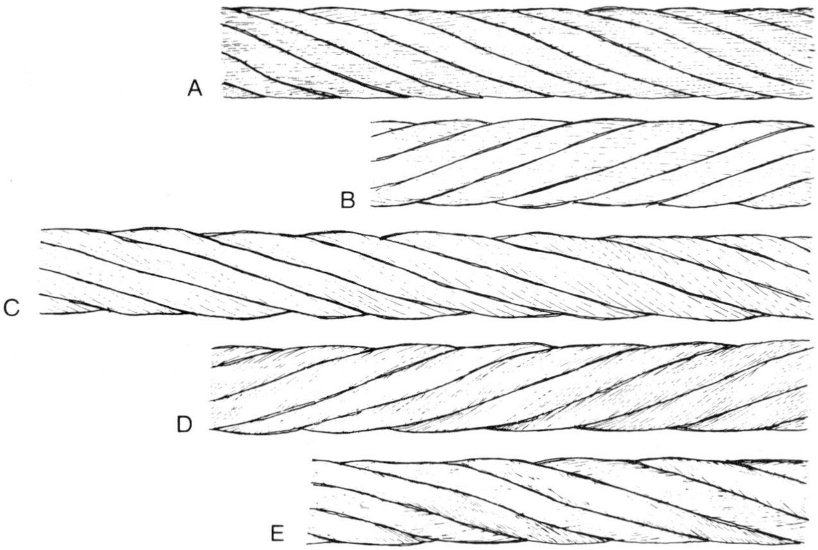

Fig. 1-3: A comparison of typical wire rope lays: (A) Right regular lay; (B) left regular lay; (C) right lang lay; (D) left lang lay; and (E) right alternate lay.

and the service life of the rope favors lang lay construction on applications where fatigue or abrasion are controlling factors. To illustrate this point, Fig. 1-4 compares a regular lay with a lang lay rope, each of which has been worn to the same amount of reduction in their respective diameters. Hence, it is not the total of the rope's worn surface area that governs the life span of rope and equipment. It is, rather, the inherent characteristics of properly used lang lay ropes that gives them a significant advantage in resistance to both abrasion and fatigue.

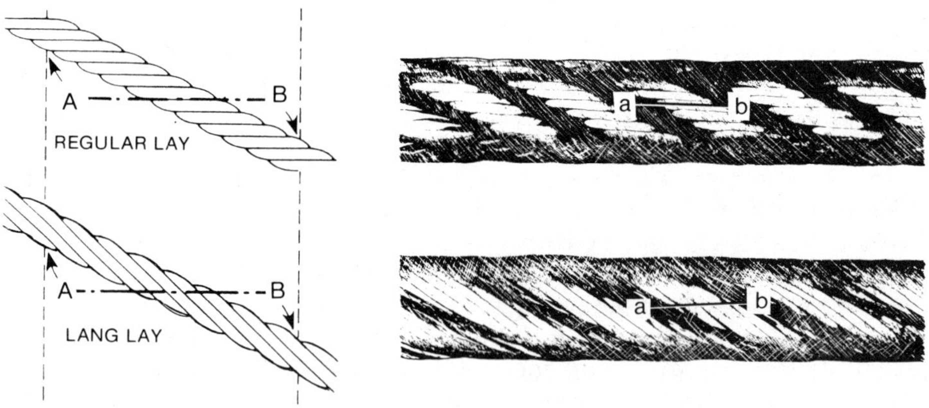

Fig. 1-4: A comparison of wear characteristics between lang lay and regular lay ropes. The line a-b indicates the rope axis.

However, lang lay ropes have some disadvantages. They are more susceptible to damage resulting from handling abuses, bending over extremely small sheaves, pinching in undersize sheave grooves, crushing when improperly wound on drums, and they are subject to excessive rotation. In fact, this latter tendency for the rope and the strands to unwind in the same direction requires that lang lay ropes should be secured at both ends to prevent unlaying or spinning out.

In conjunction with wire rope, the term "lay" can be used in another meaning; that is, the *lay length* or *length of lay* is the length of rope required for one strand to make one complete revolution around the rope. Figure 1-5 illustrates the method of measuring the length of the lay.

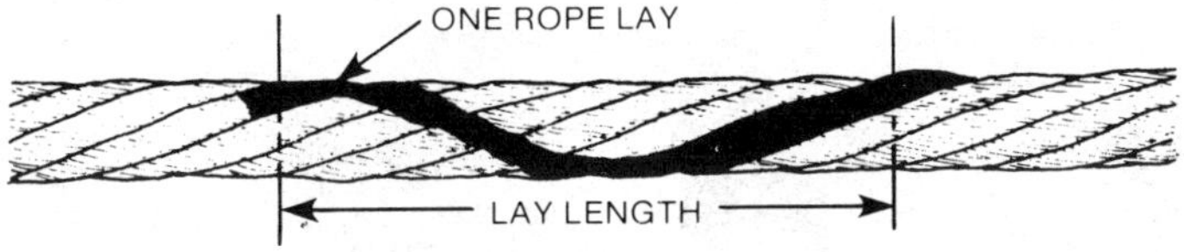

Fig. 1-5: Method of determining the lay length of a six-strand wire rope.

Wire Rope Construction

Preforming is a wire rope manufacturing process wherein the strands and their wires are shaped, during fabrication, to the spiral form that they will ultimately assume in the finished rope or strand. As previously noted, wire rope strands are made up of a number of wires. The wire arrangement in the strands will determine the rope's functional characteristics, i.e., its capacity to meet the operating conditions to which it will be subjected. There are many basic design constructions around which standard wire ropes are built; some of these are shown in Fig. 1-6.

Round strand ropes are grouped into classifications according to the number of wires per strand. As the number increases, the rope develops greater ability to resist bending fatigue—at the expense of resistance to abrasion.

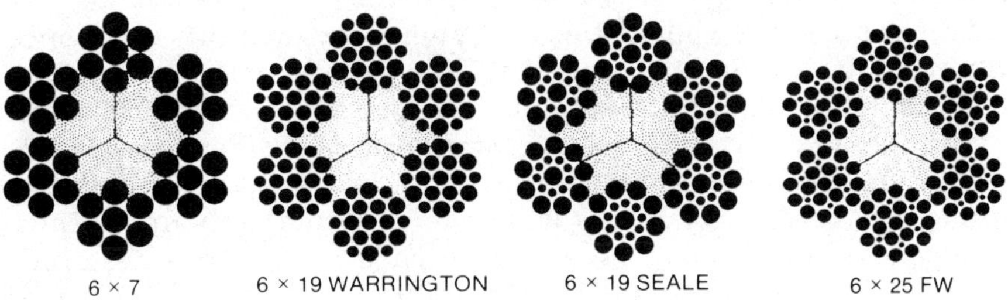

Fig. 1-6: Basic constructions around which standard wire ropes are built.

Within the round strand classification, the wires may be arranged in certain geometric patterns, called "constructions." The number and pattern of the wires defines the rope's characteristics.

The wires in the strands may be all the same size or a mixture of sizes. There are four common arrangements (Fig. 1-7):

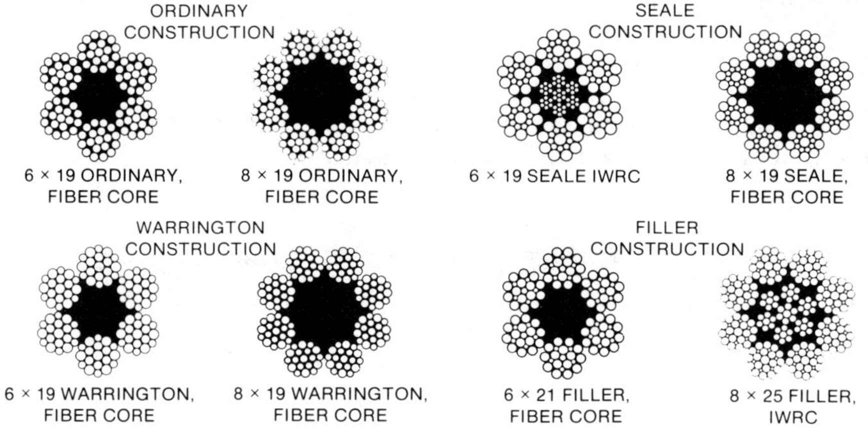

Fig. 1-7: Wire rope constructions.

1. *Ordinary,* where the wires are all the same size.
2. *Seale,* where larger diameter wires are used on the outside of the strand to resist abrasion and smaller wires are used inside to provide flexibility.
3. *Warrington,* where alternate wires are large and small to combine great flexibility with resistance to abrasion.
4. *Filler,* where very small wires fill in the valleys between the outer and inner rows of wires to provide good abrasion and fatigue resistance.

As the number of wires per strand increases, combinations of Filler, Seale, and Warrington patterns are used to provide certain characteristics. This combination of rope classifications and strand constructions produces varying degrees of resistance to bending fatigue and abrasion. The effect of varying constructions can be best shown by using, as examples, the most commonly used rigging rope, the 19 wire strand (Fig. 1-8).

Wire ropes are usually identified by a nomenclature that is referenced to: (1) the number of strands in the rope; (2) the number (nominal or exact) and arrangement of wires in each strand; and (3) a descriptive word or letter indicating the type of construction, i.e., geometric arrangement of wires. Wire rope nomenclature also defines length, size (i.e., diam.), type, direction of lay, grade of rope, type of core, and whether it is preformed (p/f) or nonpreformed (np/f). If the direction and type of lay are omitted from the rope description, it is presumed to be a right regular lay. In addition, if no mention is made as to preforming, this will be presumed as a requirement for preforming. On the other hand, an order for a rigging rope requires an explicit statement since p/f

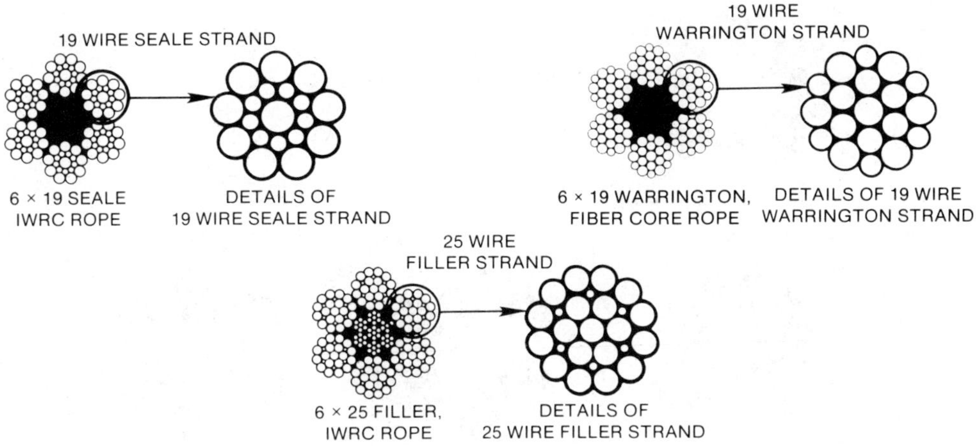

Fig. 1-8: Basic 19 wire strand construction.

and np/f ropes are used extensively. An example of a complete description would appear thus:

600 ft 3/4″ 6×25 FW Left lang lay
Improved plow IWRC

The wire rope described above would be made *preformed* (p/f).

The most widely used classifications of wire rope are listed and described in Table 1-2. When a center wire is replaced by a strand, it is considered as a single wire, and the rope classification remains unchanged.

Cross sections of wire rope shown in Fig. 1-9 are among the most commonly used in rigging applications, and they are arranged in their respective classification groups. There are ropes available for specific functions (Fig. 1-10). For example, where rotation is a problem, such as when a two-part line keeps winding up, it is a good idea to employ a rotation resistant rope. In the design of such rope, the lay of alternate strand layers is in opposite directions. The outer strands form a right lay and the inner strands a left lay. Thus, the tendency of one layer to twist in one direction is counteracted by the tendency of the layer to twist in the opposite direction. Rotation resistant wire ropes require special care in use and storage since they are easily damaged. Core slippage is the most common damage that must be guarded against.

Grades

To meet the demand for varying degrees of strength, toughness, abrasion resistance, and corrosion resistance, wire ropes are manufactured in the following grades. The most common for rigging applications are:

1. Grade 120/130 Special Improved Plow Type II—Used for special installations where maximum rope strength is required.

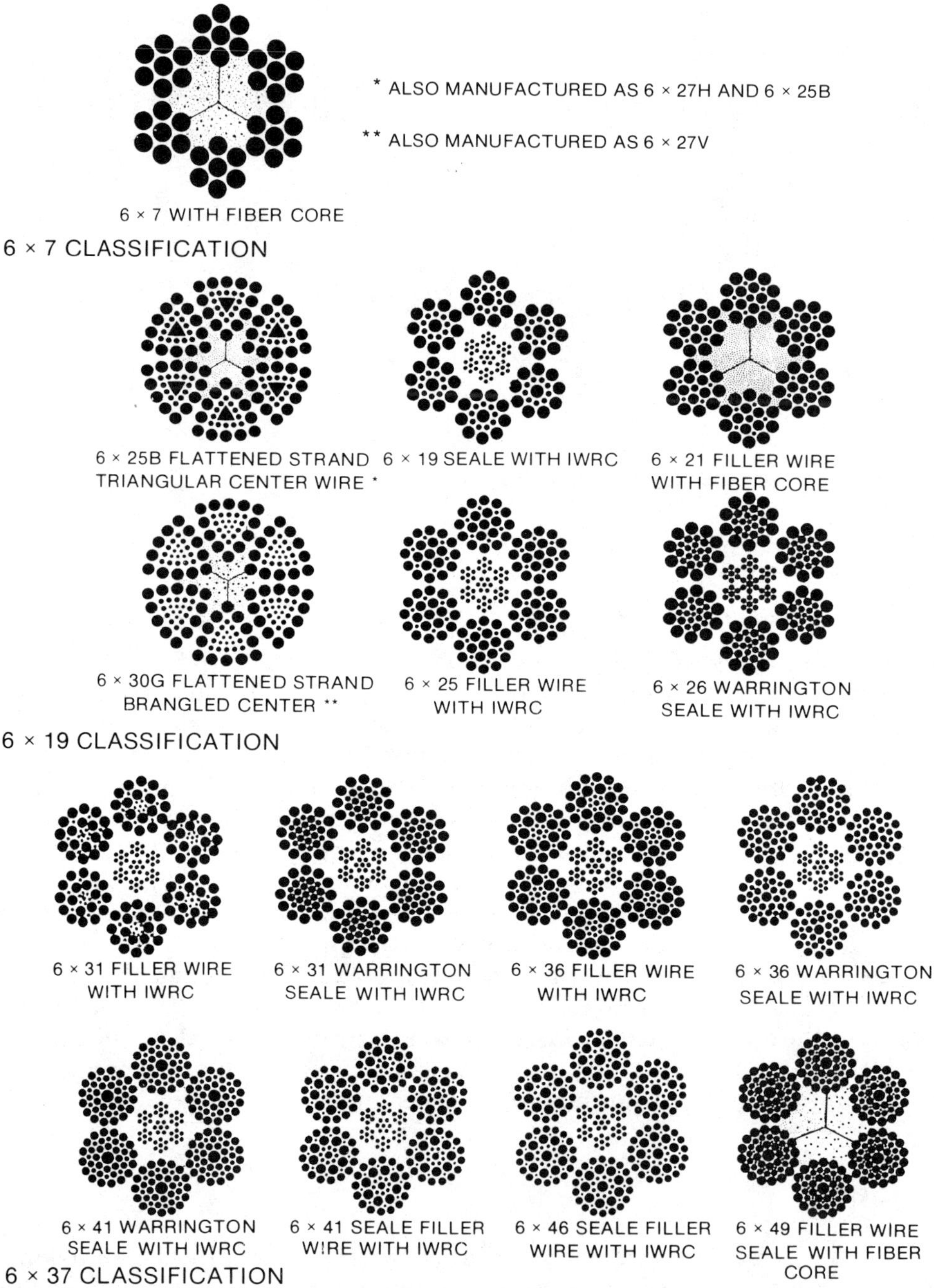

Fig. 1-9: Cross sections of some commonly used wire rope constructions.

Table 1-2: Wire Rope Classifications
(Based on the Nominal Number of Wires in Each Strand)

Classification	Description
6 × 7	Containing 6 strands that are made up of 3 through 14 wires, of which no more than 9 are outside wires.
6 × 19	Containing 6 strands that are made up of 15 through 26 wires, of which no more than 12 are outside wires.
6 × 37	Containing 6 strands that are made up of 27 through 49 wires, of which no more than 18 are outside wires.
6 × 61	Containing 6 strands that are made up of 50 through 74 wires, of which no more than 24 are outside wires.
6 × 91	Containing 6 strands that are made up of 75 through 109 wires, of which no more than 30 are outside wires.
6 × 127	Containing 6 strands that are made up of 110 or more wires, of which no more than 36 are outside wires.
8 × 19	Containing 8 strands that are made up of 15 through 26 wires, of which no more than 12 are outside wires.
19 × 7 and 18 × 7	Containing 19 strands, each strand is made up of 7 wires. It is manufactured by covering an inner rope of 7 × 7 left lang lay construction with 12 strands in right regular lay. (The rotation-resistant property that characterizes this highly specialized construction is a result of the counter torques developed by the two layers.) When the steel wire core strand is replaced by a fiber core, the description becomes 18 × 7.

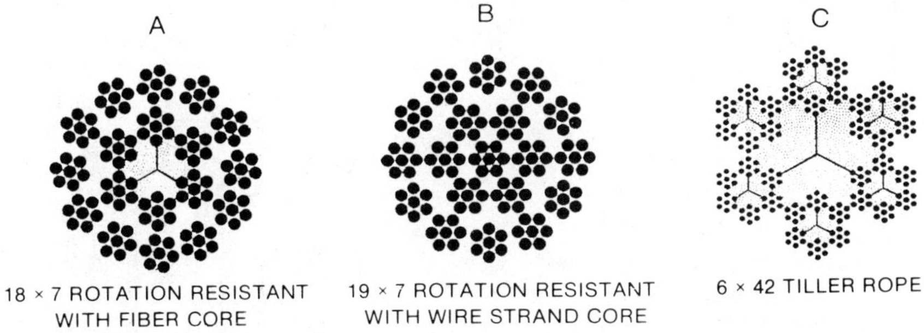

Fig. 1-10: Special function ropes: (A and B) The two rotation resistant constructions are identical except for the cores—one of which is wire strand and the other fiber. The wire strand core increases the number of strands from 18 to 19. The tiller rope construction in C is good for situations where a severe bending condition exists.

2. Grade 115/125 Special Improved Plow Type I—Used for special applications where breaking strengths somewhat higher than obtained with Grade 110/120 are desired and where other conditions such as sheave and drum diameters are favorable to its use.
3. Grade 110/120 Improved Plow—Has a good combination of tensile strength, wearing qualities, and excellent fatigue resistant properties. These steels have high strength, are tough, ductile, and highly wear resistant and are drawn to 110/120 long tons per square inch or 246,000 to 268,000 psi.
4. Grade 100/110 Plow—Has lower tensile strength and resistance to wear than Grade 110/120 but retains high fatigue resistant properties. Can be used when strength is secondary to wear resistance.

Characteristics of Wire Rope

There are several wire rope "characteristics" that should be understood. They are as follows:

Size. Whatever its grade, wire rope is usually measured by its diameter. The correct and incorrect methods of measuring the diameter of wire rope are shown in Fig. 1-11. To measure wire rope correctly, place it in the caliper so that the outermost points of the strands will be touching the jaws of the caliper. Because of friction and tension, the diameter of used wire rope will be 1/64″ to 1/8″ (0.4 to 3.18 mm) less than when it was new.

It is important to check the diameter of the rope *before* installation. This is to make certain that the rope diameter meets the specified requirements for the given machine or equipment. With an *undersize* diameter rope, stresses will be higher than designed for and the probability of breaking the rope will be increased; an *oversize* diameter rope will wear out prematurely. This happens because of abuse to the rope caused by pinching in the grooves of the sheave and drum.

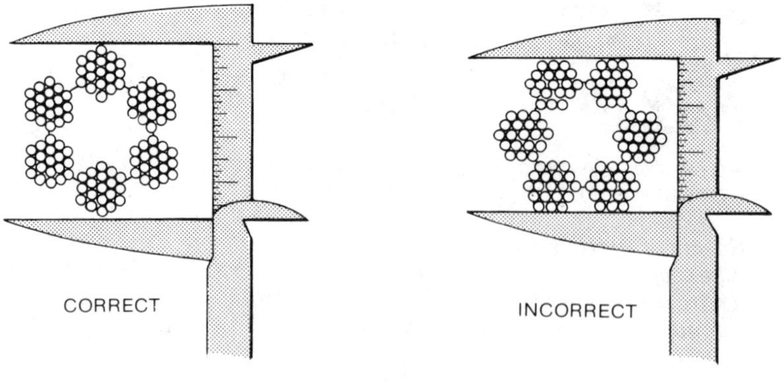

Fig. 1-11: Measuring wire rope.

Weight. No rule of thumb can be given for determining the weight of any particular size or type of rope construction. However, approximate weights for certain sizes are given in Table 1-3. The use of this table presupposes the knowledge of the type of steel that is used in the wire's production.

Breaking Strength. The strength of wire rope is determined by its method of fabrication, its size, its grade, and the ingredients of the individual wires. Whether or not the wires are made of traction steel, mild plow steel, improved plow steel, and extra improved plow steel will have a substantial bearing on the strength of the finished product. "Breaking strength" is the ultimate or maximum strength of a wire rope. Table 1-3 also gives the breaking strength of 6 × 19 wire ropes of various grades of steel.

Table 1-3: Breaking Strength of 6 × 19 Wire Rope *
(Divide by factor of safety in Table 1-4 to obtain safe load)

Diameter, inches	Mild plow steel, pounds	Plow steel, pounds	Improved plow steel, pounds	Extra improved plow steel, pounds	Weight per foot, pounds
1/4	4,140	4,780	5,480	6,040	0.10
5/16	6,440	7,420	8,520	9,380	0.16
3/8	9,240	10,620	12,200	13,420	0.23
7/16	12,500	14,380	16,540	18,180	0.31
1/2	16,260	18,700	21,400	23,600	0.40
9/16	20,400	23,600	27,000	29,800	0.51
5/8	25,200	29,000	33,400	36,600	0.63
3/4	36,000	41,400	47,600	52,400	0.90
7/8	48,600	56,000	64,400	70,800	1.20
1	63,200	72,800	83,600	92,000	1.60
1-1/8	79,600	91,400	105,200	115,800	2.00
1-1/4	97,600	112,400	129,200	142,000	2.50

**Note:* For ropes of 6 × 7 type, use 96.5% of the tabulated strengths.
For ropes of 6 × 37 class, use 95% of the tabulated strengths.
For ropes of 18 × 7 type, use 91.5% of the tabulated strengths.
For ropes with independent wire-rope center, use 107% of the tabulated strengths and 110% of the tabulated weights.

Safety Factors. In order to ensure against failure of wire rope in service with possible disastrous consequences, the actual strain on the rope should be only a fraction of the breaking load. The breaking strength divided by the actual total stress on the rope is known as the "factor of safety" and should not be less than the values given in Table 1-4.

Keep in mind that useful formulas for determining the safe working loads of several grades of wire rope have constants not to be confused with safety factors. For example, the basic formula for the safe working load (SWL), in short tons of 2,000 pounds each, of extra improved plow steel wire rope is diameter (D) squared times 10.

Table 1-4: Minimum Safety Factors for Wire Rope

Slings	8 to 10
Use of rotation resistance rope	8
Overhead electric hoists (small)	7
Overhead traveling cranes (small)	6-1/2
Industrial truck cranes	6
Locomotive cranes	6
Derricks	6
Grab buckets	6
Overhead traveling cranes (large)	5-1/2
Hoisting tackle	5
Shovels	5
Derrick guys	5
Stack guys	4
Tow ropes	4
Ferry ropes	4
Cableway track ropes	4
Contractor's suspension bridges	3

Let us find the SWL of 1″ 6 × 19, extra improved plow steel wire rope.

$$\text{SWL} = D^2 \times 10 = \text{short tons (STON)}$$
$$1 \times 1 \times 10 = 10 \text{ STON}$$

Now we come to a figure relatively constant in rigging operations, especially for new wire rope, and that is the safety factor (SF). It is 5. We use it with the SWL to find the breaking strength or strain (BS).

$$\text{BS} = \text{SWL} \times 5$$
$$\text{BS} = 10 \times 5 = 50 \text{ STON}$$

To show that these useful formulas err on the side of safety, the manufacturer listed this rope's BS as 51.7 STON.

What did the 10 tell us? Simply that we were working with extra improved plow steel wire rope. Of course, we use the safe working load given by the manufacturer when we know it. Otherwise, we use the formula. Because improved plow steel is used widely by the military and because plow steel is used to a lesser extent, we will likely be working with one of these two grades. When we do not know which of these two grades we are working with, we will use the formula for plow steel. Of course, if we happen to work with mild plow steel, we will use the formula for it. That is why it is included.

The formulas for improved plow steel, plow steel, and mild plow steel, 6 × 19 wire rope, are as follows:

Improved plow steel	Plow steel	Mild plow steel
SWL = $D^2 \times 8$ = STON	SWL = $D^2 \times 7$ = STON	SWL = $D^2 \times 6$ = STON
BS = SWL × SF = STON	BS = SWL × SF = STON	BS = SWL × DF = STON

The wire rope breaking strength given here is determined in a standard test wherein fittings are attached to the ends of the rope and the rope is pulled in a straight line. If, however, the rope passes over a curved surface (such as a sheave or pin), its strength is decreased. The amount of such reduction will depend on the severity of the bend as expressed by the D/d ratio, in which D is the tread diameter of the sheave or pin and d is the nominal diameter of the rope. For example, a rope bent around a pin of its own diameter will have only 50% of the strength attributed to it in the standard test. This is called "50% efficiency" (Fig. 1-12). Even at D/d ratios of 40, there may be a loss of up to 5%. At smaller D/d ratios, the loss in strength increases quite rapidly. For this reason, *never* bend a wire rope over a pin of its own diameter (or anywhere close to it), or it will be severely damaged, preventing the rope's further use.

The angle of bend need not be 180°, 90°, or even 45°; relatively small bends can cause considerable loss. All discussion of strength presupposes a gradually applied load not in excess of 1″/minute.

Resistance to Bending Fatigue. To describe this, a close analogy can be made with a paper clip. If it is repeatedly bent back and forth at one point, it will eventually break. The reason for this is a phenomenon called "metal fatigue." To some degree, the same thing happens when a wire rope is bent

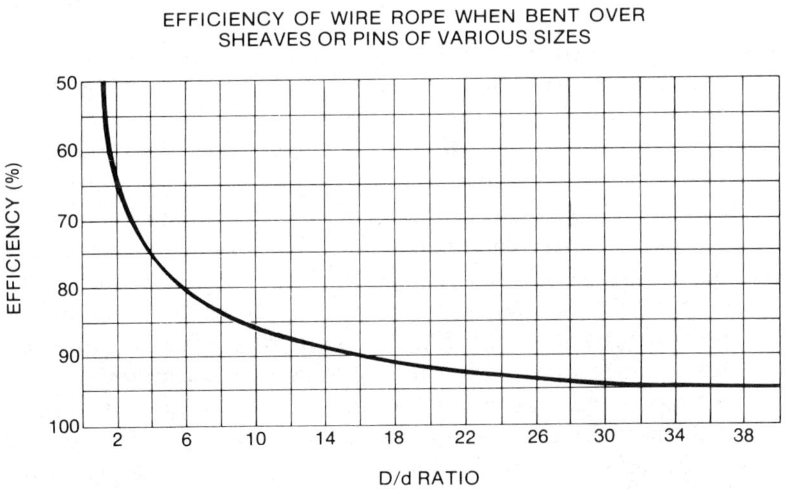

Fig. 1-12: Derived from standard test data, this curve relates rope strength efficiency to various D/d ratios. The curve is based on static loads only and applies to 6 × 19 and 6 × 17 class ropes.

around sheaves, drums, and rollers. The sharper—or more acute—the bend, the quicker the fatigue factor does its work. Accelerating the rate of travel also speeds up fatigue; close-coupled *reverse bending* will speed it up at an even greater rate.

But fatigue can be greatly reduced if sheaves and drums have, at the very least, the recommended minimum diameter. In most construction work, the ratios of the sheave or drum to rope diameter for cranes and derricks usually require the winding drum barrel diameter be not less than 18 rope diameters. While the ratio for the upper block is also 18 minimum, the lower block ratio may not be less than 16.

Ratios of a wire rope block or sheave to rope diameter are given using the pitch diameter, which is the diameter to the center of the rope on the sheave; in other words, the tread plus rope diameter. These ratios are often referred to as D/d ratios, in which D is the pitch tread diameter of the sheave and d equals the nominal diameter of the wire rope. Heavy-duty wire rope construction blocks normally have a D/d ratio of about 18 to 20, while snatch blocks have a slightly smaller D/d ratio. For example, to find the proper size heavy-duty construction block to use with 1″ diameter wire rope, a sheave with an 18 to 20″ pitch tread diameter will be needed.

When considering the size of a sheave, it is important to note that as its size increases so does its weight; this is also true of the weight and size of its mounting. This dead load increase reflects a direct reduction in lifting capacity; conversely, a large crane or derrick structure will be needed to attain the previous capacity. For this reason, crane designers and rigging engineers are motivated to use the smallest D/d ratios permitted by code unless particular considerations dictate otherwise.

When considering the bending fatigue of wire rope, keep in mind this general overall rule: The greater the number of wires in each strand, the greater the resistance of the rope to bending fatigue.

Table 1-5 shows the effect of rope construction and D/d ratios on service. Figure 1-13 gives the service life curve for various D/d ratios.

Resistance to Abrasion. Abrasion is one of the most common destructive conditions to which wire rope is exposed. It will occur whenever a rope either rubs against or is dragged through any soil or other material. It happens whenever a rope passes around a sheave or drum. And, it takes place within the rope itself whenever it is loaded or bent. Abrasive action weakens the rope simply by removing metal from both inside and outside wires.

When excessive wear is encountered in an operation, the problem usually stems from faulty sheave alignment, incorrect groove diameters, an inappropriate fleet angle, or improper drum winding. There may, however, be other causes. If, on investigation, none of these common conditions are found to be causative factors, the solution may lie in changing to a more suitable rope construction. In making such a change, it is helpful to remember that *larger outer wires and lang lay ropes are more abrasion resistant than regular lay ropes.*

Table 1-5: Relative Bending Life Factors

Rope Construction	Factor	Rope Construction	Factor
6 × 7	.57	6 × 31 WS	1.09
18 × 7	.67	6 × 36 WS	1.31
6 × 19 S	.80	8 × 25 FW	1.39
6 × 30 Style G	.80	6 × 41 SFW	1.39
6 × 25 Style B	.80	6 × 43 FWS	1.54
6 × 21 FW	.92	6 × 49 SWS	1.54
6 × 25 FW	1.00	6 × 42 Tiller	2.00

If a change in construction is being considered as a means of obtaining longer service on a rope influenced principally by bending stresses, the table of factors may be useful. For example: A change from a 6 × 25 FW with a factor of 1.00 to a 6 × 36 WS with a factor of 1.31 would mean the service life could be expected to increase 1.31 times or 31%. It must be pointed out, however, that these factors apply only for bending stresses. Other factors which may contribute to rope deterioration have not been considered.

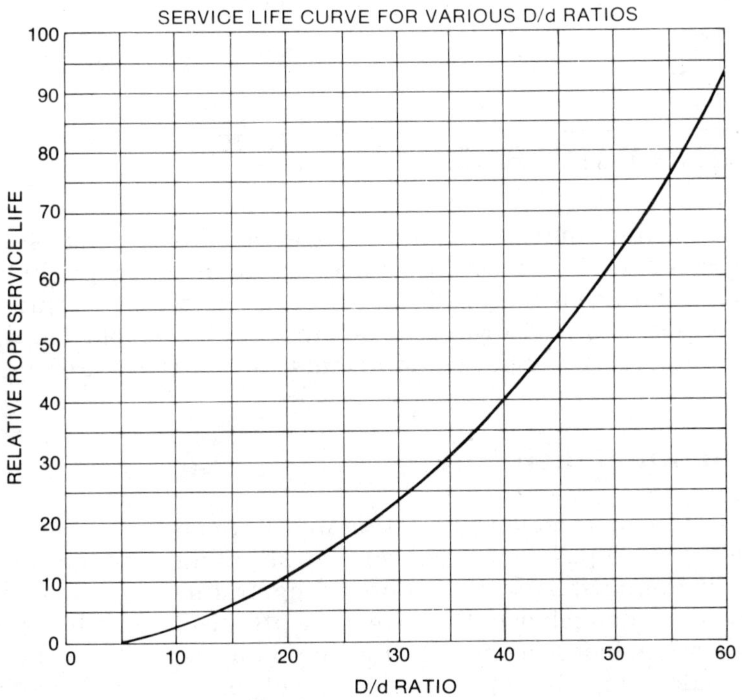

Fig. 1-13: This service life curve only takes into account bending and tensile stresses. Its applicability can be illustrated by the following example: A rope working with a D/d ratio of 26 has a relative service life of 17. If the same rope works over a sheave that increases its D/d ratio to 35, the relative service life increases to 32. In short, this rope, used on a larger sheave, increases its service life from 17 to 32—or 88%.

Resistance to Crushing. Wire rope can be crushed: (1) by its own pressure against a sheave; (2) from improperly sized grooves; and (3) from overwinding on a drum.

The pressure of rope against a sheave is determined by the sheave diameter and the load. The pressure of rope to a drum is influenced in great measure by the support of the groove; smooth drums have a more adverse effect than those that are grooved.

Overwinding is also a cause of wear even when the winding is done in an orderly (thread-winding) manner. Irregular or scramble winding is an even greater cause of damage.

Obviously, in each of these cases, reducing the load will ease the condition. If, however, this is not feasible, offending sheaves should be replaced with sheaves that have larger tread diameters. Unsuitable drums and/or winding conditions should be corrected. Otherwise, the rope will have to be replaced by one with a construction better designed to resist the abuse.

If the original rope has a fiber core, the replacement should have a steel core because a steel core rope will provide greater physical support. And here it is well to remember that regular lay ropes are better able to resist crushing than lang lay ropes.

HANDLING WIRE ROPE

The rigger must know much about handling wire rope and must use these skills properly in his/her work. These include the principles of unreeling or uncoiling, seizing, welding, cutting, coiling, avoiding kinks, determining the minimum size of sheaves and drums, locating drums and blocks, winding on drums, cleaning, lubricating, reversing ends, and storing wire rope.

Unreeling and Uncoiling

Wire rope is shipped in cut lengths, either in coils or on reels. Great care should be taken when the rope is removed from the shipping package since it can be permanently damaged by improper unreeling or uncoiling. Looping the rope over the head of the reel or pulling the rope off a coil while it is lying on the ground will create loops in the line. Pulling on a loop will, at the very least, produce imbalance in the rope and may result in open or closed kinks (Fig. 1-14). Once a rope is kinked, the damage is permanent. To correct this condition, the kink must be cut out and the shortened pieces used for some other purpose.

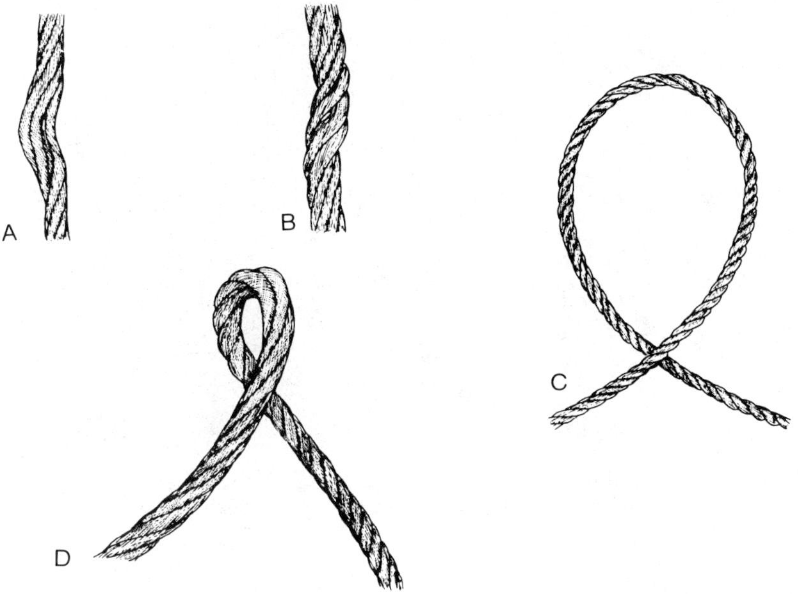

Fig 1-14: Improper handling will help create (A) open or (B) closed kinks. The open kink will open the rope lay; the closed kink will close it. (C) The starting loop. Do not allow the rope to form a small loop. If, however, a loop forms and is removed at the point shown, a kink will be avoided. (D) The kink. Here the looped rope has been put under tension, the kink has formed, the rope is permanently damaged and is of little value.

Unwinding wire rope from its reel also requires careful and proper procedure. There are three methods to perform this step correctly:

1. The reel is mounted on a shaft supported by two jacks or a roller payoff (Fig. 1-15). Since the reel is free to rotate, the rope is pulled from the reel by a worker holding the rope end and walking away from the reel as it unwinds. A braking device should be employed so that the rope is kept taut and the reel is restrained from overrunning the rope. This is necessary particularly with powered dereeling equipment.
2. Another method involves mounting the reel on an unreeling stand (Fig. 1-16). It is then unwound in the same manner as described above (1). In this case, however, greater care must be exercised to keep the rope under tension sufficient to prevent the accumulation of slack—a condition that will cause the rope to drop below the lower reel head.
3. In another accepted method, the end of the rope is held while the reel itself is rolled along the ground. With this procedure the rope will pay off properly; however, the end being held will travel in the direction the reel is being rolled. As the difference between the diameter of the reel head and the diameter of the wound rope increases, the speed of travel will increase.

Fig. 1-15: The wire rope is mounted on a shaft supported by jacks. This permits the reel to rotate freely, and the rope can be unwound either manually or by a powered mechanism.

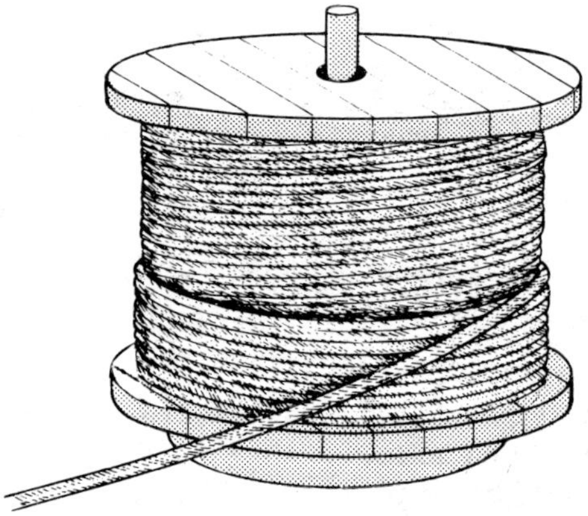

Fig. 1-16: A vertical unreeling stand.

When rereeling wire rope from a horizontally supported reel to a drum, it is preferable for the rope to travel from the top of the reel to the top of the drum; or, from the bottom of the reel to the bottom of the drum (Fig. 1-17). Rereeling in this manner will avoid putting a reverse bend into the rope as it is being installed. If a rope is installed so that a reverse bend is induced, it may cause the rope to become livelier and, consequently, harder to handle.

When unwinding wire rope from a coil, there are two suggested methods for carrying out this procedure in a proper manner:

1. One method involves placing the coil on a vertical unreeling stand. The stand consists of a base with a fixed vertical shaft. On this shaft there is a "swift," consisting of a plate with inclined pins positioned so that the coil may

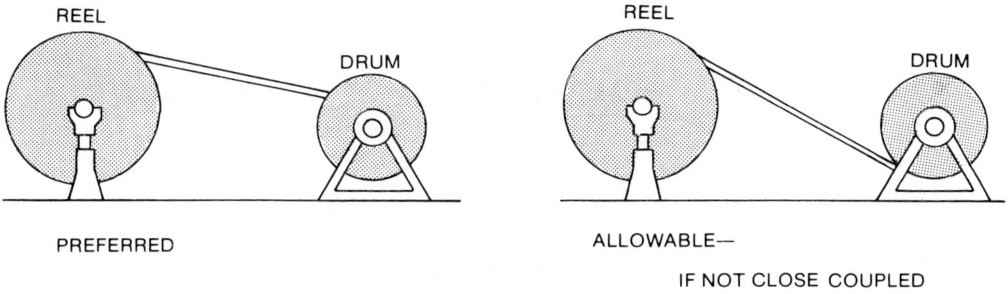

Fig. 1-17: Winding wire rope from reel to drum.

be placed over them. The whole swift and coil then rotate as the rope is pulled off. This method is particularly effective when the rope is to be wound on a drum.

2. The most common as well as the easiest uncoiling method is merely to hold one end of the rope while rolling the coil along the ground like a hoop (Fig. 1-18).

Seizing Wire Rope

Before cutting wire rope, it is necessary to apply seizing on both sides of the place where the cut is to be made. In a wire rope, carelessly or inadequately seized ends may become distorted and flattened, and the strands may loosen. Subsequently, when the rope is put to work, there may be an uneven distribution of loads to the strands—a condition that will significantly shorten the life of the rope.

Fig. 1-18: Uncoiling wire rope.

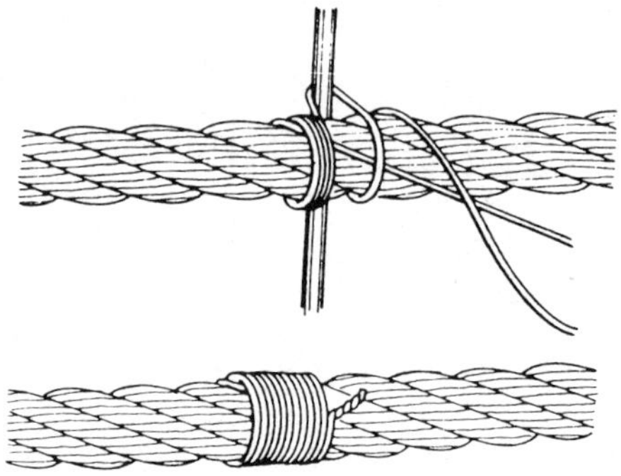

Fig. 1-19: One method of seizing wire rope.

There are two widely accepted methods of applying seizing:

Method A (Fig. 1-19). Lay one end of the seizing wire in the groove between two strands; wrap the other end *tightly* in a close helix over a position of the groove using a seizing iron (a round bar 1/2″ to 5/8″ diameter × 18″ long). Both ends of the seizing wire should be twisted together tightly. Seizing widths should not be less than the rope diameter.

Method B. After taking the required number of turns (Fig. 1-20, step 1),

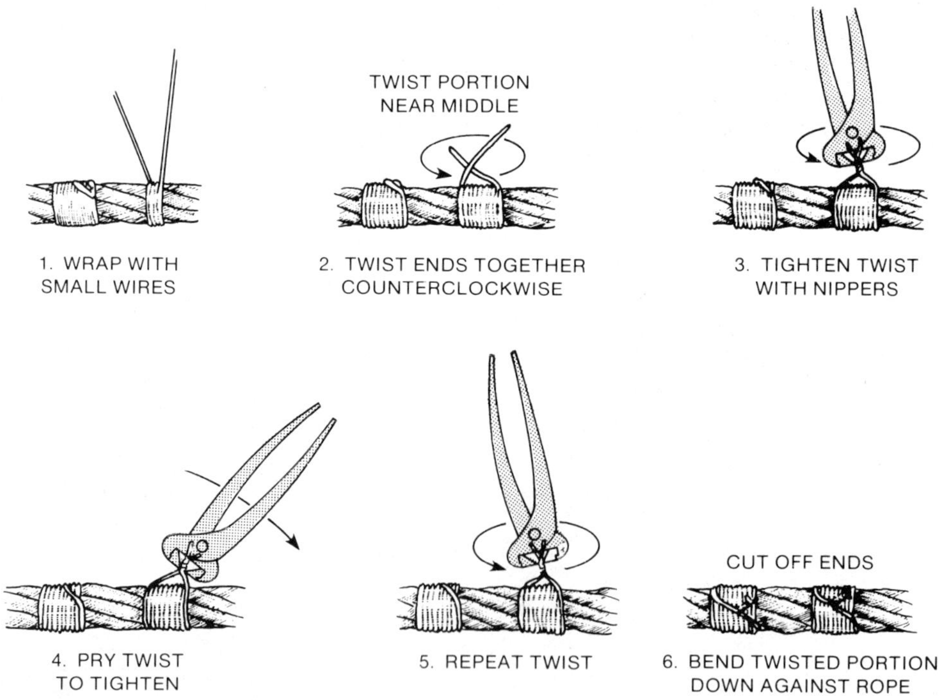

Fig. 1-20: Another way to seize wire rope.

twist the ends of the wires counterclockwise as in step 2. Grasp the ends with endcutting nippers and twist up the slack (step 3). Do not try to tighten seizing by twisting. Draw up on the seizing as in step 4. Twist up the slack. Repeat steps 4 and 5 if necessary. Cut the ends and pound them down on the rope as in step 6.

The seizing itself should be a soft or annealed wire or strand. The seizing wire diameter and the length of the seize will depend on the diameter of the wire rope. But, the length of the seizing should never be less than the diameter of the rope being seized. For preformed ropes, one seizing on each side of the cut is normally sufficient. But for those that are not preformed, a minimum of two seizings is recommended. Seizings should be spaced six rope diameters apart. Table 1-6 lists seizing lengths and the seizing wire diameters suggested for use with some commonly used wire ropes.

Table 1-6: Seizing—Suggested Diameters and Lengths

Rope Diameters		Seizing Wire Diameters		Seizing Lengths	
inches	mm	inches	mm	inches	mm
1/8–5/16	3.5–8.0	.032	0.813	1/4	6.0
3/8–9/16	9.5–14.5	.048	1.21	1/2	13.0
5/8–15/16	16.0–24.0	.063	1.60	3/4	19.0
1-1-5/16	26.0–33.0	.080	2.03	1-1/4	32.0
1-3/8–1-11/16	35.0–43.0	.104	2.64	1-3/4	44.0
1-3/4–2-1/2	45.0–64.0	.124	3.15	2-1/2	64.0
2-9/16–3-1/2	65.0–89.0	.124	3.15	3-1/2	89.0

Welding. Wire rope ends may also be bound together or seized by fusing or welding the wires. This is a satisfactory method if carefully done because it does not increase the size of the rope and requires little time to complete. Before welding the rope, cut a short piece of the core out of the end so that a clean weld will result and the core will not be burned deep into the rope. The area heated should be kept to a minimum, and no more heat should be applied than essential to fuse the metal.

Cutting

Wire rope may be cut with a wire rope cutter, a cold chisel, a hacksaw, bolt clippers, or an oxyacetylene cutting torch. As mentioned previously, wire rope must be properly seized before being cut (Fig. 1-21). To use the wire rope cutter (Fig. 1-22), insert the wire rope in the bottom of the cutter with the blade of the cutter coming between the two central seizings. Push the blade down against the wire rope and strike the top of the blade sharply with a sledge hammer sev-

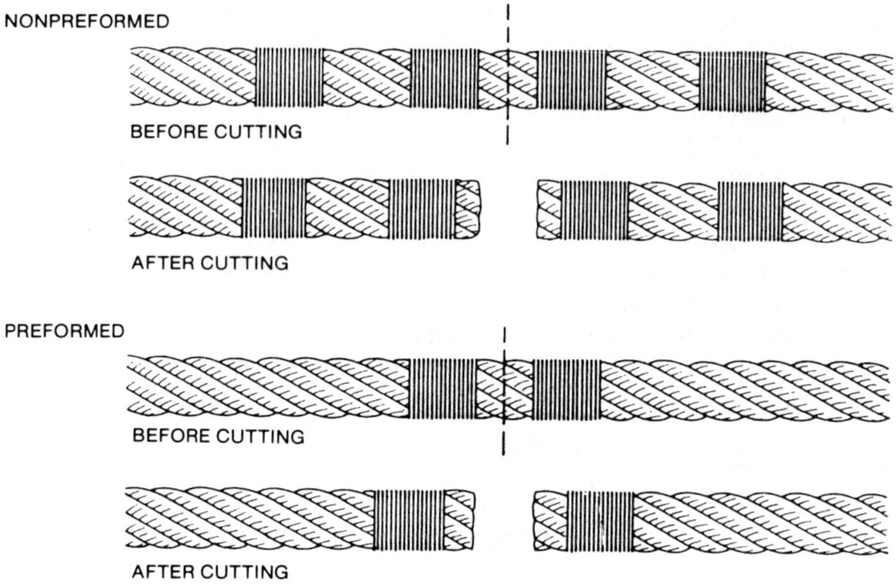

Fig. 1-21: Seizings, whether on nonpreformed or preformed wire rope, are applied before cutting.

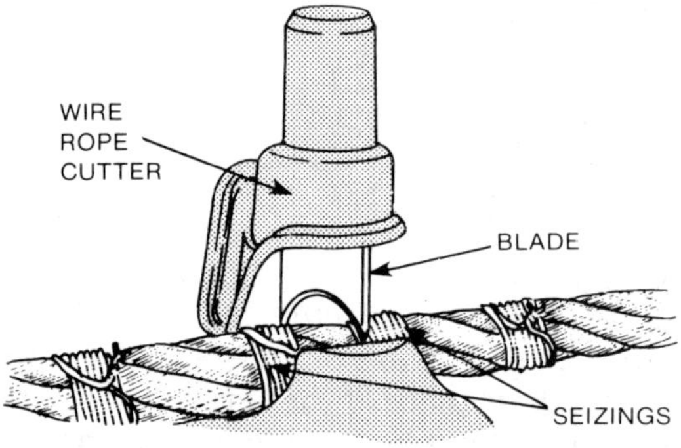

Fig. 1-22: Wire rope cutter.

eral times. The bolt clippers can be used only on wire rope of fairly small diameter, but the oxyacetylene torch can be used on wire of any diameter. Cutting with the hacksaw and cold chisel is slower than cutting with the other tools and equipment.

Installation of Wire

As stated earlier, all wire ropes operating over sheaves and drums are subjected to cyclic bending stresses, hence the rope wires will eventually fatigue.

The magnitude of these stresses depends—all other factors being constant—upon the ratio of the diameter of the sheave or drum to the diameter of the rope. Frequently, fatigue from cyclic, high-magnitude bending stress is the principal reason for shortened rope service.

To illustrate, in order to bend around a sheave, the rope's strands and wires must move relative to one another. This movement compensates for the difference in diameter between the underside and the top side of the rope, the distance being greater along the top side than it is on the underside next to the groove. Proper rope action (and service) is adversely affected if shifting the wires cannot compensate for this situation. Also, there can be additional motion retardation because of excessive pressure caused by a sheave whose groove diameter is too small or by a lack of lubrication. Changing the bending direction from one sheave to another should be scrupulously avoided as this reverse bending still further accelerates wire fatigue.

The relationship between sheave diameter and rope diameter is a critical factor that is used to establish the rope's fatigue resistance or relative service life. It is expressed in the tread D/d ratio mentioned earlier, in which D is the tread diameter of the sheave and d is the diameter of the rope. As a "field" rule of thumb, the minimum recommended sheave and drum diameter is 20 times the diameter of the rope. For example, determine the minimum sheave diameter for 5/8″ rope.

$$20 \times 5/8 = 12\text{-}1/2'' \text{ sheave}$$

If a 12-1/2″ sheave is not on hand, use the next larger size. Never use a smaller size.

Whenever possible, drums and blocks with sheaves should be arranged so as to avoid reverse bends. Figure 1-23 shows incorrect and correct arrangements.

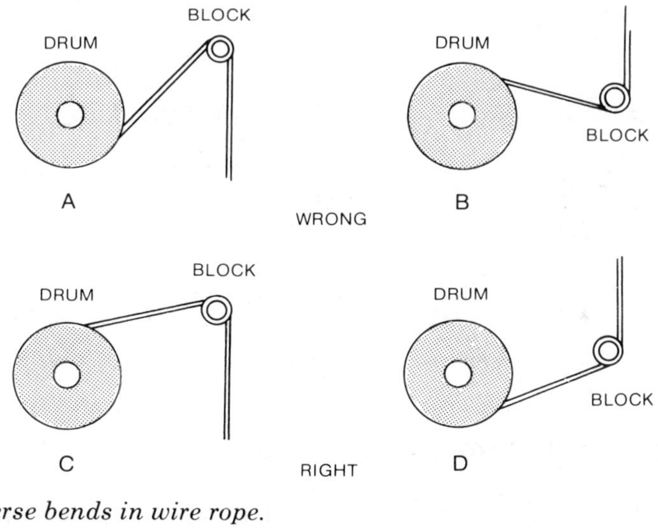

Fig. 1-23: Reverse bends in wire rope.

A reverse bend occurs when the rope bends in one direction around one drum, block, or sheave and bends in the opposite direction around the next (Fig. 1-24). Where a reverse bend must be used, the drums, blocks, or sheaves causing the reversal should have a larger diameter than ordinarily used and should be spaced as far apart as possible so that there will be more time allowed between the bending motions.

After a new wire rope has been installed and the ends secured in the correct manner, the mechanism should be started carefully and then permitted to run through a cycle of operation at very slow speed. During this trial operation, a very close watch should be kept on all working parts—sheaves, drums, rollers—to make certain that the rope runs freely and without any possible obstructions as it makes its way through the system. If no problems appear in running the rope, the next step should include several run-throughs of the normal operational cycle under light load and at reduced speed. This procedure allows the component parts of the new rope to make a gradual adjustment to the actual operating conditions.

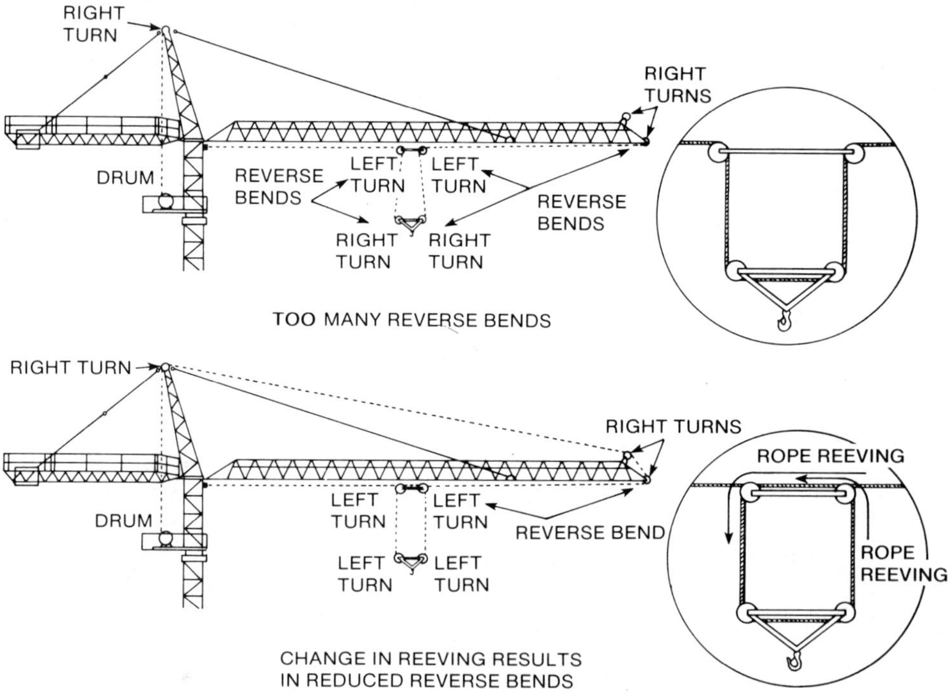

Fig. 1-24: Elimination of reverse bends in wire rope.

Winding on a Drum

Drums are the means by which power is transmitted to the rope and thence to the object to be moved. For the wire rope to pick up this power efficiently and to transmit it properly to the working end, installation must be carefully controlled. The proper way to wind wire rope on a drum depends on whether the drum is grooved or plain (smooth).

Grooved Drums. If the drum is grooved, the winding conditions should be closely supervised to assure adherence to the following recommended procedures:

1. The end of the rope must be secured to the drum by such means as will give the end attachment at least as much strength as is specified by the equipment manufacturer.
2. Adequate tension must be maintained on the rope while it is being wound so that the winding proceeds under continuous tension.
3. The rope must follow the groove.
4. There should be at least three dead turns remaining on the drum when the rope is unwound during normal operation. Two dead turns are a mandatory requirement in many codes and standards.

If the wire rope is carefully wound and, as a result, jumps the grooves, it will be crushed and cut where it crosses from one groove to the other. Another, almost unavoidable problem is created at the drum flange; as the rope climbs to a second layer there is further crushing and the wires receive excessive abrasion. Riser and filler strips may help remedy this condition.

Plain or Smooth Drums. Installation of a wire rope on a plain (smooth) face drum requires a great deal of care. The starting position should be at the drum end so that each turn of the rope will wind tightly against the preceding turn (Fig. 1-25). Use a piece of wood or a sledge handle to drive one turn against

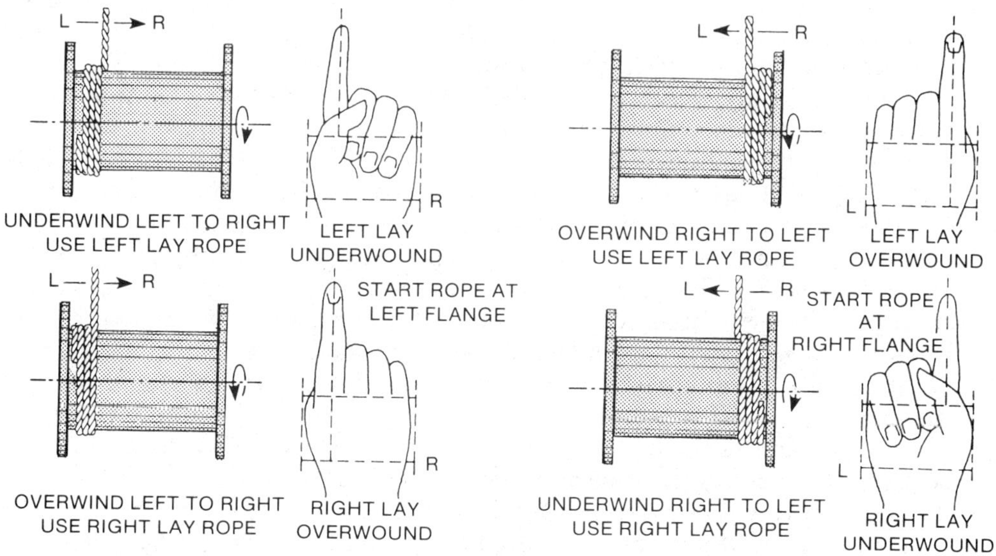

Fig. 1-25: By holding the right or left hand with index finger extended, palm up or palm down, the proper procedure for installing left and right lay rope on a smooth drum can be easily determined.

the other; *never* use your fingers. Close supervision should be maintained all during installation. This will help make certain that:

1. the rope is properly attached to the drum
2. appropriate tension on the rope is maintained as it is wound on the drum
3. each turn is guided as close to the preceeding turn as possible, so that there are no gaps between turns
4. there are at least two dead turns on the drum when the rope is fully unwound during normal operating cycles.

Loose and uneven winding on a plain- (smooth-) faced drum can and usually does create excessive wear, crushing, and distortion of the rope. The results of such abuse are lower operating performance and a reduction in the rope's effective strength. Also, for an operation that is sensitive in terms of moving and spotting a load, the operator will encounter control difficulties as the rope will pile up, pull into the pile, and fall from the pile to the drum surface. The ensuing shock can break or otherwise damage the rope.

The proper direction of winding the first layer on a smooth drum can be determined by standing behind the drum and looking along the path the rope travels, then following one of the procedures illustrated in Fig. 1-25. The diagrams show the correct relationship that should be maintained between the direction of lay of the rope (right or left), the direction of rotation of the drum (overwind or underwind), and winding from left to right or right to left.

Drums—Multiple Layers. Many installations are designed with requirements for winding more than one layer of wire rope on a drum. Winding multiple layers presents some further problems.

The first layer should wind in a smooth, tight helix which, if the drum is grooved, is already established. The grooves allow the operator to work off the face of the drum and permit the minimum number of dead turns.

A smooth drum presents an additional problem, initially, as the wire rope must be wound in such a manner that the first layer will be smooth and uniform and will provide a firm foundation for the layers of rope that will be wound over it. The first layer of rope on the smooth drum should be wound with tension sufficient to assure a close helix, each turn being wound as close as possible to the preceding turn, and most, if not all, of the entire layer being used as dead turns. The first layer then acts as a helical groove which will guide the successive layers. Unlike wire ropes operating on groove drums, *the first layer should not be unwound* from a smooth-faced drum with multiple layers.

After the rope has wound completely across the face of the drum (either smooth or grooved), it is forced up to a second layer at the flange. The rope then winds back across the drum in the opposite direction, lying in the depression between the turns of the rope on the first layer. Advancing across the drum on the second layer, the rope, following the "grooves" formed by the rope on the first layer, actually winds back one turn in each revolution of the drum. The rope must then cross two rope "grooves" in order to advance across the drum for each turn. The point at which this occurs is known as the *crossover.* Crossover is unavoidable on the second and all succeeding layers. Figure 1-26 illustrates the winding of a rope on the second layer from left to right and from right to left—the direction is shown by the arrows. At these crossover points, the rope is subjected to severe abrasion and crushing as it is pushed over the two rope "grooves" and rides across the crown of the first rope layer. The scrubbing of the rope, as this is happening, can easily be heard. There is, however, a special drum grooving available that will greatly minimize the damage that can occur at crossover points.

Severe abrasion can also be reduced by applying the rule for the correct rope lay (right or left lay) to the second layer rather than to the first layer. It is for this reason that the first layer of a smooth drum should be wound tight and used as dead turns.

Fleet Angle. The achievement of even winding on a smooth faced drum is closely related to the magnitude of the D/d ratio, the speed of rotation, load on the rope, and the fleet angle. Of all these factors, the one that exerts perhaps the greatest influence on winding characteristics is the fleet angle.

The schematic drawing (Fig. 1-27) shows an installation where the wire rope runs from a fixed sheave, over a floating sheave, and then on to the surface of a smooth drum. The fleet or lead angle (Fig. 1-27) may be defined as the included angle between two lines: one line drawn through the middle of the fixed sheave and the drum and perpendicular to the axis of the drum and a second line drawn from the flange of the drum to the base of the groove in the sheave. (The drum flange represents the farthest position to which the rope can travel across the drum.) There are left and right fleet angles, measured to the left or right of the centerline of the sheave, respectively.

It is necessary to restrict the fleet angle on installations where wire rope passes over the lead or fixed sheave and onto a drum. For optimum efficiency and service characteristics, the angle here should not exceed 1-1/2° for a smooth drum, nor 2° for a grooved drum. Fleet angles larger than these suggested limits can cause such problems as bad winding on smooth drums and the rope rubbing against the flanges of the sheave grooves. Larger angles also create situations where there is excessive crushing and abrasion of the rope on

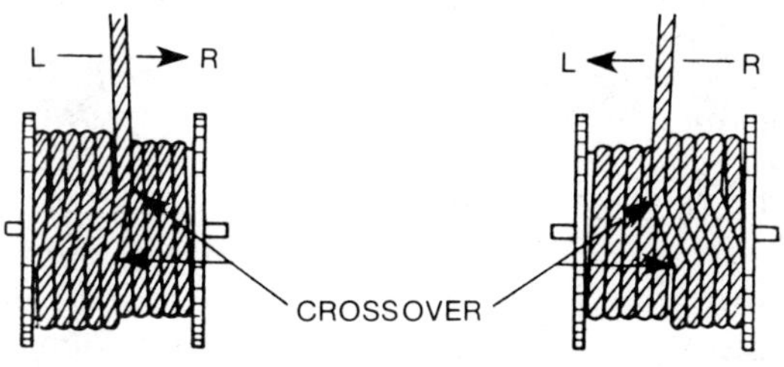

Fig. 1-26: After the first layer is wound on a drum, the point at which the rope winds back for each turn is called the crossover.

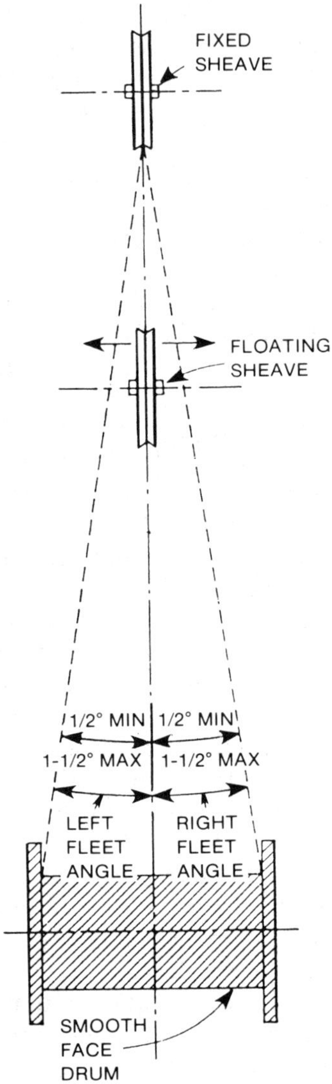

Fig. 1-27: This illustration of wire rope running from a fixed sheave, over a floating sheave, and then on to a smooth drum graphically defines the fleet angle.

the drum. Conversely, small fleet angles—less than 1/4° for grooved drums or 1° for smooth drums—should also be avoided since too small an angle will cause the rope to pile up.

Figure 1-27 shows how to determine the distance between the drum and the sheave in order to obtain the proper fleet angle. For example, let us assume that the width of a smooth drum is 60″ and the desired fleet angle is 2°. On the graph, follow the 60″ line up to the intersection of the 2° line. Then, follow a horizontal line to the left from the point of intersection to the proper position on the "distance from drum to sheave" scale. As can be seen, it would be necessary to have a distance of 70′ between the drum and the first sheave in order to obtain a 2° fleet angle on a 60″ smooth drum.

ATTACHMENTS

To obtain maximum strength at any joining of wire rope ends, each end should be made up into an eye splice (Fig. 1-28), an end fitting should be attached, or clips or clamps may be used to make an eye. The wire can then be joined to another wire, a hook, a ring, or a link of chain directly or by using a shackle.

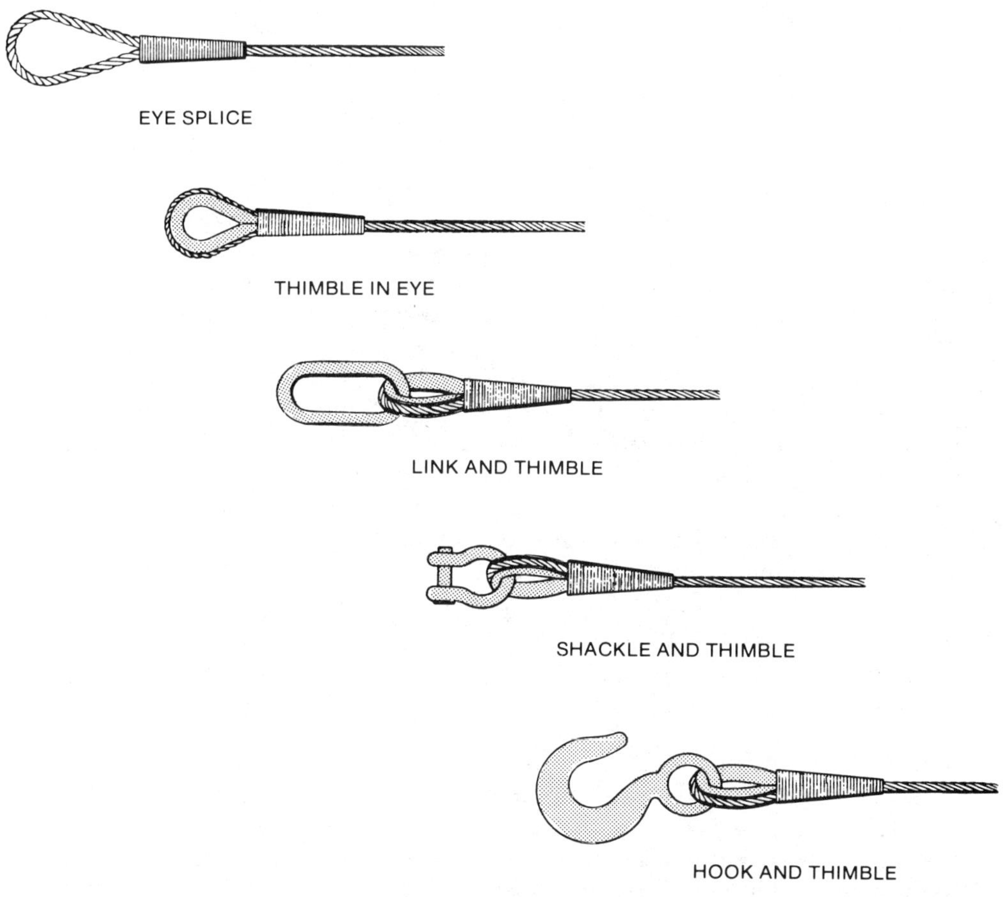

Fig. 1-28: Use of eyes in wire rope ends.

End Fittings

For a number of applications, such as tight openings in drums or other complicated reeving systems, there may be a need for making special end preparations. When these are required, there are about four basic designs (and combinations) from which to choose (Fig. 1-29).

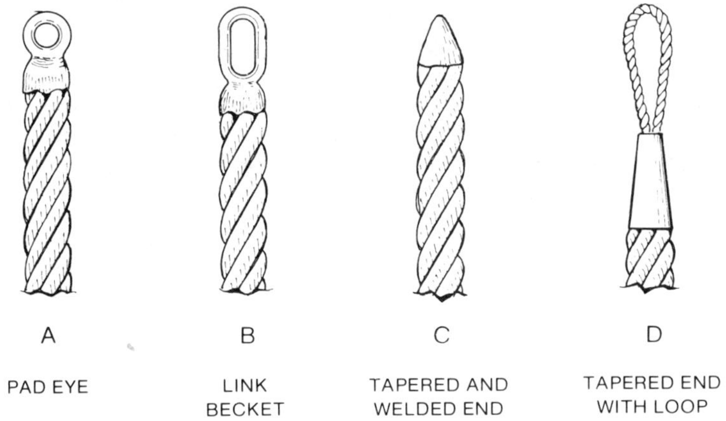

Fig. 1-29: Beckets, or end preparations, are used on wire rope ends when another rope is needed to pull the operating rope into place. Four commonly used beckets are illustrated.

Becket loops are used when another rope is needed to pull the new rope into place. The rope end must be fastened to the mechanism so that force and motion are transferred efficiently. End fittings thus become items of great importance for transferring these forces. Each basic type of end fitting has its own individual characteristics. Thus, one type will usually fit the needs of a given installation better than the others (Fig. 1-30).

Efficiency of End Attachments. It should be noted that not all end attachments will develop the full strength of the wire rope used. To lessen the possibility of error, the wire rope industry has determined terminal efficiencies for various types of end attachments. Table 1-7, listing these efficiencies, permits calculation of the holding power of the more popular end fittings for any size, grade, and construction of wire rope.

Socketing

Wire rope terminals that are improperly attached can lead to serious—possibly unsafe—conditions. To perform properly, all wire rope elements must be held securely by the terminal. If this is not accomplished, the strands will "loaf on the job," and there is every likelihood that a strand will become "high." A high strand condition is illustrated in Fig. 1-31. In the case shown, selective abrasive wear of the loose strand will necessitate early removal of the rope.

Table 1-7: Terminal Efficiencies (Approximate)— (Efficiencies are based on nominal strengths)

Method of Attachment	Efficiency Rope with IWRC*	Efficiency Rope with FC**
Wire Rope Socket-Spelter or Resin Attachment	100%	100%
Swaged Socket	95%	(Not established)
Mechanical Spliced Sleeve		
1″ dia. and smaller	95%	92-1/2%
1-1/8″ dia. thru 1-7/8″	92-1/2%	90%
2″ dia. and larger	90%	87-1/2%
Loop or Thimble Splice—Hand Spliced (Tucked) (Carbon Steel Rope)		
1/4″	90%	90%
5/16″	89%	89%
3/8″	88%	88%
7/16″	87%	87%
1/2″	86%	86%
5/8″	84%	84%
3/4″	82%	82%
7/8″ thru 2-1/2″	80%	80%
Loop or Thimble Splice—Hand Spliced (Tucked) (Stainless Steel Rope)		
1/4″	80%	
5/16″	79%	
3/8″	78%	
7/16″	77%	
1/2″	76%	
5/8″	74%	
3/4″	72%	
7/8″	70%	
Wedge Sockets*** (Depending on Design)	75% to 90%	75% to 90%
Clips*** (Number of clips varies with size of rope)	80%	80%

*IWRC = Independent Wire Rope Core **FC = Fiber Core

***Typical values when applied properly. Refer to fittings manufacturers for exact values and method.

Basket Sockets. Basket or spelter sockets (Fig. 1-32A) are efficient and permanent terminal attachments for wire rope. They are the most reliable of all terminal fittings. When properly attached, these standard drop forged sockets are recommended for all standing ropes and whenever service conditions are severe.

A basket socket ordinarily is attached to the end of the rope with molten zinc or babbitt metal and is a permanent end fitting. If properly made up, this fitting is as strong as the rope itself. If molten lead is used instead of zinc, the strength of the connection must be assumed to be reduced to one-fourth the

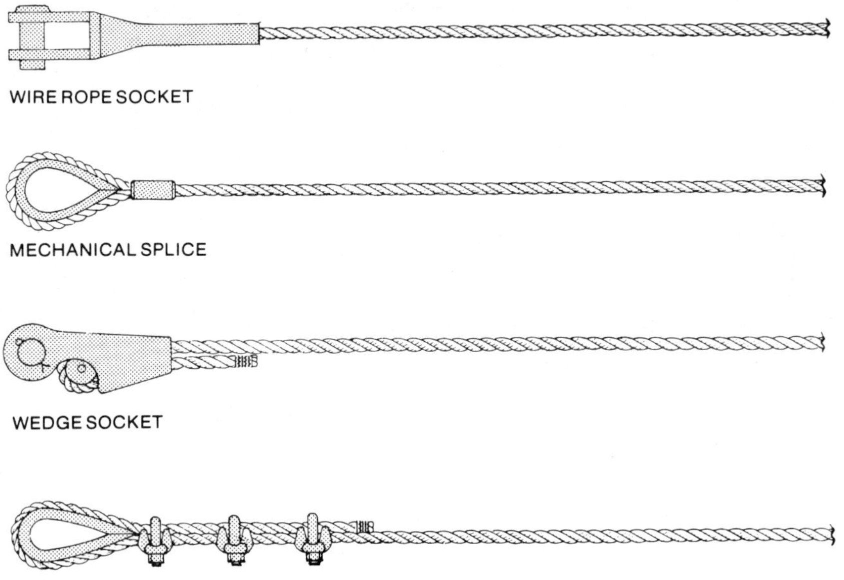

Fig. 1-30: Types of wire rope end fittings.

Fig. 1-31: An example of "high strand." The excessive wear of a single strand is caused by improper socketing.

strength of a zinc connection. In either case, the socket should be installed by the manufacturer or in the shop; *never* in the field.

Swaged Sockets. These sockets (Fig. 1-32B) also make efficient and permanent terminal attachments for wire rope. They are made by compressing a steel sleeve over the rope with a hydraulic press. Properly made, they provide an efficiency of 100%.

Careful inspection of the wires leading into these socketed terminals is most important because the strength of this section (from 5′ to 10′ above the load attachment) is bound to be greatly reduced because of fatigue within the wires. On visual inspection, one broken wire is sufficient to cause it to be condemned. It is, therefore, advisable to cut this damaged section off periodically even though it may look sound. About one-third of the estimated life of the rope is a good interval.

When inspecting any existing socketed connection, examine the rope closely for corrosion at the socket base. This corroded wire is very susceptible to fatigue.

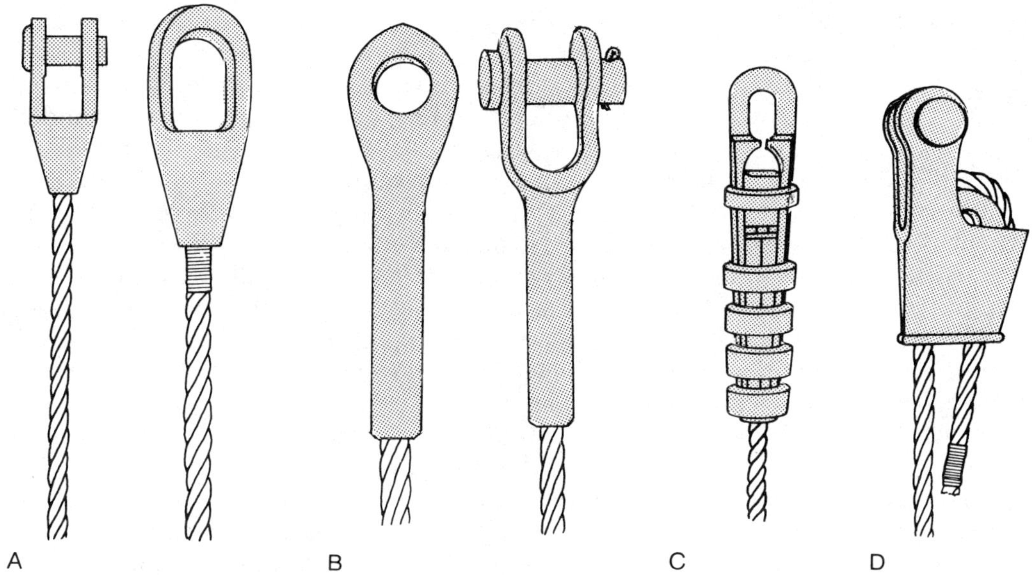

Fig. 1-32: (A) Basket socket; (B) swaged socket; (C) cappel socket; and (D) wedge socket.

Cappel Sockets. These sockets (Fig. 1-32C), when properly installed and frequently inspected, also give 100% efficiency. Their efficiency depends entirely upon the wedges being kept tight. Since the wedges are interlocked and cannot move independently of each other, the unit affords a holding power equal to the strength of any type of wire rope. They also afford the facility of frequent inspection of the whole of that part of the rope upon which the grip is exerted.

Wedge Sockets. Wedge sockets (Fig. 1-32D) are among the simplest devices for anchoring a wire rope for any purpose. They are intended for "on the job" attachment and for quick rope replacement. (Remember, basket, swaged, and cappel sockets are either manufacturer or shop installed.) Principal advantages are simplicity, ease and speed of applying and detaching. They are also used where conditions are such that spliced eyes could not be reeved and would have to be made after the rope was in place.

The efficiency of a wedge socket is low—approximately 70% of the strength of the rope. Care must be taken that moving loads do not force the wedges out and accidental slackening of the cable does not release the socket. To prevent this they should be positive locking. This can be accomplished by proceeding as follows:

1. Inspect the wedge and socket; all rough edges or burrs that might damage the rope should be removed.
2. If the end of the rope is welded, the welded end should be cut off. This will allow the distortions of the rope strands, caused by the sharp bend around the wedge, to adjust themselves at the end of the line. If the weld is not cut off, the

distortions will be forced up the working line. This may result in the development of high strands and wavy rope.

3. Place the socket in an upright position and bring the rope around in a large, easy to handle loop. Care must be taken to make certain that the live-loaded-side of the rope is in line with the ears (Fig. 1-33).

4. The dead end of the rope should extend from the socket for a distance approximately nine times the rope diameter. The wedge is now placed in the socket, and a wire rope clip is placed around the dead end by clamping a short, extra piece of rope to the tail. (*Do not clamp to live part.*) The U-bolt should bear against the tail; the saddle of the clip should bear against the short extra piece.

5. Secure the ears of the socket to a sturdy support and carefully take a strain on the live side of the rope. Pull the wedge and rope into position with tension sufficiently tight to hold them in place.

6. After final pin connections are made, increase the loads *gradually* until the wedge is properly seated. Avoid sudden shock loads.

The foregoing is the recommended procedure. If variations are made to suit special conditions, they should be carefully evaluated beforehand. For example, wedge socket anchorages should not be used on nonrotating ropes (because of the small radius of wedge sockets and the possibility of severe core slippage), unless extreme care is used in their installation and the rope end is very tightly seized. Also, do not clamp the dead end of the cable to the live side of the line with the clip as this will seriously weaken the attachment. The clip will ultimately take the load and could deform and break the cable (Fig. 1-34).

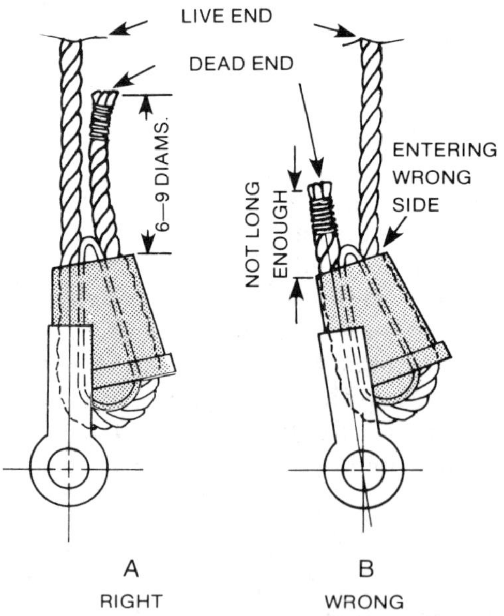

Fig. 1-33: The wedge socket is a very popular end attachment; it is easily installed and quickly dismantled, but it must be applied correctly (A).

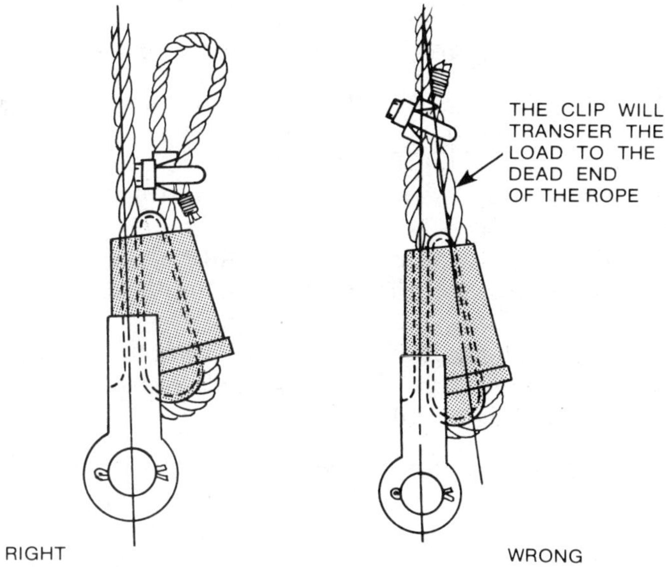

Fig. 1-34: Proper method of securing dead end of rope when using a wedge socket.

Wire Rope Clips

Wire rope clips are widely used for attaching wire rope to haulages, hoists, and for joining two ropes. Clips are available in two basic designs: the *U-bolt* and *fist grip*. The efficiency of both types is the same.

How to Apply U-Bolt Clips. When using U-bolt clips, extreme care must be exercised to make certain that they are attached correctly, i.e., the *U-bolt* must be applied so that the "U" section is in contact with the dead end of the rope (Fig. 1-35). Also, the tightening and retightening of the nuts must be accomplished as required.

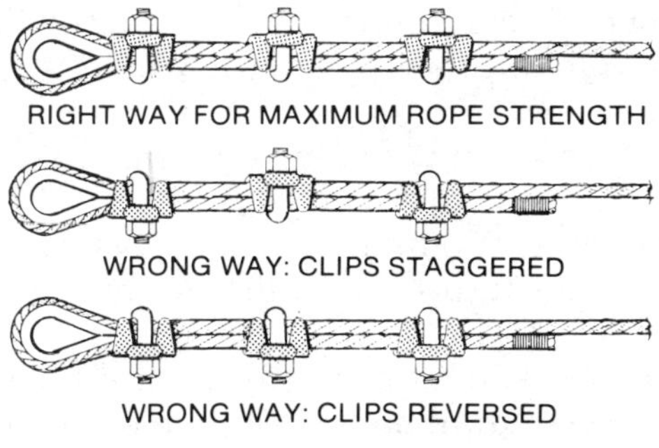

Fig. 1-35: The correct way to attach U-bolts is shown at the top. The "U" section is in contact with the rope's dead end.

Using Table 1-8 as a reference, the following is the recommended method of applying U-bolt clips to obtain maximum holding power:

1. Turn back the specified amount of rope from the thimble. Apply the first clip one base width from the dead end of the wire rope (U-bolt over dead end—live end rests in clip saddle). Tighten nuts evenly to recommended torque.
2. Apply the next clip as near the loop as possible. Turn on nuts firm but do not tighten.
3. Space additional clips, if required, equally between the first two. Turn on nuts—take up rope slack—tighten all nuts evenly on all clips to recommended torque.
4. Apply the initial load and retighten the nuts to their approximate recommended torque. Rope will stretch and shrink in diameter when loads are applied. Inspect periodically and retighten.

A termination made in accordance with the above instructions and using the number of clips given in Table 1-8 has an approximate 80% efficiency rating. This rating is based upon the catalog breaking strength of wire rope. If a pulley is used in place of a thimble for turning back the rope, add one additional clip.

The number of clips suggested in Table 1-8 is based upon using right regular or lang lay wire rope, 6 × 19 class or 6 × 37 class, fiber core or IWRC, IPS, or XIPS. If Seale construction or similar large outer wire type construction in the 6 × 19 class is to be used for sizes 1″ and larger, add one additional clip. The number of clips stated also applies to right regular lay wire rope, 8 × 19 class, fiber core, IPS, sizes 1-1/2″ and smaller; and right regular lay wire rope, 18 × 7 class, fiber core, IPS or XIPS, sizes 1-3/4 and smaller.

For other classes of wire rope not mentioned above, it may be necessary to add additional clips to the number shown. If a greater number of clips are used than shown in the table, the amount of rope turnback should be increased proportionately. The above is based on the use of clips on new rope.

How to Apply Fist Grip Clips. The recommended method of applying fist grip clips (they are also called double-saddle or Crosby-saddle clips) is as follows:

1. Turn back the specified amount of rope from the thimble. Apply the first clip one base width from the dead end of the wire rope. Tighten nuts evenly to recommended torque.
2. Apply the next clip as near the loop as possible. Turn on nuts firmly but do not tighten.
3. Space additional clips, if required, equally between the first two. Turn on nuts—take up rope slack—tighten all nuts evenly on all clips to recommended torque.
4. Apply the initial load and retighten nuts to the recommended torque. Rope will stretch and shrink in diameter when loads are applied. Inspect periodically and retighten.

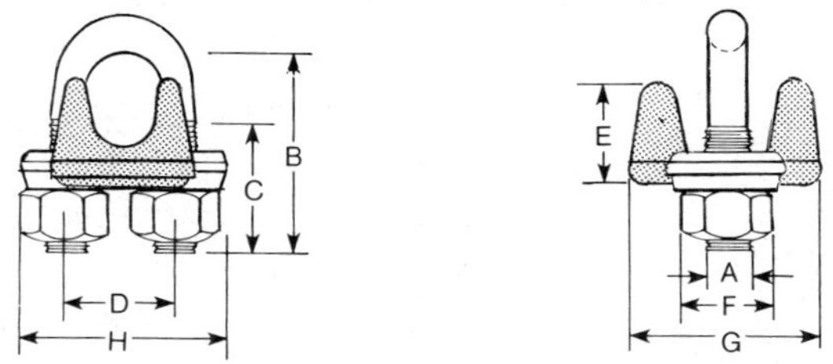

Table 1-8: U-Bolt Clip Details

Clip Size	A	B	C	D	E	F	G	H	Min. no. of clips	Amount of rope to turn back	Torque in lb/ft	Weight lb/100
1/8	.22	.72	.44	.47	.41	.38	.81	.94	2	3-1/4	4.5	5
3/16	.25	.97	.56	.59	.50	.44	.94	1.16	2	3-3/4	7.5	9
1/4	.31	1.03	.50	.75	.66	.56	1.19	1.44	2	4-3/4	15	18
5/16	.38	1.38	.75	.88	.72	.69	1.31	1.69	2	5-1/4	30	30
3/8	.44	1.50	.75	1.00	.91	.75	1.63	1.94	2	6-1/2	45	42
7/16	.50	1.88	1.00	1.19	1.03	.88	1.81	2.28	2	7	65	70
1/2	.50	1.88	1.00	1.19	1.13	.88	1.91	2.28	3	11-1/2	65	75
9/16	.56	2.25	1.25	1.31	1.22	.94	2.06	2.50	3	12	95	100
5/8	.56	2.38	1.25	1.31	1.34	.94	2.06	2.50	3	12	95	100
3/4	.63	2.75	1.44	1.50	1.41	1.06	2.25	2.84	4	18	130	150
7/8	.75	3.13	1.63	1.75	1.59	1.25	2.44	3.16	4	19	225	240
1	.75	3.50	1.81	1.88	1.78	1.25	2.63	3.47	5	26	225	250
1-1/8	.75	3.88	2.00	2.00	1.91	1.25	2.81	3.59	6	34	225	310
1-1/4	.88	4.25	2.13	2.31	2.19	1.44	3.13	4.13	6	37	360	460
1-3/8	.88	4.63	2.31	2.38	2.31	1.44	3.13	4.19	7	44	360	520
1-1/2	.88	4.94	2.38	2.59	2.53	1.44	3.41	4.44	7	48	360	590
1-5/8	1.00	5.31	2.63	2.75	2.66	1.63	3.63	4.75	7	51	430	730
1-3/4	1.13	5.75	2.75	3.06	2.94	1.81	3.81	5.28	7	53	590	980
2	1.25	6.44	3.00	3.38	3.28	2.00	4.44	5.88	8	71	750	1340
2-1/4	1.25	7.13	3.19	3.88	3.94	2.00	4.56	6.38	8	73	750	1570
2-1/2	1.25	7.69	3.44	4.13	4.44	2.00	4.69	6.63	9	84	750	1790
2-3/4	1.25	8.31	3.56	4.38	4.88	2.00	5.00	6.88	10	100	750	2200
3	1.50	9.19	3.88	4.75	5.34	2.38	5.31	7.63	10	106	1200	3200

A termination made in accordance with the above instructions and using the number of clips detailed in Table 1-9 has an approximate 80% efficiency rating. This rating is based upon the catalog breaking strength of wire rope. If a pulley is used in place of a thimble for turning back the rope, add one additional clip.

The number of clips given in Table 1-9 is based upon using right regular or lang lay wire rope, 6 × 19 class or 6 × 37 class, fiber core or IWRC, IPS, or EIPS. If Seale construction or similar large outer wire type construction in the 6 × 19 class is to be used for sizes 1″ and larger, add one additional clip. The number of clips suggested also applies to right regular lay wire rope, 8 × 19 class, fiber core, IPS, sizes 1-1/2″ and smaller; and right regular lay wire rope, 18 × 7 class, fiber core, IPS or EIPS, sizes 1-1/2 and smaller.

For other classes of wire rope not mentioned above, it may be necessary to add additional clips to the number suggested. If a greater number of clips are used than shown in the table, the amount of rope turnback should be increased proportionately. The above is based on the use of fist grip clips on new wire rope.

Splicing Wire

Only a few tools are required for splicing wire rope. In addition to the tools shown in Fig. 1-36, a hammer and cold chisel will be helpful for cutting ends of strands. Two slings of marline and two sticks should be used for untwisting the wire, and a pocket knife may be needed for such operations as cutting the hemp core.

Long Splice. Splicing wire rope requires skill and care. The beginner should follow the instructions carefully and expect to have difficulty while he/she is learning. Later results will be more gratifying. For strength as well as appearance, the long splice is best. The recommended number of feet re-

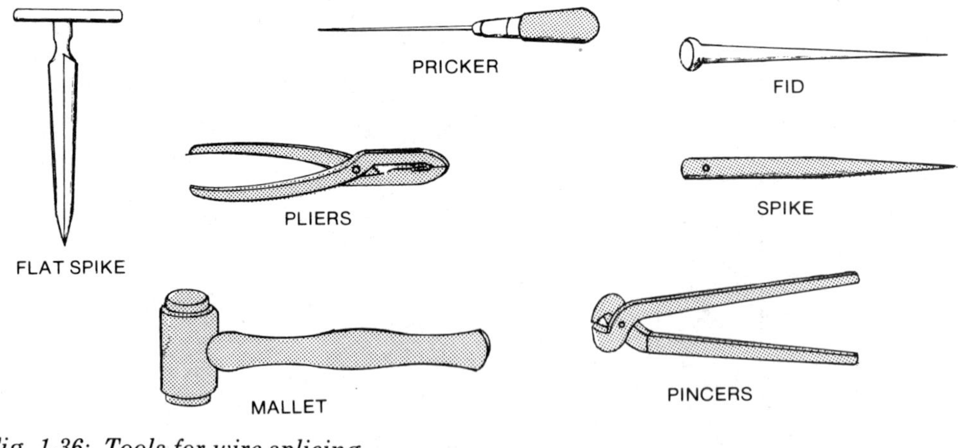

Fig. 1-36: Tools for wire splicing.

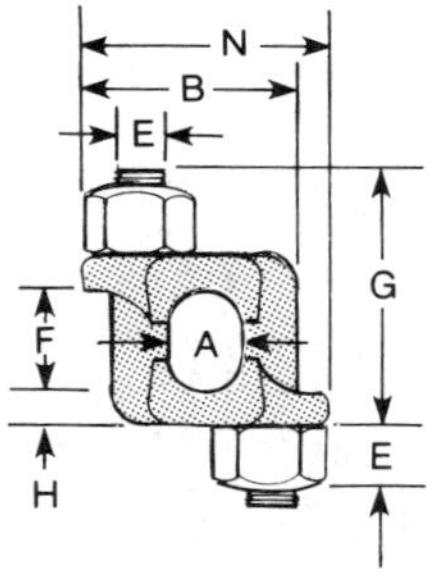

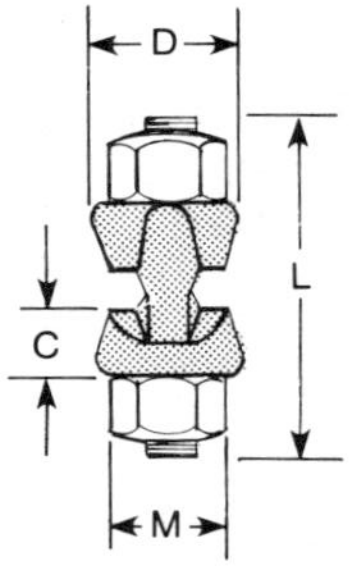

Table 1-9: Fist Grip Clips Details

Clip Size	A	B	C	D	E	F	G	H	L Approx.	M	N	Min. no. of clips	Amount of rope to turn back	Torque in lb/ft	Weight lb/100
3/16–1/4	.25	1.25	.34	.94	.38	.50	1.28	.22	1.63	.69	1.47	2	4	30	21
5/16	.31	1.34	.44	1.06	.38	.63	1.47	.19	1.94	.69	1.56	2	5	30	26
3/8	.38	1.59	.50	1.06	.44	.75	1.81	.25	2.38	.75	1.88	2	5-1/2	45	37
7/16	.50	1.88	.56	1.25	.50	1.00	2.19	.28	2.75	.88	2.19	2	6-1/2	65	60
1/2	.50	1.88	.56	1.25	.50	1.00	2.19	.28	2.75	.88	2.19	3	11	65	60
9/16	.63	2.28	.69	1.50	.63	1.25	2.69	.28	3.50	1.06	2.63	3	12-3/4	130	110
5/8	.63	2.28	.69	1.50	.63	1.25	2.69	.28	3.50	1.06	2.63	3	13-1/2	130	110
3/4	.75	2.69	.88	1.81	.75	1.50	2.94	.31	3.75	1.25	3.06	3	16	225	140
7/8	.88	2.97	.97	2.13	.75	1.75	3.31	.38	4.13	1.25	3.14	4	26	225	220
1	1.00	3.06	1.19	2.25	.75	2.00	3.72	.41	4.63	1.25	3.53	5	37	225	270
1-1/8	1.13	3.44	1.28	2.38	.88	2.25	4.19	.44	5.25	1.44	3.91	5	41	360	300
1-1/4	1.25	3.56	1.34	2.50	.88	2.50	4.25	.50	5.25	1.44	4.03	6	55	360	410
1-3/8	1.50	4.13	1.56	3.00	1.00	3.00	5.56	.56	7.00	1.63	4.66	6	62	500	680
1-1/2	1.50	4.13	1.56	3.00	1.00	3.00	5.56	.56	7.00	1.63	4.66	6	66	500	680

quired to make a long splice in wire rope is approximately 40 times the diameter in inches; for example, a long splice in a 3/4″ (19.05 mm) rope requires 40 times 3/4″, or approximately 30′ (15′ [4.57 m] of each rope). The directions given here are for making a 30′ (9.14 m) splice in a 3/4″ (19.05 mm) regular lay, round strand, hemp center wire rope; when a longer or shorter splice is required, the lengths are increased or decreased proportionately. Other strand combinations differ only when there is an uneven number of strands.

1. Measure off the desired length (15′ [4.57 m]) on each of the ends to be spliced (Fig. 1-37A) and mark them.
2. Unlay the strands of both ends for the distance marked. Cut off the hemp cores and bring the bunches of strands together so that each strand of the left-hand rope is between two strands of the right-hand rope (Fig. 1-37B).
3. Start with any strand. Unlay the strand from its own rope and replace it immediately with the adjacent strand from the other rope, laying the new strand tightly in the groove left by the unlaid strand and making the twist agree exactly. Continue this until all of the new strand (15′ [4.57 m]) except 24″ (0.61 m) has been laid into the rope. Cut off the unlaid strand, which has been replaced, leaving an end of about 24″ (0.61 m) (Fig. 1-37C). A good rule is to make these ends equal to one-fifteenth of the total length of the splice.

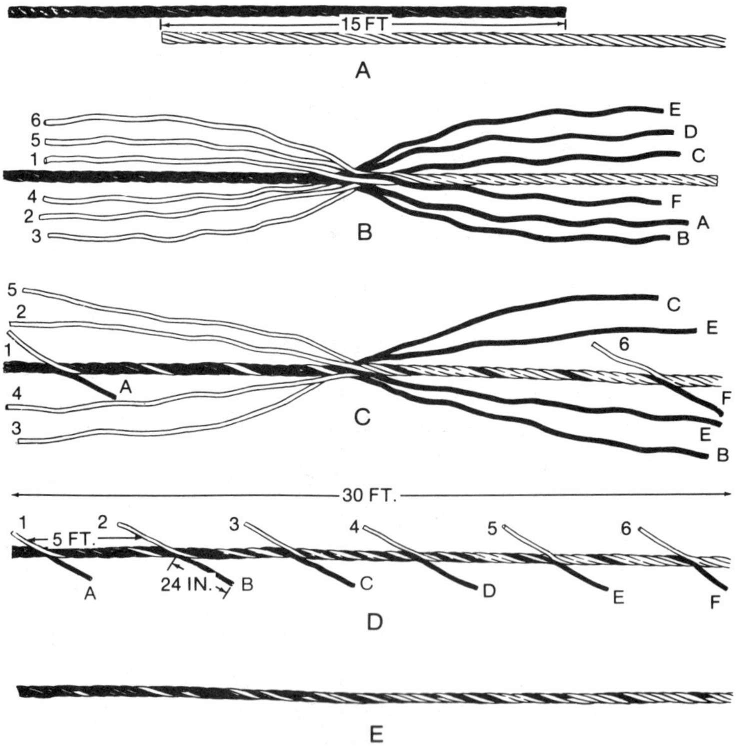

Fig. 1-37: Steps in making a long splice.

4. Repeat step 3 on the other rope, unlaying the strand next to the strand which has been worked into the first rope.
5. Repeat step 3 with the next strand of the first rope, stopping 5′ (1.52 m) short of the ends of the first pair of strands. Continue in this manner, unlaying successive strands, working on alternate ends, and leaving 5′ (1.52 m) intervals, until all strands are relaid (Fig. 1-37D).
6. The pairs of ends are tucked into the rope without enlarging the diameter of the rope (Fig. 1-37E). The two ends should be laid in side-by-side, not crossed over each other (Fig. 1-38A).
7. Straighten out all ends and wrap them with friction tape or twine to make each strand approximately the size of the hemp center. This prevents the ends from working out after being tucked. Fasten two rope clamps (Fig. 1-38A) or similar means on each side of the point where the two strands intersect, and twist them in opposite directions, sufficiently opening the lay of the rope to permit cutting the hemp core at the point of intersection.
8. Pull out the hemp core for a short distance. Drive a spike under two adjacent strands (Fig. 1-38B), and work the loose strand end into the space previously occupied by the hemp core.
9. Rotate the spike to open the rope farther. Continue pulling out the hemp

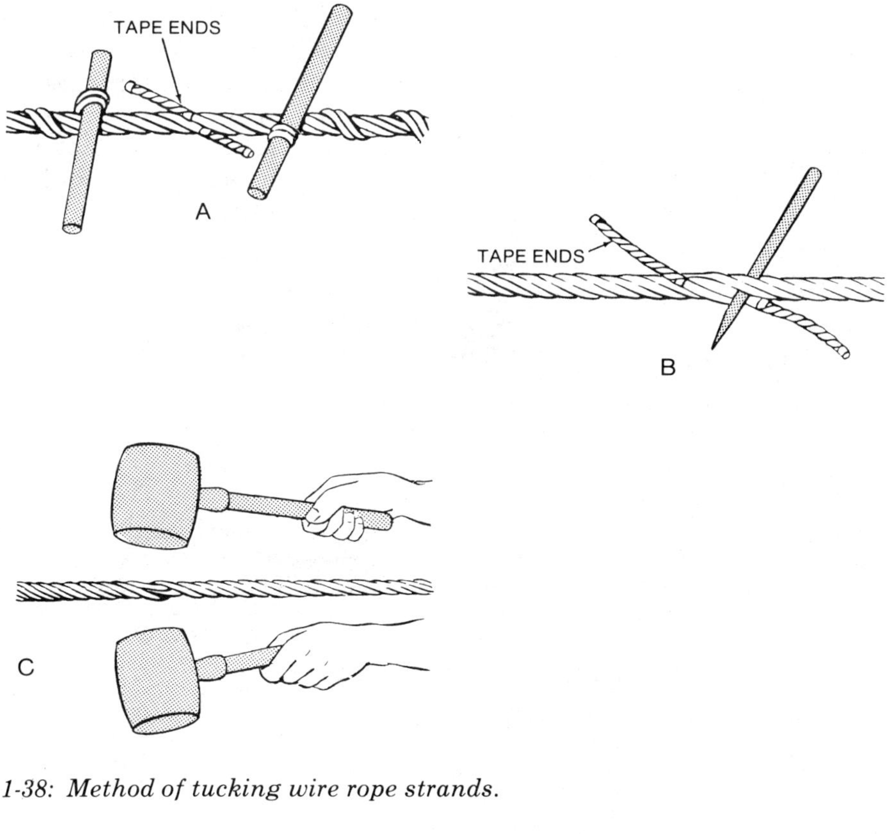

Fig. 1-38: Method of tucking wire rope strands.

core and working in the strand until all of the strand is in the center of the rope. Cut off the hemp core exactly where the strand ends, and push the end of the core back into place.

10. Repeat steps 8 and 9 for the other strand at the same intersection, working in the opposite direction.

11. At the points where tucks are made, tap the rope lightly with a copper or wood mallet (Fig. 1-38C) to fix strands firmly in place.

12. Shift the clamps to the next intersection of strands and repeat the operations.

When splicing a round strand lang lay rope, it is advisable to make a slightly longer splice than for the same size rope of regular lay, because of the tendency to untwist. Up to the point of tucking the ends, the procedure previously described is followed. Then, instead of laying the strands side-by-side where they pass each other, cross them over to increase the holding power of the splice. At the point where they cross, untwist the strands for a length of about 3″ (76.2 mm) so they cross over each other without materially increasing the diameter of the rope. Then, finish the tucks in the usual way, steps 7 to 12.

In splicing wire ropes having five or seven strands, the odd tuck is made at the center of the splice.

Short Splice. The short splice is frequently used for making an eye in the end of a wire rope, either with or without a thimble in the eye. This gives rise to the names "loop splice" and "thimble splice" (Fig. 1-39). The recommended extra length in feet to allow for a short splice is approximately three times the diameter in inches. This is measured from the end of the loop to the end of the rope. A short splice in a 3/4″ (19.05 mm) rope requires 3 times 3/4 or approximately 2-1/4′ (0.69 m). A short splice is not as strong as a well made long splice, developing only from 70 to 90% of the strength of the rope.

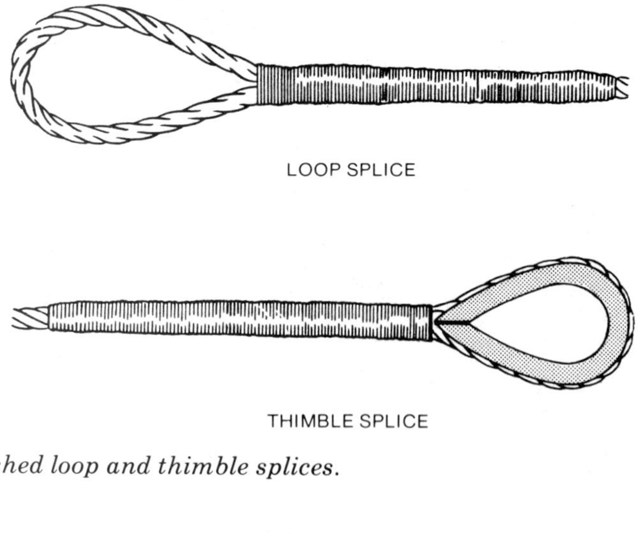

Fig. 1-39: Finished loop and thimble splices.

Loop or Tucked Eye Splice. To make a loop splice, proceed as follows:

1. Apply a seizing at each of the two points which will be ends of the loop. This will prevent the strands in the loop from untwisting and becoming displaced while the splice is being made.
2. Clamp the loop in a vise and proceed with the splice in the same manner as for the thimble splice described below.

Thimble or Flemish Eye Splice. To make a thimble or mechanical Flemish splice, proceed as follows:

1. Measure off the length of rope to be used and tie the back of the thimble to the rope at the desired point. Bend the rope around the thimble and press it tightly against the sides of the thimble by placing it in a vise. Using wire, bind the rope in position around the thimble as near the point of the thimble as possible (Fig. 1-40A).
2. Hold the thimble and rope in the vise and untwist the strands of the running end of the rope back to the thimble. Bind the end of each strand with friction tape or soft wire. Bend the strands back out of the way (Fig. 1-40B), and cut the hemp center or core close to the thimble. It is advisable to untwist the main rope slightly. This can be done by means of a hemp rope and a short length of pipe.
3. If a right lay rope is being spliced, stand on the same side as the short end of the rope and drive the spike from right to left, under two adjacent strands, a and f, on top and in front of the point of the thimble (Fig. 1-40C). Push the top of the spike to make a larger opening and insert strand 1 in this opening (Fig. 1-40D) between the spike and the thimble. The strand should enter the opening in the same direction as the spike.
4. Draw this strand tight and pull it back toward the thimble (Fig. 1-40E), at the same time rotating the spike back, forcing the strand up close to the thimble.
5. Insert the spike from left to right (Fig. 1-40F) and insert strand 2 under strand a. This time the strand is pushed through the opening in the opposite direction (Fig. 1-40G) to that of the spike. Pull tight and force strand 2 back toward the thimble, as with strand 1.
6. Drive the spike under strand b (Fig. 1-40H) and insert strand 3 through the opening (Fig. 1-40I) in the same manner as strand 2, and force it back toward the thimble.
7. Proceed in the same manner, tucking strand 4 between d and c, under c, and out between c and b.
8. Before working strand 5, give strand 4 two wraps around c (Fig. 1-40J). Do not remove the spike. Push it forward about a half turn to make the opening larger. Before making these wraps, the remainder of strand 4 should be untwisted so that when it is wrapped around c it will be flattened instead of round.

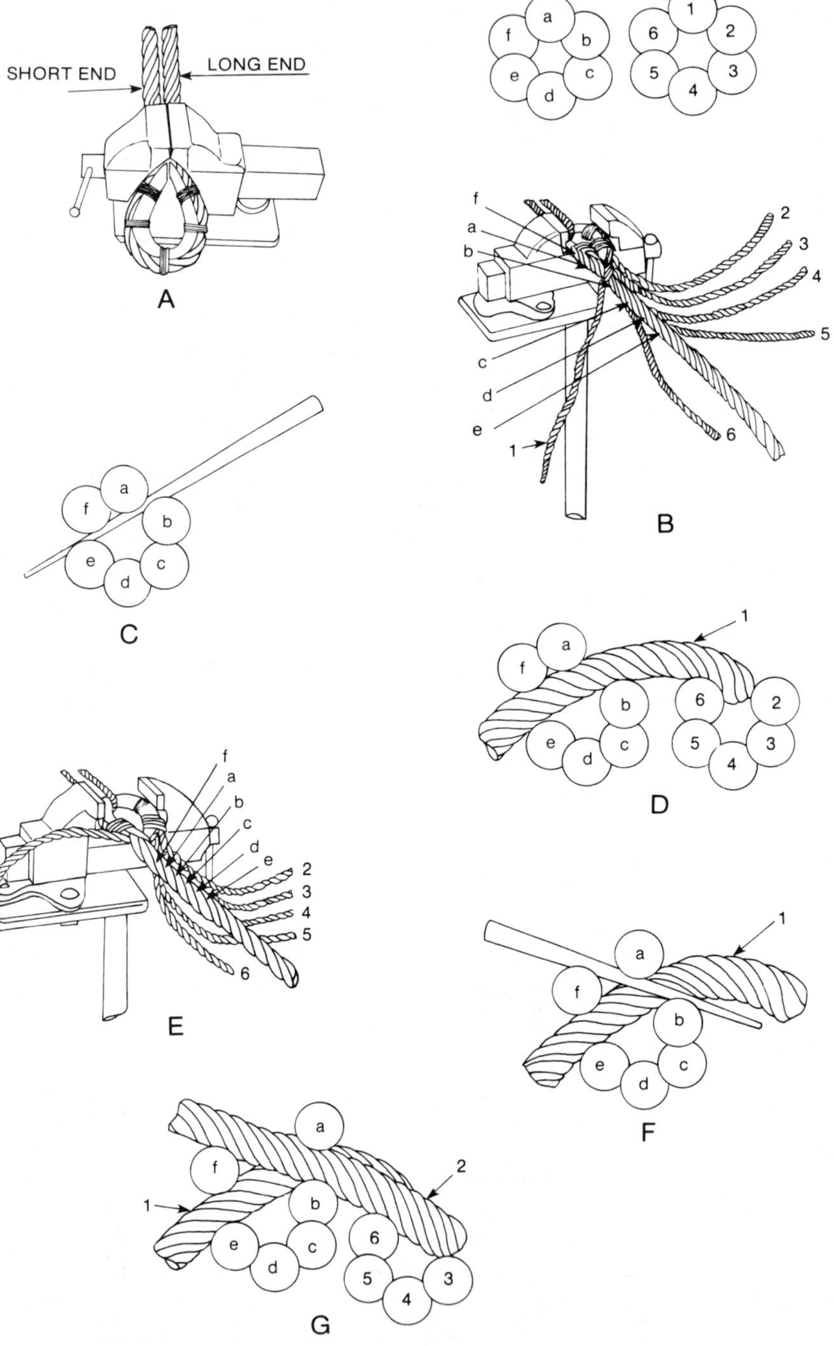

Fig. 1-40: Steps in making a thimble splice.

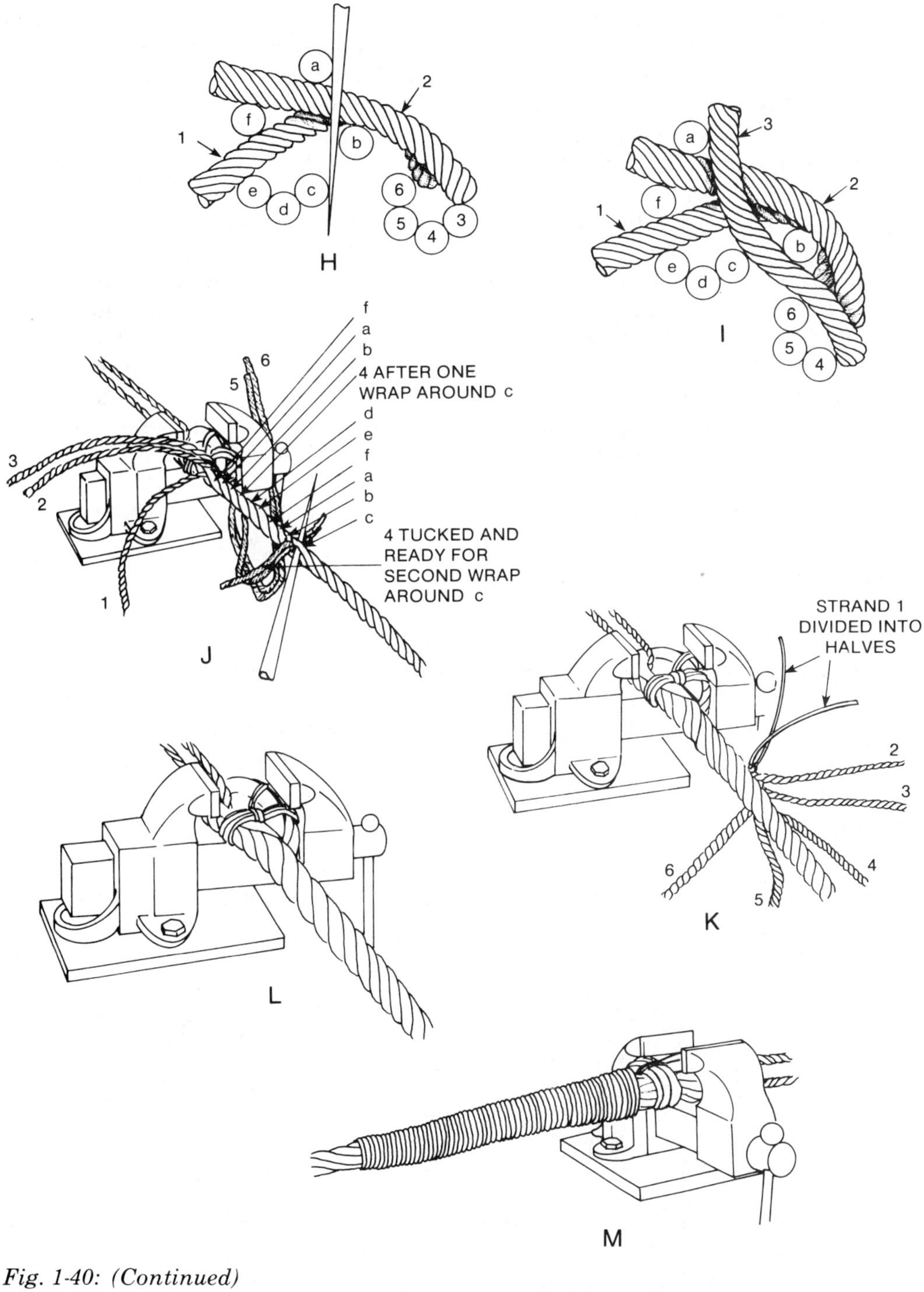

Fig. 1-40: (Continued)

9. Go back to strand 5, and proceed with it in the same way, giving it one tuck under and two wraps around d. Follow with strand 6, giving it a tuck under and two wraps around e. Go back to strand 3, which has been given one tuck, and give it two wraps around b, in the same manner strands 4, 5, and 6 were wrapped. Follow the same procedure with strands 2 and 1.

10. Strands 1, 2, 3, 4, 5, and 6 now have been tucked once under and wrapped twice around strands f, a, b, c, d, and e, respectively. Divide the wires (Fig. 1-40K) in strand 1 into two bundles. Bend one bundle back out of the way and continue wrapping the other bundle (half strand) once around f.

11. Divide strand 2 and wrap half of it around strand a and proceed in the same manner with strands 3, 4, 5, and 6, wrapping half of each once around strands b, c, d, and e, respectively.

12. Clip all projecting ends of the strands close to the splice (Fig. 1-40L). Any inequalities in the splice should now be removed by hammering with a wood or plastic mallet. In order to avoid injuring the hands on the cut ends, and to hold the splice secure, a seizing (Fig. 1-40M) is applied over the entire length of the splice.

Every wire rope manufacturer attaches a different name to his/her particular type of eye splice, but they are usually variations of the three basic types:

Group I: —Flemish eye (rolled eye)
—Flemish eye + serving (Fig. 1-41)
—Flemish eye + pressed metal sleeve (Fig. 1-42)
Group II: —Tucked eye (hand spliced eye or Liverpool splice)
—Tucked eye + serving (Fig. 1-43)
—Tucked eye + pressed metal sleeve (Fig. 1-44)
Group III: —Fold back eye + pressed metal sleeve (Fig. 1-45)

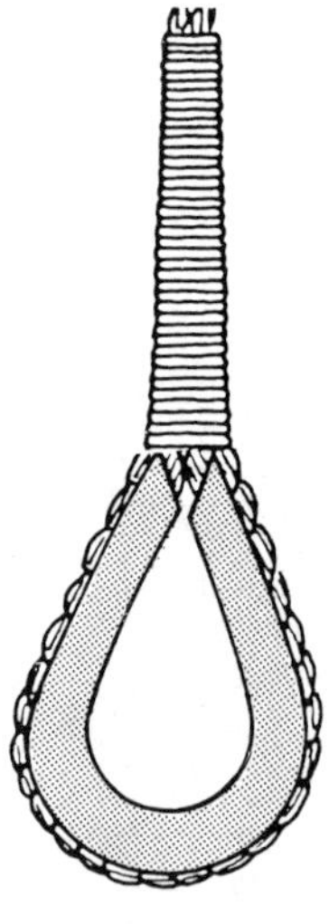

Fig. 1-41: Flemish eye + serving.

Fig. 1-42: Flemish eye + pressed metal sleeve.

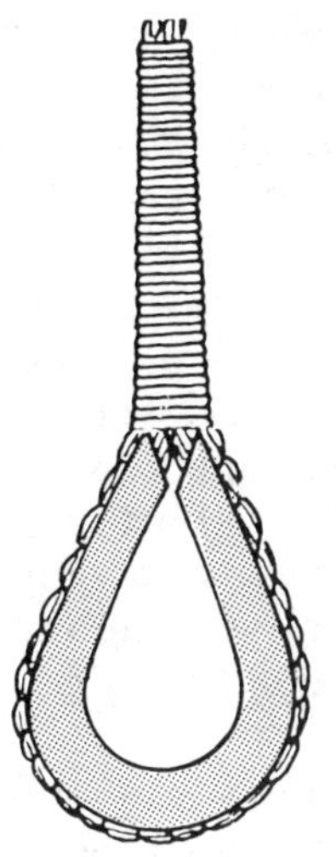

Fig. 1-43: Tucked eye + serving.

Fig. 1-44: Tucked eye + pressed metal sleeve.

Fig. 1-45: Fold back eye + pressed metal sleeve.

A comparison of Figs. 1-41 through 1-44 shows that it is almost impossible to determine what kind of splice you are working with when it is covered by serving wire or pressed metal sleeves. All personnel must be told exactly what type the company uses and all splices must be identical on all slings.

The best and most secure splices are those from Group I, and the best of all and the one recommended for all rigging and hoisting use is the Flemish eye + pressed metal sleeve. It is an excellent attachment. When properly made it develops almost 100% of the catalogue breaking strength. The strand ends of the spliced eye are secured against the live portion of the rope by means of a steel or aluminum sleeve set in place under pressure. Of the various types of mechanical splices, this is the most dependable. The strength of the splice is not wholly dependent on the sleeve since it has the hand splicing to fall back on should the sleeve fail and the tapered end of the sleeve allows the splice to pass over obstructions without snagging.

The splices from Group II (with the exception of the tucked eye + pressed metal sleeve, which is almost 100% efficient) are less secure than those from Group I. The tucked eye unlike the Flemish eye can develop only 70% of the strength of the rope and tends to come free as the rope unwinds. As the ropes untwist, the tucks in the eye begin to pop free. On short lengths of cable or sling, one or two turns will seriously weaken the eye and, consequently, they should not be used for hoisting loads. This, of course, refers to the splice connection only. The addition of wire seizings protects the splice and the rigger's hands, but adds no strength.

All Group II eye splices should have at least five tucks, and the completed splice should be carefully and tightly wrapped with a wire serving to cover the whole of the splice. The efficiencies of the tucked or hand spliced eyes (without pressed metal sleeves) are as given in Table 1-10.

Table 1-10: Efficiencies of Tucked or Hand Spliced Eyes

Rope Diameter	Efficiency
1/4″ and smaller	95%
5/16″—3/4″	90%
7/8″—1″	85%
1-1/8″—1-1/2″	80%
1-5/8″—2″	75%
2-1/8″ and larger	70%

The Group III eye splices, the fold back eyes + pressed metal sleeves, are made by bending the rope to the eye dimension required and securing the free or dead end of the rope against the live portion of the rope by means of a steel or aluminum sleeve set in place under pressure. Some manufacturers unlay or separate the dead end strands and relay them into the grooves between the strands around the body of the live side of the rope. To hold these strands in place, steel ferrules are pressed onto the splice by a hydraulic press.

It is strongly recommended that these splices never be used for overhead hoisting, since improper swaging or split sleeves will result in complete failure without warning (Fig. 1-46).

It is very easy to confuse the excellent mechanically spliced eye with this fold back eye, so be certain that all spliced eyes on the job are of one type or the other, preferably of the mechanically spliced type.

Collet connections (Fig. 1-47) are similar to the fold back eye except that instead of the sleeve being swaged on, it takes the form of a split collet, sleeve

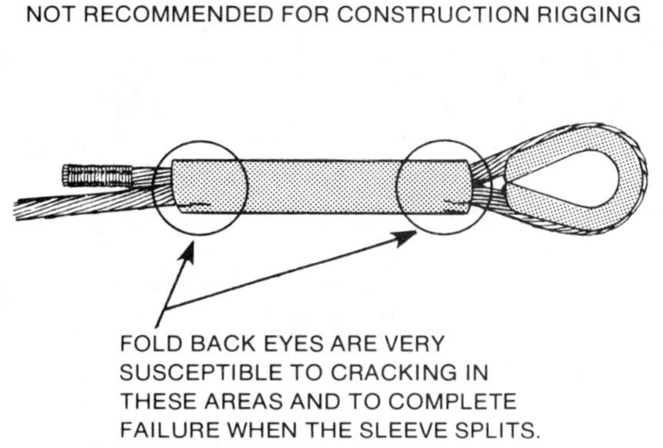

Fig. 1-46: Failure of fold back eye splice.

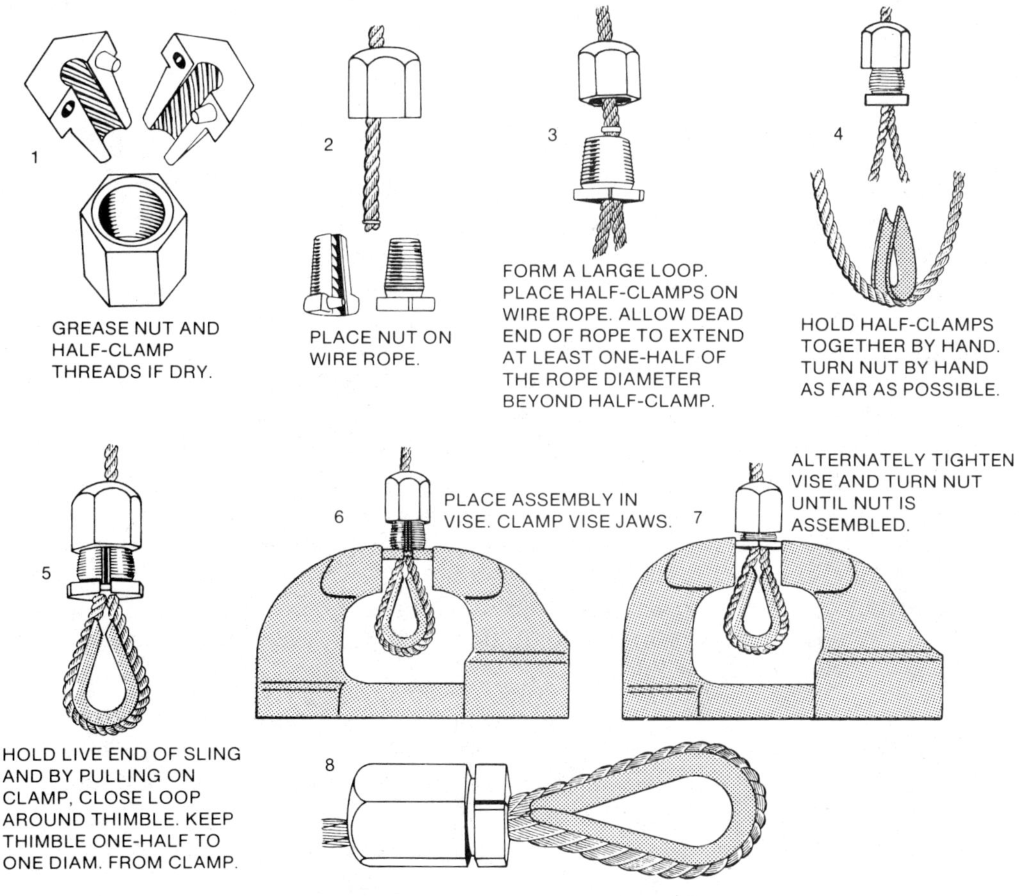

Fig. 1-47: Making a collet connection.

and nut arrangement. The nut is placed on the rope, a loop of the eye size required is made. The collets are placed on the rope and the dead end of the rope is extended at least one-half of the rope diameter beyond them. The collets are held together by hand and the nut turned on as far as possible by hand. The loop is closed around the thimble by holding the live end of the rope and pulling on the clamp. The thimble is kept one-half to one-rope diameter from the clamp. The assembly is then placed in a vise and the nut is torqued.

CARE AND MAINTENANCE OF WIRE ROPE

There are certain maintenance procedures and care precautions that must be followed for the *safe* use of wire rope. They include such things as cleaning and lubrication, inspection, avoiding unnecessary abuses, and proper storage.

Cleaning and Lubrication

Used wire rope should be carefully cleaned of any accumulation of dirt, grit, or other foreign material. Rust should be removed at regular intervals by wire brushing. Scraping or "live" steaming will remove most of the dirt or grit which may accumulate on the wire. The rope should always be carefully cleaned just prior to lubrication. The object of cleaning at that time is to remove all foreign material and old lubricant from the valleys between the strands and from the spaces between the outer wires to permit the newly applied lubricant free entrance into the rope.

Lubrication. During fabrication, wire ropes receive lubrication, the kind and amount depending on the rope's size, type, and anticipated use. This in-process treatment will provide the finished rope with ample protection for a reasonable time if it is stored under proper conditions. But, when the rope is put into service, the initial lubrication may be less than needed for the full useful life of the rope. Because of this possibility, periodic applications of a suitable rope lubricant are necessary. The following are the important characteristics of a good wire rope lubricant:

1. It should be free from acids and alkalis.
2. It should have sufficient adhesive strength to remain on the ropes.
3. It should be of viscosity capable of penetrating the interstices between wires and strands.
4. It should not be soluble in the medium surrounding it under the actual operating conditions.
5. It should have a high film strength.
6. It should resist oxidation.

When it is normal for the wire rope to operate in dirt, rock, or other abrasive material, the lubricant should be selected with great care to make certain that it will penetrate and, at the same time, will not pick up any of the material through which the rope must be dragged. In every case where the wire rope is being stored for any length of time, it should be cleaned and lubricated prior to storage.

As a general rule, the most efficient and most economical means to do field lubrication/protection is by using some method or system that continuously applies the lubricant while the rope is in operation. Many techniques are used; these include the continuous bath, dripping, pouring, swabbing, painting, or where circumstances dictate, automatic systems can be used to apply lubricants either by a drip or pressure spray method (Fig. 1-48).

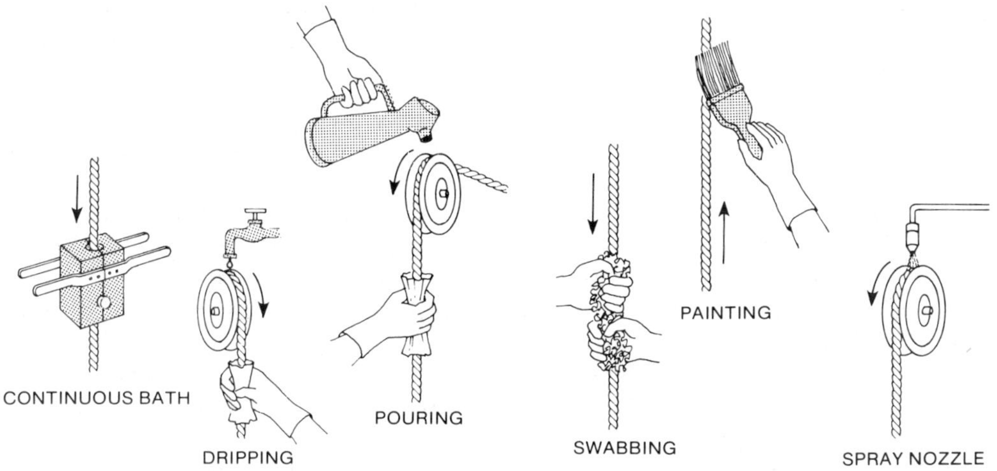

Fig. 1-48: Lubricant application methods in general use today include continuous bath, dripping, pouring, swabbing, painting, and spraying. The arrows indicate the direction in which the rope is moving.

Inspection

Wire rope should be inspected at frequent intervals and frayed, kinked, worn, or corroded rope replaced. The frequency of inspection is determined by the amount of use of the rope. A rope that is used 1 or 2 hours a week requires less frequent inspection than a rope which is in use 24 hours a day.

The common causes of wire rope failures are the use of rope which is of incorrect size, construction, or grade; allowing rope to be dragged over obstacles; not properly lubricating the rope; operating rope over sheaves and drums of inadequate size; overwinding or crosswinding on drums; operating over sheaves and drums out of alignment; using rope when frozen; allowing rope to bind in sheaves; permitting to jump sheaves; subjecting rope to moisture or acid fumes; permitting rope to untwist; and allowing rope to become kinked.

Inspection Procedure. The following is a fairly comprehensive listing of critical inspection factors. It is not, however, presented as a substitute for an experienced inspector. It is rather a user's guide to the accepted standards by which ropes must be judged.

Abrasion. Rope abrades when it moves through an abrading medium or over drums and sheaves. Most standards require that rope is to be removed if the outer wire wear exceeds 1/3 of the original outer wire diameter. This is not easy to determine, and discovery relies upon the experience gained by the inspector in measuring wire diameters of discarded ropes.

Rope Stretch. All ropes will stretch when loads are initially applied. But, as rope deteriorates from wear, fatigue, etc. (excluding accidental damage), continued application of a load of constant magnitude will produce varying amounts of rope stretch. A "stretch" curve plotted for stretch vs. time (Fig. 1-49) displays three discrete phases:

1. *Phase 1*—Initial stretch, during the early (beginning) period of rope service, caused by the rope adjustments to operating conditions (constructional stretch).
2. *Phase 2*—Following break-in, there is a long period—the greatest part of the rope's service life—during which a slight increase in stretch takes place over an extended time. This results from normal wear, fatigue, etc. On the plotted curve—stretch vs. time—this portion would almost be a horizontal straight line inclined slightly upward from its initial level.
3. *Phase 3*—Thereafter, the stretch begins to increase at a quicker rate. This means that the rope is reaching the point of rapid deterioration—a result of prolonged subjection to abrasive wear, fatigue, etc. This second upturn of the curve is a warning indicating that the rope should soon be removed.

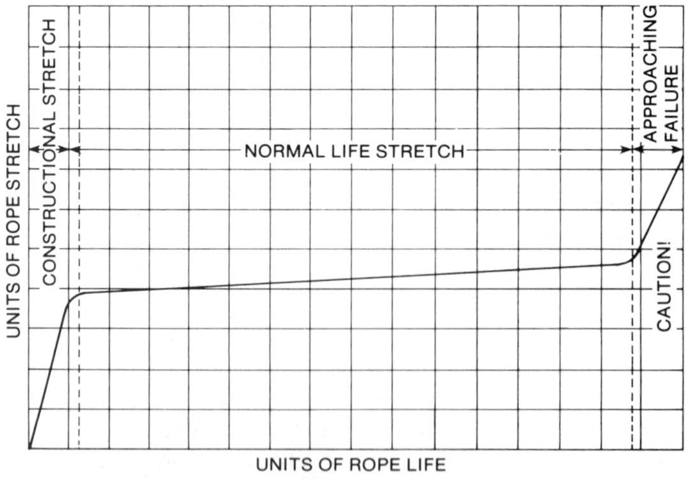

Fig. 1-49: This curve is plotted to show the relationship of wire rope stretch to the various stages of a rope's life.

Reduction in Rope Diameter. Any marked reduction may be attributed to:

1. Excessive external abrasion.
2. Internal or external corrosion.
3. Loosening or tightening of rope lay.
4. Inner wire breakage.
5. Rope stretch.
6. Ironing or milking of strands.

In the past, whether or not a rope was allowed to remain in service depended to a great extent on the rope's diameter at the time of inspection. Currently this practice has undergone significant modification. Previously, a decrease in the rope's diameter was compared with published standards of minimum diameters. The amount of change in diameter is, of course, useful in assessing a rope's condition. But, comparing this figure with a fixed set of values is, for the most part, useless. These long-accepted minima are not, in themselves, of any serious significance since they do not take into account such factors as: (a) variations in compressibility between IWRC and Fiber Core; (b) differences in the amount of reduction in diameter from abrasive wear, or from core compression, or a combination of both; and (c) the actual original diameter of the rope rather than its nominal value. As a matter of fact, all ropes will show a significant reduction in diameter when a load is applied. Therefore, a rope manufactured close to its nominal size may, when subjected to loading, undergo a greater reduction in diameter than that stipulated in the minimum diameter table. Yet, under these circumstances, the rope would be declared unsafe although it may, in actuality, be safe.

As an example of the possible error at the other extreme, we can take the case of a rope manufactured near the upper limits of allowable size. If the diameter has reached a reduction to nominal or slightly below that, the tables would show this rope to be safe. But it should, perhaps, be removed.

Today, evaluations of the rope diameter are first predicated on a comparison of the original diameter—when new and subjected to a known load—with the current reading under like circumstances. Periodically, throughout the life of the rope, the actual diameter should be recorded when the rope is under equivalent loading and in the same operating section. This procedure, if followed carefully, reveals a common rope characteristic: After an inital reduction, the diameter soon stabilizes. Later, there will be a continuous, albeit small, decrease in diameter throughout its life.

Core deterioration, when it occurs, is revealed by a more rapid reduction in diameter and, when observed, it is time for removal. Deciding whether or not a rope is safe is not always a simple matter. A number of different but interrelated conditions must be evaluated. It would be dangerously unwise for an inspector to declare a rope safe for continued service simply because its diameter had not reached the minimum arbitrarily established in a table if, at the same time, other observations lead to an opposite conclusion. Because criteria for removal are varied, and because diameter, in itself, is a vague criterion, the table of minimum diameters has been deliberately omitted from this text.

Corrosion. Corrosion, while difficult to evaluate, is a more serious cause of degradation than abrasion. Usually, it signifies a lack of lubrication. Corrosion will often occur internally before there is any visible external evidence on the rope surface. Pitting of wires is a cause for immediate rope removal. Not only does it attack the metal wires, but it also prevents the rope's component parts from moving smoothly as it is flexed. Usually, a slight discoloration because of rusting merely indicates a need for lubrication.

Severe rusting, on the other hand, leads to premature fatigue failures in the wires, necessitating the rope's immediate removal from service. When a rope shows more than one wire failure adjacent to a terminal fitting, it should be removed immediately. To retard corrosive deterioration, the rope should be kept well lubricated. In situations where extreme corrosive action can occur, it may be necessary to use galvanized wire rope.

Kinks. Kinks are permanent distortions caused by loops drawn too tightly. Ropes with kinks must be removed from service.

"Bird Caging." Bird caging results from torsional imbalance that comes about because of mistreatments such as sudden stops, the rope being pulled through tight sheaves, or wound on too small a drum. This is cause for rope replacement unless the affected portion can be removed.

Localized Conditions. Particular attention must be paid to wear at the equalizing sheaves. During normal operations, this wear is not visible. Excessive vibration or whip can cause abrasion and/or fatigue. Drum crossover and flange point areas must be carefully evaluated. All end fittings, including splices, should be examined for worn or broken wires, loose or damaged strands, cracked fittings, worn or distorted thimbles, and tucks of strands.

Heat Damage. After a fire or the presence of elevated temperatures, there may be metal discoloration or an apparent loss of internal lubrication. Fiber core ropes are particularly vulnerable. Under these circumstances the rope should be replaced.

Protruding Core. If, for any reason, the rope core protrudes from an opening between the strands, the rope will be unfit for service.

Scrubbing. Scrubbing refers to the displacement of wires and strands as a result of rubbing around or against an object. This, in turn, causes wear and displacement of wires and strand along one side of the rope. Corrective measures should be taken as soon as this condition is observed.

Fatigue Failure. Wires that break with square ends and show little surface wear have usually failed as a result of fatigue. Such failures can occur on the crown of the strands or in the valleys between the strands where adjacent strand contact exists. In almost all cases, these failures are related to bending stresses or vibration.

If diameter of the sheaves, rollers, or drum cannot be increased, a more flexible rope should be used. But, if the rope in use is already of maximum flexibility, the only remaining course that will help prolong its service life is to move the rope through the system by cutting off the dead end. By moving the rope through the system, the fatigued sections are moved to less fatiguing areas of the reeving. This technique is most frequently used in rotary drilling.

Damaged End Attachments. Cracked, bent, or broken end fittings must be eliminated. The cause should be sought out and corrected. In the case of bent hooks, the throat openings—measured at the narrowest point—should not exceed 15% over normal nor should twisting be greater than 10°.

Peening. Continuous pounding is one of the causes of peening. The rope strikes against an object such as some structural part of the machine, it beats against a roller, or it hits itself. Often, this can be avoided by placing protectors between the rope and the object it is striking. Another common cause of peening is continuous passage—under high tension—over a sheave or drum. Where peening action cannot be controlled, it is necessary to have more frequent inspections and to be ready for earlier rope replacement. Figure 1-50 shows the external appearance of two ropes, one of which has been abraded and the other peened. Also shown are the cross sections of both wires in these conditions.

Broken Wires. The number of broken wires on the outside of a wire rope are an indication of: (a) its general condition, and (b) whether or not it must be considered for replacement (Fig. 1-51). Frequent inspection will help determine the elapsed time between breaks. Ropes should be replaced as soon as the wire breakage reaches the numbers given in Table 1-11. Such action must be taken without regard to the type of fracture.

On occasion, a single wire will break shortly after installation. However, if no other wires break at that time, there is no need for concern. On the other hand, should more wires break, the cause should be carefully investigated.

On any installation, valley breaks, i.e., where the wire ruptures between strands, should be given serious attention. When two or more such conditions

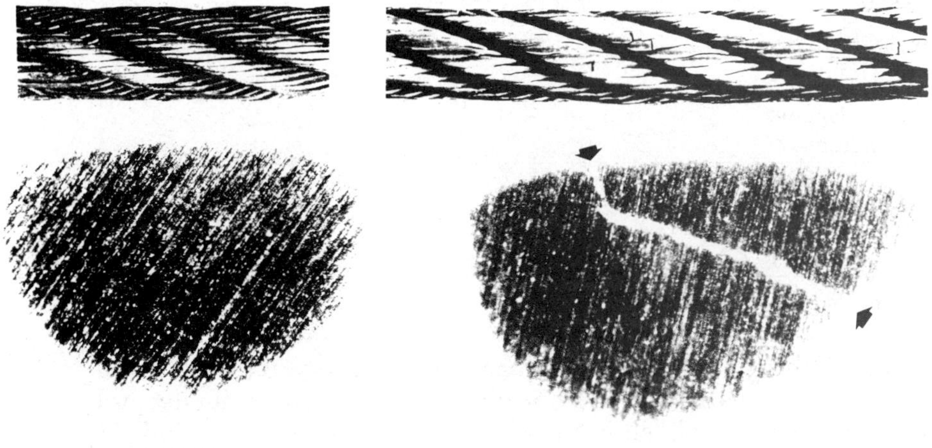

ABRASION PEENING

Fig. 1-50: These plan views and cross sections show the effects of abrasion and peening on wire rope. Note that a crack has formed as a result of heavy peening.

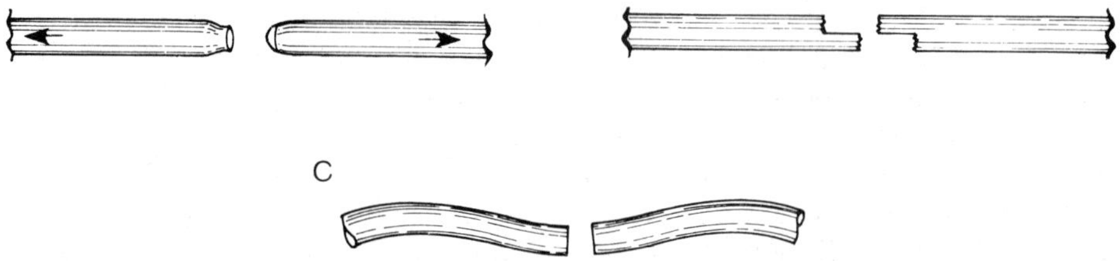

Fig. 1-51: (A) A wire that has broken under a tensile load in excess of its strength is recognized by the "cup and cone" configuration at the fracture point. The necking down of the wire at point of failure shows that failure occurred while the wire retained its ductility. (B and C) A fatigue break is usually characterized by squared-off ends perpendicular to the wire either straight across or Z-shaped.

Table 1-11: When to Replace Wire Rope—Based on Number of Broken Wires

	Number Broken Wires In Running Ropes		Number Broken Wires In Standing Ropes	
Equipment	**In One Rope Lay**	**In One Strand**	**In One Rope Lay**	**At End Connection**
Overhead & Gantry Cranes	12	4	Not Specified	
Portal, Tower & Pillar Cranes	6	3	3	2
Crawler, Locomotive & Truck Cranes	6	3	3	2
Derricks	6	3	3	2
Base Mounted Drum Hoists	6	3	3	2
Floating Cranes and Derricks	6	3	3	2
Personnel Hoists	6*	3	2*	2
Material Hoists	6*	Not Specified	Not Specified	

*Also remove for 1 valley break.

are found, the rope should be replaced immediately. It is well to remember that once broken wires appear in a normal rope operating under normal conditions, a good many more will show up within a relatively short period. Attempting to squeeze the last measure of service from a rope beyond the allowable number of broken wires (Table 1-11) will create an intolerably hazardous situation.

A diagnostic guide to some of the most prevalent rope abuses is given in Table 1-12. Some of the more common abuses are illustrated in Fig. 1-52.

Table 1-12: Diagnostic Guide to Common Wire Rope Abuses

Abuse	Symptoms	Possible Causes
Fatigue	Wire break is transverse—either straight across or Z shape. Broken ends will appear grainy.	Check for rope bent around too small a radius; vibration or whipping; wobbly sheaves; rollers too small; reverse bends; bent shafts; tight grooves; corrosion; small drums & sheaves; incorrect rope construction; improper installation; poor end attachments. All running rope, if left in service long enough, will eventually fail by fatigue.
Tension	Wire break reveals predominantly cup and cone fracture with some 45° shear breaks.	Check for overloads; sticky, grabby clutches; jerky conditions; loose bearing on drum; fast starts, fast stops; broken sheave flange; wrong rope size & grade; poor end attachments. Check for too great a strain on rope after factors of deterioration have weakened it.
Abrasion	Wire brake mainly displays outer wires worn smooth to knife edge thinness. Wire broken by abrasion in combination with another factor will show a combination break.	Check for change in rope or sheave size; change in load; overburden change; frozen or stuck sheaves; soft rollers, sheaves, or drums; excessive fleet angle; misalignment of sheaves; kinks; improperly attached fittings; grit & sand; objects imbedded in rope; improper grooving.
Cut, gouged, or rough wire	Wire ends are pinched down, mashed, and/or cut in a rough diagonal shear-like manner.	Check on all the above conditions for mechanical abuse, or either abnormal or accidental forces during installation.
Torsion or twisting	Wire ends show evidence of twist and/or cork-screw effect.	Check on all the above conditions for mechanical abuse, or either abnormal or accidental forces during installation.
Mashing	Wires are flattened and spread at broken ends.	Check on all the above conditions for mechanical abuse, or either abnormal or accidental forces during installation.
Corrosion	Wire surfaces are pitted with break showing evidence either of fatigue, tension or abrasion.	Indicates improper lubrication or storage.

Table 1-12: (Cont'd)

Abuse	Symptoms	Possible Causes
Abrasion plus fatigue	Reduced cross section is broken off square thereby producing a chisel shape.	A long term condition normal to the operating process.
Abrasion plus tension	Reduced cross section is necked down as in a cup and cone configuration. Tensile break produces a chisel shape.	A long term condition normal to the operating process.

A THIS ROPE WAS DAMAGED BY BEING ROLLED OVER SOME SHARP OBJECT.

B THIS EFFECT OF DRUM CRUSHING IS EVIDENCE OF BAD WINDING CONDITIONS.

C AN OCCURRENCE THAT IS CALLED A POPPED CORE.

D THIS IS A TYPICAL BIRD CAGE CONDITION.

E THIS IS THE APPEARANCE OF A TYPICAL TENSION BREAK; A RESULT OF OVERLOADING.

F HERE THE STRAND WIRES WERE SNAGGED.

G

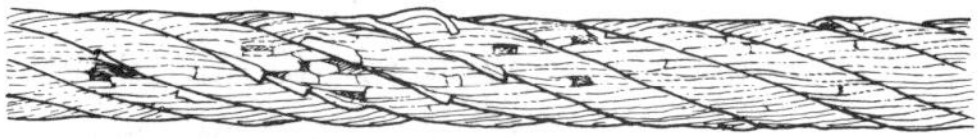

H A) SERIOUS WEAR RESULTING FROM EXCESSIVE BENDING, AND B) LOCALIZED WEAR BROUGHT ABOUT BY POOR CUT-OFF PRACTICE.

I THIS IS AN ILLUSTRATION OF A SERIOUS CONDITION WHERE THE ROPE SLIDES OVER OR AGAINST ITSELF.

J AN ILLUSTRATION OF VALLEY TYPE FATIGUE BREAKS. FLEXING THE ROPE EXPOSES BROKEN WIRES HIDDEN IN VALLEYS BETWEEN STRANDS.

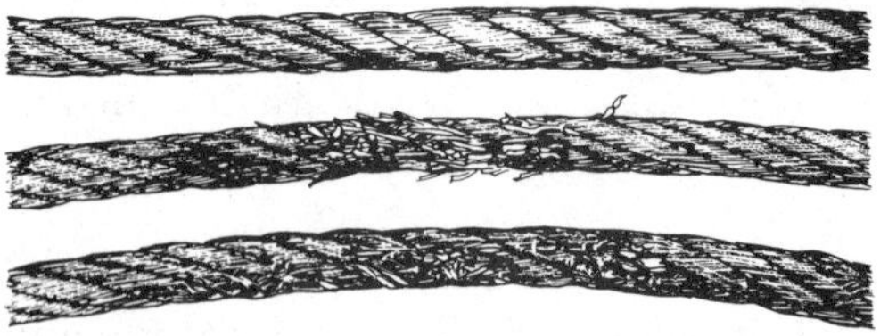

K THESE DAMAGES WERE THE RESULT OF BAD DRUM WINDING.

Fig. 1-52: Examples of wire rope abuse.

Reversing Ends. To lengthen the service of a wire rope, it is sometimes advisable to reverse ends or cut back the ends to get more service from wire rope. Reversing ends is more satisfactory than just cutting ends, because frequently the wear and fatigue on a rope are more severe at certain points than at others. Reversing distributes other, stronger parts of the rope to the points that cause wear and fatigue. To reverse ends, remove the drum end and put it in the attachment. Then, fasten the end taken from the attachment to the drum. Cutting back the ends has a similar effect, but there is not as much change involved. In reversing ends, cut off short lengths of both ends to remove the sections that have sustained the greatest local fatigue.

Inspection of Sheaves and Drums. Under normal conditions, machines receive periodic inspections, and their overall condition is recorded. Such inspections usually include the drum, sheaves, and any other parts that may come into contact with the wire rope and subject it to wear. As an additional precaution, rope-related working parts should be reinspected prior to the installation of a new wire rope.

The very first item to be checked when examining sheaves, rollers, and drums is the condition of the grooves (Figs. 1-53, 1-54, and 1-55). To check the size, contour, and amount of wear, a *groove gauge* is used. As shown in Fig. 1-53, the gauge should contact the groove for about 150° of arc.

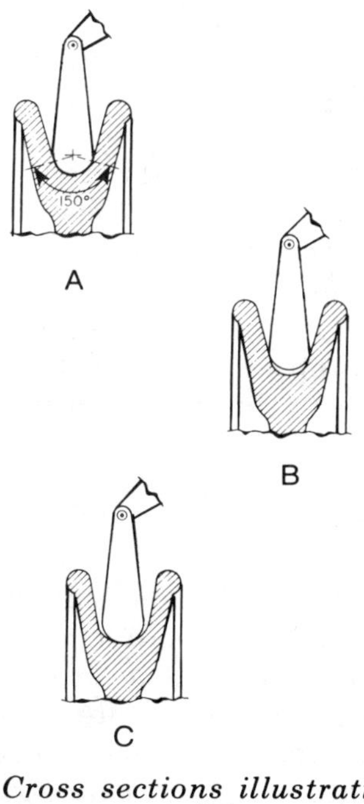

Fig. 1-53: Cross sections illustrating three sheave-groove conditions revealed by the metric arrangement of wires in the strand: (A) Correct; (B) too tight; and (C) too loose.

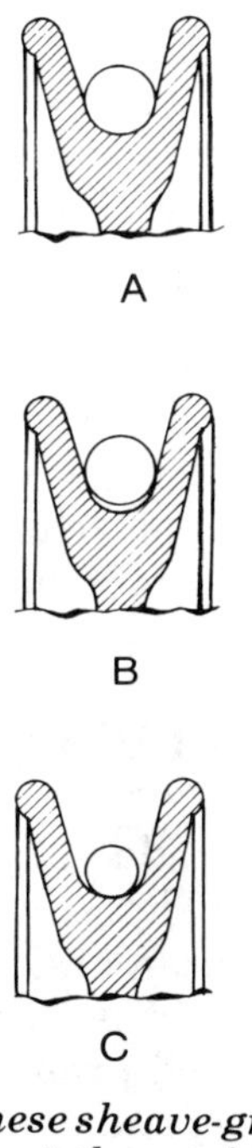

Fig. 1-54: These sheave-groove cross sections represent three wire rope seating conditions: (A) A new rope in a new groove; (B) a new rope in a worn groove; and (C) a worn rope in a worn groove.

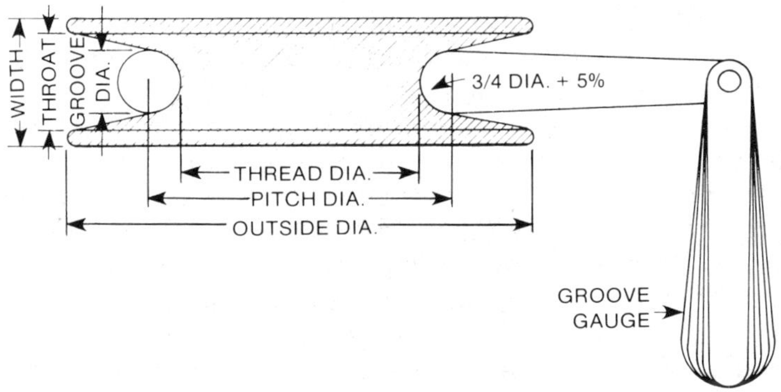

Fig. 1-55: Illustrating the various dimensions of a sheave, and the use of a groove gauge.

Two types of groove gauges are in general use and it is important to note which of these is being used. The two differ by their respective percentage *over nominal.*

For new or remachined grooves, the groove gauge is nominal plus the full oversize percentage. The gauge carried by most wire rope representatives today is used for worn grooves and is made nominal plus 1/2 the oversize percentage.

This latter gauge is intended to act as a sort of "no-go" gauge. Any sheave with a groove smaller than this *must* be regrooved or, in all likelihood, the existing rope will be damaged.

When the sheave is regrooved, it should be machined to the dimensions for "new and machined" grooves given in Table 1-13. This table lists the requirements for new or remachined grooves, giving the groove gauge diameter in terms of the nominal wire rope diameter plus a percentage thereof. Similarly, the size of the "no-go" gauge is given, against which worn grooves are judged. Experience has clearly demonstrated that the service life of the wire rope will be materially increased by strict adherence to these standards.

Storage of Wire Rope

Wire rope should be coiled on a spool for storage and should be properly tagged as to size and length. It should be stored in a dry place to reduce corrosion and kept away from chemicals and fumes which might attack the metal. Prior to storage, wire rope should always be cleaned and lubricated. If the lubricant film is applied properly and the wire is stored in a place protected from the weather, corrosion will be virtually eliminated. Rusting, corrosion of the wires, and deterioration of the fiber core sharply decrease the strength of the rope. It is not possible to estimate accurately the loss in strength from these effects.

Table 1-13: Minimum Sheave- and Drum-Groove Dimensions*

Nominal Rope Diameter		Groove Radius			
		New		Worn	
inches	mm	inches	mm	inches	mm
1/4	6.5	.137	3.48	.129	3.28
5/16	8.0	.167	4.24	.160	4.06
3/8	9.5	.201	5.11	.190	4.83
7/16	11	.234	5.94	.220	5.59
1/2	13	.271	6.88	.256	6.50
9/16	14.5	.303	7.70	.288	7.32
5/8	16	.334	8.48	.320	8.13
3/4	19	.401	10.19	.380	9.65
7/8	22	.468	11.89	.440	11.13
1	26	.543	13.79	.513	13.03
1-1/8	29	.605	15.37	.577	14.66
1-1/4	32	.669	16.99	.639	16.23
1-3/8	35	.736	18.69	.699	17.75
1-1/2	38	.803	20.40	.759	19.28
1-5/8	42	.876	22.25	.833	21.16
1-3/4	45	.939	23.85	.897	22.78
1-7/8	48	1.003	25.48	.959	24.36
2	51	1.070	27.18	1.019	25.88
2-1/8	54	1.137	28.88	1.079	27.41
2-1/4	58	1.210	30.73	1.153	29.29
2-3/8	61	1.273	32.33	1.217	30.91
2-1/2	64	2.338	33.99	1.279	32.49
2-5/8	67	1.404	35.66	1.339	34.01
2-3/4	71	1.481	37.62	1.409	35.79
2-7/8	74	1.544	39.22	1.473	37.41
3	77	1.607	40.82	1.538	39.07
3-1/8	80	1.664	42.27	1.598	40.59
3-1/4	83	1.731	43.97	1.658	42.11
3-3/8	87	1.807	45.90	1.730	43.94
3-1/2	90	1.869	47.47	1.794	45.57
3-3/4	96	1.997	50.72	1.918	48.72
4	103	2.139	54.33	2.050	52.07
4-1/4	109	2.264	57.51	2.178	55.32
4-1/2	115	2.396	60.86	2.298	58.37
4-3/4	122	2.534	64.36	2.434	61.82

*Values given are applicable to grooves in sheaves and drums; they are not generally suitable for pitch design since this may involve other factors.

Table 1-13: (Cont'd)

Nominal Rope Diameter		Groove Radius			
		New		Worn	
inches	mm	inches	mm	inches	mm
5	128	2.663	67.64	2.557	64.95
5-1/4	135	2.804	71.22	2.691	68.35
5-1/2	141	2.929	74.40	2.817	71.55
5-3/4	148	3.074	78.08	2.947	74.85
6	154	3.198	81.24	3.075	78.16

2

Fiber Rope

INTRODUCTION

Basically, fiber rope is made of either vegetable or synthetic fibers. Vegetable fibers include manila, sisal, hemp, cotton, and flax. Synthetic fibers include nylon, dacron (polyester), polyethylene, and polypropylene. While manila is still in use, most of the fiber rope used by the rigger is of the synthetic type. Of course, most of the rope used in rigging, as mentioned earlier, will be made of wire.

Manila Rope

Of the vegetable or natural fibers, the only type available in quantity today is manila rope, which is made from the leaf stems of the stalk of the abaca plant in the banana family. Varying in length from 4′ to 15′ (1.22 to 4.57 m) in their natural state, the fibers have the length and quality to give manila rope relatively high elasticity, strength, and resistance to wear and deterioration, when compared to other vegetable fibers such as sisal, hemp, agave, coir, and cotton.

Manila fiber varies in color from a yellow white to dark brown. The quality of the manila fiber varies with the color. The lighter shades are soft and clean and are used in the best rope. The darker shades, used in poorer quality rope, are coarse and harsh. Most manila rope used in rigging work is of the so-called *Number 1 grade*, which is very light in color.

Most rigging manila rope consists of three strands, each containing a number of threads. The threads are usually twisted right, the strands to the left, and the rope to the right (Fig. 2-1A). This alternate winding helps to eliminate the rope's natural tendency to untwist. In the manufacture of fiber rope, the threads and strands are laid into the rope either loosely (soft-laid) or tightly (hard-laid). Soft-laid rope is slightly stronger than hard-laid but is less resistant to abrasion. For general rigging use, a medium-laid rope is recommended.

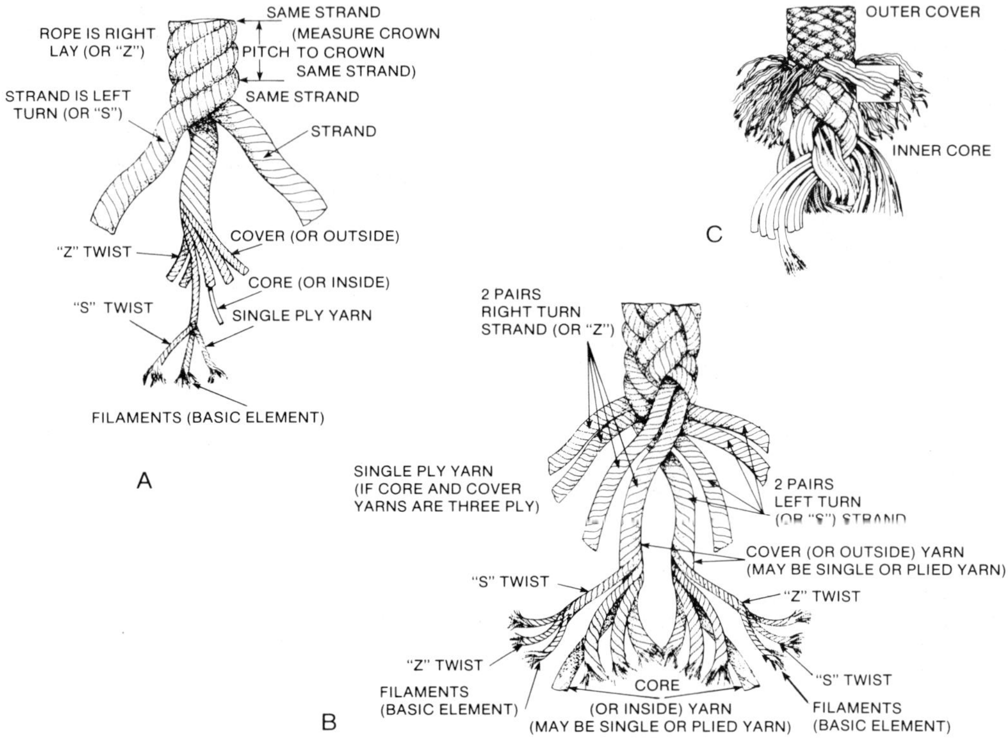

Fig. 2-1: Three popular rope windings: (A) Three-strand rope; (B) eight-strand plaited rope; and (C) double-braided rope.

Synthetic Fiber Ropes

While natural fibers are relatively short and must be overlapped, those employed in synthetic types of rope are continuous for the entire length of the rope. The synthetic thread may be a single filament (monofilament) or made up of several filaments (multifilament). The filament(s) may be laid up in any of the three ropes shown in Fig. 2-1. The synthetics may be of nylon, dacron, polyethylene, and polypropylene. New synthetic fibers are now in the experimental stages.

Nylon Rope. Nylon rope is two and one-half times stronger than manila and is the most widely used of the synthetic rope family. It is chalky white in color, has a smooth surface, is soft and pliant, and has a feeling of elasticity.

Some properties of nylon ropes are:

1. High breaking strength (wet or dry).
2. Light weight per unit of strength.
3. Excellent elasticity and tensile recovery.

4. Superior absorption of impact and shock loads.
5. Excellent flex and abrasion resistance.
6. Good flexibility.
7. Excellent resistance to rot.
8. Good sunlight and weather resistance.
9. High melting point.

Due to its continuous fiber construction, nylon rope is strong and resistant to creep under sustained load. Its elasticity recommends its use where high energy absorption is needed or where shock loading is a factor. The high degree of stretch can be a serious disadvantage if headroom for lifting is restricted and slings must be as short as possible. Under actual service conditions, nylon rope will stretch to about 16% under working loads and over 40% under breaking loads. The initial loading of a nylon rope produces a permanent elongation of approximately 8%, but recovery is complete from subsequent stretching under load. Therefore, nylon ropes should be broken in prior to field use.

Nylon rope tolerates heat without loss of strength or physical properties up to 300°F (melting point 482°F). Nylon ropes absorb moisture and lose approximately 10% of their strength when wet, as well as becoming very slippery. Full recovery occurs when the rope dries out. Figure 2-2 compares nylon and manila ropes under a heavy sustained load.

Store nylon rope away from heat and exposure to sunlight. Nylon is highly resistant to alkalies, but can be degraded by most acids, paints, and linseed oil. All contact with chemicals should be avoided. If contact is suspected, wash the rope thoroughly in cold water and then carefully examine the fibers for evidence of weakening.

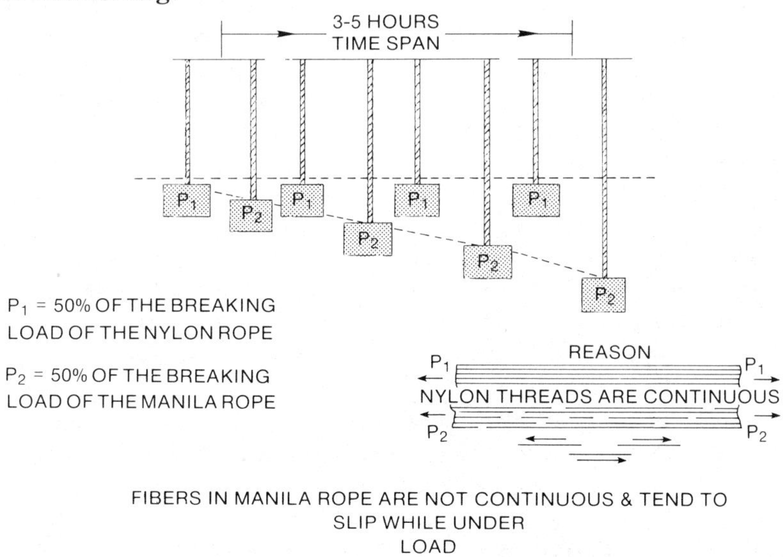

Fig. 2-2: Comparison of nylon and manila ropes under heavy, sustained load.

Braided Nylon Rope. Braided nylon rope provides the highest possible strength since the load is divided equally between a braided sheath and a braided core. The rope is soft and flexible and does not twist or kink. Twisted three strand rope transmits a turning, twisting motion under load while braided nylon does not.

Braided nylon presents approximately 50% more surface area for wear and grip, is stronger for size and more stable, exhibits less stretch when working, and has less permanent stretch and greater flexibility than three strand nylon ropes.

The braided sheath and core construction is available in synthetic combinations in addition to nylon:

1. Nylon core/nylon cover—high strength, more stretch, good wearing qualities.
2. Polypropylene core/nylon cover—lower strength, lower stretch.
3. Polypropylene core/polyester cover—very low stretch, high strength, and good abrasion resistance.

For example, a nylon core/nylon cover rope with a nominal diameter of 7/8″ carries a safe working load of 4,800 pounds, while a polypropylene core/nylon cover rope will carry a safe working load of 4,150 pounds. A polypropylene core/polyester cover rope will support a safe working load of 3,800 pounds. The safety factor for braided synthetic combinations is 5.

Polyester Rope. Polyester rope (trade names: Dacron, Terylene) is nearly identical to nylon in appearance but has little or no elastic feeling. Size for size, it is heavier and not as strong as nylon rope, although similar in construction. This is a continuous filament rope and resists creep under sustained loads. The low stretch properties of polyester rope represent a considerable advantage where headroom is limited since, after the initial permanent stretch of approximately 6%, subsequent loads produce a temporary lengthening of approximately 5.9% (compared to 16% for nylon). Polyester ropes should be broken in before field use.

Abrasion resistance is similar to nylon rope. Like nylon, polyester rope is unaffected by temperatures up to 300°F. It absorbs much less water than nylon and thus does not lose strength when wet. Polyester rope resists sun and weather damage as well as attack by rot and mildew; however, precautionary storage practices are recommended. Polyester is more resistant to chemicals than nylon, but exposure and contact should be avoided. Slings should be washed frequently in cold water.

Polypropylene Rope. Polypropylene and polyester ropes are almost identical in appearance; they are available in various colors and are smooth and pliant and somewhat slippery, particularly polypropylene.

A singular advantage of polypropylene rope is its ability to float on water; however, it is not as strong as nylon or polyester and ranks between natural fiber ropes and the more sophisticated nylon and polyester.

Stretch properties of polypropylene ropes vary with the type of construction; however, they generally exceed those of polyester. Energy absorption is approximately half that of nylon.

Polypropylene ropes have a heat tolerating capacity comparable to nylon and polyester; however, some deterioration occurs in sunlight. Chemical resistance is generally good as is resistance to rot and mildew. An important feature is polypropylene rope's safety around electricity as it is a nonconductor because it absorbs no water.

Polyethylene Rope. Polyethylene ropes are low in strength compared with the other synthetic fiber ropes. They tolerate heat only up to 250°F and undergo some deterioration in sunlight. They are, however, resistant to chemicals except sulfuric acid.

CHARACTERISTICS OF FIBER ROPE

The three characteristics of any rope—whether it be fiber or wire—are size, strength, and weight. The latter, in the case of fiber rope, varies with type of fiber, weather conditions, use, fiber arrangement, added preservatives, and other factors. For this reason, when the weight of the rope is an important factor, it is wise to check the manufacturer's property specifications.

Size

Wire and fiber rope sizes are designated by inches of diameter and circumference. Wire rope is always designated by diameter; fiber rope is usually designated by diameters up to 7/8″ (22.23 mm). Then, it is designated by circumference up to 12″ (3.66 m) or more. Since this is only a "usual" standard, most tables give both the diameter and circumference of fiber rope. By the way, the circumference is approximately three times the size of the diameter; that is, a rope with a 3″ circumference would have a diameter of 1″.

Strength

Table 2-1 gives the working strength of a three-strand rope of various materials when subjected to direct pull. A safety factor of 5 has been used for synthetic fiber ropes, 10 for natural fiber ropes. Since the factor of safety (FS) is the breaking strength (BS) divided by the safe working load (SWL), the breaking strength of rope can be found as follows:

$$BS = FS \times SWL$$

Or, in the case of 1/2″ nylon rope:

$$BS = 5 \times 1460 \text{ or } 7300 \text{ pounds.}$$

In the case of the same size manila rope, the breaking strength would be:

$$10 \times 265 \text{ or } 2650 \text{ pounds.}$$

The safety factor should always be used in determining the size of rope because the breaking strength becomes reduced after use and exposure to weather conditions. In addition, a safety factor is required because of shock loading, knots, sharp bends, and other stresses which the rope may have to withstand during its use. Some of these stresses reduce the strength of rope as much as 50%. If the tables are not available, the safe working capacity may be closely approximated by an old sailor's rule of thumb. The rule of thumb for the safe working load, in tons, for fiber rope is equal to the square of the rope diameter in inches ($SWL = D^2$). The safe working capacity, in tons, of a 1/2″ diameter fiber rope would be 1/2″ squared or 1/4 ton. The rule of thumb will allow a safety factor of approximately 4.

CARE OF FIBER ROPES

The useful life of fiber rope will be shortened considerably and its strength decreased rapidly by improper storage, exposure to dampness, or damage to the strands. Loose ends of fiber rope should always be tied in a knot to prevent unlaying of the rope, or the ends should be whipped. When fiber rope is to be cut, put two whippings on the rope 1 or 2″ apart and make the cut between the whippings.

Storage

When fiber rope is being stored away for either short or long periods, care should be taken to prevent deterioration of the rope.

1. Do not store in wet or damp places.
2. Dry carefully before storing.
3. If possible, always store fiber rope on gratings or in some other manner that will allow circulation of air through the coil.

Dampness

Fiber rope deteriorates rapidly from continued dampness. Rope tends to contract when wet and, unless allowed to do so, may be injuriously strained.

Table 2-1: Properties of Fiber and Synthetic Ropes

Properties	Manila 1st Grade	Filament Nylon	Filament Dacron	Polyeth-ylene	Polyprop-ylene
Safety Factor Used	10	5	5	5	5
Strength, dry (Safe Working Loads)					
1/4 in. dia.	60#	360#	240#	210#	250#
3/8	135	800	520	420	540
1/2	265	1,460	900	720	900
5/8	440	2,180	1,380	1,080	1,320
3/4	540	3,120	1,900	1,400	1,600
7/8	770	4,280	2,520	1,900	2,150
3 in. Circ.	900	5,400	2,300	2,400	2,500
3-1/2 in.	1,200	6,700	4,200		3,400
4	1,500	9,300	5,600		4,400
4-1/2	1,850	11,300	7,000		5,600
5	2,250	14,000	8,200		7,000
5-1/2	2,650	16,650	10,000		8,400
6	3,100	19,600	12,000		10,000
Strength, wet Dry strength = 100	100+	85-90	100	105	105
Weight, dry Dry manila rope = 100	100	84	104	66	63
Flexing, Resistance to repeated	F			G	E
Abrasion, Resistance to	F	G	G	G	G
Acids, Resistance to	P	F	G	E	G
Alkalis, Resistance to		G	G	E	G
Oils, Resistance to	G	*E			
Sun light, Resistance to	P	P	E	P	
Solvents, Resistance to		E		E	G
Mildew, decay, organisms, Resistance to	F	E	E	E	
High Temperature, Use at	P				E
Low Temperature, Use at	F	F		G	G
Wet, Use when	P	G		**G	F
Stretch under load	A	H	A+	H	
Dielectric strength, dry	G			E	E
Splicing, Ease of	G		G		
Slippery to handle	N	Y		Y	F
Work Life	A	G		G	

E = excellent A = average G = good F = fair P = poor H = high Y = yes N = no *Except linseed oil **Floats on water.

1. Slacken dry, taut lines before they are exposed to rain or damp weather.
2. Do not cover fiber rope unless it is absolutely necessary, as covering holds in any moisture present and prevents discovery of deterioration.

Damage to Strands

The strength of fiber rope decreases rapidly with use because of breakage and slipping of the fibers. The fibers will slip a small amount under each strain in spite of the twisting, and no attempt should be made to load a rope to its maximum after it has been used for some time. Breakage of fibers should be avoided as much as possible.

1. Clean muddy rope by washing in water.
2. Avoid pulling rope over sharp edges. Place a board between the rope and the sharp edge to reduce breakage of the strands.
3. Sand has an abrasive action on the inner fibers of rope. Avoid dragging rope through sand or dirt. If sand gets on the rope, carefully wash the rope in water.
4. Wherever possible, use knots that can be untied easily to eliminate the necessity for cutting the rope.
5. Repair broken strands in rope as soon as possible.

Coiling and Uncoiling

New manila rope is coiled, bound, and wrapped in burlap as a protective covering. Since the covering protects the rope during storage and prevents tangling, it should not be removed until the rope is to be used. To open, strip back the burlap wrapping and look inside the coil for the end of the rope. It should be at the bottom of the coil. If it is not, turn the coil over so that the end will be at the bottom. Pull the end of the rope up through the center as shown in Fig. 2-3; as the rope comes up through the coil, it will unwind in a counterclockwise direction.

Once line is removed from the manufacturer's coil, it may be made up (disposed of for storage or ready for use) either by winding on a drum or by coiling down.

Coiling down a rope means laying it up in circles, roughly one on top of the other (Fig. 2-4). Always coil down right-laid line right-handed or clockwise. When a line is coiled down, the top end is ready to run off. Trying the bottom end will cause the line to kink. If for some reason the bottom end must go first, it is necessary to turn over the coil to free it for running.

Coiling up a rope can be accomplished as shown in Fig. 2-5. Loop the rope over the left hand a number of times (Fig. 2-5A) until about 15′ (4.57 m) of rope

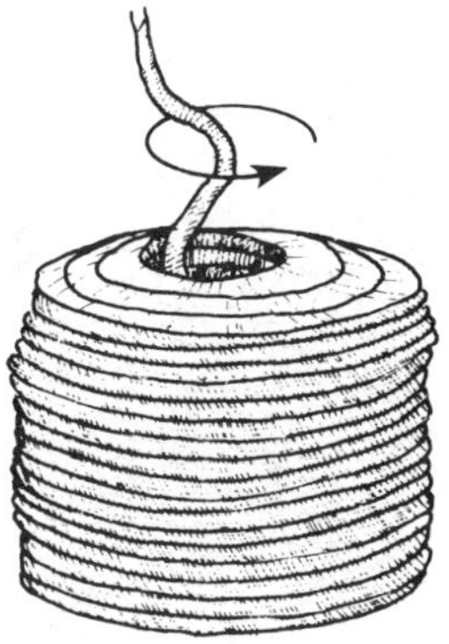

Fig. 2-3: Method of uncoiling rope.

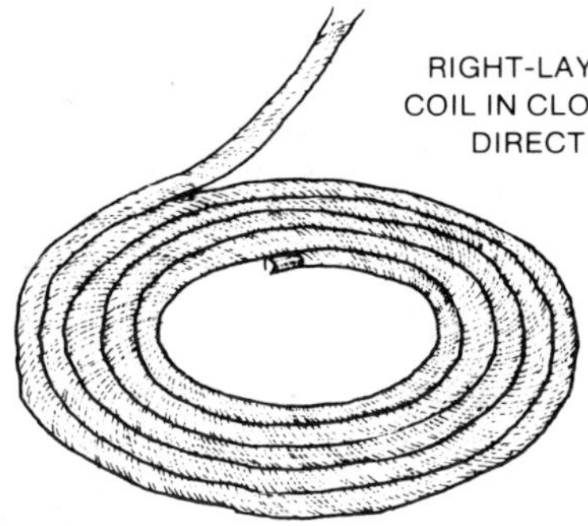

Fig. 2-4: Coiling of a fiber rope after being utilized.

remains. In coiling the rope, careful watch should be kept for kinks, which of course should be removed. Then, starting about 1′ (0.3 m) from the top of the coil, the rope should be wrapped about six times around the loops by rolling them in the left hand (Fig. 2-5B). Then, the left hand is extended through the coil (Fig. 2-5C) and the bight pulled back through the loops. Two half-hitches are tied around the bight, leaving a short end for carrying or for tying to a peg or supporting bar (Fig. 2-5D).

Most nylon and synthetic rope comes on reels. To uncoil it, place a reel on stands or jacks. Nylon is handled differently from natural fiber line. Coil three-strand nylon clockwise one week and counterclockwise the next week because continuously coiling three-strand nylon down one way tends to take the lay out of the strands.

Whipping Ends of Rope. The raw, cut end of a rope, as mentioned earlier, has a tendency to untwist and should always be knotted or fastened in some manner to prevent this untwisting. Whipping is one method of fastening the end of the rope to prevent untwisting. This method is particularly satisfactory because it does not increase the size of the rope very much. The whipped end of a rope will still thread through blocks or other openings. Before cutting a rope, place two whippings on the rope 1″ or 2″ (25.4 or 50.8 mm) apart and make the cut between the whippings (Fig. 2-6E). This will prevent the cut ends from untwisting immediately after they are cut. A rope is whipped by wrapping the end tightly with a small cord. Make a bight near one end of the cord and lay

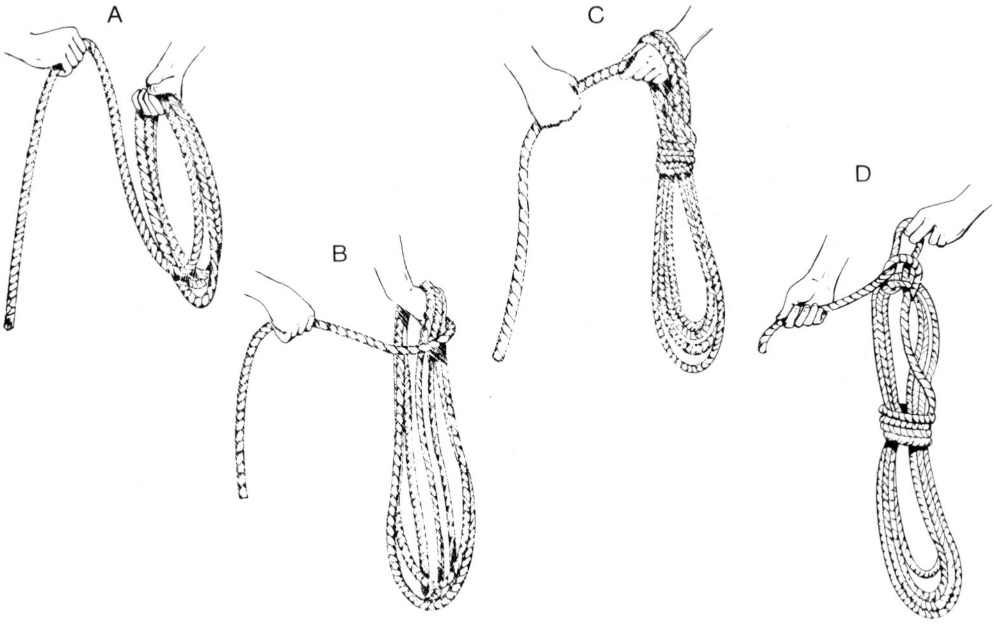

Fig. 2-5: Another method of coiling up a rope.

both ends of the cord along one side of the rope in the groove between two strands (Fig. 2-6A). The bight should project beyond the end of the rope about 1/2″ (12.7 mm). Start wrapping the standing part (b) of the cord tightly around the rope and cord (Fig. 2-6B) at the end of the whipping which will be farthest from the end of the rope. The whipping should end (Fig. 2-6C) about 1/2″ (12.7 mm) from the end of the rope and should be about as long as, or one and one-half times longer than, the diameter of the rope. Wrap the cord in the same direction as the twist of the strands in the rope. Continue wrapping the cord around the rope, keeping it tight, to within about 1/2″ (12.7 mm) of the end. At this point, slip the cord end (b) through the bight (c) of the cord (Fig. 2-6C). Pull the free end of the cord (a) until the bight of the cord is pulled under the whipping and the cord (b) is tightened (Fig. 2-6D). Cut off the ends (a and b) at the edge of the whipping, leaving the rope end whipped as shown in Fig. 2-6E.

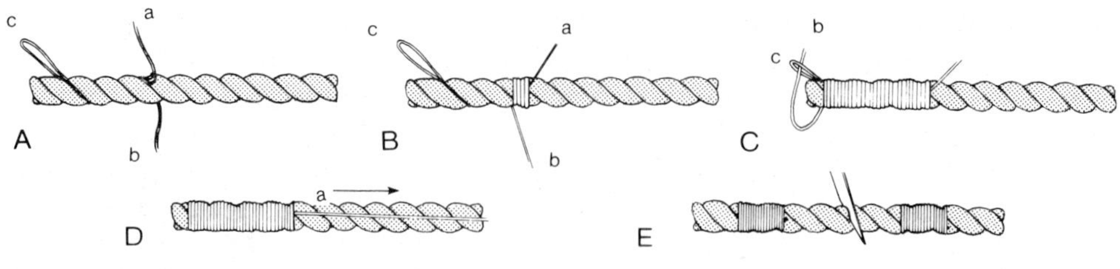

Fig. 2-6: Whipping the end of a rope.

After putting a whipping on nylon or synthetic rope, regardless of fabrication, heat the end of the line to seal it and make the line more stable. Another point on whipping nylon applies to putting a whipping on either an eye splice or an end-for-end splice—one to join two lines of unequal size. Put a strain on the line until it stretches and then put the whipping on the line. When the strain is taken off, the line returns to its normal length and, in doing so, makes the whipping more stable.

Inspection

The outside appearance of fiber rope is not always a good indication of its internal condition. The rope will soften with use and deteriorate according to the manner in which it is handled. Dampness, the amount of strain to which the rope has been subjected, the fraying and breaking of strands, and chafing on rough edges, all weaken the rope considerably. Overloading of a rope may cause heavy damage to material and serious injury to personnel. For this reason rope should be inspected carefully at regular intervals to determine its exact condition. Since the surface appearance of the rope is not a good indication of its condition, the strands should be untwisted slightly to open the rope so that the inside can be examined. Mildewed rope will have a musty odor and inner fibers of the strands will have a dark stained appearance. Broken strands or broken yarns ordinarily are easy to find. The presence of dirt and sawdust-like material inside the rope, caused by chafing, are indications of damage. In rope having a central core, the core should not break away in small pieces upon examination. If it does so, it indicates that the rope has been overstrained. Since any weak point in the rope weakens the entire rope, it should be examined at a number of places. If the rope appears to be satisfactory in all other respects, pull out a couple of fibers and try to break them. Sound fibers should offer considerable resistance to breakage. When any unsatisfactory conditions are found, the rope should be destroyed or cut up in short pieces. This is to prevent the use of the rope for hoisting. The short pieces can be used for miscellaneous service, but none of them should be left long enough to permit their use in hoisting.

KNOTS, BENDS, AND HITCHES

A good knot must be easy to tie, must hold without slipping, and must be easy to untie. The choice of the best knot, bend, or hitch to use depends largely on the job it has to do. This text explains why a given one is used and also gives the efficiency or strength of many of the knots, bends, and hitches. Always follow this rule: Never tie a knot on which you are not willing to stake your life.

Each of the three terms—knot, bend, and hitch—has a specific definition. In a knot, a line is usually bent or tied to itself, forming an eye or a knob or securing a cord or line around an object, such as a package. In its noun form, a bend ordinarily is that used to join two lines together. In its verb form, bend means the act of joining; bent is the past tense of bend. A hitch differs from a knot and a bend in that it ordinarily is tied to a ring, around a spar or stanchion, or around another line. In other words, it is not merely tied back on itself to form an eye or to bend two lines together.

Tying a knot, bend, or hitch in a line weakens it because the fibers are bent sharply, causing the line to lose varying degrees of its efficiency or strength. A general rule to follow, then, is to use a knot, bend, or hitch for temporary work. On the other hand, use a splice for permanent work because it retains more of the line's strength. A rigger, when working with fiber rope, must know how to tie knots, bends, and hitches, as well as how to splice a rope. But taking a look at some of the more common methods of tying rope, let us define these terms:

1. *Running end.* The running end of a rope is that end which is free or working, not anchored to some fixture of at least semipermanence.
2. *Standing part.* The standing part of a rope is the whole of the rope save for the running end.
3. *Bight.* A bight is a bind or U-shaped curve in a rope.

Knots at the End of Rope

There are basically two knots that are tied at the end of rope:

1. *Overhand knot.* The simplest and most commonly used knot is the overhand (Fig. 2-7A). It has many and varied uses, from preventing the unraveling of a rope, to forming a knob at the end to keep the rope from pulling through a particular hole or opening. It is used to increase a person's grip on a rope and sometimes serves as a part of another knot.
2. *Figure eight knot.* When a larger knot than that formed by an overhand knot is desired, then a figure eight (Fig. 2-7B) is employed. It too is used to keep a rope from slipping through a hole or through a fastening or loop in another rope.

Knots for Joining Two Ropes

Here are a few knots that can be used for joining two ropes:

1. *Square knot.* The square knot is used for joining two ropes of the same size (Fig. 2-8). In the square knot, the end and standing part of one rope comes out on the same side of the bight formed by the second rope. The greater the strain placed upon the knot, the greater it tightens. This knot should not be used if the

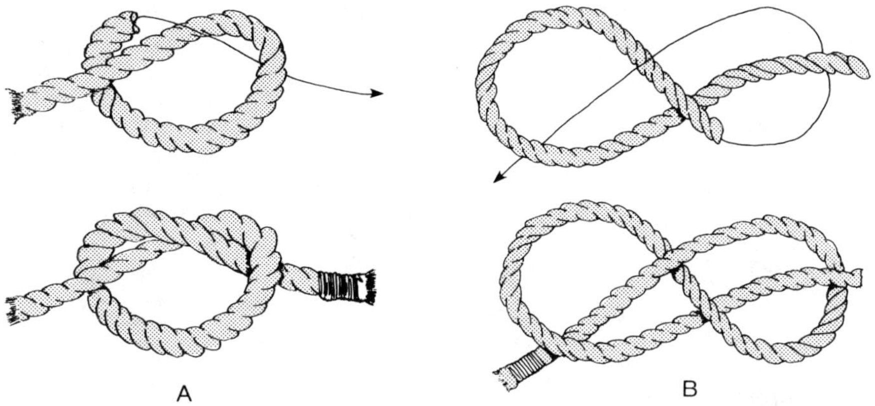

Fig. 2-7: Overhand and figure eight knots.

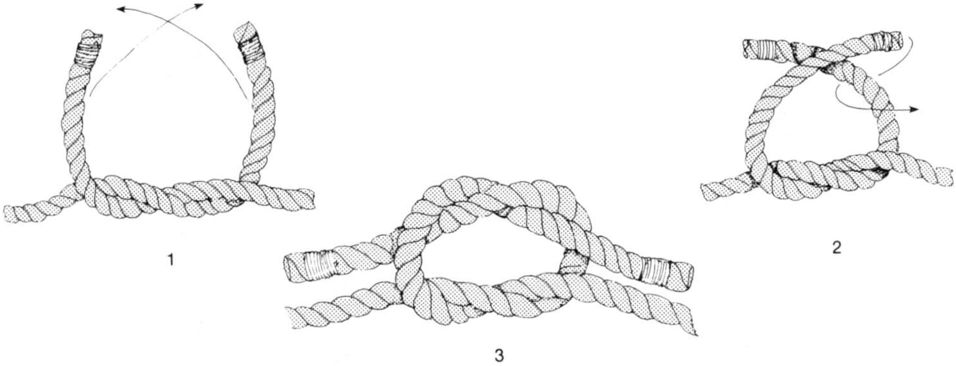

Fig. 2-8: Steps in making a square knot.

ropes are wet or of uneven size. If they are, the knot will slip. The square knot can be easily untied by grasping the two ends of the bights and pulling the knot apart.

2. *Single sheet bend.* The single sheet bend has two principal uses (Fig. 2-9): tying together two ropes of unequal size or tying a rope to an eye. This knot will draw extremely tight, is stronger than a square knot, but will loosen or slip when ropes are slackened.

3. *Double sheet bend.* For joining ropes of any size, for joining wet ropes, or for tying a rope to an eye, the double sheet bend (Fig. 2-10) is more effective than the single sheet bend. It will not slip or draw tight under heavy loads and is quite secure when used in a spliced eye.

4. *Carrick bend.* The carrick bend (Fig. 2-11) is used for heavy loads and for joining large diameter rope. It will not draw tight under a heavy load and is easily untied if the ends are seized to their own standing part.

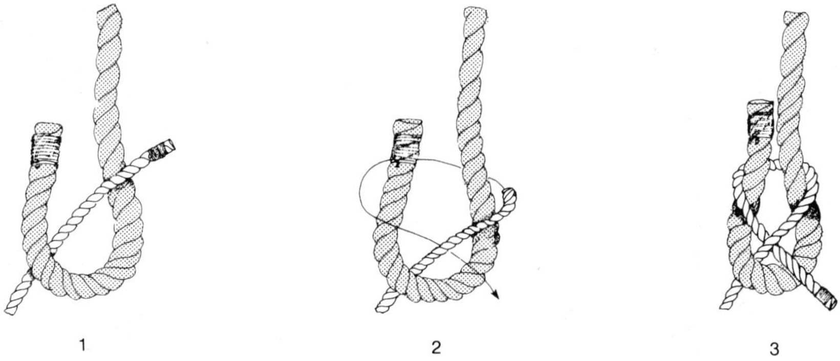

Fig. 2-9: Single sheet bend.

Fig. 2-10: Double sheet bend.

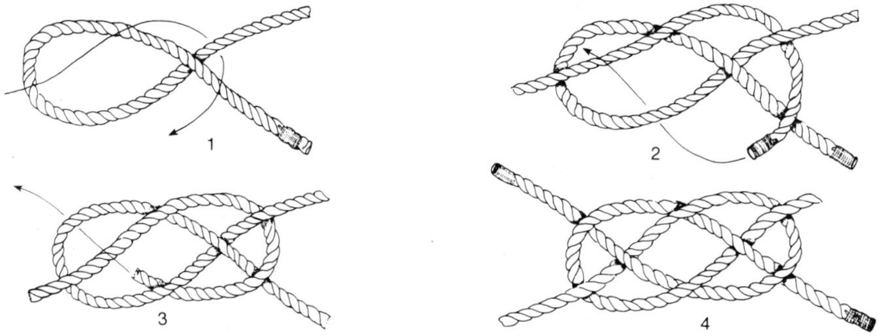

Fig. 2-11: Carrick bend.

Knots for Making Loops

Other knots with which the rigger should be familiar are those used for making loops:

1. *Bowline.* Of all the knots for making loops, probably the most common and versatile is the bowline (Fig. 2-12). It is frequently used for the lowering of

personnel and materials. The bowline forms a loop which may be of any length and which will not tighten or slip under strain. It is easily untied if each running end is seized to its own standing part.

2. *Double bowline.* Note in Fig. 2-13 that the double bowline forms three nonslipping loops. This knot is often used for slinging a person so that when he/she sits in the sling, one loop is used to support the back and the other to support the legs. A notched board passed through the two loops makes a comfortable seat known as a boatswain's (bosun's) chair.

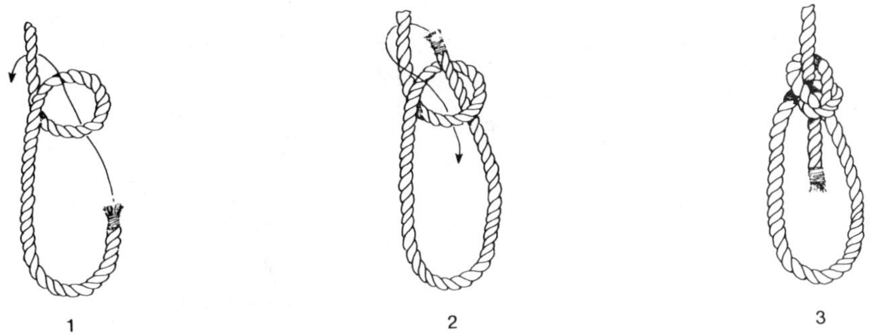

Fig. 2-12: Bowline.

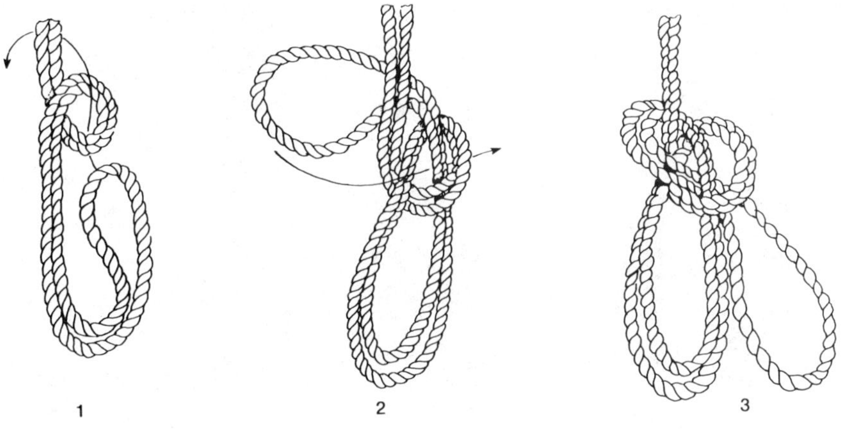

Fig. 2-13: Double bowline.

3. *Spanish bowline.* A Spanish bowline (Fig. 2-14) can be tied at any point in the rope, either at a place where the line is double or at an end which has been doubled back. This knot is frequently used in rescue work or to give a double grip for lifting pipe or other round objects in a sling.

4. *French bowline.* The most frequent use for the French bowline (Fig. 2-15) is in rescue operations of an injured person. When so used, one loop is used as a seat and the other loop is put around the body under the arms. The weight of the person keeps the two loops tight so that it is almost impossible for him/her to fall out. It is particularly good for unconscious personnel. The French bowline is sometimes used when a person is working alone and needs hands free. The two loops of the knot can be adjusted to the size needed.

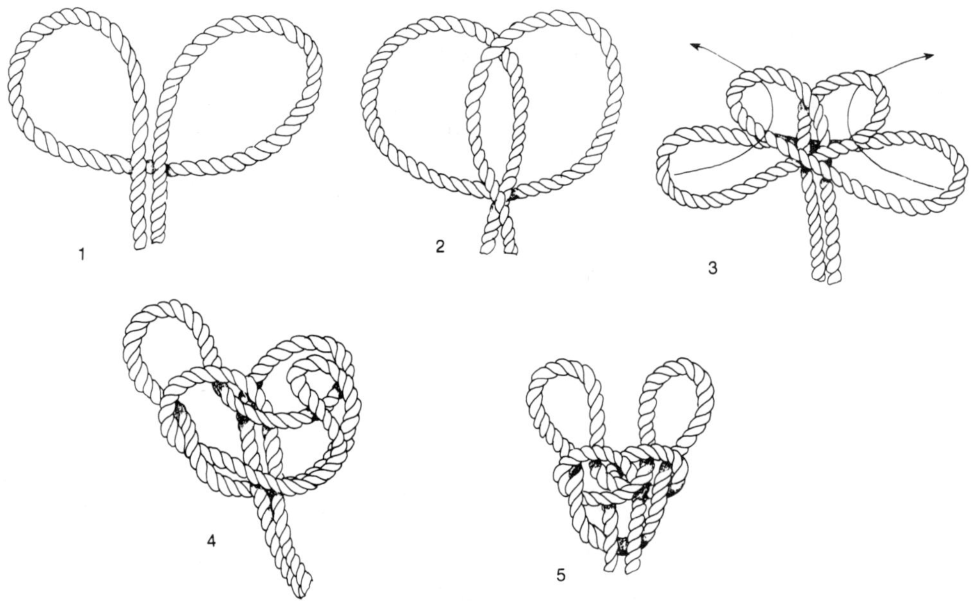

Fig. 2-14: Spanish bowline.

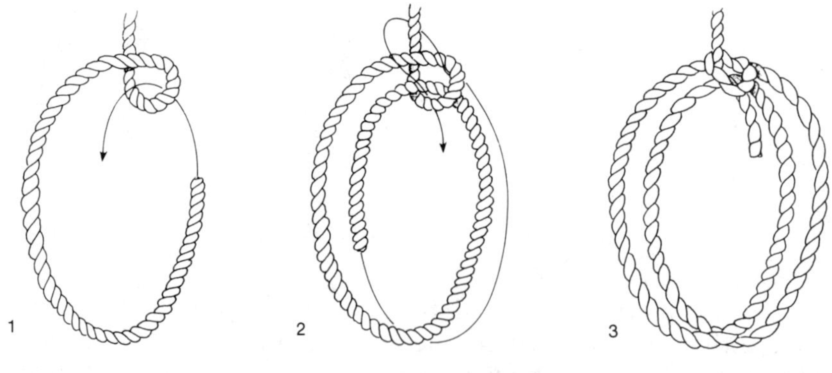

Fig. 2-15: French bowline.

Hitches

Of the better known hitches, the following may be used by riggers:

1. *Half hitch.* To tie a rope to a larger rope or to a timber, the half hitch (Fig. 2-16A) should be employed. *It is not a secure hitch* because to hold it requires a steady pull on the standing part of the rope. It is, however, used for securing the free end of a rope and is an aid and the foundation for many knots.
2. *Two half hitches.* Securing the running end of a rope to its standing part is the useful role of two half hitches (Fig. 2-16B). If the two hitches are slid together along the standing part of the rope, the hitches become a *clove hitch.* Actually, the clove hitch (Fig. 2-16C) is one of the most widely used of all hitches. It is used to fasten a rope to almost any object. It forms the basis for other knots. The clove hitch can be tied at any point in a rope. If there is not constant tension, however, there may be a slackening of the rope.
3. *Timber hitch.* The timber hitch (Fig. 2-16D) is used for moving heavy timber or poles. The pressure of the coils, one over the other, holds the timber securely; the more tension, the tighter the hitch. It will not slip but will readily loosen when the strain is relieved.
4. *Timber hitch and half hitch.* Combining a timber hitch with a half hitch (Fig. 2-16E) allows heavy timbers to be held while being lifted or dragged. A timber hitch alone may become untied when the rope is slack or if a sudden strain is put on it. Be careful to check ropes for abrasive wear wherever this hitch is used.

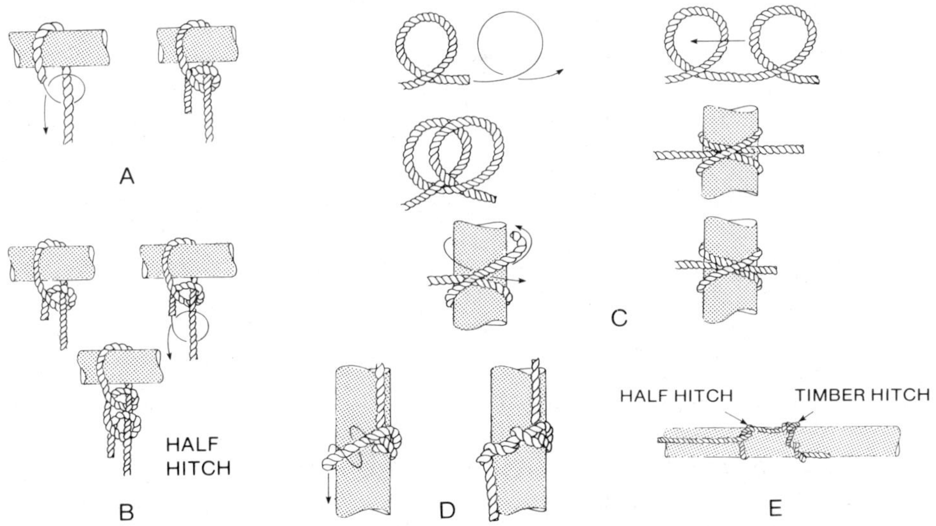

Fig. 2-16: (A) Half hitch; (B) two half hitches; (C) clove hitch; (D) timber hitch; and (E) combination timber hitch and half hitch.

5. *Catspaw.* The catspaw (Fig. 2-17) is a hitch that can be made at the end of a rope for fastening the rope to a hook, or it can be used for fastening an endless sling to a hook. It will not slip off and need not be kept taut to make it hold.

6. *Sheepshank.* A sheepshank (Fig. 2-18) is a method used to shorten rope or to take a load off a weak spot in the rope. Unless the eyes are fastened to the standing part on each end, it is only a temporary knot.

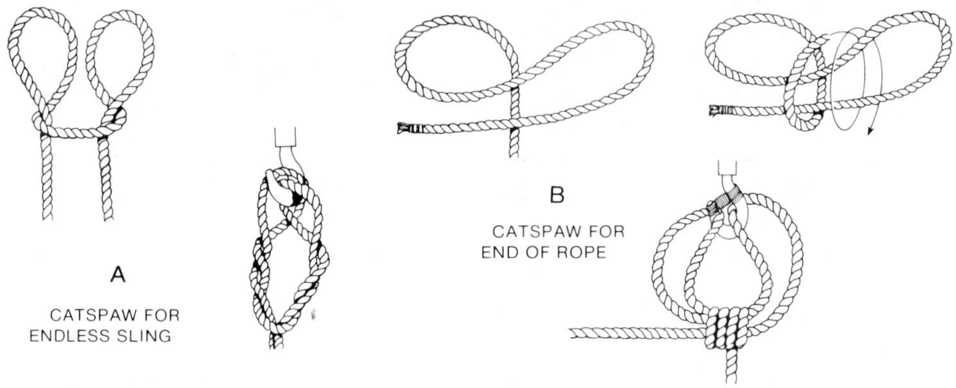

Fig. 2-17: Two types of catspaws.

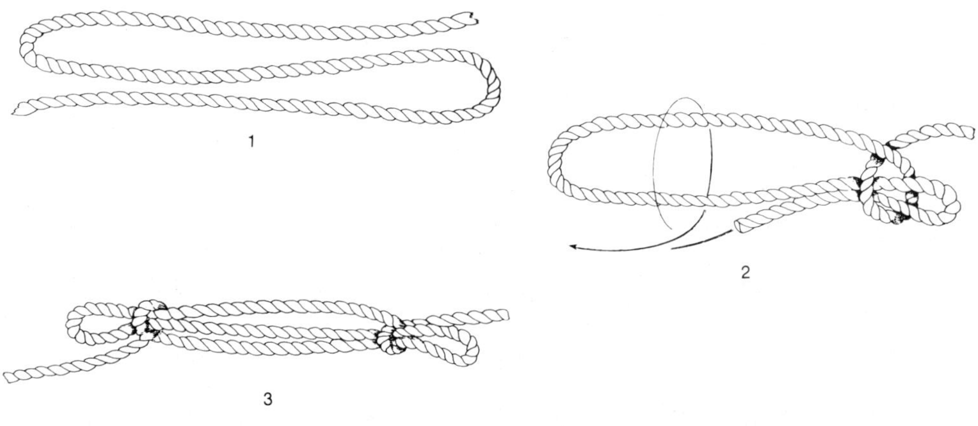

Fig. 2-18: Steps in making a sheepshank.

7. *Scaffold hitch.* The scaffold hitch (Fig. 2-19) is used to support the end of a scaffold plank with a single rope and prevents the plank from tilting. It is not the recommended way to scaffold but will serve in an emergency. Here again, as in other knots or hitches that come in direct contact with square sided timbers or planks, watch for the effects of abrasion.

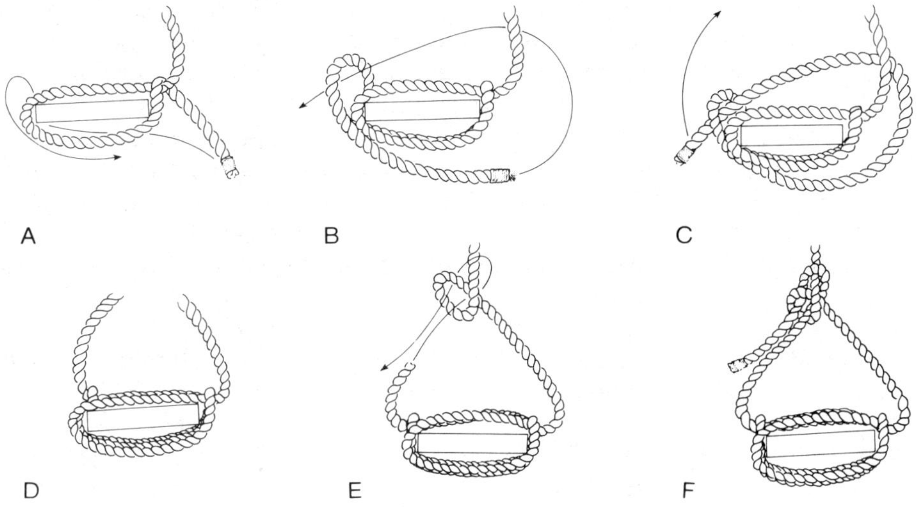

Fig. 2-19: Steps in making a scaffold hitch.

Splices

Fiber rope is spliced by unlaying strands of both ends and interweaving these strands. Learning to splice fiber rope is not difficult and once learned can be of material advantage. While there are four general types of splices in fiber rope, all four types are similar. If one type of splice is mastered, all splices are easily accomplished. No special tools are necessary for splicing fiber rope, but it will be found helpful to have a spike available. The circumstances of the situation will dictate the type of splice to be used. In the following paragraphs, the methods of making a short splice, an eye or side splice, a long splice, and a crown or back splice are given. All of these generally consist of three principal steps, which are: unlaying the strands of the rope; placing the rope ends together; and interweaving the strands and tucking them into the rope. A somewhat similar procedure is followed in renewing a broken strand of rope.

Short Splice. The short splice is as strong as the rope in which it is made and will hold as much as a long splice. However, the short splice causes an increase in the diameter of the rope for a short distance and can be used only where this increase in diameter will not affect operations with the rope. It is called the short splice because the minimum reduction in rope length takes place in making the splice.

To make the short splice, the following steps are necessary:

1. Untwist the strands at the end of each rope (Fig. 2-20A) for approximately seven turns. Tie one or two turns of light cord around each rope at the point where the strands begin separating.

2. Lay back the strands of each rope so that they do not cross each other. Place the rope ends together with the untwisted strands intermingled. Each strand of one end should lie between two strands of the other end of rope.

3. Cut the light cord binding off the left side of the junction (Fig. 2-20B), and tie strands d, e, and f from the left rope to the right rope with a temporary cord binding. This will leave strands a, b, and c of the rope on the right clear for tucking. Take strand a from the rope on the right and bring it up over the nearest strand of the rope on the left, down under the next strand, and back up again. It should be pushed over and under the strands so that it crosses each strand at nearly right angles.

4. Take strand b of the rope from the right side and bring it up over the nearest strand of the rope on the left side, and under the second strand of the rope on the left side. Repeat this operation with strand c of the rope on the right side (Fig. 2-20C).

5. Cut the temporary binding off strands d, e, and f and cut the cord with which the rope on the right was bound. Tuck each of these three strands of the rope on the left over and under one strand of the rope on the right in the same manner as in step 3 (Fig. 2-20D).

6. Take each of the strands from the right hand rope and give them two more tucks, going over the nearest strand and under the next strand of the rope on the left. Repeat this with each of the three strands from the left hand rope, taking two tucks with each strand.

7. Divide the fibers of each strand into two parts and take two more tucks with each of the half strands.

8. Cut off all loose ends (Fig. 2-20E) and roll the splice on a hard surface.

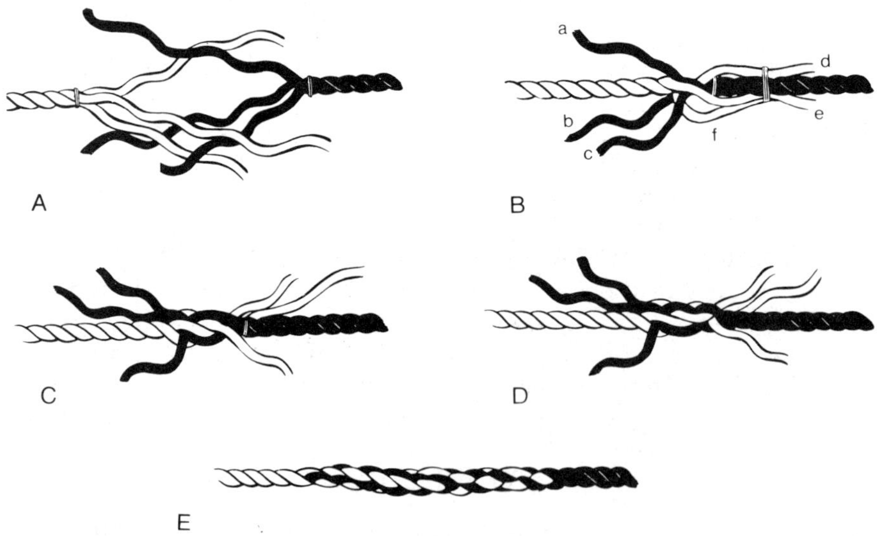

Fig. 2-20: Making the short splice.

Eye or Side Splice. The eye or side splice is used for making a permanent loop in the end of a rope. The loops can be used for fastening the rope to a ring or hook and can be made up with or without a thimble. A thimble is used to reduce wear. To make the eye splice, proceed as follows:

1. Unlay five turns of the end of the rope.
2. Bend the rope into a loop or around the ring or thimble as desired, bringing the end of the unlaid portion back against the standing part of the rope. Push strand a of the unlaid end under one strand (Fig. 2-21A) of the standing part of the rope and up over the next strand.
3. In the same manner as in step 2 above, pass strand b of the unlaid end of the rope under the strand of the standing part next to the one under which strand a was tucked (Fig. 2-21B). Tuck strand c in the same manner, so that each of the three unlaid strands has been tucked once under a different strand of the standing part of the rope.
4. Take each strand of the unlaid end of the rope and make two more tucks under one and over one strand of the standing part of the rope. A fid can be used to hold the strands of the standing part open for tucking (Fig. 2-21C).
5. Cut off the ends of the strands close to the standing part of the rope and roll the splice to smooth it (Fig. 2-21D).

Long Splice. The long splice is as strong as the rope itself. It has a neater appearance than a short splice. The long splice is used when the larger diame-

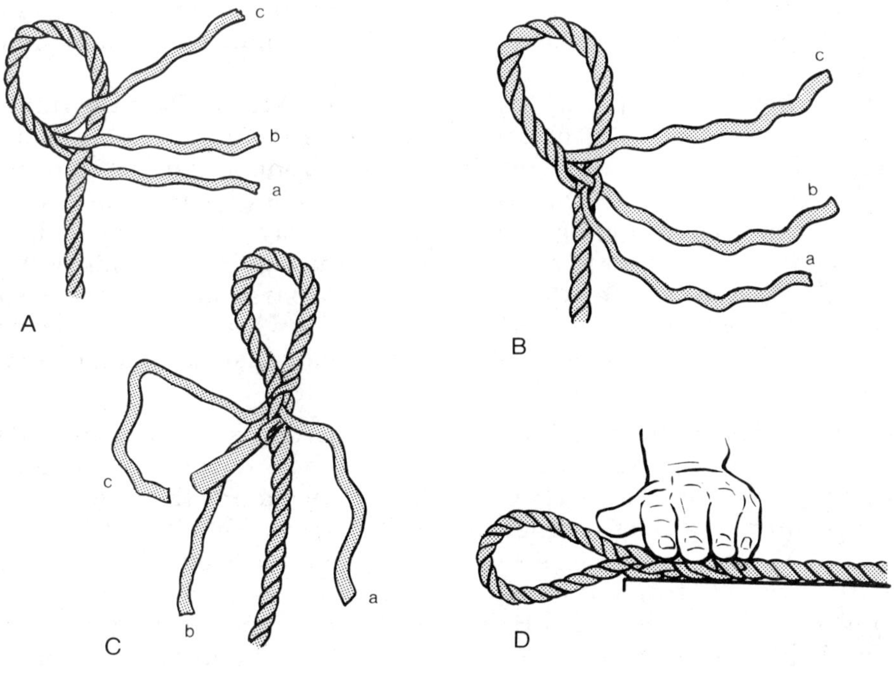

Fig. 2-21: Eye or side splice.

ter of the short splice would prevent the use of the rope. To make a long splice in a three-strand rope, proceed as follows:

1. Unlay the strands for 15 turns from each end. Tie a light cord (Fig. 2-22A) around each rope at this point to prevent the ropes from unlaying further.
2. Bring the two parts of rope together with each strand of the right-hand rope between two strands of the left-hand rope (Fig. 2-22B) and the untouched portions butting against each other.
3. At point a where the two parts of rope are placed together, tie two strands from the right-hand rope temporarily to the left-hand rope with a piece of cord, and cut the first cord placed around the right-hand rope. Take one strand of this rope and unlay it to the right (Fig. 2-22C). Lay the adjoining strand from the left-hand rope carefully in the place from which the right-hand strand has been removed.
4. Continue unlaying the strand from the right-hand rope and laying the strand from the left-hand rope in its place until all but about five turns of the strand from the left have been put into the right-hand rope.
5. Cut the cord securing the two remaining strands of the right-hand rope to the left-hand rope. Cut the cord holding the left-hand strands from unlaying. Unlay one strand of the left-hand rope and insert in its place the adjoining strand from the right-hand rope in the same manner as previcusly described.
6. There are now (Fig. 2-22D) three pairs of strands at three different places in the splice, one pair at the center (a) and one each at the right (b) and left (c) ends of the splice. These pairs of strands must be tucked into the rope. Cut off the long ends of the strands so that only about five turns of each strand project out of the main rope.
7. Starting with any pair of strands, bring the strand from the right (Figs. 2-22E and 2-22F) over the nearest strand from the left and under the next strand (Fig. 2-22G). Then, give the strand from the left one tuck in the same manner (Figs. 2-22H and 2-22I). Each strand should now be given two more tucks in a direction almost at right angles to the direction of twist of the rope (Fig. 2-22J).
8. Tuck the second pair of strands and the third pair of strands in the same manner as in step 7 above. After all three pairs of strands have been tucked, cut off the ends of the strands and smooth the places where the tucks are made by rolling the rope against a hard surface. The finished splice is shown in Fig. 2-22K.

Crown or Back Splice. Where the end of a rope is to be finished off to prevent unlaying and a slight enlargement of the end is not objectionable, a crown splice (Fig. 2-23) may be used to finish the end. In making the crown splice, first unlay or untwist six turns of the rope and tie a crown knot (Fig. 2-24) on the end (Fig. 2-23A). Then, give each strand three or four tucks by bringing the strand over the nearest strand of the standing part of the rope and under the next strand for each tuck (Figs. 2-23B and 2-23C), using a fid to separate the strands. Trim off the ends of the strands close to the standing part of the rope and roll the finished crown splice (Fig. 2-23D) underfoot to smooth the tucks.

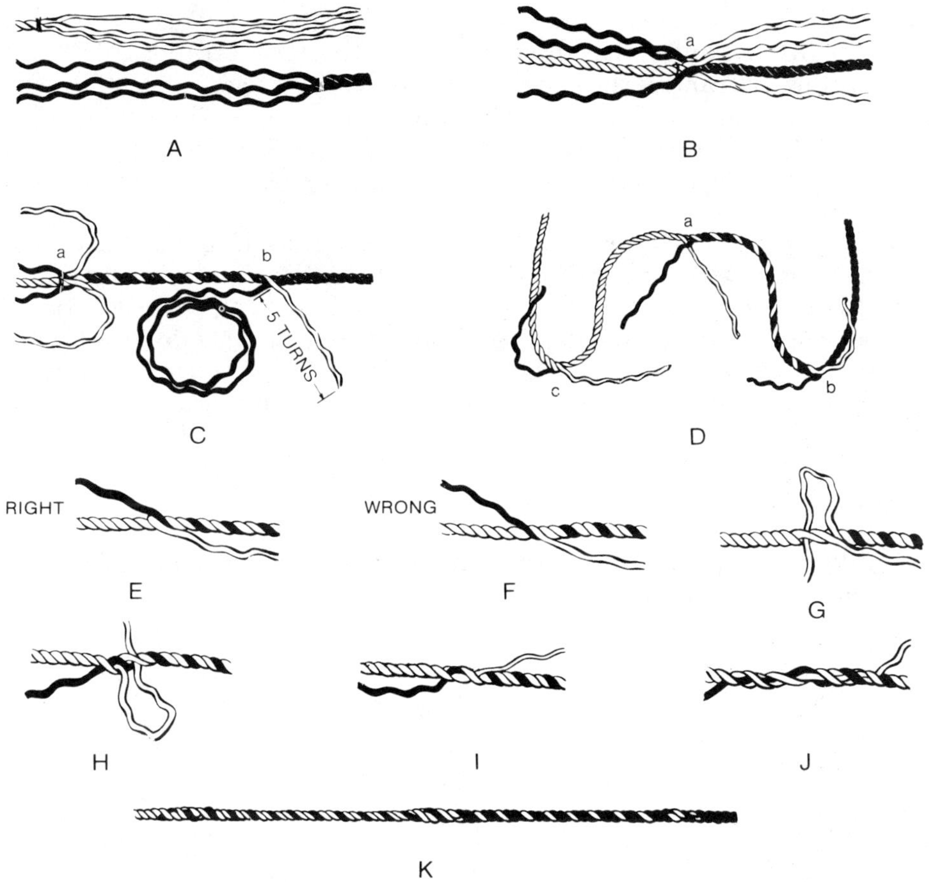

Fig. 2-22: Long splice.

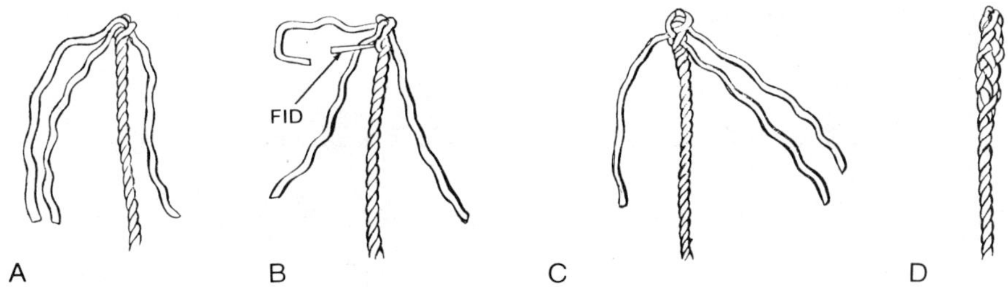

Fig. 2-23: Crown or back splice.

Renewing Rope Strands. When one strand of a rope is broken, it is not possible to tie the ends together as this would shorten the strand. However, the strand can be renewed (Fig. 2-25) by inserting a piece longer than the break and tying the ends together. To accomplish this, proceed as follows:

1. Unlay each end of the broken strand (a) for five or six turns around the rope (Fig. 2-25B) by pulling it out from between the other strands.
2. Obtain a strand of the same size from another rope, preferably a used rope, since this has already been stretched and will fit better. This strand will need to be about twenty turns long. Insert this strand (b) into the rope (Fig. 2-25C) in the space left by unlaying the broken strand (a), leaving the ends sticking out of the rope.
3. Tie the ends of the broken strand (a) to the ends of the strand (b) which has been inserted with an overhand knot at each end of the new strand.
4. At one knot, tuck the strand from the right over the nearest strand in the left side of the rope and under the second strand, using a fid to hold the second strand up while tucking. Tuck the strand from the left over one strand of the rope on the right and under another strand, again using a fid to help make the tuck.
5. Repeat step 4 for the other knot.
6. Tuck each of the two strand ends two more times in the same manner as in step 4.
7. Cut off the ends of the strands close to the rope and smooth the tucked ends by rolling (Fig. 2-25D).

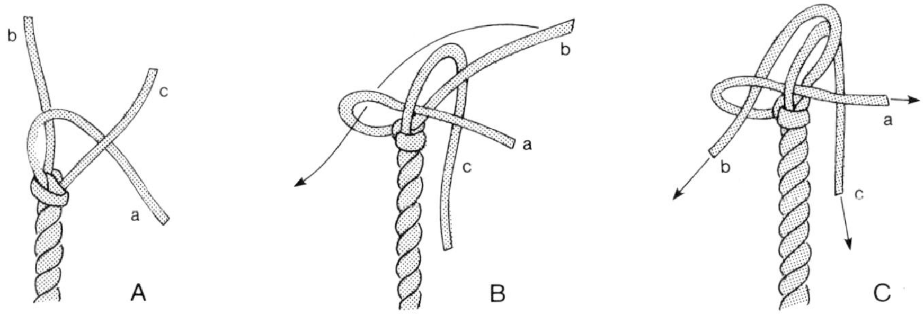

Fig. 2-24: Making a crown knot.

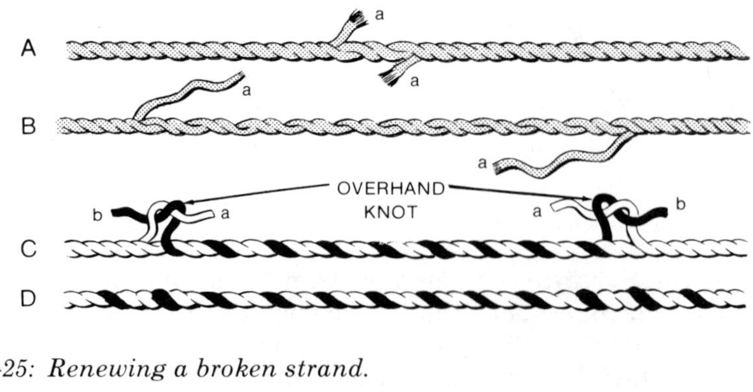

Fig. 2-25: Renewing a broken strand.

3

Chains

INTRODUCTION

The manufacture and use of chain predate the Christian era and have been found in the relics of the Phoenician traders. Originally they were used almost exclusively for heavy hoisting, but today they have, for that purpose, given way to wire rope. In fact, a good general rigging rule to follow is **never** use a chain if the job can be accomplished with a wire rope. In fact, it is best **never** to use chain for general construction rigging operations.

The reason for this statement makes good safety sense. Chains are made up of a series of links fastened through each other. Each link is made of a rod of wire bent into an oval shape and welded at one or two points. The weld ordinarily causes a slight bulge on the side or end of the link. The chain size refers to the diameter in inches of the rod used to make the link. As shown in Fig. 3-1, a chain link consists of two sides, either of which in failing would cause the link to open and drop the load. A wire rope, on the other hand, is frequently composed of 114 individual wires, all of which must fail before it parts. A chain just does *not* have this kind of reserve strength.

When a wire rope is fatigued from severe service, the wires break one after the other over a relatively long period of time and thus afford an opportunity

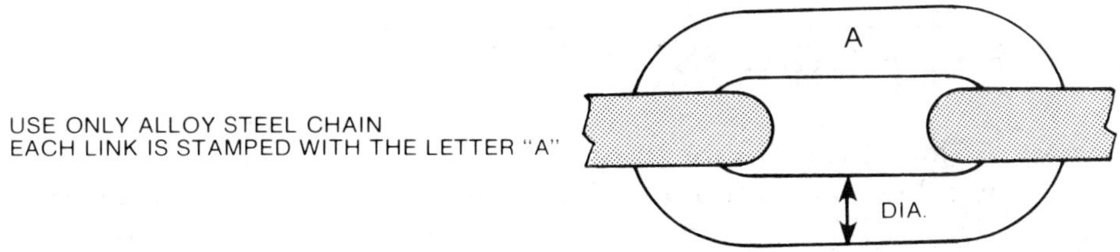

Fig. 3-1: Identification of alloy steel chain.

to discover the condition. If severely overloaded, the wires and strands will break progressively over a period of perhaps several seconds, and with resultant noise, before complete failure occurs. This may afford the person handling the load a brief time in which to jump to safety before it is dropped.

Chains will usually stretch under excessive loading so that the individual links will be bent slightly. Bent links are a warning that the chain has been overloaded and might fail suddenly under load. Wire, on the other hand, will fail a strand at a time, giving warning before complete failure occurs. If a chain is equipped with the proper hook, the hook should start to fail first, indicating that the chain is overloaded. However, if the overloading is great enough, it will ultimately fail with less warning than wire rope. Should the weld be defective, a chain may break with little, if any, warning.

There are certain jobs in the field for which chains are better suited than wire rope. Chains withstand rough handling, do not knit, are easily stored, have dead flexibility, and when used as slings, grip the load well. They are much more resistant to abrasion and corrosion than wire rope and are particularly well suited as slings for lifting rough loads such as heavy castings, which would quickly weaken or destroy wire rope slings due to the sharp bends over the edges of the castings. Chains can also be used on ratchet hoists and for binder chains on trucks and other vehicles. They can also be employed on chain blocks if they are equipped with proof-tested chains supplied by the manufacturer.

A number of types (Fig. 3-2) and grades of chains are available.

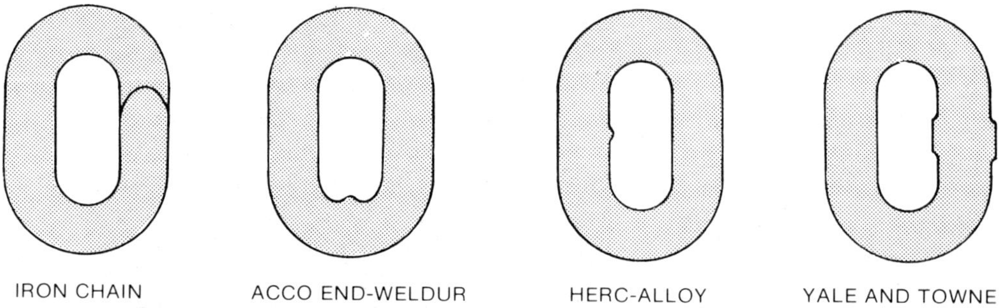

Fig. 3-2: Types of chains.

GRADES OF CHAINS

It is very difficult to determine the grade of a chain by examination. However, all chain suitable for construction use and in particular for hoisting is marked by an identifying letter stamped on the links. For the highest degree of safety, ensure that only chain stamped with the letter "A" is used on hoisting operations. The letter "A" designates heat treated alloy steel, which is the strongest chain commonly made (Fig. 3-1).

The alloy steel chains must never be welded or exposed to excessive temperatures as the heat treatment will be affected and the chain will lose its strength.

STRENGTH OF CHAINS

Table 3-1 gives the maximum safe working loads of new alloy steel chain when used as a single vertical sling. Maximum safe working loads of inclined slings, etc., are given on page 113.

Table 3-1: Maximum Safe Working Load "A" Type Alloy Steel Chain Single Vertical Sling

Chain Size (Inches)	Capacity (Pounds)
1/4	3,250
3/8	6,600
1/2	11,250
5/8	16,500
3/4	23,000
7/8	28,750
1	38,750
1-1/8	44,500
1-1/4	57,500
1-3/8	67,000
1-1/2	80,000
1-3/4	100,000

Any change in the above factors, such as twisting of the chain, deterioration of the chain or its fittings by overload, use, rust, shock loading, or inclined (offset) loading, will lessen the load that the chain will safely carry.

Alloy steel chain can be used in environments up to 500°F without reducing the safe working load, but at temperatures above this level the load limit decreases as per Table 3-2.

INSPECTION OF CHAINS

All chains used regularly should be thoroughly inspected on a link-by-link basis at least once a month by a competent inspector. Under no circumstances

Table 3-2: Effect of Heat on Safe Working Load

Chain Temperature	Reduction of Safe Working Load Limit while Heated*	Permanent Reduction in Safe Working Load Limit**
500°F	None	None
600°F	10%	None
700°F	20%	None
800°F	30%	None
900°F	40%	10%
1000°F	50%	15%

* While chain is at temperature in first column.

** When chain is used at room temperature after being used at temperature shown in first column.

should a chain be used for hoisting purposes unless it has been closely examined for defects or wear.

Whenever a chain is subjected to shock or impact loads, it must be immediately inspected before being put back into service.

It is recommended that every chain carry a small metal identification tag bearing a serial number and its safe working load (Fig. 3-3). A log book should also be kept for the chain identifying its characteristics and setting up an inspection schedule.

When inspecting chain remember to give every link a close examination. One bad link is all it needs to fail. The following guidelines may be of assistance in carrying out the inspection:

1. Clean the chain thoroughly in a solvent solution.
2. Lay it out on a clean surface or hang it up in a well lit area. Use a magnifying glass to aid in the inspection.

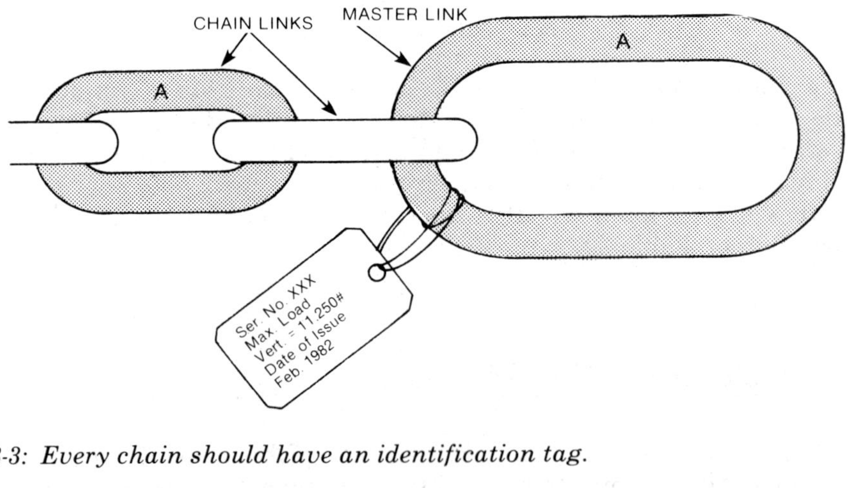

Fig. 3-3: Every chain should have an identification tag.

3. Look for elongated or stretched links. When the links are severely stretched they tend to close up so that in some cases stretch may be indicated by links that bind or a chain that will not hang perfectly straight. Elongation should be determined by measuring all new chains in sections of 1 to 3′ with a caliper and remeasuring them during the inspection. If the inspection reveals a stretch of more than 3%, take the chain out of service (Fig. 3-4).

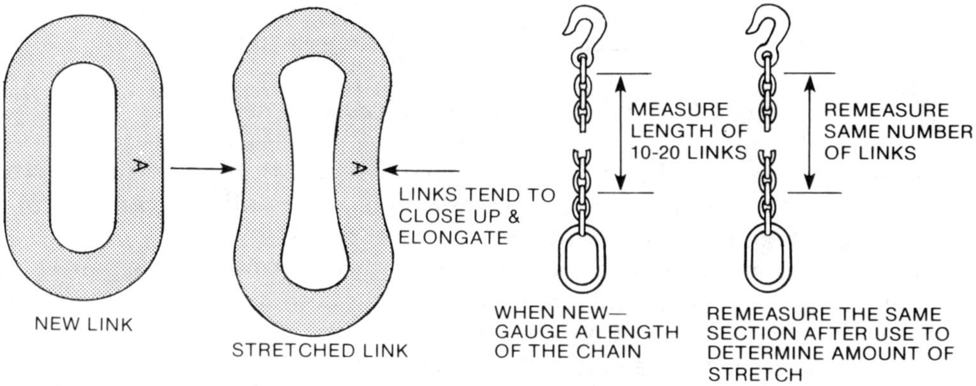

Fig. 3-4: Look for chain stretch during inspections.

4. Look for bent, twisted, or damaged links that often occur when the sling is used to lift a load having unprotected sharp edges (Fig. 3-5).
5. Look for cracked links. The presence of any crack, regardless of its size, means the chain is unsafe and must be removed from service. Where a crack is suspected, the link may be soaked in thin oil and then wiped dry. A coating of powdered chalk or other white material is applied to the surface and allowed to remain there for several hours. If a crack exists, the oil pocketed in it will be drawn out and discolor the white coating.
6. Look for gouges, chips, scores, or cuts in each link. If they are deep or large in area, then the chain should be removed from service. If the depth of these

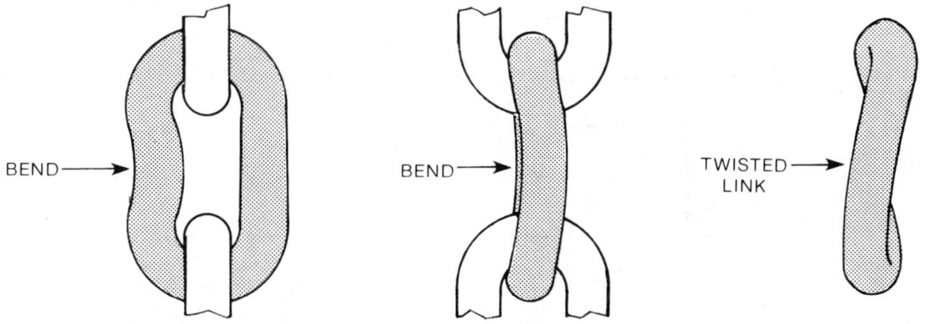

Fig. 3-5: Inspect all links for bends, twists, and damage.

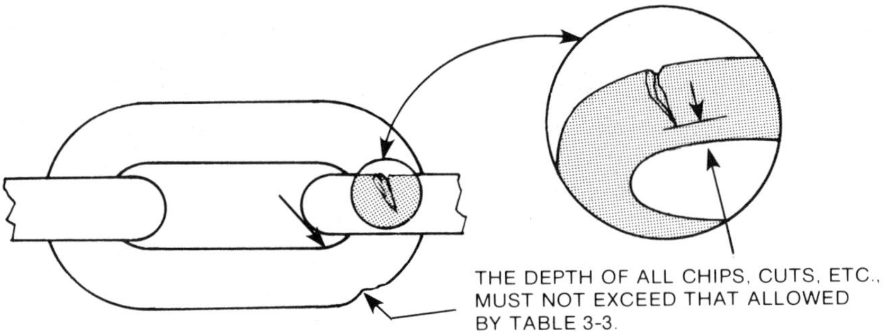

Fig. 3-6: Inspect all links for gouges, chips, and cuts.

defects is such that the link size is reduced below that listed in the table for wear, then the chain is unsafe. In addition, watch all sharp nicks and cuts because cracks usually initiate at or near them (Fig. 3-6).

7. Look for small dents, peen marks, and bright polished surfaces on the links. These usually indicate that the chain has been work-hardened or fatigued.
8. Look for lifted fins at welds. They are evidence of severe overloading and indicate that the whole chain should be destroyed.
9. Look for severe corrosion resulting in measurable material loss or severe pitting.
10. Be particularly careful in determining link wear at the point where they bear on each other. A caliper should be used for the measurement and the degree of wear at the worst section, or the most badly worn link must be determined during the inspection. Table 3-3 gives the percentages by which the rated loads must be reduced to compensate for wear and the point at which the chain must be removed from service (Fig. 3-7).

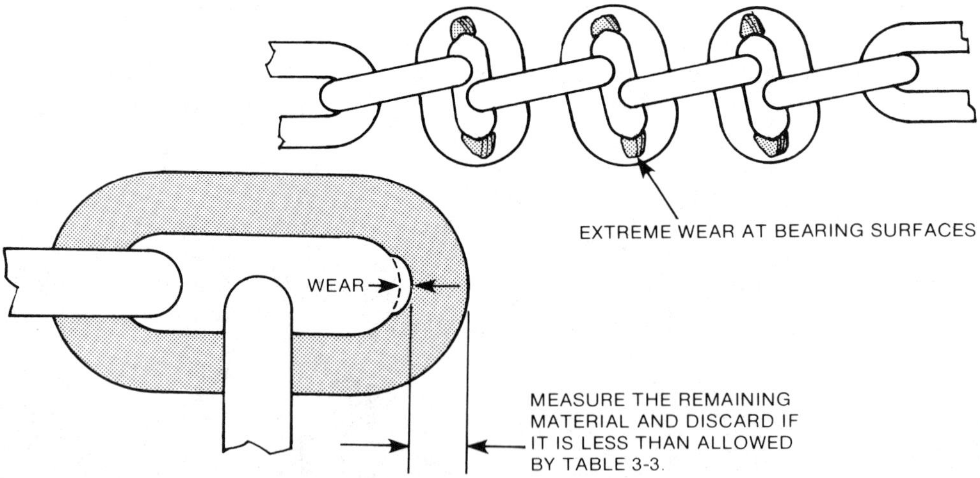

Fig. 3-7: Inspect all links for wear at the bearing surfaces.

Table 3-3: Correction Table to Compensate for Wear

Original Nominal Chain Stock Diameter (Inches)	Reduce Rated Capacity by Following % When Diameter at Worn Section is as Follows (Inches) 5%	10%	Remove from Service When Diameter is (Inches)
1/4 = 0.250	0.244	0.237	0.233
3/8 = 0.375	0.366	0.356	0.335
1/2 = 0.500	0.487	0.474	0.448
5/8 = 0.625	0.609	0.593	0.559
3/4 = 0.750	0.731	0.711	0.671
7/8 = 0.875	0.853	0.830	0.783
1 = 1.000	0.975	0.949	0.895
1-1/8 = 1.125	1.100	1.070	1.010
1-1/4 = 1.250	1.220	1.190	1.120
1-3/8 = 1.375	1.340	1.310	1.230
1-1/2 = 1.500	1.460	1.430	1.340
1-5/8 = 1.625	1.590	1.540	1.450
1-3/4 = 1.750	1.710	1.660	1.570
1-7/8 = 1.875	1.830	1.780	1.680
2 = 2.000	1.950	1.900	1.790

CHAIN CARE AND USE

Use only alloy steel chains and never exceed their rated safe working loads. When hoisting heavy metal objects using chains for slings, padding should be inserted around the sharp corners of the load to protect the chain links from being cut. The padding may be either planks or heavy fabric. Chains should not be permitted to twist or kink when under strain. Know the weight of all loads to avoid accidental overload. Avoid impact loading. Make sure all attachments and fittings are of a type, grade, and size suitable for service with the chain used.

Never shorten a chain by twisting or knotting it (Fig. 3-8). Also, links of chain should never be fastened together with bolts or wire because such connections weaken the chain and limit its safe working capacity. Worn or damaged links should be cut out of the chain and replaced with a cold shut link. This practice, however, should be discouraged because the strength of a worn or damaged link is not equal to that of the original links and the strength of the entire chain is thus greatly reduced. Remember that a chain is only as strong as its weakest link. Also, do not attempt to shorten one leg of bridle sling by twisting it. This also causes a serious reduction in strength.

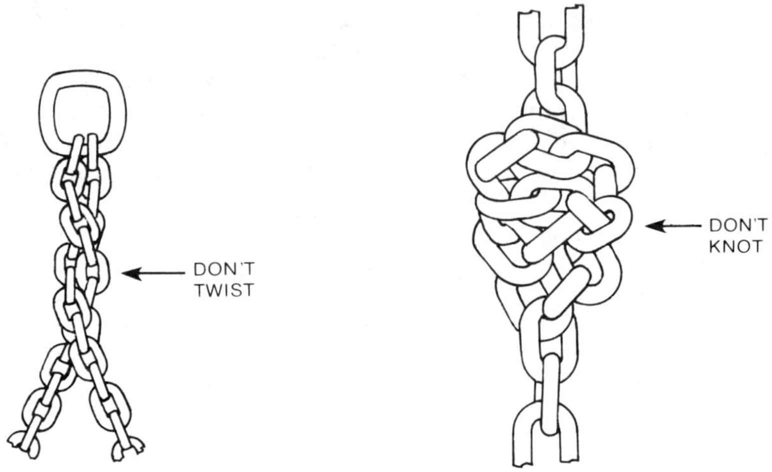

Fig. 3-8: Never twist or knot a chain.

Painting a chain to prevent rusting is not advisable because the paint will interfere with the freedom of action of the links. A light coat of lubricant can be applied to prevent rusting. Chains should be stored in a dry and well ventilated place to prevent rusting.

Chains must be inspected frequently, depending on the amount of use. During an inspection, check *each* link of the chain very carefully. If you overlook one link, that one may be the defective link. It will be necessary to wipe off most of the grease to make the inspection properly. Damaged or worn sections can sometimes be returned to the manufacturer for repair and reconditioning. Only specialists can repair alloy steel chain—do not attempt it yourself. Also, never reweld alloy steel chain links. They must be replaced by the manufacturer.

Look for links that may be elongated owing to overloading. If the links bind, condemn the chain. A free-hanging chain that is not perfectly straight is indicative of binding links. If there is any question as to the stretching of the chain, remove it from service. As mentioned previously, small dents similar to peen marks on the surface of the links or even a bright or polished surface usually indicates that the chain has been work-hardened or fatigued. Also, watch for any sharp nicks and cuts, as it is in these spots that cracks will usually start. Never hammer a chain to straighten the links or to force the links into position.

4

Slings

INTRODUCTION

Slings require special attention because they are almost always subjected to severe wear, abrasion, impact loading, crushing, kinking, and overloading. They also merit special attention because seemingly insignificant changes in sling angle drastically affect the loading. When using slings, exercise extreme caution because unknown loads will be developed under less than ideal circumstances, in less than perfect equipment.

Failure to provide blocking or protective pads will permit sharp corners to cut slings. Pulling slings from under loads will result in abrasion and kinking. Dropping loads on slings or running equipment over them will cause crushing. Sudden starts and stops when lifting loads will increase the stresses in them. Also, improper storage will result in deterioration.

Because of the severe service expected of slings, errors in determining load weight, and the effect of sling angle on the loading, it is recommended that all safe working loads be based on a factor of safety of *at least* 5:1.

SLING TYPES AND CONFIGURATIONS

The term "sling" includes a wide variety of configurations for all fiber ropes, wire ropes, chains, and webs. The most commonly used types in rigging will be considered here because improper application can affect the safety of the lift.

There are three basic types of slings: single slings, endless slings, and braided slings. Let us take a look at each, as well as the configurations which use them.

Single Slings

A single sling can be made of wire rope, fiber rope, metal mesh, or chain. Each end of a single sling (Fig. 4-1) is made into an eye or has an attached hook. In some instances, the ends of a wire rope are spliced into eyes around thimbles and one eye is fastened to a hook with a shackle (Fig. 4-2). With this type of single sling, the shackle and hook can be removed when desired. A single sling can be used in several different ways for hoisting. It is advisable to have four single slings of wire rope available at all times. These can be used singly or in combination (configuration) as may be necessary.

Single Vertical Hitch. This is a method (Fig. 4-3) of supporting a load by a single vertical part or leg of the sling. The total weight of the load is carried by a single leg, the angle of the lift is 90°, and the weight of the load can equal the maximum safe working load of the sling and fittings. This sling configuration must not be used for lifting loose material, lengthy material, or anything that will be difficult to balance. Use them only on items equipped with lifting eye bolts or shackles such as concrete buckets. They provide absolutely no control over the load because they permit rotation.

Bridle Hitch. Two, three, or four single slings (Figs. 4-4, 4-5, and 4-6) can be used together to form a bridle hitch for hoisting an object that has the neces-

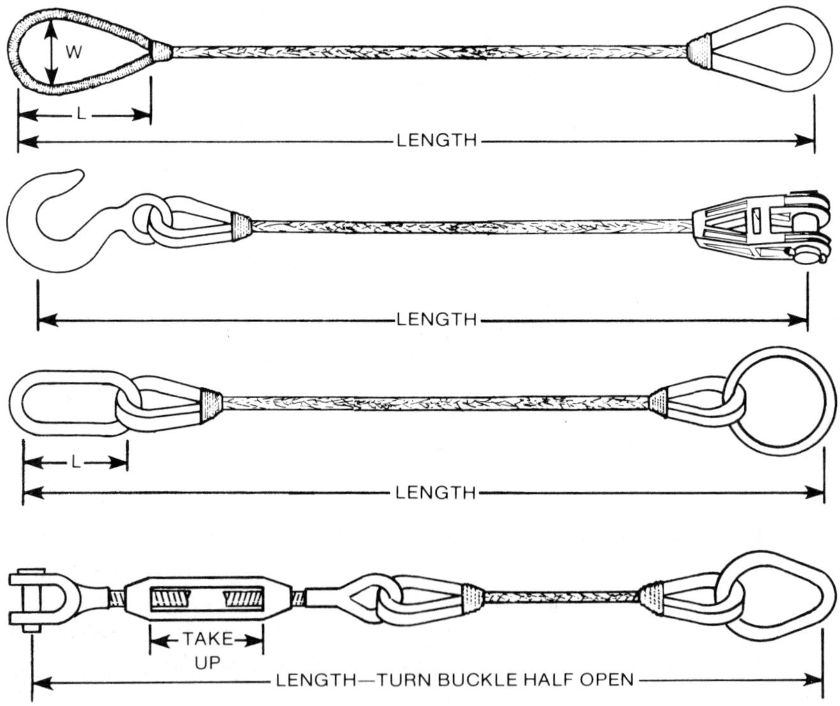

Fig. 4-1: Various single slings and how they are measured.

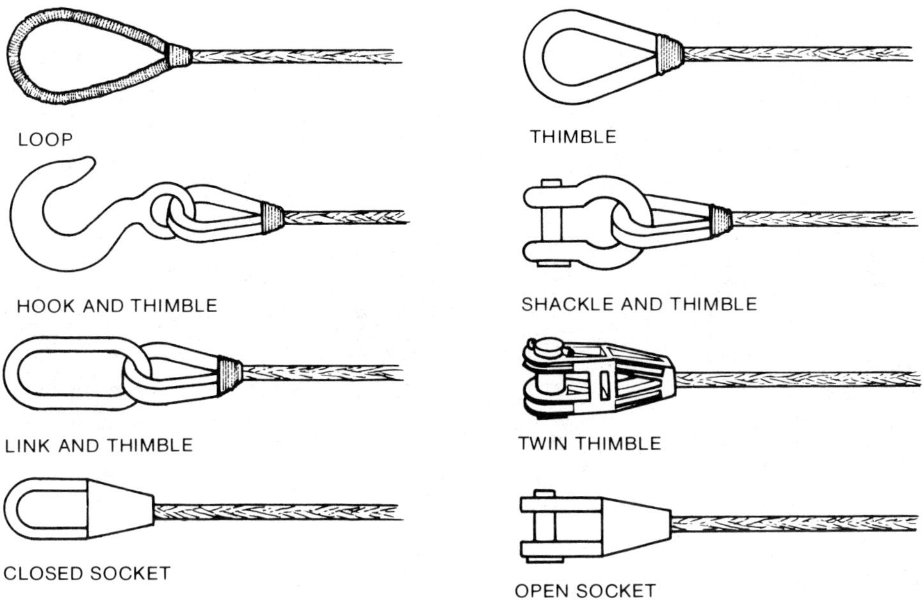

Fig. 4-2: Various fittings used on single slings.

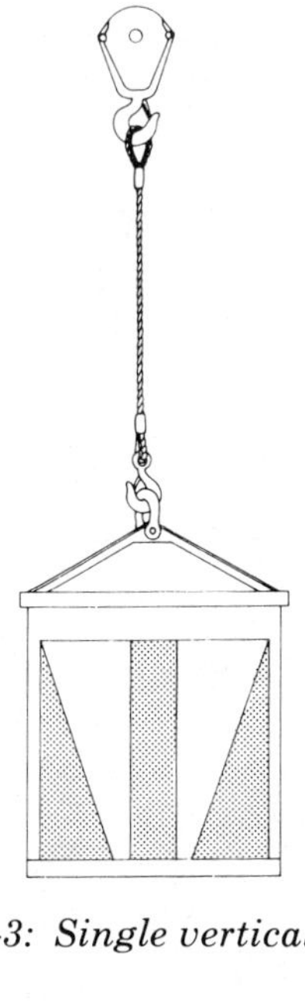

Fig. 4-3: Single vertical hitch.

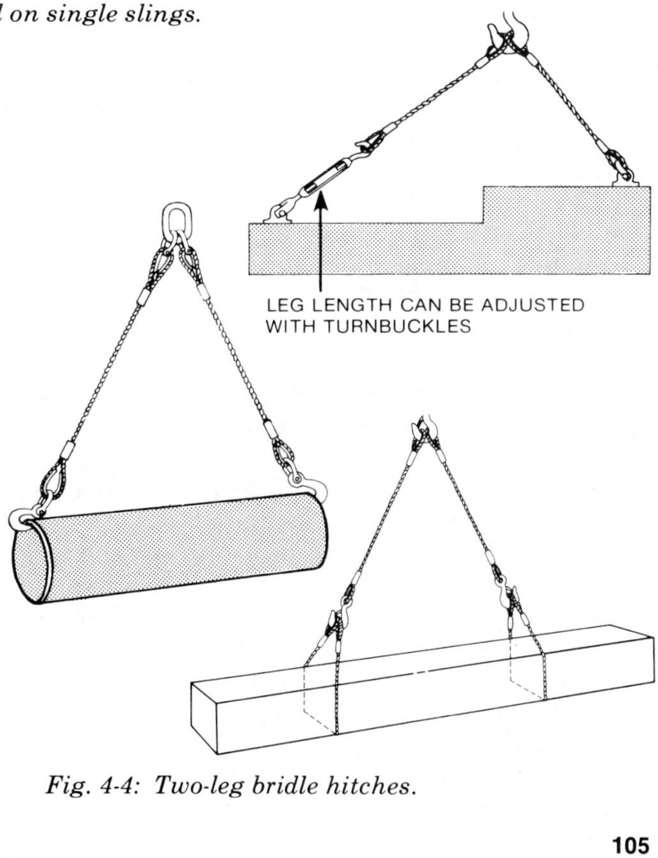

Fig. 4-4: Two-leg bridle hitches.

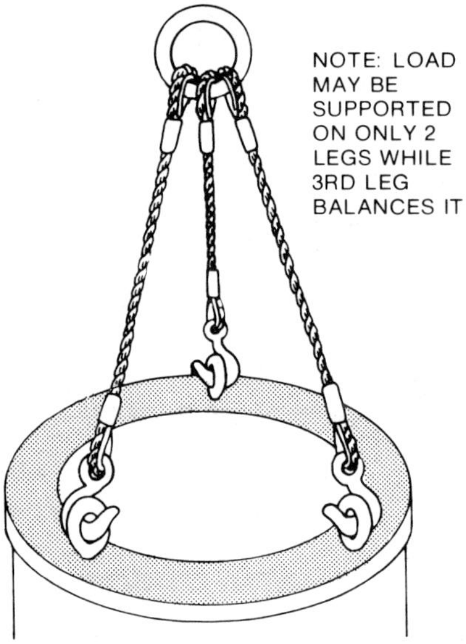

Fig. 4-5: Three-leg bridle hitch.

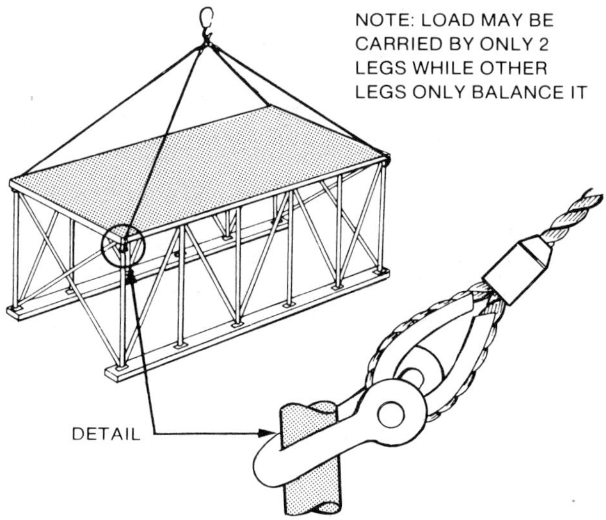

Fig. 4-6: Four-leg bridle hitch.

sary lifting lugs or attachments. They can be used with a wide assortment of end fittings. They provide excellent load stability when the load is distributed equally among the legs, when the hook is directly over the center of gravity of the load and the load is raised level. In order to distribute the load equally, it may be necessary to adjust the leg lengths with turnbuckles or shackles. The use of a bridle sling requires that the sling angles be carefully determined to be sure that the individual legs are not overloaded.

Unless the load is flexible, it is wrong to assume that a three or four leg hitch will safely lift a load equal to the safe load on one leg multiplied by the number of legs, because there is no way of knowing that each leg is carrying its share of the load. With slings having more than two legs and a rigid load, it is possible for two of the legs to take practically the full load while the others only balance it.

Single Basket Hitch. This is a method of supporting a load by hooking one end of a sling to a hook, wrapping it around the load, and securing the other end to the hook (Fig. 4-7). It cannot be used on any load that is difficult to balance because the load can tilt and slip out of the sling. On loads having inherent stabilizing characteristics, the load on the sling will be automatically equalized with each leg supporting half the load. Be sure that the load does not turn or slide along the rope during a lift because both the load and rope will become damaged. Remember that the capacity of a basket hitch is affected by the sling angle.

Double Basket Hitch. The double basket hitch (Fig. 4-8) consists of two single basket hitches passed under the load. They must be placed under the

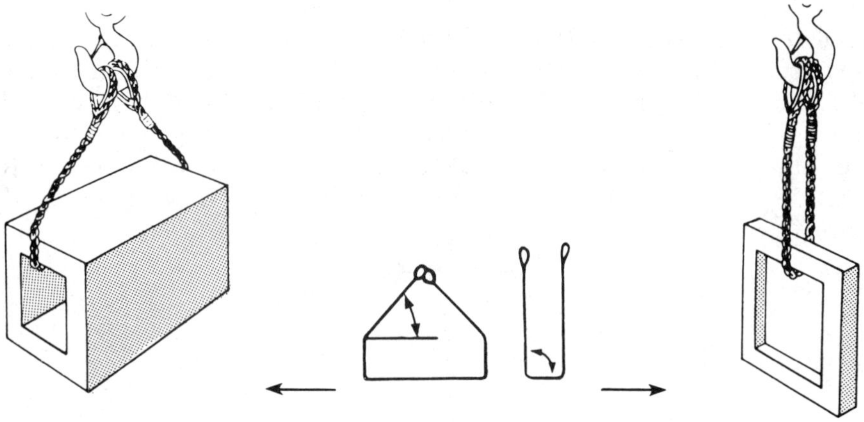

Fig. 4-7: Single basket hitch.

load so that it is properly balanced. The legs of the hitches must be kept far enough apart to provide balance but not so far apart that excessive angles are developed or to create a tendency for the legs to be pulled in toward the center (Fig. 4-8A). On smooth surfaces, both sides of the hitches should be snubbed against a step or change of contour to prevent the rope from slipping as the load is applied (Figs. 4-8B and C). Another way of preventing slippage is to keep an angle of 60° or more between the load and the sling (Fig. 4-8D).

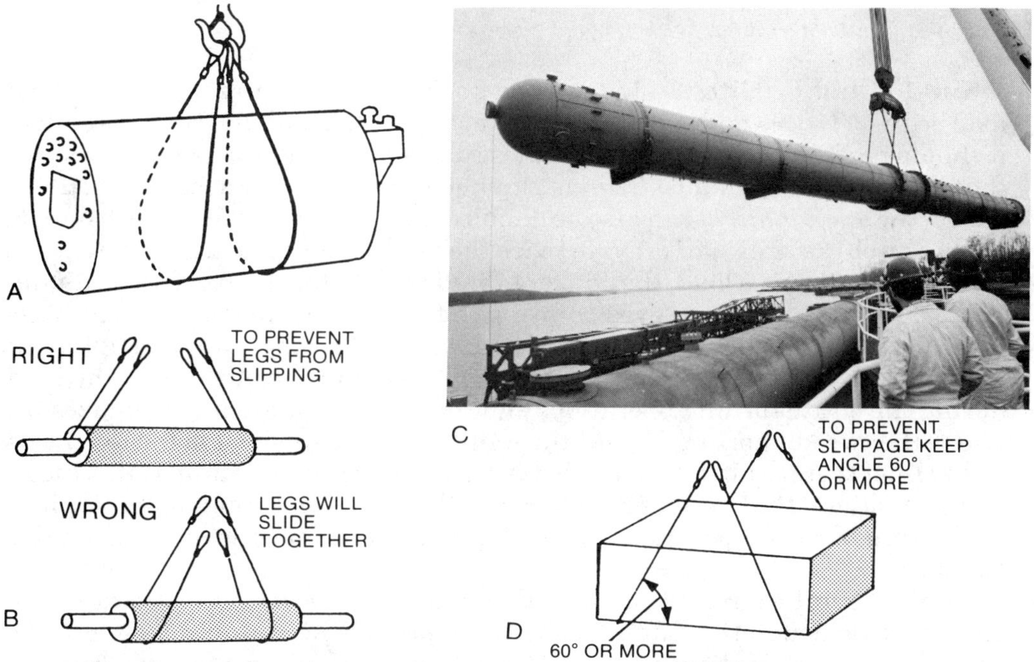

Fig. 4-8: Double basket hitches (photo courtesy of CBI Industries, Inc.).

Double Wrap Basket Hitch. This is a basket hitch that is wrapped completely around the load rather than just supporting as does the ordinary basket hitch. The double wrap basket hitch (Fig. 4-9) can be used in pairs like the double basket hitch. This method is excellent for handling loose material, pipe, rod, or smooth cylindrical loads because the rope or chain exerts a full 360° contact with the load and tends to draw it together. It does, however, decrease the sling's capacity.

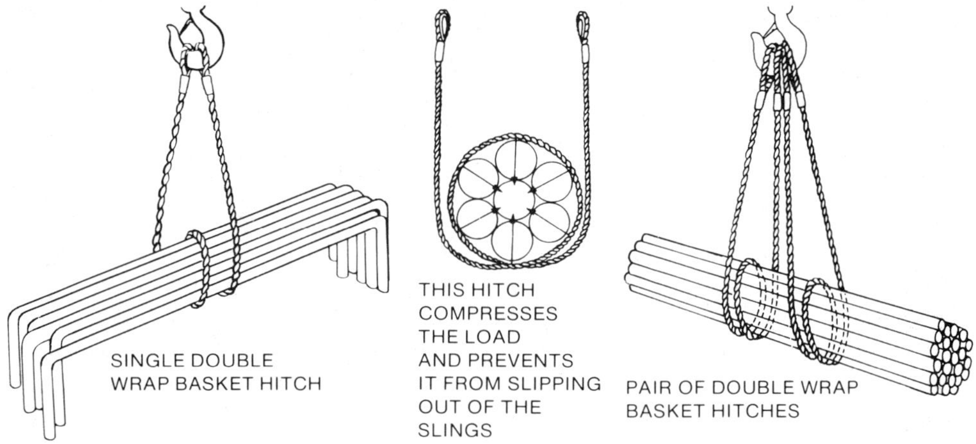

Fig. 4-9: Double wrap basket hitch.

Single Choker Hitch. This forms a noose in the rope that tightens as the load is lifted. It does not provide full 360° contact with the load. Because of this, it should not be used to lift loose bundles from which material can fall or loads that are difficult to balance. The single choker (Fig. 4-10), or as it is sometimes called, the *anchor hitch,* can also be doubled up (not to be confused with double choker hitch) as shown to provide twice the capacity or to turn a load. When it is necessary to turn a load, the choker is made by placing both eyes of the sling on top of the load with the eyes pointing in the direction opposite to the direction of turn. The center of the sling is passed around the load, through both eyes and up to the hook. This hitch provides complete control over the load during the entire turning operation, and the load automatically equalizes between the two supporting legs of the sling. Because the load is turned into a tight sling, there is no movement between the load and the sling. If it is incorrectly made and the two eyes are placed on the crane hook, the supporting legs of the sling may not be equal in length and the load may be imposed on one leg only.

Double Choker Hitch. The double choker hitch consists of two single chokers attached to the load and spread to provide load stability (Fig. 4-11). They, like the single choker, do not completely grip the load; however, because the load is less likely to tip, they are better suited for handling loose bundles, pipes, rods, etc.

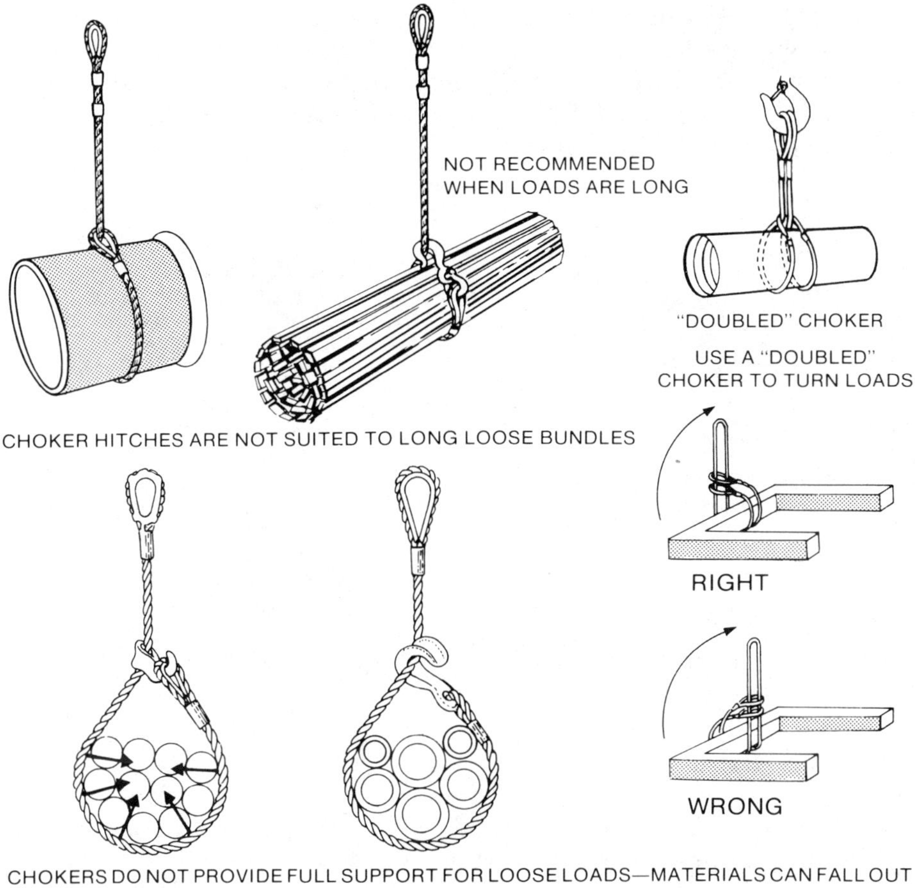

Fig. 4-10: Single choker hitches.

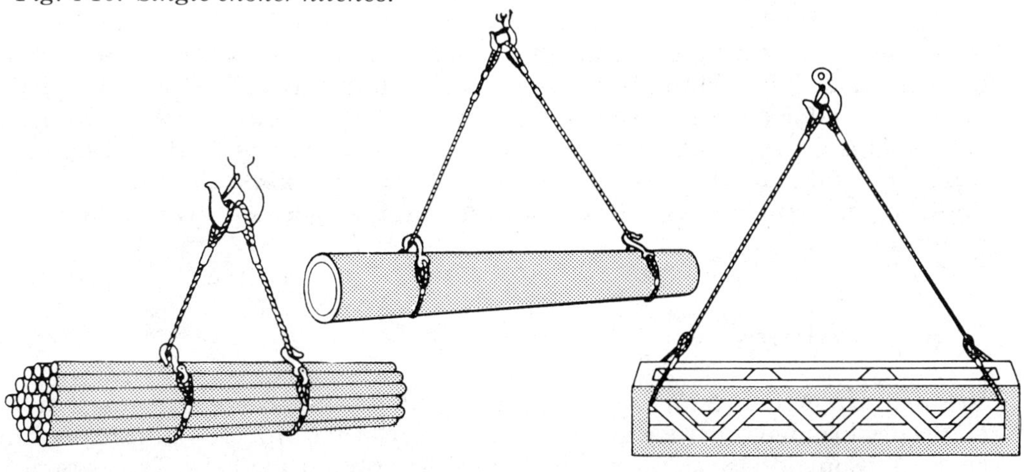

Fig. 4-11: Double choker hitches.

Double Wrap Choker Hitch. This hitch (Fig. 4-12) is one in which the rope or chain is wrapped completely around the load before being hooked into the vertical part of the sling. This hitch is in full contact with the load and tends to draw it tightly together. It can be used either singly on short, easily balanced loads or in pairs on longer loads.

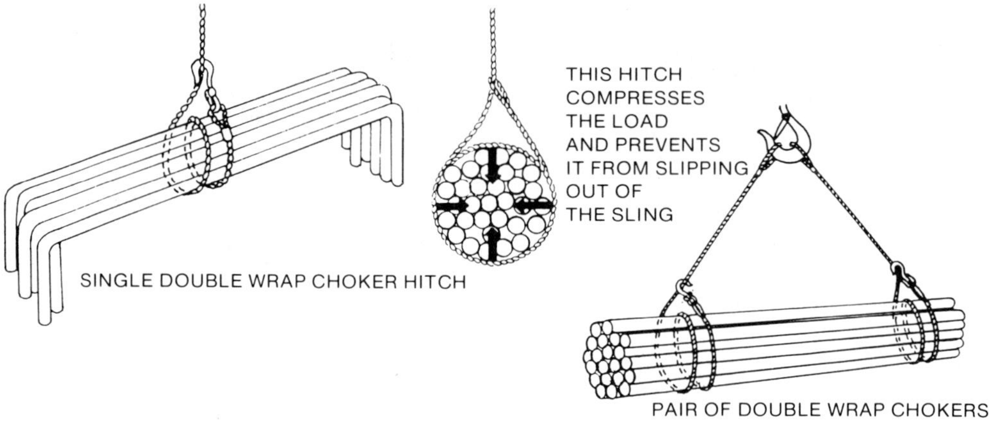

Fig. 4-12: Double wrap choker hitches.

Endless Slings

The endless sling, or as it is sometimes called, the *grommet sling,* is made by splicing the ends of a piece of wire rope (Fig. 4-13) or fiber rope together or by inserting a cold shut link in a chain. Cold shut links should be welded after insertion in the chain. There is only one tuck in the entire circumference where the two ends enter the rope. These slings are simple to handle and can be used in a number of configurations, as vertical hitches, basket hitches, choker hitches, and all combinations of these basic configurations (Fig. 4-14). They are very flexible but tend to wear and deteriorate more rapidly than the other slings. This results because they are not normally equipped with fittings and thus are deformed when bent over hooks, bearing against themselves at the bight. When using endless slings, be sure that the splice is always clear of the hooks and load.

Braided Slings

Braided slings (Fig. 4-15) are usually fabricated from six or eight small diameter ropes braided together to form a single rope that provides a large bearing surface, tremendous strength, and flexibility in all directions. They are very

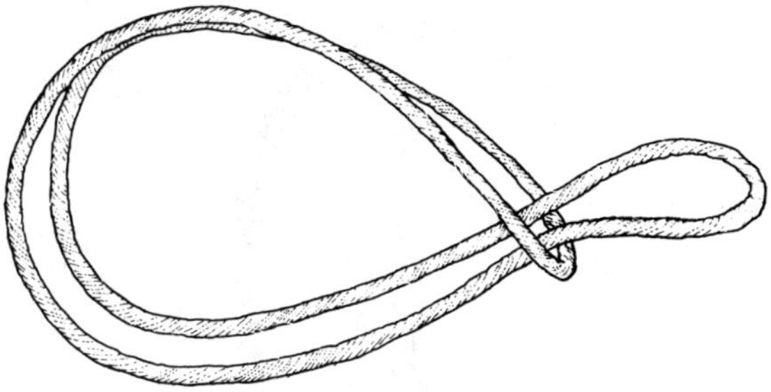

Fig. 4-13: Endless sling.

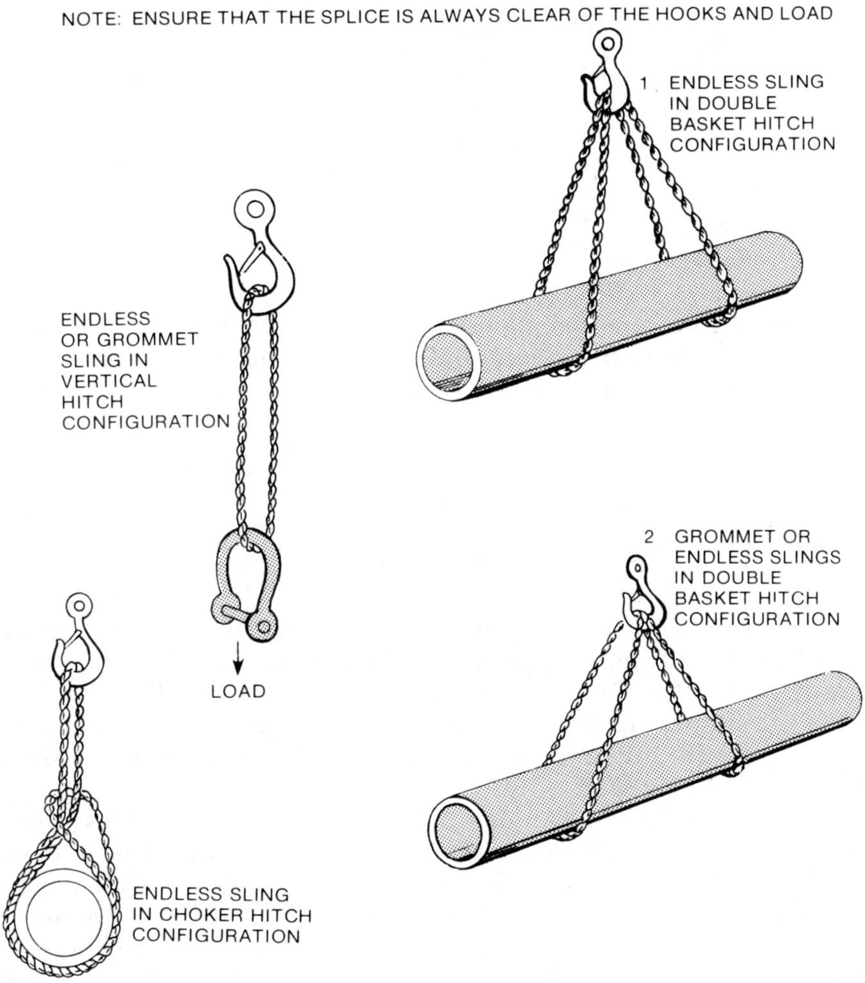

Fig. 4-14: Endless slings and their use in various configurations.

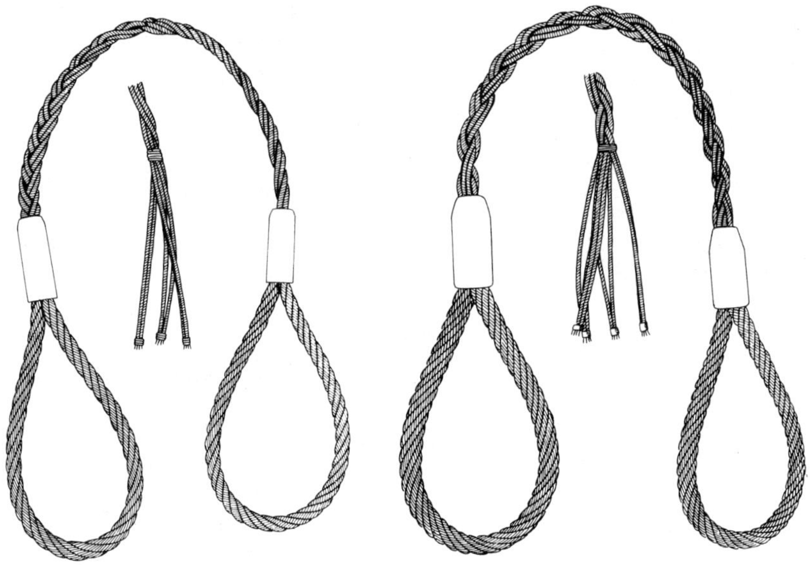

Fig. 4-15: Braided slings.

easy to handle and almost impossible to kink. The braided sling can be used in all the standard configurations and combinations but is especially useful for basket hitches where low bearing pressure is desirable or where the bend is extremely sharp.

SLING ANGLES

The rated capacity of any sling depends on its size, its configuration, and the angles formed by the legs of the sling and the horizontal. A sling with two legs that is used to lift a 1000 pound (453.5 kg) object will have a 500 pound (226.75 kg) load in each leg when the sling angle is 90°. The load in each leg will increase as the angle is decreased and at 30° the load will be 1000 pounds (453.5 kg) in each leg (Fig. 4-16).

If possible, keep the sling angles greater than 45°; sling angles approaching 30° should be considered extremely hazardous and should be avoided at all costs. The sharp increase in loading at low angles is clearly shown in Fig. 4-17.

Some load tables list sling angles as low as 15°, but the use of any sling at an angle less than 30° is extremely dangerous. This is not only because of the high loads associated with them, but also because of the effect on the load of an error in sling angle measurement of as little as 5°. Table 4-1 illustrates the effect of a 5° error in sling angle measurement on the sling load.

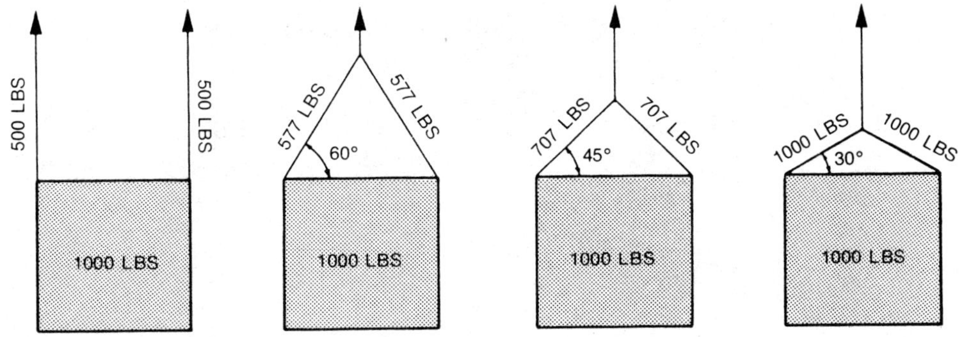

Fig. 4-16: Effect of sling angle on sling load.

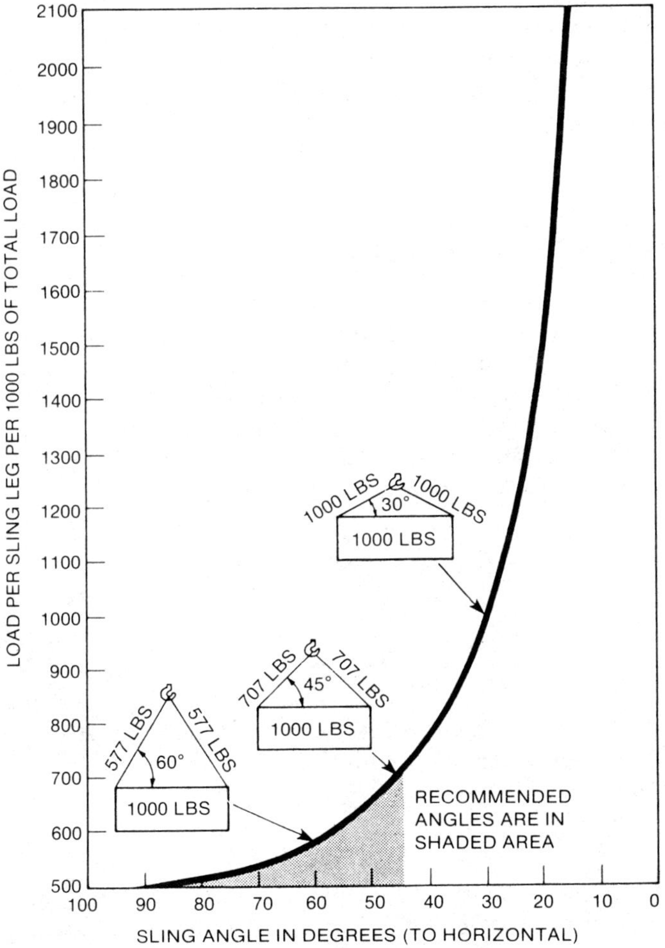

Fig. 4-17: Graph of the sling angle effect on sling load.

Table 4-1: Effect of Sling Angle Measurement Error on Loads

Assumed Sling Angle	Assumed Load (Pounds per Leg)	Actual Angle (is 5° Less than Assumed Angle)	Actual Load (Pounds per Leg)	Error in Percent
90°	500	85°	502	0.4
75°	518	70°	532	2.8
60°	577	55°	610	5.7
45°	707	40°	778	9.1
30°	1,000	25°	1,183	18.3
15°	1,932	10°	2,880	49.0

There is almost a 50% error in the assumed load at the 15° sling angle. This illustrates the degree of caution needed in not only ensuring the angle is greater than 45°, but also in measuring it accurately. The easiest and most accurate way to determine the angle is by measuring it with a large plywood measure graduated in degrees.

SAFE WORKING LOADS

Appendix D lists the safe working loads of fiber rope slings and wire rope slings under various conditions. There are a number of formulas used for obtaining the safe working loads of slings and sling configurations. Most of these formulas are based on the safe working load of a single vertical hitch. In the Occupational Safety and Health Administration (OSHA) tables, load ratings, or rated capacities, for a wire rope, single vertical hitch are generally based on the formula:

$$C_v = \frac{B \times E}{S}$$

in which:

C_V = Capacity of single vertical hitch, in pounds or tons
B = Breaking strength of wire rope, in pounds or tons
E = Efficiency of end fitting, in percent
S = Safety factor

Example:

Determine the rated capacity of a single vertical hitch made from 1″ 6 × 19 IWRC (improved plow steel) wire rope whose breaking strength is 112.1 tons. The end fitting is a hand-tucked spliced-eye that has an efficiency of 80%. As stated earlier in the chapter, the safety factor should be 5.

Solution:

$$C_v = \frac{B \times E}{S} = \frac{112.1 \times 0.80}{5} = 17.936\text{, or } 17.9 \text{ tons}$$

It would be a very difficult task to remember all load size and sling angle combinations, but the following rules of thumb work well for estimating the loads in the most common sling configurations. Each of the rules of thumb for a given sling configuration, material and size, is based on the safe working load of the single vertical hitch of that sling. The efficiencies of whatever end fittings are used will also have to be taken into account when determining the capacity of the combination.

Bridle Hitch

To determine the capacity of the two-leg bridle hitch (Fig. 4-18), measure the length of the sling legs (L) and measure the head room between the hook and the load (H). When the sling legs are not of equal length, use the smallest H/L ratio. The safe work load (SWL) of the two-leg bridle hitch is found by the following rule of thumb formula:

$$\text{SWL of Bridle Hitch} = \text{SWL of Single Vertical Hitch} \times \frac{H}{L} \times 2$$

Example:

If the safe work load of a single vertical hitch is 17.9 tons and the length of the two-leg bridle hitch is 15′, while the head room between the hook and the load is 12′, what is the safe working load of the configuration?

Solution:

$$SWL_{BH} = SWL_{VH} \times \frac{H}{L} \times 2 = \frac{17.9 \times 12 \times 2}{15} = 28.64 \text{ tons}$$

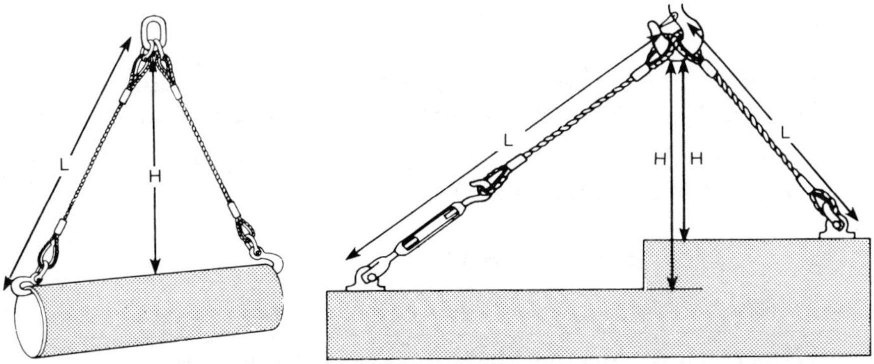

Fig. 4-18: Determination of capacity of two-leg bridle hitches.

The formula for a two-leg bridle hitch can also be used for the three- and four-leg hitches (Fig. 4-19). It is *wrong* to assume that a three- or four-leg hitch will safely lift a load equal to the safe load on one leg multiplied by the number of legs because there is no way of knowing that each leg is carrying its share of the load. With slings having more than two legs and a rigid load, it is possible for two of the legs to take practically the full load while the others only balance it.

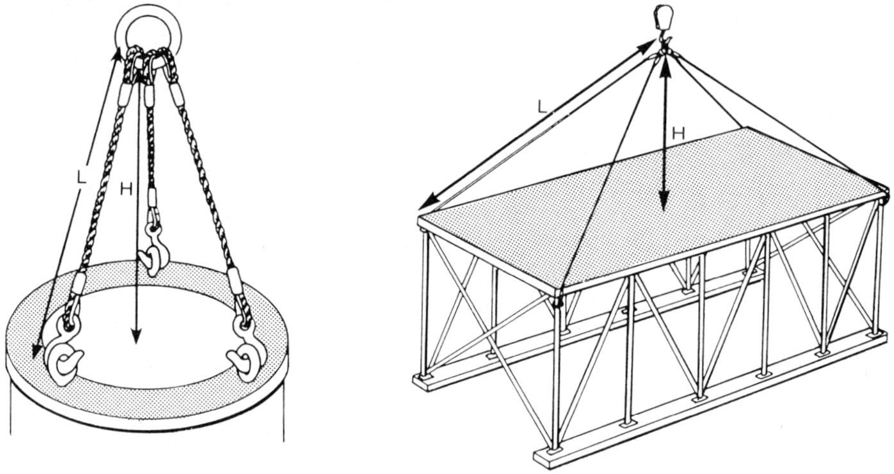

Fig. 4-19: Determination of capacity of three-leg and four-leg bridle hitch.

Single Basket Hitch

To determine the safe working capacity of a single basket hitch with vertical legs (Fig. 4-20A), use the following rule of thumb formula:

$$SWL_{SBH} = SWL_{VH} \times 2$$

in which:

SWL_{SBH} = Safe working load for a single basket hitch
SWL_{VH} = Safe working load for a single vertical hitch

The rule of thumb formula for the safe working capacity of a single basket hitch with inclined legs (Fig. 4-20B) is:

$$SWL_{SBH} = SWL_{VH} \times \frac{H}{L} \times 2$$

in which:

SWL_{SBH} = Safe working load for a single basket hitch
SWL_{VH} = Safe working load for a single vertical hitch

L = Length of one leg of hitch
H = Head room between block and load

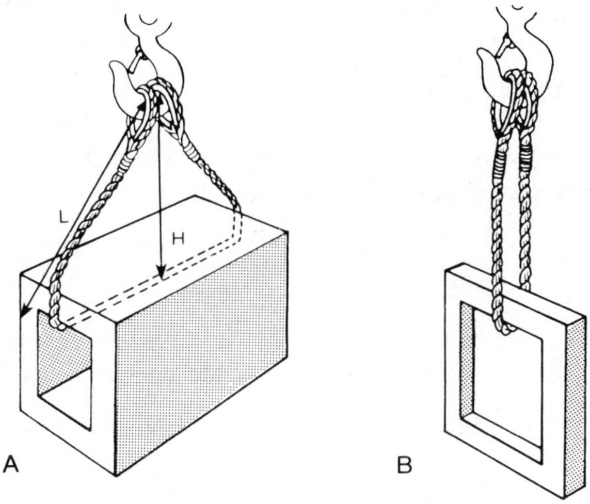

Fig. 4-20: Determination of capacity of single basket hitch.

Double Basket Hitch

The safe working capacities of double basket hitches (Fig. 4-21) are determined in basically the same manner as the single basket hitches, except that the multiplier is four instead of two. That is,

For vertical legs:

$$SWL_{DBH} = SWL_{VH} \times 4$$

For inclined legs:

$$SWL_{DBH} = SWL_{VH} \times \frac{H}{L} \times 4$$

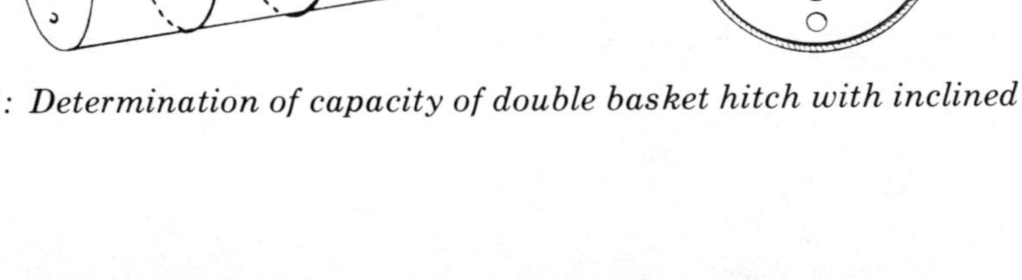

Fig. 4-21: Determination of capacity of double basket hitch with inclined legs.

Depending on the configurations, the safe working loads for *double wrap basket hitches* are the same as for the single basket hitch or the double basket hitch.

Single Choker Hitch

For sling angles of 45° or more (Fig. 4-22A), the rule of thumb formula for the safe working capacity of the single choker hitch is:

$$SWL_{CH} = SWL_{VH} \times 3/4$$

in which:

SWL_{CH} = Safe working load for a single choker hitch
SWL_{VH} = Safe working load for a single vertical hitch

Sling angles of less than 45° are *not* recommended; however, if they must be used, the formula is:

$$SWL_{CH} = SWL_{VH} \times \frac{Y}{Z}$$

in which:

SWL_{CH} = Safe working load for a single choker hitch
SWL_{VH} = Safe working load for a single vertical hitch
Y = Head room between hook and load
Z = Length of one leg of hitch (Fig. 4-22B)

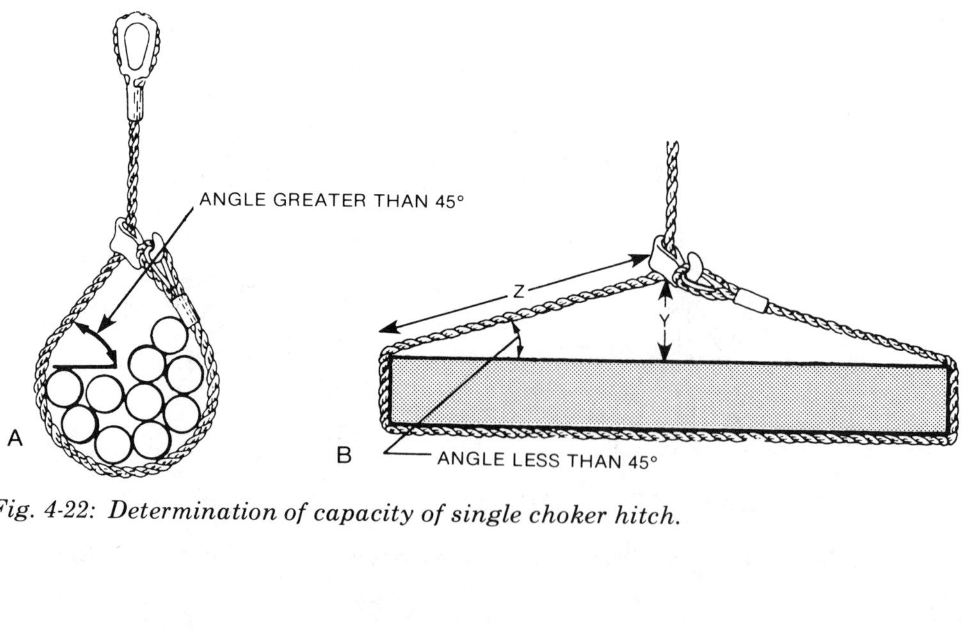

Fig. 4-22: Determination of capacity of single choker hitch.

Double Choker Hitch

For sling angles of 45° or more, formed by the choker (Fig. 4-23A), the formula is:

$$SWL_{CH} = SWL_{VH} \times \frac{3}{4} \times \frac{H}{L} \times 2$$

in which:

SWL_{CH} = Safe working load for a double choker hitch
SWL_{VH} = Safe working load for a single vertical hitch
H = Head room between hook and load
L = Length of sling

Sling angles of less than 45° (formed by the choker) are *not* recommended; however, if they must be used, the formula for such a double choker hitch arrangement (Fig. 4-23B) is:

$$SWL_{CH} = SWL_{VH} \times \frac{H}{L} \times \frac{Y}{Z} \times 2$$

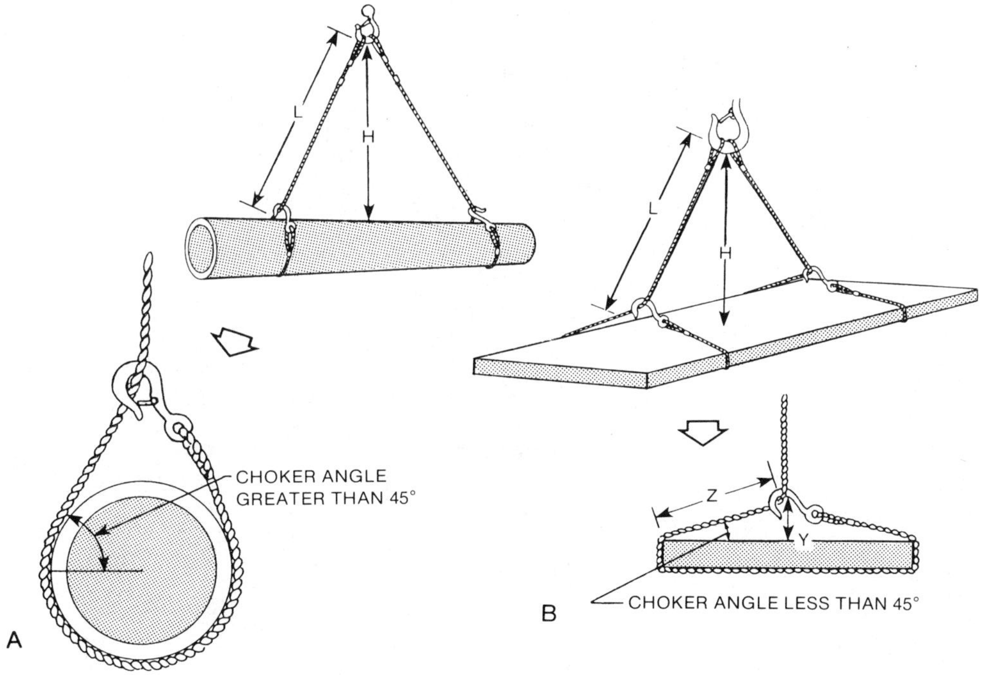

Fig. 4-23: Determination of capacity of double choker hitch.

in which:

SWL_{CH} = Safe working load for a double choker hitch
SWL_{VH} = Safe working load for a single vertical hitch
H = Head room above main hook and load
L = Length of sling from main hook and load
Z = Length from secondary hook to load
Y = Head room from secondary hook to load.

Depending on the configuration, the safe working loads of a *double wrap choker hitch* are the same as for the single choker hitch or the double choker hitch.

To measure and cut wire rope for chokers, proceed as follows (Fig. 4-24):

1. Measure and mark the splicing end and put on wire stop.
2. Measure and mark the crown and eye marks.
3. Measure the required length between the crown marks. Add twice the circumference of the wire rope for the take-up allowance at the top as shown in Table 4-2, and then mark off the eye, wire stop, and splicing end.

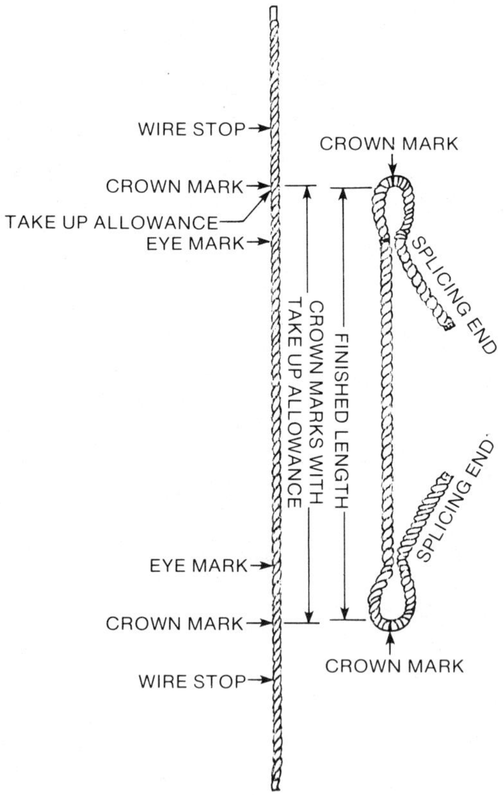

Fig. 4-24: Making a choker.

Table 4-2: Measurements for Cutting Wire Rope

Dia. of Wire Rope	Take-up Allowance	Splicing End	Eye
3/8″	2″	12″	8″
1/2″	3″	14″	10″
5/8″	4″	16″	12″
3/4″	5″	18″	14″
7/8″	6″	24″	16″
1	6-1/4″	30″	18″
1-1/8″	6-1/2″	36″	20″
1-1/4″	7-1/4″	44″	22″

Endless Slings

Depending on the exact configuration, the safe working loads for endless slings are two times the values for the previously mentioned configurations.

SELECTING SLINGS

As stated at the beginning of this chapter, slings can be made of fiber rope, metal mesh, chain, or wire rope. Each material has its advantages and each has its use. Let us look at the various sling materials.

Fiber Rope

Fiber rope slings are preferred for some applications because they are pliant, they grip the load well, and they do not mar the surface of the load. They should be used only on light loads, however, and must not be used on objects that have sharp edges capable of cutting the rope or in applications where the sling will be exposed to high temperatures, severe abrasion, or acids.

The choice of the rope type and size will depend upon the application, the weight to be lifted, and the sling angle. Before lifting any load with a fiber rope sling, be sure to inspect the sling carefully because they deteriorate far more rapidly than wire rope slings and their actual strength is very difficult to estimate.

The safe working loads given in Appendix D are approximate only and are based on a 5:1 factor of safety. They refer to new unknotted ropes and they assume that the loads are symmetrical and balanced, that the sling legs are of equal length, that the rope splices and fittings used to couple the slings to the load are strong enough, and that padding is used to protect the slings when lifting objects with sharp corners.

Chain Slings

Chain slings find application in those areas where the primary requirements are ruggedness, abrasion resistance, and high temperature resistance. They should **not** be used as slings in construction work, however, if it is possible to use wire rope or other materials because of their susceptibility to sudden failure.

When chain slings must be used, they should be padded where they bear on sharp edges of metal parts, otherwise some of the links will be subjected to bending stresses for which they were not designed. The chain may also damage the material being lifted if it is not padded.

The only chain suitable for overhead lifting is fabricated from alloy steel and identified by a letter, usually A, marked on each link.

Wire Rope Slings

The use of steel wire rope slings for lifting materials provides several advantages over other materials. Great strength and a minimum of weight are combined with flexibility. Warning occurs before failure by the breaking of outer wires. Reserve strength is provided in that the inner wires are protected by the outer wires and possess sufficient strength to carry the load if a reasonable factor of safety is allowed for the sling.

Properly fabricated wire rope slings are the safest type available for general construction usage. They do not wear as rapidly as fiber rope slings and they are not susceptible to the weakest link ailment of chains. Wire rope slings show, by inspection, their true condition, and the appearance of broken wires clearly indicates the extent of fatigue, wear, abrasion, and the like.

However, they must be carefully selected for the application and service in which they are to be placed as they also have limitations. The procedures and precautions outlined in Chapter 1 must be followed regardless of the sling type or application.

All wire rope slings should be made of improved plow steel and should have independent wire rope cores to reduce the possibility of the rope being crushed in service. Do not use slings from lang lay wire rope.

It is recommended that all eyes in wire rope slings be equipped with thimbles and that the eyes be formed with the Flemish splice and secured by swaged or pressed mechanical sleeves or fittings. With the exception of socketed connections, this is the *only* method that produces an eye that is as strong as the rope itself and provides reserve strength should the mechanical sleeve or fitting fail or loosen.

Appendix D is provided for information only on the more commonly used types of wire rope slings. For actual safe working loads of slings consult with the sling manufacturer. These tables are based on safety factors of 5:1 and assume that the slings are in excellent condition. Any deterioration, such as described in Chapter 1, reduces the capacity of the sling.

Metal (Wire or Chain) Mesh

Wire or metal mesh slings are well adapted for use where the loads are abrasive, hot, or tend to cut other types of slings. They resist abrasion and cutting, grip the load firmly without stretching, and can withstand temperatures of up to 550°F (288°C). They have smooth, flat bearing surfaces, they conform to irregular shapes, do not kink or tangle, and resist corrosion (Fig. 4-25).

They are available in three mesh sizes: 10 gauge mesh (heavy duty), recommended for general purpose lifting because it combines strength and abrasion resistance with flexibility; 12 gauge mesh, for medium duty applications; and 14 gauge mesh, for very light duty.

For handling materials that would damage the wire mesh or for handling loads with finishes that the mesh would damage, the slings can be coated with rubber or plastic. For their safe working loads, see Appendix D.

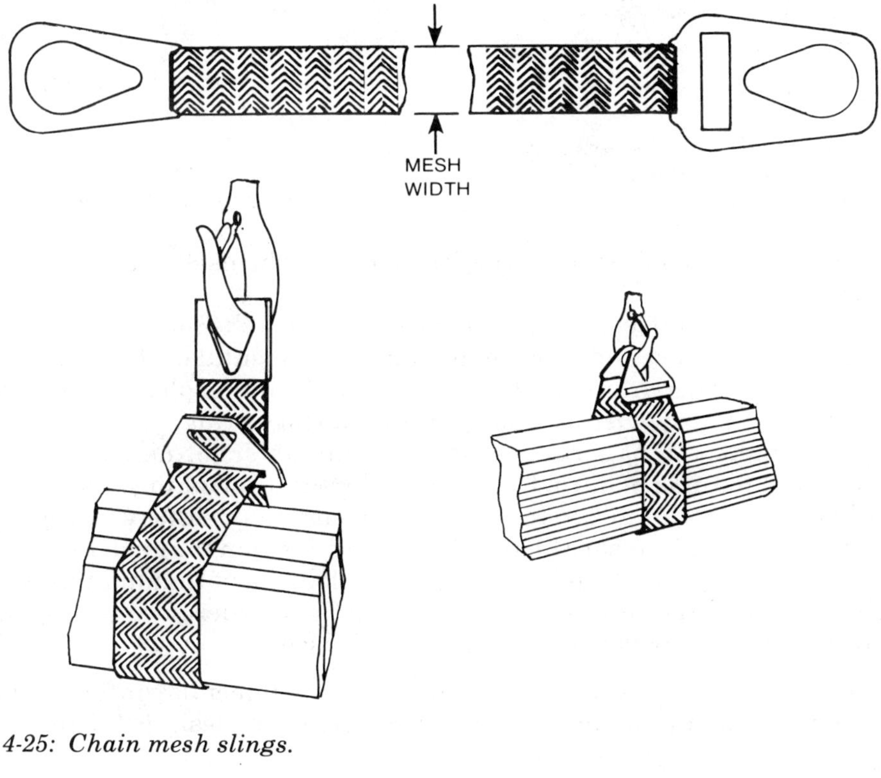

Fig. 4-25: Chain mesh slings.

SYNTHETIC WEBBING SLINGS

In recent years, synthetic webbing slings have become very popular in construction rigging jobs. Such slings offer a number of advantages:

1. Their relative softness and width mean they have much less tendency to mar or scratch finely machined, highly polished or painted surfaces and have less tendency to crush fragile objects than do fiber rope, wire rope, or chain slings (Figs. 4-26 and 4-27).

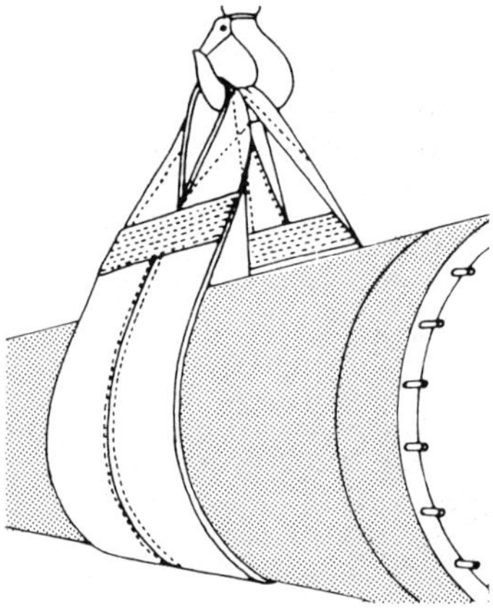

Fig. 4-26: Synthetic web slings do not damage or crush like wire ropes or chain.

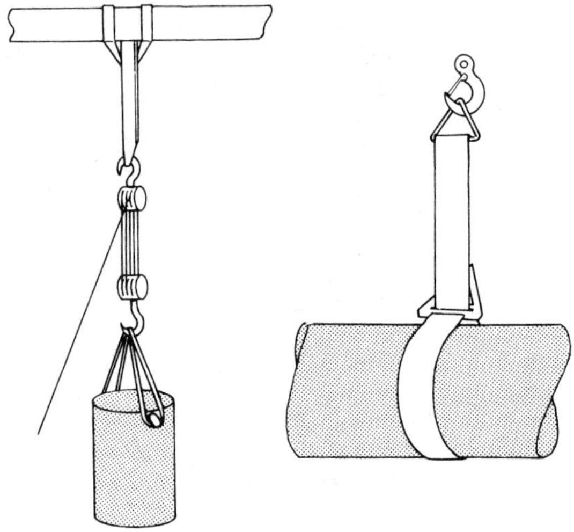

Fig. 4-27: Other uses of synthetic web slings.

2. Because of their flexibility, they tend to mold themselves to the shape of the load (Fig. 4-28).
3. They are not affected by moisture and certain chemicals.
4. They do not rust and thus do not stain ornamental cast materials.
5. They are nonsparking and can be used safely in explosive atmospheres.
6. They minimize twisting and spinning during lifting.
7. Their light weight permits ease of rigging, their softness precludes hand cuts, and the danger of harm from a bump by a free-swinging sling is minimal.
8. They are elastic and stretch under load more than either wire rope or chain and are thus able to absorb heavy shocks and cushion the load. In cases where sling stretching must be minimized, it should be kept in mind that synthetic web slings will stretch 10% of their rated capacity; therefore, a sling of larger load capacity or a polyester sling should be used.

Synthetic webbing slings are available in a number of configurations that find application in the construction trade. Some are designed with a reinforc-

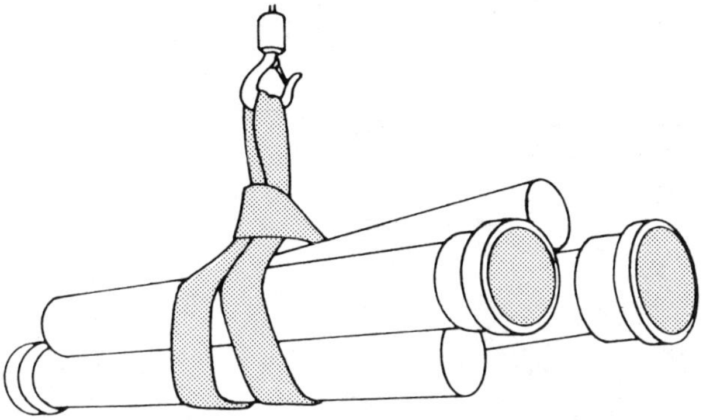

Fig. 4-28: Pipe handling illustrates the tendency of webbing slings to mold themselves to the load. This allows handling irregularly shaped loads securely.

ing core of inner load bearing yarns which carry 80% of the load and are protected from wear by the outer fibers.

Endless or Grommet Sling

In this type of sling, both ends of one piece of webbing are lapped and sewn to form a continuous piece. They can be used as vertical hitches, bridle hitches, in choker arrangements, or as basket slings. Because load contact points can be shifted with every lift, wear is evenly distributed and sling life is extended (Fig. 4-29).

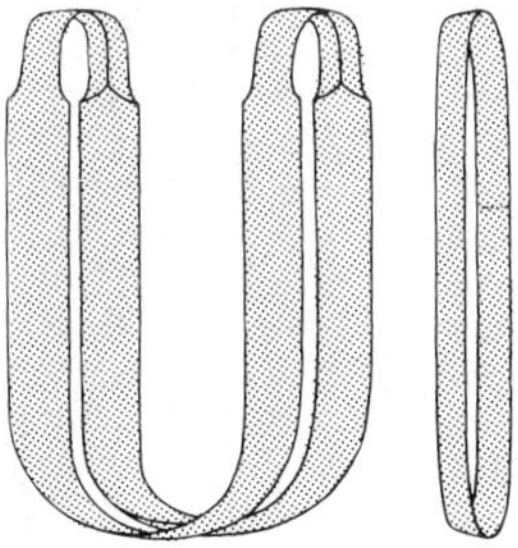

Fig. 4-29: Endless or grommet sling.

Standard Eye and Eye Sling

The webbing is assembled and sewn to form a flat body sling with an eye at each end and the eye openings in the same plane as the sling body. The eyes may either be full web width or may be tapered by being folded and sewn to a width narrower than the webbing width (Fig. 4-30).

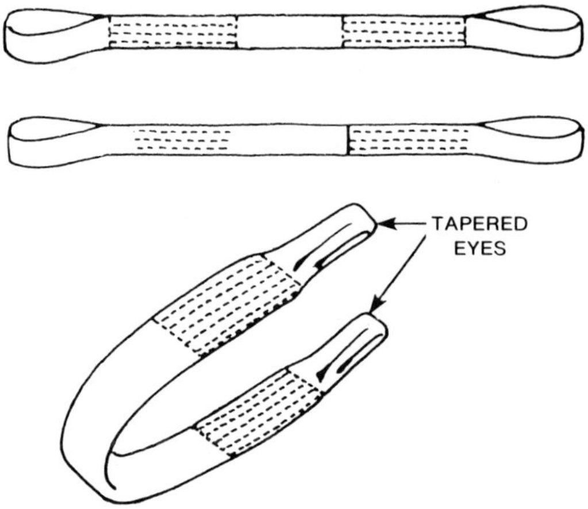

Fig. 4-30: Standard eye and eye slings.

Twisted Eye

This is an eye-and-eye type with twisted terminations at both ends. The eye openings are at 90° to the plane of the sling body. This configuration is also available with either full width or tapered eyes (Fig. 4-31).

In place of the sewn eyes, the slings are also available with metal end fittings. The most common are triangle and choker hardware. Combination hardware consists of a triangle for one end of the sling and a triangle/rectangle (choker attachment) for the other end. With this arrangement, both choker and basket as well as straight hitches may be rigged. They help reduce wear in the sling eyes and thus lengthen sling life (Fig. 4-32).

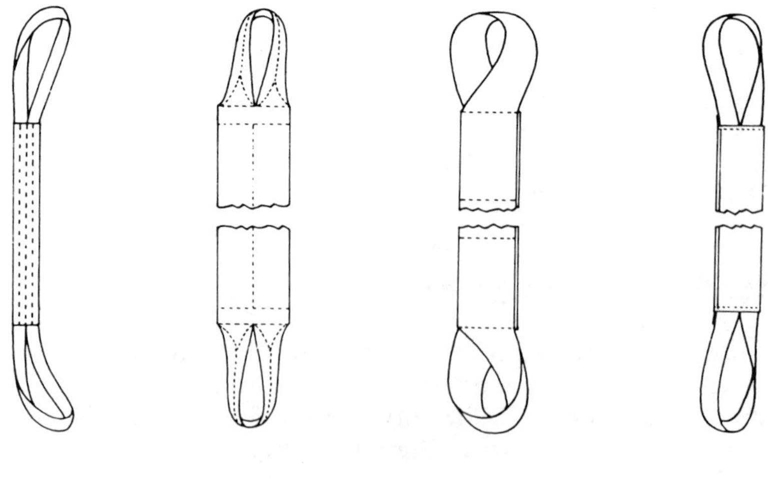

Fig. 4-31: Twisted eye slings.

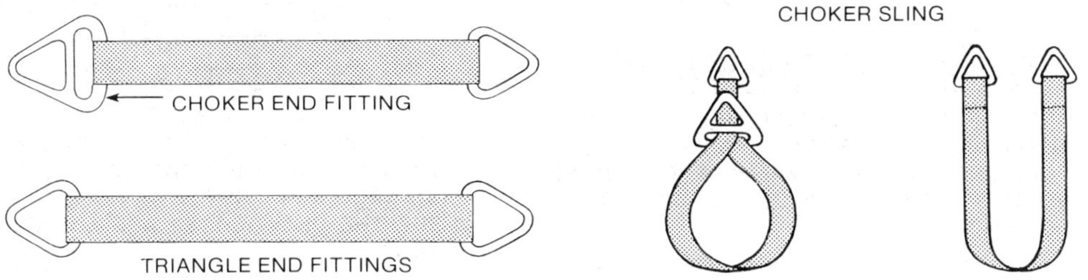

Fig. 4-32: Metal end fittings.

Protection of Slings

Despite their inherent toughness, synthetic webbing slings can be cut by repeated use around sharp-cornered objects and they eventually show signs of abrasion when rough-surfaced products are continually hoisted. There are, however, protective devices offered by most sling manufacturers that minimize these effects. The most common of these (Fig. 4-33) are:

1. Buffer strips of leather, nylon, or other materials sewn on the body of a sling protect against wear. Leather pads are the most resistant to wear and cutting, but are subject to weathering and gradual deterioration. They are not recommended in lengths over 6′ (1.83 m) due to the different stretch characteristics of leather and webbing. On the other hand, nylon web wear pads are more resistant to weathering, oils, grease, and most alkalies, and they stretch in the same ratio as the sling body.
2. Edge guards are also available and consist of strips of webbing or leather sewn around each edge of the sling. This is necessary for certain applications where the sling edges are subject to damage.
3. Sleeve or sliding tube type wear pads are available for slings used to handle material having sharp edges. They can be positioned on the sling where required, do not move when the sling stretches, adjust to the load, and cover both sides of the sling.
4. Reinforcing strips can be sewn into the sling eyes that double or triple the eye thickness and greatly increase the life and safety.
5. Coatings can be applied to provide added abrasion and chemical resistance. These treatments also increase the coefficient of friction, affording a better grip when loads with slippery surfaces are to be handled. These coatings can be brightly colored for safety or load rating purposes.
6. Some manufacturers offer a cotton-faced nylon webbing for hoisting granite and other rough surfaced material.

The synthetic webbing sling capacities quoted in Appendix E are approximate only and are based on nylon webbing having breaking strengths of 8000 pounds per inch of webbing width and 6000 pounds per inch of width and Dacron having a breaking strength of 5000 pounds per inch of webbing width.

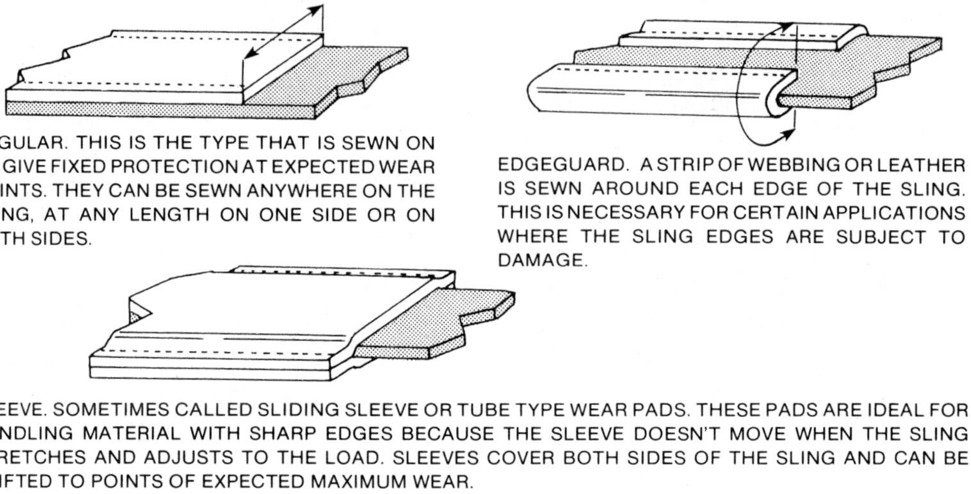

Fig. 4-33: Web and edge protectors.

The capacities are also based on a 5:1 factor of safety and assume that the end fittings are of adequate strength.

Although safe working loads for bridle hitches in the choker or double basket configuration hitches are provided, they are generally used only with extreme caution, because as the sling angle decreases, one edge of the web will take all the load with risk of tearing (Fig. 4-34).

The importance of inspection of the slings must not be underestimated even though, in general, if they look good they probably are good. Their damage is usually easy to detect; worn eyes and worn or distorted fittings, cuts, holes, punches, tears, frayed material, broken stitching and acid, caustic or heat burns—all are immediately evident and indicate that the slings should be replaced. Do not attempt to repair them yourself.

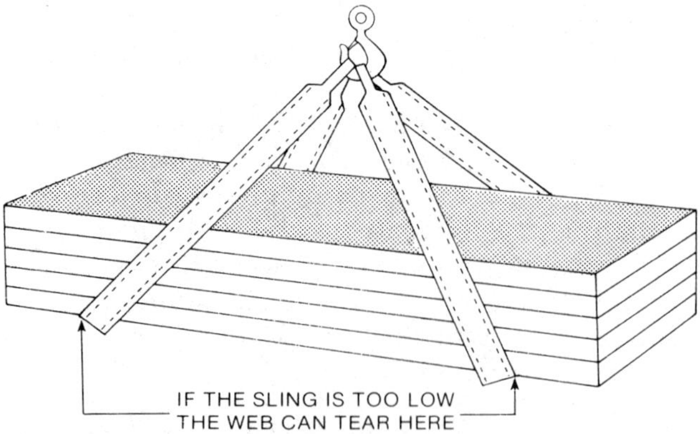

Fig. 4-34: Effect of low sling angle on webbing.

SLINGS AND SAFETY

The American Insurance Association records show that sling accidents can be attributed to five major factors. They are as follows:

Overloading

By establishing the weight of the load, the environment to be lifted in, and the load characteristics, the right type and capacity of sling can be selected. All load weights, when possible, should be marked on the loads to avoid any mistake that might be made in selecting the proper sling. Chain identification tags should not be removed from the chain since an untagged chain could be used for loads greater than those for which it is designed. With the wide variety of configurations, materials, and attachments in use on manufactured and job made slings, it is very important that slings be properly identified prior to use. Manufacturers' specifications, material properties, and test data should be utilized and appropriate records kept. OSHA has numerous tables for various types of slings and these must also be observed.

Sling Leg Angles

Spreading the sling legs reduces the angle of lift and increases the load on each leg. For example, reducing the angle which the sling makes with the horizontal from 90° to 30° causes the load on each branch of a two-leg sling to be doubled. An overload condition can be prevented by selecting the angle specified for working-load limits for slings as listed in Table 4-3.

Table 4-3: Sling Leg Angles

Angle Sling Makes with the Horizontal (Degrees)	Load per Leg for Each 1000 pounds Load Lifted (Pounds)	Angle Sling Makes with the Horizontal (Degrees)	Load per Leg for Each 1000 pounds Load Lifted (Pounds)
5	5736	50	653
10	2878	55	611
15	1931	60	577
20	1460	65	552
25	1183	70	538
30	1000	75	518
35	872	80	508
40	778	85	502
45	707	90	500

Faulty Hook-Up

Attaching the load off center reduces the load lifting capabilities of slings and frequently causes failure of a leg or fitting. Point loading also leads to many mishaps with slings. If a load begins to slip, the crane or lift operator should lower the load so that adjustments can be made in order to lift it without endangering personnel and equipment.

Impact Loading

Sudden strains on the sling are to be avoided, since such loading can snap a sling or cause the links to kink and stretch. No sling is manufactured to withstand this type of abuse.

Wear, Abrasion, and Distortion

A twisted sling can cause extensive damage by twirling a load. If this occurs, the load should be lowered and the sling untwisted. If the sling becomes kinked, the load should be lowered, the tension reduced, and the kink removed. Abrasions can significantly reduce a sling's load capacity through elongation and kinks from bending and pulling stresses. In fact, undue exposure to the weather, extremes of temperatures, and foreign materials (dirt, sand, etc.) can result in a decrease of a sling's wear quality through general deterioration. Remember, examine all slings before use, and discard any that are defective. Do not use a sling that contains a severe kink.

5

Rigging Hardware

INTRODUCTION

The hardware used in and with rigging installations and applications is just as important as the ropes, chains, etc., that are used with it. Not only must the rigger know what hardware to use and how to use it, but he/she must know how the safe working loads compare with the rope or chain, etc. It is extremely important that all fittings be of adequate strength for the application.

In Chapter 1, drums were discussed in relationship to ropes. In this chapter, a more detailed look at drums and drum assemblies as rigging hardware is undertaken.

In order to obtain the most from drums and drum assemblies, it is necessary to make certain that:

- They have adequate power and operational characteristics to perform all hoisting, holding, and lowering functions when operated under all conditions and configurations as recommended and approved by the manufacturer.
- They are provided with suitable clutching or power engaging devices that facilitate immediate starting and stopping of the drum motion.
- They are provided with self setting brakes that are capable of supporting all rated loads with recommended reeving.
- Their brakes and clutches are provided with adjustments to compensate for wear and maintain adequate force in springs where used.
- The drums have sufficient rope capacity with recommended rope size and reeving to perform all hoisting and lowering functions under recommended and actual service conditions. In addition, all hoist drums should be provided with adequate means to ensure even spooling of the rope on the drum.
- At least two or three full wraps of rope remain on the drum in all service conditions.

• The drum end of the rope is anchored by a clamp, securely attached to the drums with an arrangement approved by the manufacturer.

• The drums are provided with rims and flange guards of size sufficient to prevent the rope from jumping off the drum.

• Grooved drums have the correct groove pitch for the diameter of the rope. The depth of the groove must also be correct for the diameter of the rope (Fig. 5-1).

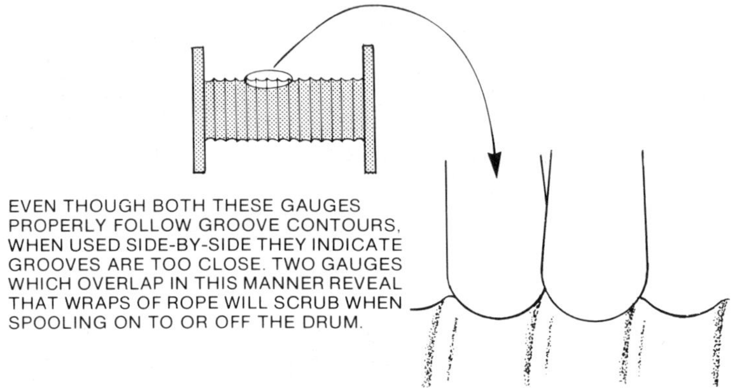

Fig. 5-1: Check drum grooves during inspections.

• The flanges on grooved drums must project either twice the rope diameter or 2″ (50.8 mm) beyond the last layer of rope, whichever of the two is greater (Fig. 5-2).

• The flanges on ungrooved drums must project either twice the rope diameter or 2-1/2″ (63.5 mm) beyond the last layer of rope, whichever of the two is the greater (Fig. 5-2).

• The fleet angle for grooved drums should lie between 1/4° and 1-1/4°, and for smooth drums it should lie between 1° and 2° (Figs. 5-3 and 5-4).

• Where the operator cannot see the drum or rope, drum rotation indicators should be provided and located to afford easy sensing by the operator.

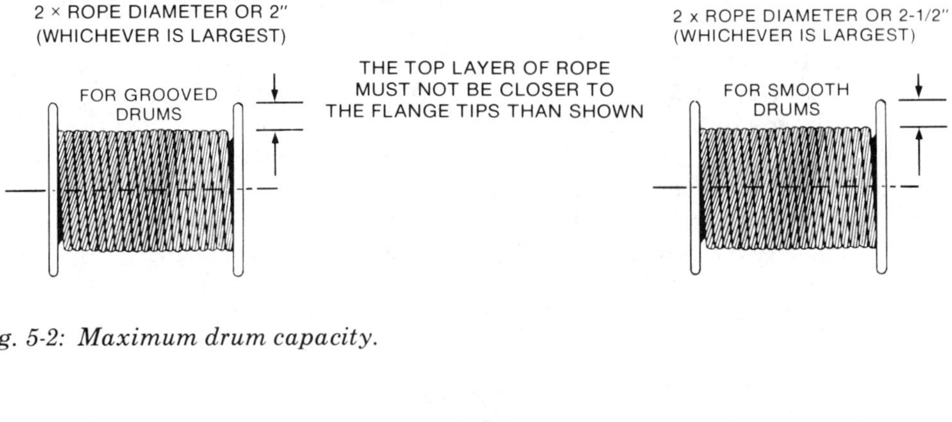

Fig. 5-2: Maximum drum capacity.

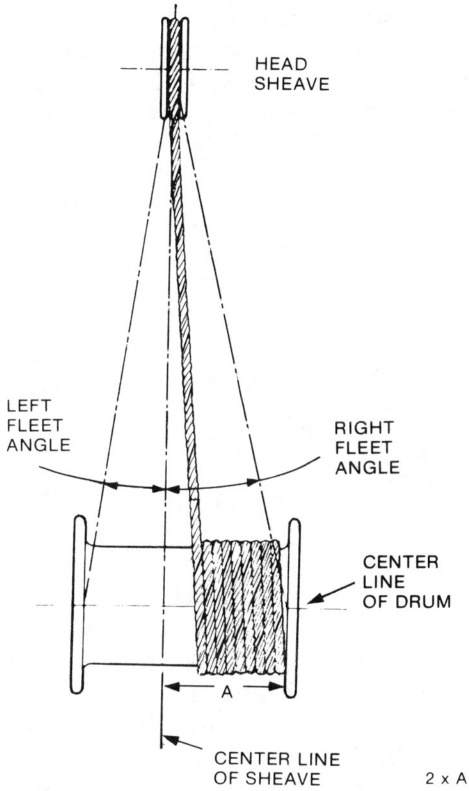

Fig. 5-3: Fleet angles.

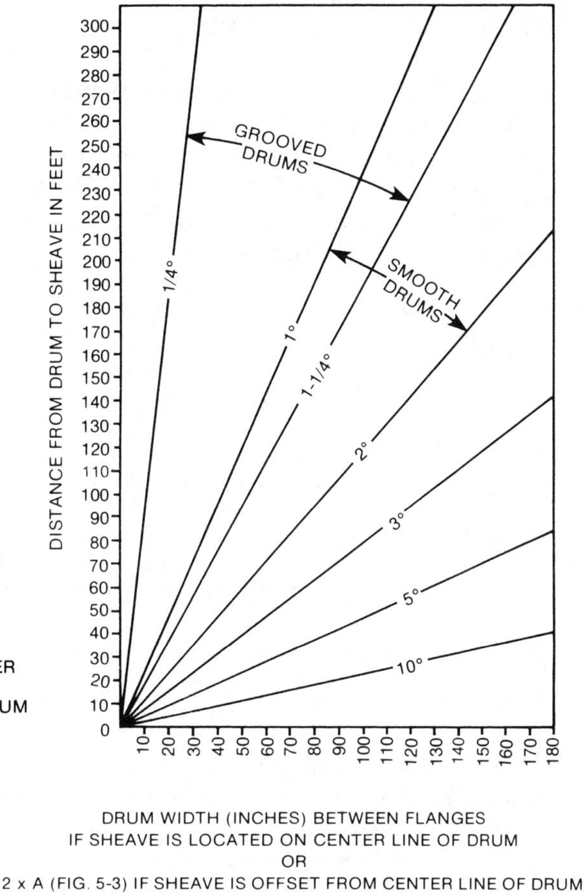

Fig. 5-4: Proper fleet angle ranges for grooved and smooth drums.

Whenever possible, a drum should be designed to hold all the rope on one smooth, even layer. Two and sometimes three layers are permitted, but more than three layers may cause crushing of the rope on the bottom layer as well as the end of any layer where pinching occurs (Fig. 5-5).

Fig. 5-5: Crushed, jammed, and flattened strands.

Winding onto a Drum

Always be sure the rope is properly installed on the drum. When winding a rope onto a machine drum, make certain that it bends in the same direction. Always reel from top to top or bottom to bottom. It is also necessary to apply a tensioning load to the rope to achieve good spooling. A simple brake such as a plank, rigged to bear against the reel flanges, will provide ample rope tension throughout the winding (Fig. 5-6).

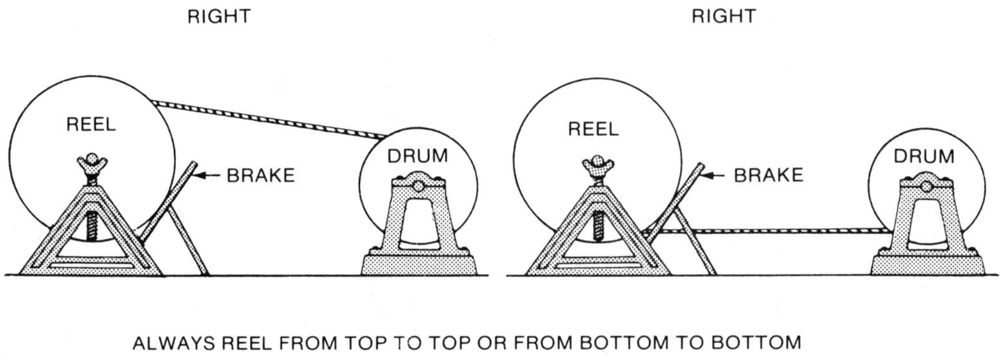

Fig. 5-6: Methods of winding rope from reel to drum or reel to reel.

To locate the rope anchorage point on a drum in relation to the direction of the lay of the rope, proceed in the following manner. Find out if the rope is to be overwound or underwound on the drum (Fig. 5-7). As mentioned in Chapter 1, if installing a *right* lay rope, visualize yourself as being behind the drum; make a fist with your *right* hand to simulate the drum and extend your forefinger to simulate the rope. For overwound rope, keep the back of your fist up (see how your extended finger is simulating the rope coming off the top of the drum). For underwound rope, turn your palm up and your finger simulates the rope coming off the bottom of the drum. Your extended finger always points to the flange to which the rope must be secured. For *left* lay ropes use exactly the same procedure but use your *left* hand (see Fig. 1-25).

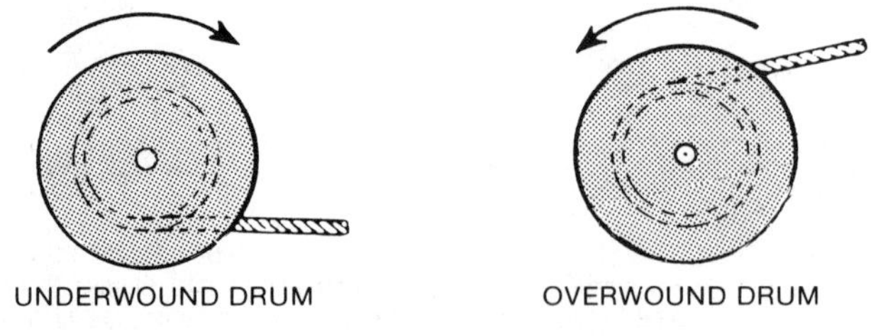

Fig. 5-7: Difference between overwound and underwound drums.

On a flat faced drum, it is most important that the rope winds in a straight helix at its proper angle. This can be assisted by means of a steel starting piece cut in a taper to fill the space between the first turn and the flange (Fig. 5-8).

Precautions should also be taken at the point where the rope enters the face of the drum at the anchorage to avoid a knuckle forming in the rope. This is particularly important for multiple-layer winding where the overwinding turn would cause hammering and rapid fatigue in the knuckle section.

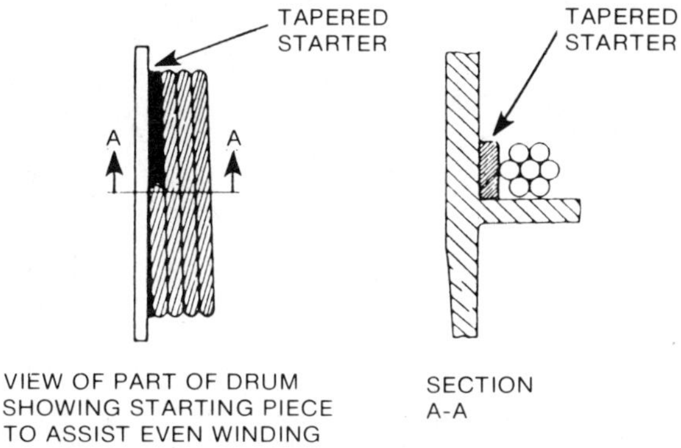

Fig. 5-8: Tapered starting piece to assist even winding.

When installing new rope, it is important that all turns of the first layer on the drum be tight and true. Open or wavy winding will result in serious damage to multiple windings. Adjacent turns should be tapped against each other with a soft metal or wooden mallet, but not so that the strands interlock. The succeeding layers of rope should wind in the groove formed between adjacent turns of the preceding layer of rope. Wire rope must never be allowed to cross-wind (Fig. 5-9).

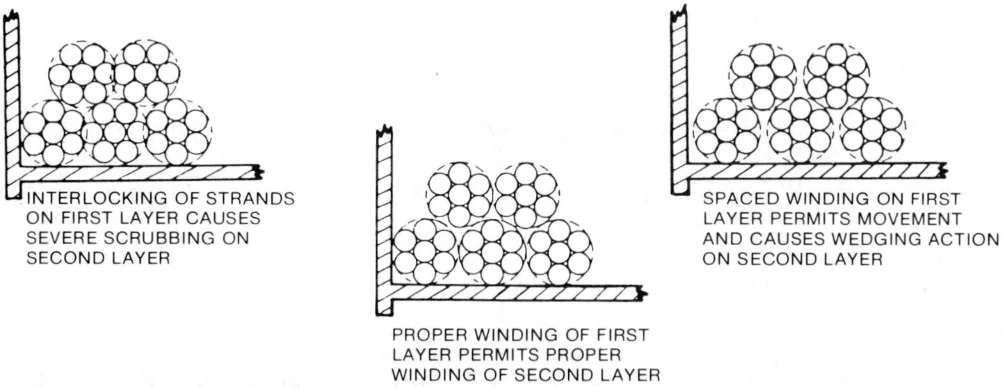

Fig. 5-9: Effect of improper first layer winding on rope life.

As far as is practical, the drum should accommodate all the rope in one smooth layer. More than three layers of rope result in reduced rope life and may cause crushing and severe abrasion of the rope on the bottom layer as well as at the end of any layer where pinching occurs. On drums with multiple-layer winding, difficulty may be experienced where, as the rope rises from the first to the second layer, wedging action and severe abrasion take place. This can be overcome by the use of a tapered steel lifter to raise the rope from the first to the second layer (Fig. 5-10).

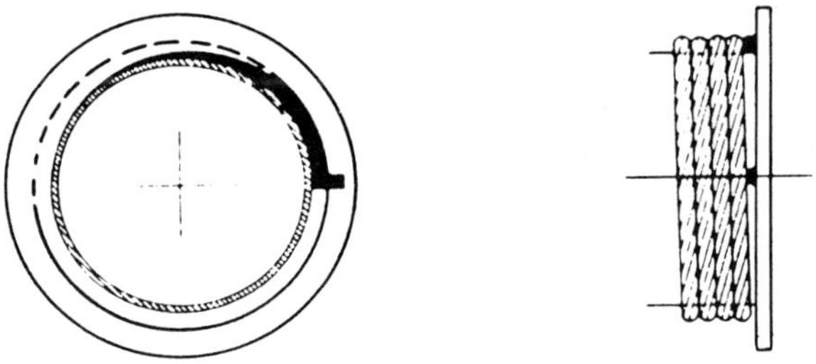

THIS TAPERED LIFTER PROVIDES A RAMP FOR THE ROPE TO RIDE UP AT THE FLANGE TO PREVENT WEDGING THE ROPE AGAINST THE FLANGE AND ADJACENT TURNS

Fig. 5-10: Tapered steel lifter to assist even winding.

The whole rope should be wound on the drum tightly and correctly as this will facilitate good winding during its full life. A poor start may mean poor winding and short life and, therefore, the extra care required for proper installation is well justified. The manner in which rope spools on a drum depends upon:

1. The rope size and construction.
2. The drum size and ratio of drum diameter to rope diameter.
3. The rope speed.
4. The amount of rope to be spooled.
5. The load on the rope.

To be certain that the rope spools evenly on the drum, either use a spooling device or keep the fleet angle to within the correct limits and keep tension on the rope at all times. Approximately 10% of the working load is recommended.

Drum Capacity

It is often necessary to know the approximate capacity of a given drum or reel for a particular rope diameter. The following method can be used to make this calculation (Fig. 5-11):

Add the diameter of the drum (B) to the depth of the flange (A). Multiply this sum by the depth of flange (A). Multiply the result by the distance between the drum flanges (C). Then multiply this result by the factor (F) listed in Table 5-1 opposite the size of rope to be installed.

Drum Capacity (feet of rope) = (B + A) × A × C × F where A, B, and C (Fig. 5-11) are in inches.

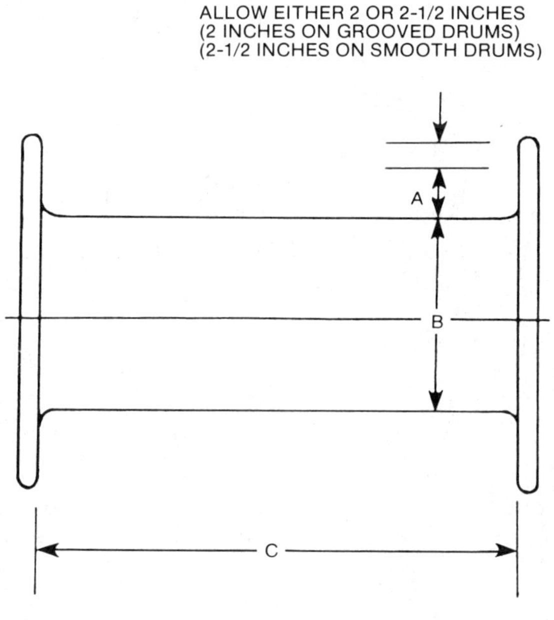

Fig. 5-11: Determination of drum capacity.

Table 5-1: Drum or Reel Capacity Factor

Nominal Rope Diameter (Inches)	F
1/4	4.160
5/16	2.670
3/8	1.860
7/16	1.370
1/2	1.050
9/16	0.828
5/8	0.672
3/4	0.465
7/8	0.342
1	0.262
1-1/8	0.207
1-1/4	0.167
1-3/8	0.138
1-1/2	0.116
1-5/8	0.099
1-3/4	0.085
1-7/8	0.074
2	0.066
2-1/8	0.058
2-1/4	0.052
2-3/8	0.046
2-1/2	0.042

Example:

If the diameter of the drum is 18″, the depth of the flange is 2″, and the distance between drum flanges is 24″, what is the drum's capacity for a 3/4″ rope?

Solution:

Using the formula for drum capacity: (B + A) × A × C × F = (18 + 2) × 2 × 24 × 0.465 = 446 feet.

The factor (F) applies to nominal rope size and level winding. Since new ropes are usually oversized by 1/32″ per 1″ of rope diameter, the result obtained by the formula must be decreased to account for oversized rope and/or random or uneven winding as follows:

1. For oversized ropes, decrease the calculated length by 0 to 6%.
2. For random wound ropes, decrease the calculated length from 0 to 8%.

Whenever possible, use grooved drums rather than smooth drums as the grooves furnish better support for the rope than do flat surfaces, and more uniform winding results in less abrasive wear on the rope. The groove surfaces on grooved drums and the complete surface on smooth drums should be perfectly smooth; those which have taken the imprint of the outer wires of previous ropes will exert a grinding action on new ropes (Fig. 5-12).

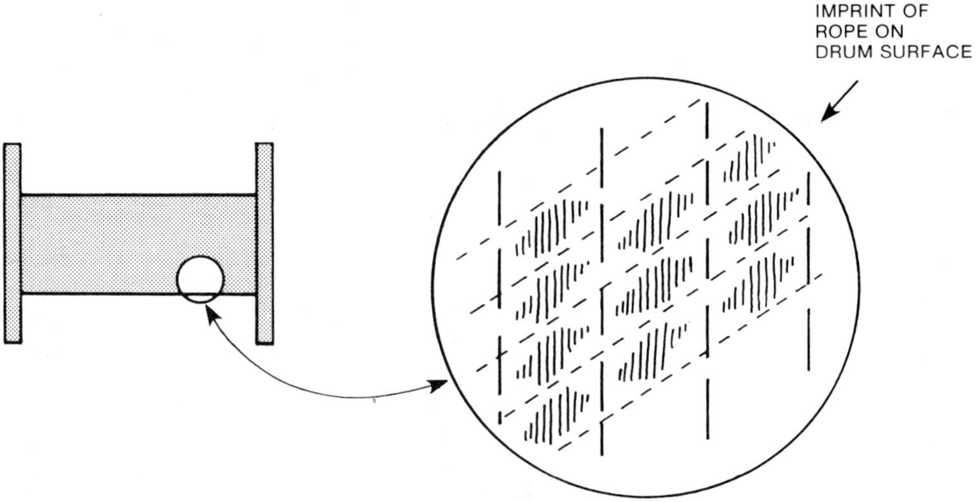

Fig. 5-12: Check for drum scoring during inspections.

This imprinting and scoring is caused by high contact pressures between the rope and drum surface. If this condition is evident, then the drum must be resurfaced and the contact pressure reduced by:

1. Decreasing the load on the rope;
2. Increasing the drum diameter; or
3. Replacing the drum with one having harder metal.

The radial contact pressure can be calculated as follows:

$$P = \frac{2L}{Dd}$$

in which:

P = Radial pressure in psi (Fig. 5-13)
L = Maximum working load of rope in pounds
D = Tread diameter of drum or sheaves (inches)
d = Rope diameter (inches)

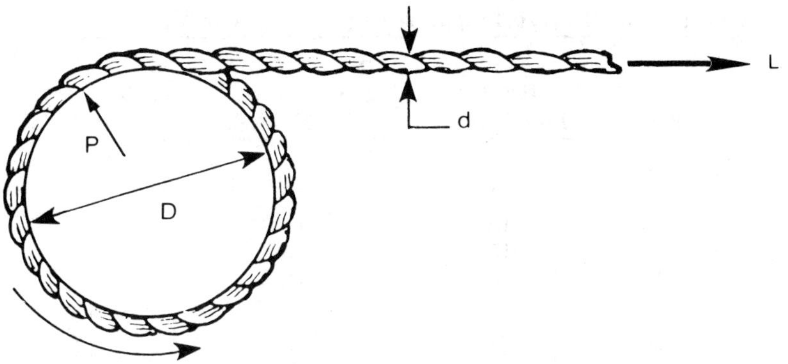

Fig. 5-13: Determination of drum contact pressures.

Example:

What is the radial contact pressure on an 18″ drum when a 7/8″ diameter 8 × 19 round strand rope is used that has a maximum working load of 11,800 pounds? Also, what type of drum material should be used?

$$P = \frac{2L}{Dd} = \frac{2\,(11{,}800)}{18 \times 7/8} = 1498 \text{ psi}$$

This contact pressure means, from Table 5-2, that the drum must be manganese steel. It is suggested that the limits given in Table 5-2 be observed for all drums and sheaves.

In order to minimize the rope bending stresses, the drum diameter should be at least as large as indicated in Table 5-3.

Table 5-2: Contact Pressure (psi) Limits for Sheave and Drum Materials

Rope Construction	Cast Iron	Cast Steel	Manganese Steel
6 × 7 Regular lay	300	550	1,500
6 × 7 Lang lay	350	625	1,700
6 × 19 Regular lay	500	900	2,500
6 × 19 Lang lay	575	1,025	2,850
6 × 37 Regular lay	600	1,075	3,000
6 × 37 Lang lay	700	1,250	3,500
8 × 19 Regular lay	600	1,075	3,000
6 × 8 Flat strand	500	900	2,500
6 × 25 Flat strand	800	1,450	4,000
6 × 33 Flat strand	975	1,800	4,900

Table 5-3: Drum and Sheave Diameters (Inches) for Various Types of Ropes

Rope Diameter (Inches)	6 × 7	6 × 19 18 × 7 N.R.	6 × 19 Warr. 6 × 16 F 6 × 27 F.S.	6 × 19 F 8 × 19 Seale	6 × 37
1/4	10	8	7	6	5
5/16	13	11	9	8	6
3/8	16	13	11	10	7
7/16	18	15	13	11	8
1/2	21	17	15	13	9
9/16	23	19	17	15	10
5/8	26	21	19	16	11
3/4	31	25	22	19	13
7/8	37	30	26	23	16
1	42	34	30	26	18
1-1/8	47	38	34	29	20
1-1/4	52	42	37	32	22
1-3/8	58	47	41	36	25
1-1/2	63	51	45	39	27
1-5/8	—	55	49	42	29
1-3/4	—	59	52	45	31
1-7/8	—	64	56	49	34
2	—	68	60	52	36
2-1/4	—	—	67	58	40
2-1/2	—	—	75	65	45

SHEAVES

The condition and contour of sheave grooves have a major influence on rope life. The grooves must be smooth and slightly larger than the rope to prevent it from being pinched or jammed in the groove. Since most ropes are made slightly larger than their nominal size, the sheave grooves for new rope should just accommodate the full oversize of the rope, as given in Table 5-4.

Table 5-4: Sheave Groove Tolerances

Nominal Rope Diameter (Inches)	Groove Oversize (Inches)	
	Min.	Max.
1/4 —5/16	1/64	1/32
3/8 —3/4	1/32	1/16
13/16—1-1/8	3/64	3/32
1-3/16—1-1/2	1/16	1/8
1-9/16—2-1/4	3/32	3/16
2-5/16 up	1/8	1/4

The bottom of the groove should have an arc of support of at least 120° to 150°, and the sides of the groove should be tangent to the arc (Fig. 5-14).

The more closely the contour of the groove approaches that of the wire rope, the greater becomes the area of contact between the two. This minimizes rope distortion and bending fatigue, as well as eases sheave rotation. If the groove diameter is too large, the rope will not be properly supported and will tend to flatten and become distorted. This accelerates bending fatigue in individual wires and can cause premature failures (Fig. 5-15).

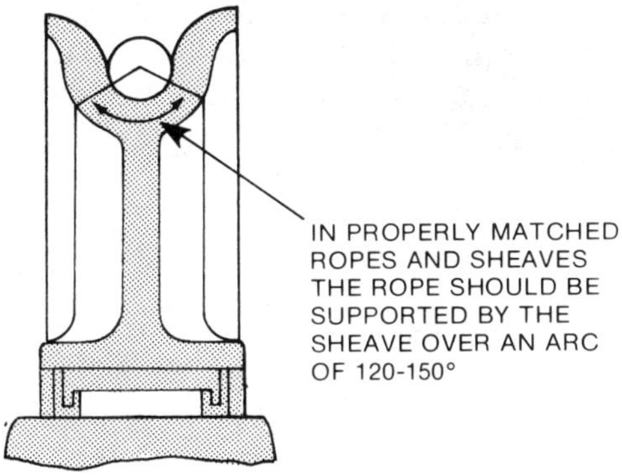

Fig. 5-14: Proper arc of support provided the rope by a sheave.

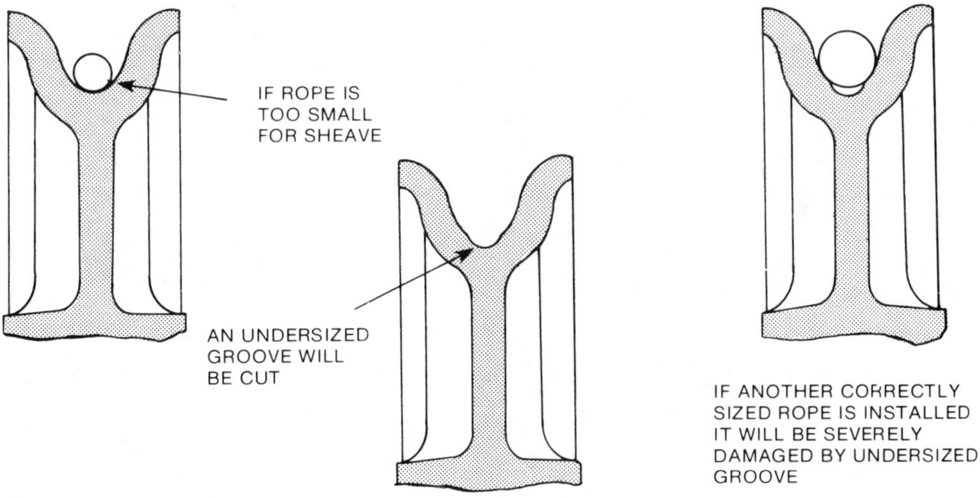

Fig. 5-15: Effect of improper match between rope and sheave.

If the sheave groove is too narrow for the rope, the operating tension will draw the rope deeply into the groove, causing it to be pinched and subjecting both the rope and sheave to severe abrasive wear. This condition can arise if new ropes are installed over old sheaves.

If the sheaves are not perfectly aligned, both the rope and sheave flanges will be subjected to severe wear and rapid deterioration will occur. A ready indication of poor alignment is rapid wear of only one of the flanges on any given sheave (Fig. 5-16).

One of the fastest ways to ruin a rope is to operate it over small sheaves. The excessive and repeated bending and straightening of the wires leads to premature failure from fatigue. Use the maximum possible diameter of sheave that the equipment will carry. The sheaves should never be of smaller diameter than indicated in Table 5-3.

The depth of the sheave grooves should be at least 1-1/2 times the rope diameter, and the tapered side walls of the grooves should not make an angle of more than 18° with respect to the centerline. The flange corners should be rounded and the rims should run true about the axis of rotation (Fig. 5-17).

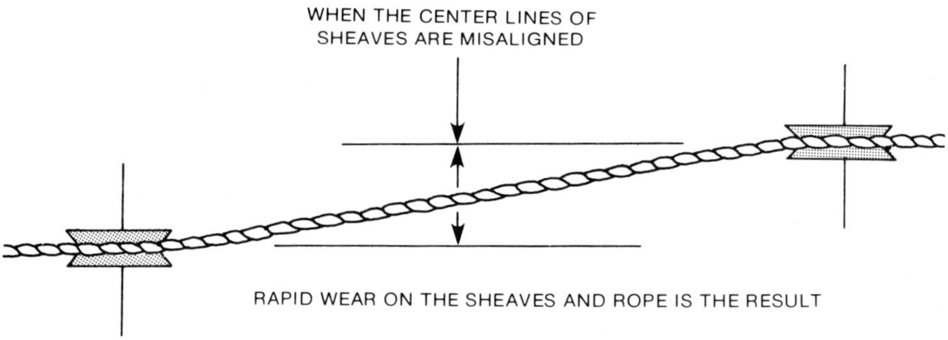

Fig. 5-16: Effect of sheave misalignment.

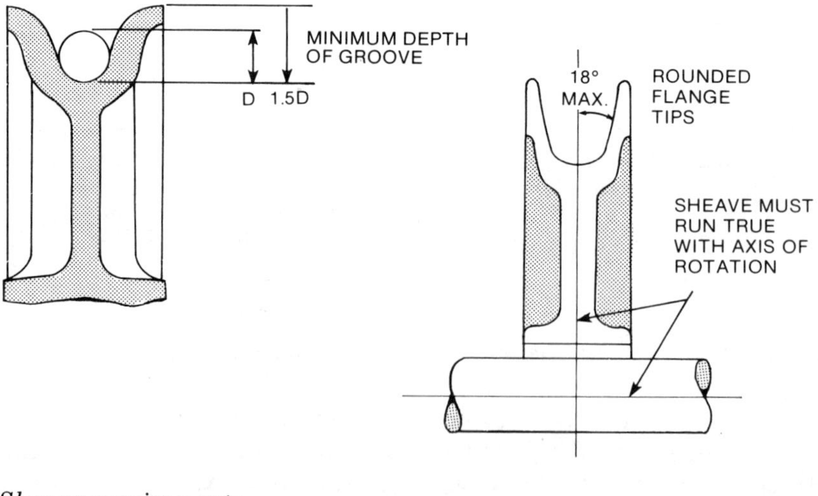

Fig. 5-17: Sheave requirements.

The bearings should either be permanently lubricated or be equipped with means for lubrication. Inadequate lubrication or a sheave that is too heavy for the load will cause the rope to slip in the sheave whenever the rope velocity changes. The momentum of the heavy sheave will cause it to continue turning after the rope has stopped. This grinding wheel action causes severe rope abrasion and will wear flat spots in the sheave that further damage the rope (Fig. 5-18).

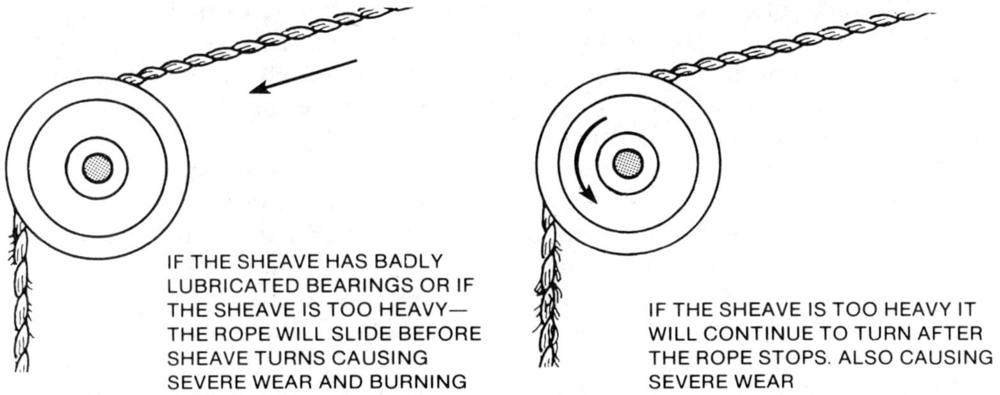

Fig. 5-18: Effect of improper sheaves or poor sheave maintenance on rope life.

If the sheaves are carrying ropes that can be momentarily unloaded, as in the case of a hoist line, then the sheave must be equipped with cable-keepers that prevent the unloaded rope from leaving the groove. The sheaves in lower load blocks should also be equipped with either cable-keepers or with close-fitting guards that prevent the ropes from becoming fouled when the block is lying on the ground with the ropes loose. When these guards are fitted, it is important that they be removed only for the purposes of maintenance, inspection, or adjustment. Failure to observe this procedure may allow a rope to jump clear and become trapped (Fig. 5-19).

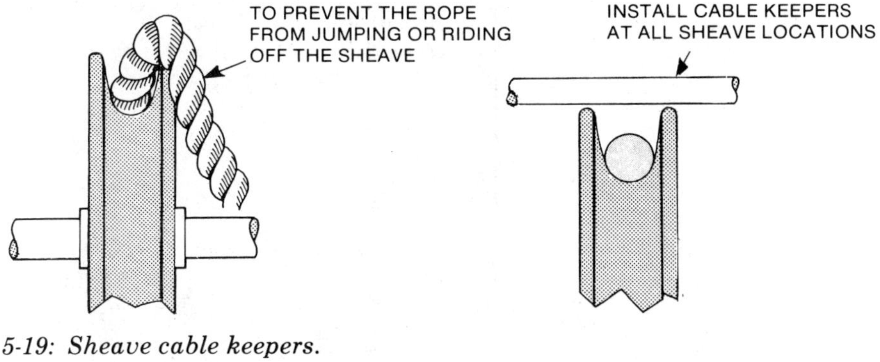

Fig. 5-19: Sheave cable keepers.

Badly worn sheaves have an adverse effect on rope life and must be examined at regular intervals. When replacement becomes necessary, only equipment supplied or approved by the equipment manufacturer should be fitted. Some designs may allow for remachining the grooves, but this operation should only be undertaken in accordance with the manufacturer's instructions, as there is a limit to the amount of metal which can be removed before the strength of the component will be affected.

Inspect the sheaves carefully for any sign of cracks in the flanges. If the flange breaks off, it will allow the rope to jump free with disastrous results. Even if only a small portion of the flange is broken, the rope is liable to be cut completely through by the rough edge of the break. Never use fiber ropes on sheaves that have seen wire rope service. Also, wire rope must never be used on equipment designed for fiber rope.

The groove surfaces on sheaves should be perfectly smooth. Those which have taken the imprint of the outer wires will exert a grinding action on the ropes (Fig. 5-20). This imprinting and scoring is caused by high contact pressures between the rope and sheave surface. If this condition is evident, then the sheave must be resurfaced and the contact pressure reduced by: (1) decreasing the load on the rope; (2) increasing the sheave diameter; or (3) replacing the sheave with one having harder metal.

The calculations for radial pressures of drums mentioned earlier apply equally well to sheaves.

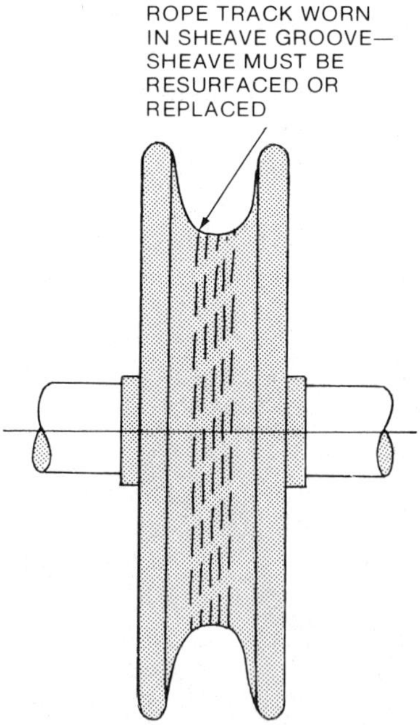

Fig. 5-20: Scored sheave groove.

HOOKS

There is such a variety of hooks available for hoisting and rigging operations that it is impossible to deal with all of them in detail (Fig. 5-21). There are, however, several considerations that apply to all hooks and observance of them will improve the safety standard.

Make sure that all hoisting hooks, excepting grab and sorting hooks, are equipped with safety catches. Inspect all hooks frequently (Fig. 5-22). Look for wear in the saddle of the hook. Look for cracks, severe corrosion, and twisting of the hook body. Be especially careful to measure the throat opening. If a hook has been overloaded or if it is beginning to weaken, the throat will open.

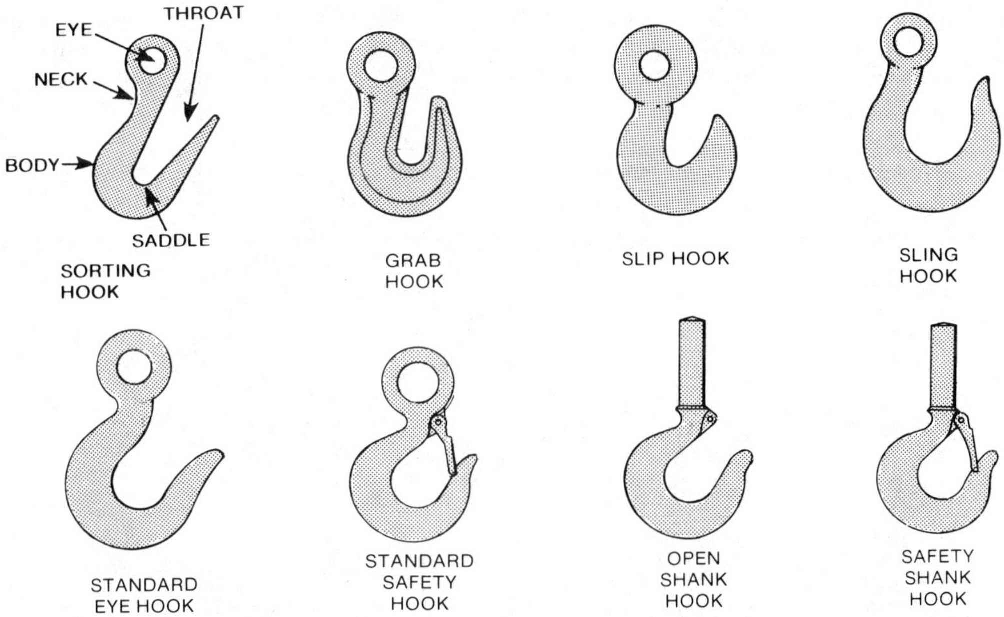

Fig. 5-21: Some of the more common hooks. The eye may be replaced with the swivel, clevis, or shank attachments.

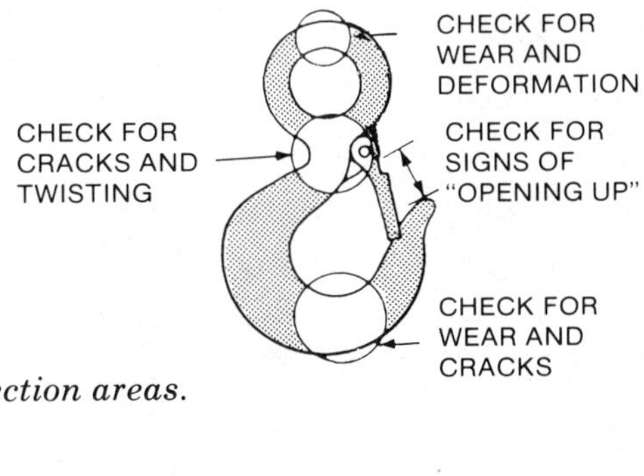

Fig. 5-22: Hook inspection areas.

If there is any evidence of opening or distortion, destroy the hook. If you discard it without destroying it, someone else may attempt to use it.

Be especially careful during the inspection to look for cracks in the saddle section and at the neck of the hook. Use only the best hooks available. These are usually forged alloy steel and generally are stamped with their rated safe working loads.

The safe working loads apply only when the load is applied in the saddle of the hook. If the hook is eccentrically loaded or if the load is applied anywhere between the saddle and the tip, the safe working load is very much reduced (Fig. 5-23). No open hook should be used in any situation where accidental dislodgement of the load could cause a risk of injury to workers. In these circumstances, a safety hook or shackle must be employed. Or if permitted by government regulations, mousing, as shown in Fig. 5-24, could be applied. To mouse a hook after the sling is on the hook, the wire or heavy twine is wrapped eight or 10 turns around the two sides of the hook. The process is completed by winding several turns of the wire or twine around the sides of the mousing and tying the ends securely.

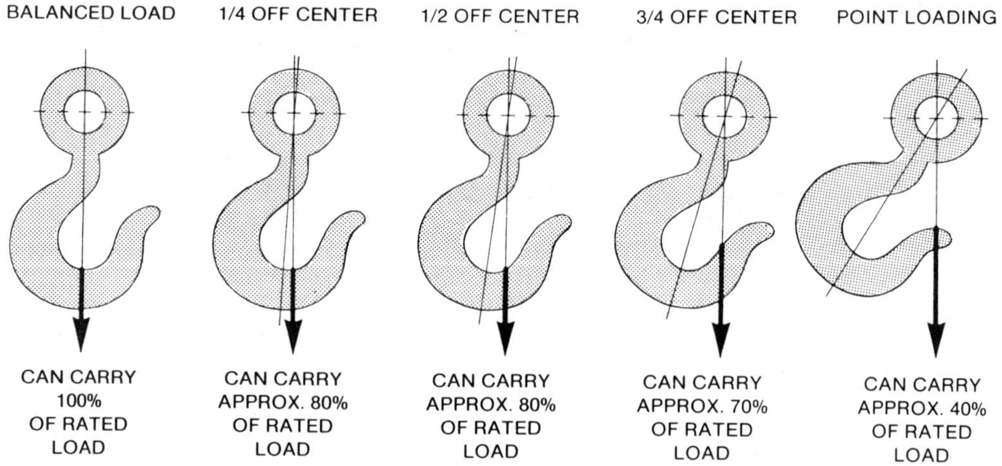

Fig. 5-23: Effect of eccentric loads on a standard eye hook capacity.

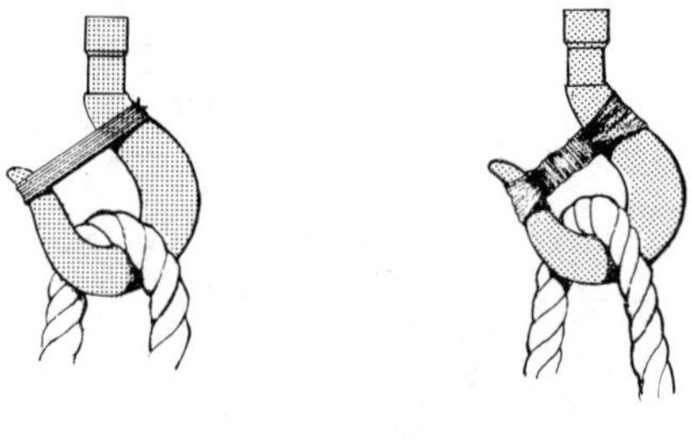

Fig. 5-24: Mousing hooks.

Commonly used choker hooks include:

1. *Standard choker hook* designed to be attached to the end of a sling. It frees the main part of the sling of fittings (Fig. 5-25).

2. *Adjustable sliding choker hook* which prevents the hook from sliding on the sling when not loaded. A spring insert has just enough tension to hold the hook in any position on the body of the sling. The hook is easily moved by hand and will choke under load (Fig. 5-26).

3. *Dual sliding choker hook* for endless slings and conventional slings when two eyes are over main hook (Fig. 5-27).

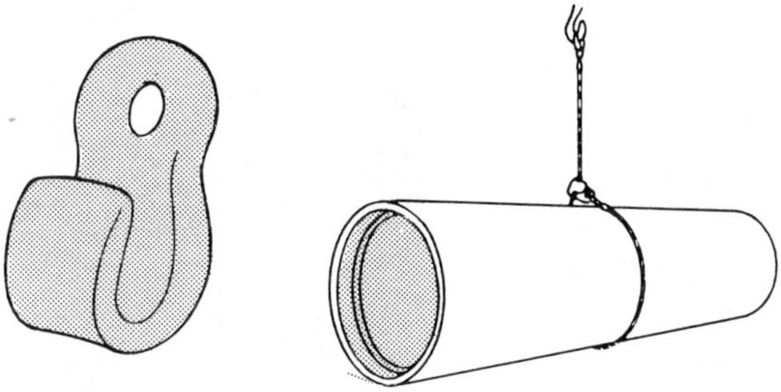

Fig. 5-25: Standard choker hook.

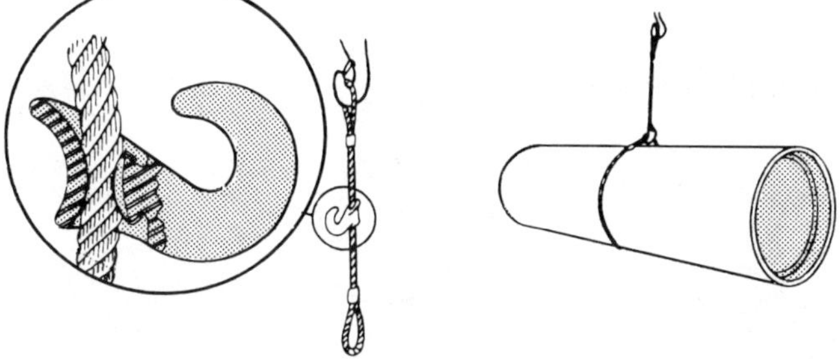

Fig. 5-26: Adjustable sliding choker hook.

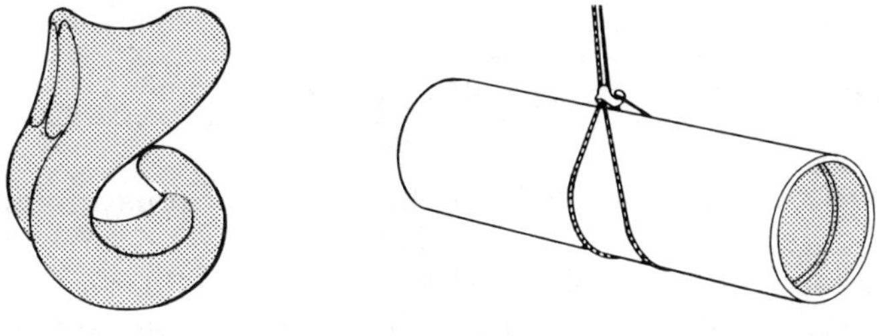

Fig. 5-27: Dual sliding choker hook.

It is recommended that all hoisting hooks, as opposed to sling hooks, be equipped with swivels and headache balls. Be sure that the headache ball is securely attached to either the hook or the rope so that there is no possibility of it sliding up and down on the load line (Fig. 5-28).

Appendix F gives typical safe working loads for some of the more common hooks. These tables are included to provide an indication of what can be expected from a hook based on its throat opening. Refer to the manufacturer's rating for specific values of specific hooks.

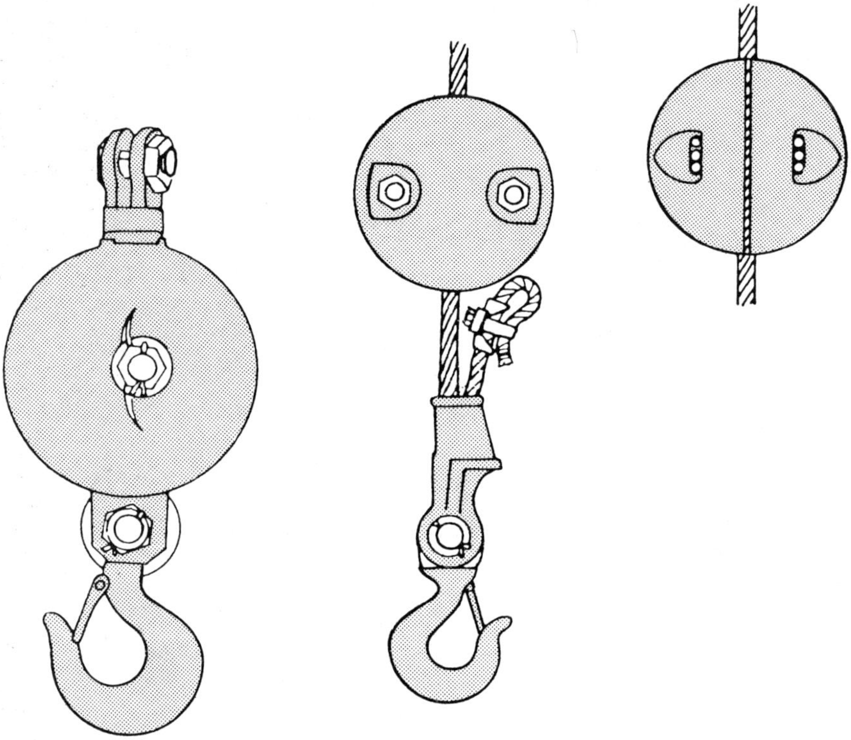

Fig. 5-28: "Headache" ball. They must be securely attached to the hook or the rope.

RINGS, LINKS, AND SWIVELS

These hardware items (Fig. 5-29), like all other hoisting and rigging fittings, must be forged alloy steel to provide the highest degree of safety.

In the absence of manufacturer's safe working loads, the tables in Appendix F can be used to get an *approximate* idea of the capacities of these rigging hardware items.

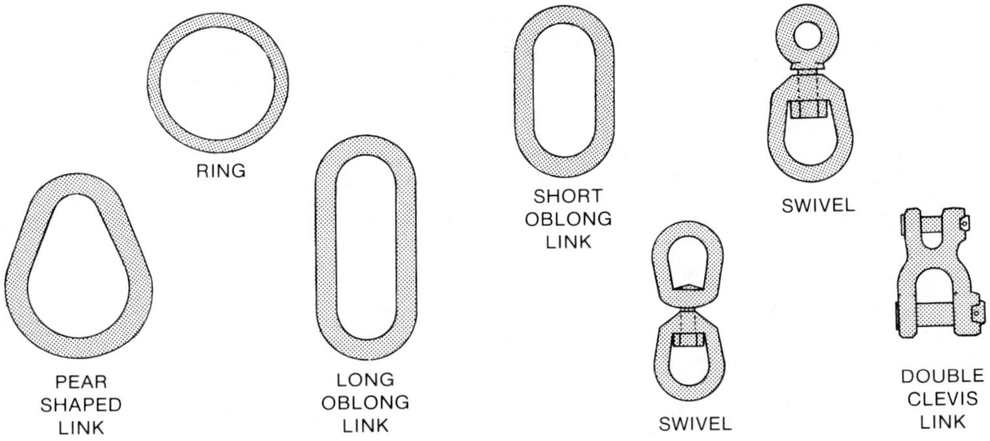

Fig. 5-29: Typical rings, links, and swivels used in rigging operations.

SHACKLES

There are two types of shackles commonly used in rigging. They are the anchor (bow type) shackle and chain ("D" type) shackle, both of which are available with screw pins or round pins (Fig. 5-30). Shackles, like most other rigging hardware, are sized by the diameter of the steel in the bow section rather than the pin size. They should only be of forged alloy steel and of weldless construction.

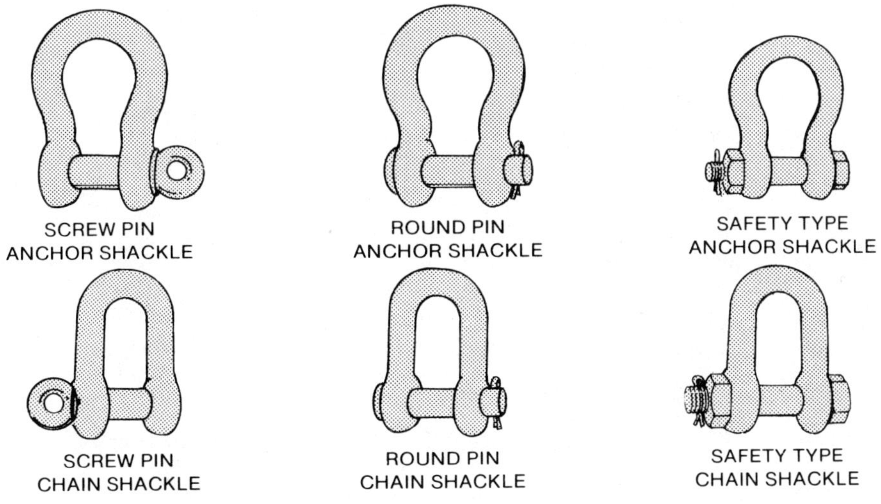

Fig. 5-30: Typical shackles.

Never replace the shackle pin with a bolt; only the proper fitted pin should be used. Bolts are not intended to take the bending that is normally applied to the pin (Fig. 5-31). Do not use a shackle if the distance between the eyes is greater or smaller than listed in Appendix F. All pins must be straight and all screw pins must be completely seated. Cotter pins must be used with all round pin shackles. Shackles worn in the crown or the pin by more than 10% of the original diameter should be destroyed (Fig. 5-32).

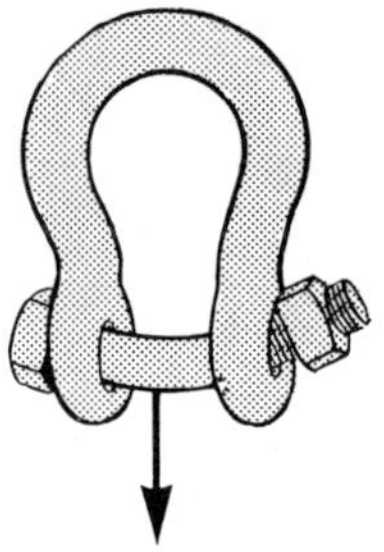

Fig. 5-31: Replacing shackle pins.

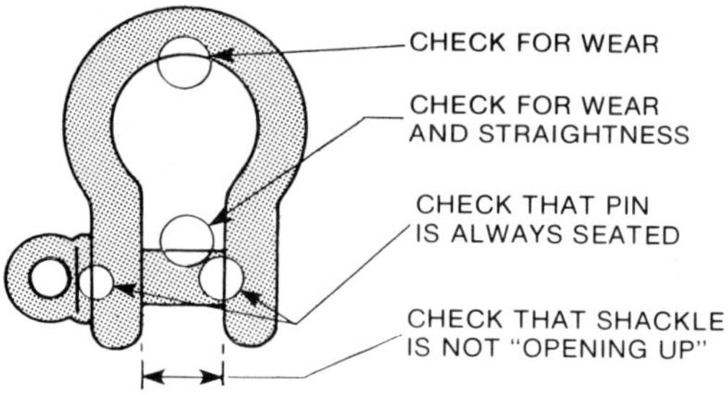

Fig. 5-32: Shackle inspection areas.

Never allow a shackle to be pulled at an angle because the capacity will be tremendously reduced. Centralize whatever is being hoisted on the pin by suitable washers or spacers (Fig. 5-33). Do not use screw pin shackles if the pin can roll under load and unscrew (Fig. 5-34).

Refer to Table 5-5 for the rated safe working loads of shackles that are of weldless construction and are made of forged alloy steel.

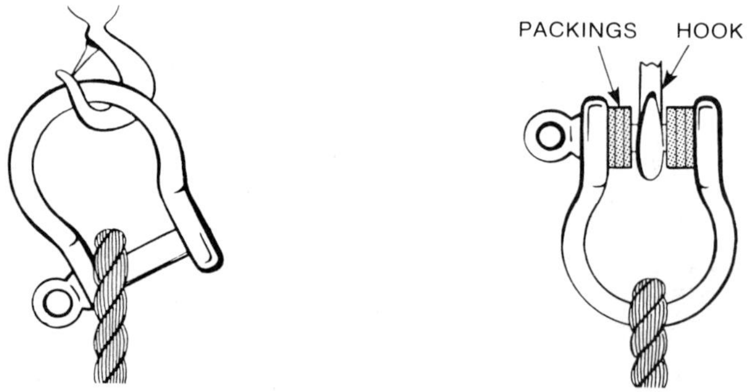

Fig. 5-33: Eccentric shackle loads.

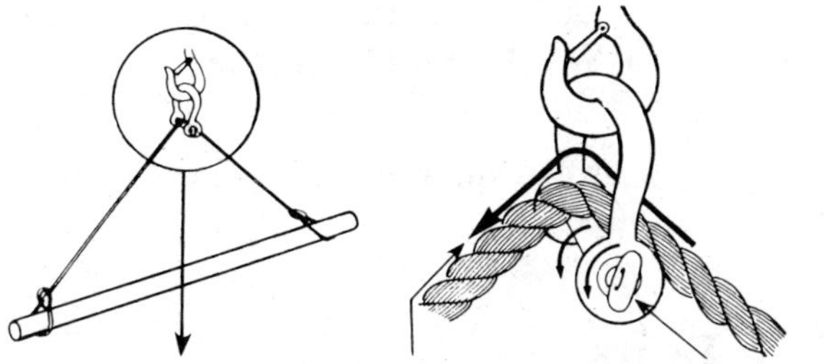

Fig. 5-34: Do not use screw pin shackles if the pin can roll under load and unscrew.

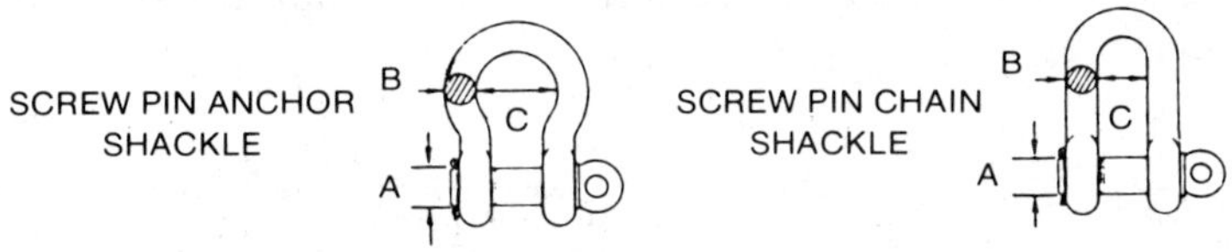

Table 5-5: Typical Strength of Shackles

Diameter of Pin A (Inches)	Diameter of Shank B (Inches)	Max. Width between Eyes C (Inches)	Safe Working Load (Tons)
3/8	3/16	1/2	1/3
7/16	1/4	5/8	1/2
1/2	7/16	3/4	3/4
5/8	1/2	7/8	1
3/4	5/8	1-1/16	1-1/2
7/8	3/4	1-1/4	2
1	7/8	1-1/2	3
1-1/8	1	1-5/8	3-1/2
1-1/4	1-1/8	1-7/8	4
1-3/8	1-1/4	2	4
1-1/2	1-3/8	2-1/8	5-1/2
1-5/8	1-1/2	2-1/4	6-1/2
1-3/4	1-5/8	2-3/4	8
2	1-3/4	2-3/4	10
2-1/4	2	3-1/4	12
2-1/2	2-1/4	4	16
2-3/4	2-1/2	4-1/8	20
3	2-3/4	4-1/2	24
3-1/4	3	5	30

Size of shackle to be identified by diameter of pin.
All shackle pins must be straight.
If the width between eyes exceeds maximum given above, the safe load given must be materially reduced.

EYE BOLTS AND WELDED LUGS

Both eye bolts and weld-on lift lugs provide an excellent means of anchorage when lifting various objects.

Lift Type Eye Bolts

It is recommended that all lift type eye bolts and ring bolts used for hoisting be of forged alloy steel or carbon steel and equipped with shoulders or collars. The plain or shoulderless eye bolt is fine for vertical loading, but as soon as it is loaded at an angle, it is subjected to bending and the load it can safely carry is severely reduced (Fig. 5-35).

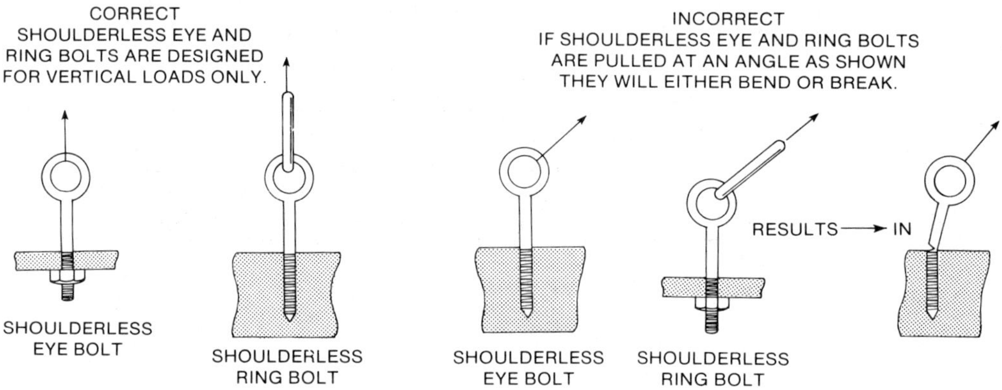

Fig. 5-35: Use of eye bolts.

Even when equipped with shoulders, the safe working loads of eye bolts and ring bolts are reduced with angular loading. When installed, the shoulder must be at right angles to the axis of the hole and must contact the working surface, and the nuts must be properly torqued. Washers may have to be used to ensure that the shoulders are firmly in contact with the working surface. The tapped hole for screwed eye bolts (body bolts) should have a minimum depth of 1-1/2 times the bolt diameter and must be a good fit for the screwed shank of the eye bolt (Fig. 5-36).

To keep the bending to a minimum, the loads should always be applied to the plane of the eye, never in the other direction. This is particularly important when bridle slings are used, because an angular pull is always developed in the eye bolts, unless a spreader bar is used as part of the sling (Fig. 5-37).

Never insert the point of a hook in an eye bolt, always use a shackle (Fig. 5-38). Do not use a sling reeved through an eye bolt or reeved through a pair of eye bolts. One single leg only should be attached to each eye bolt (Fig. 5-39).

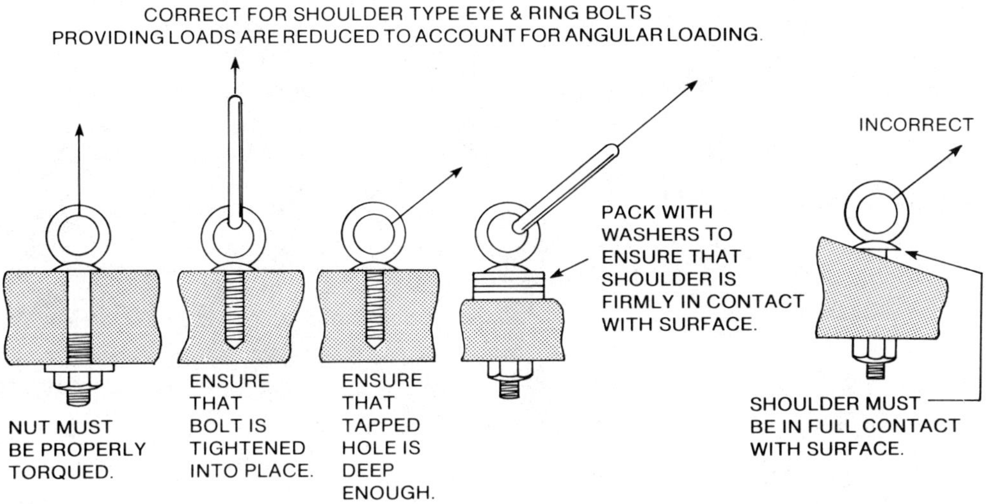

Fig. 5-36: Use of shoulder type eye and ring bolts.

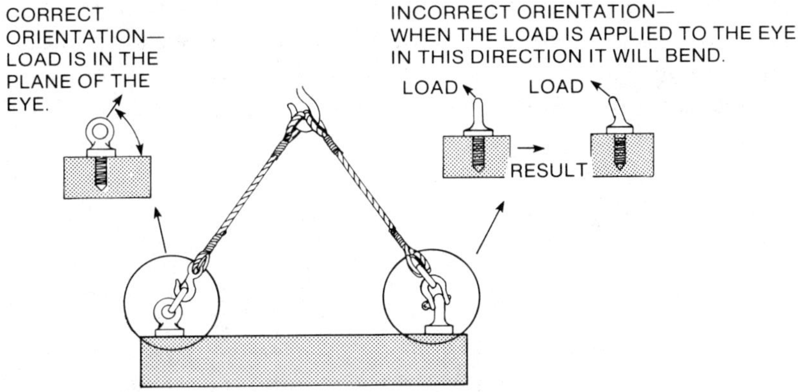

Fig. 5-37: Orientation of eye bolts.

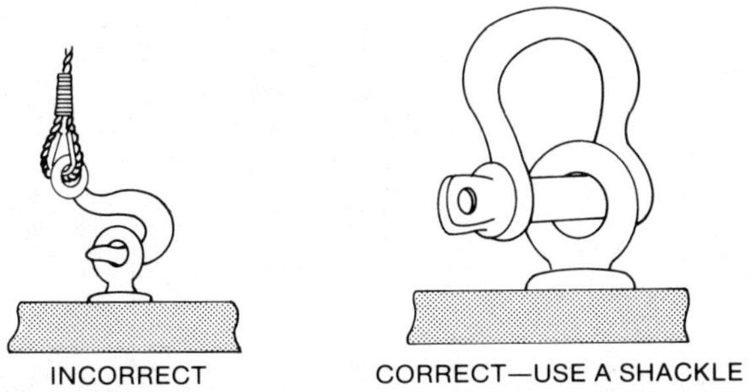

Fig. 5-38: Never insert the point of a hook in an eye bolt.

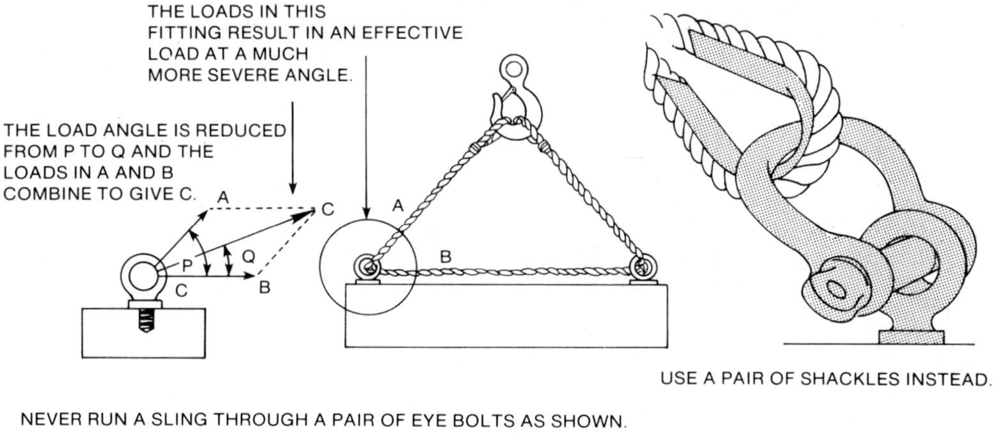

Fig. 5-39: Lifting with eye bolts.

Where eye bolts cannot be kept in line with each other and at the same time tightened, thin washers or shims may be inserted under the collars to permit the eye bolt being tightened to be turned in line with the other (Fig. 5-40). The same precautions apply to ring bolts, and the working loads are generally the same as for the eye bolts. Check Table 5-6 for the rated safe working loads of forged carbon steel eye bolts.

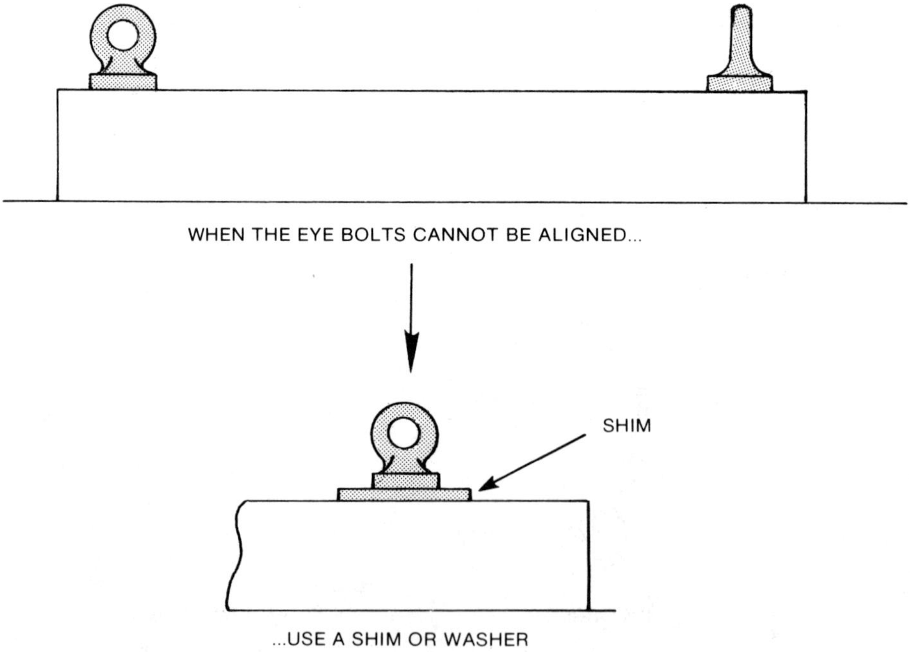

Fig. 5-40: Alignment of eye bolts.

Table 5-6: Safe Working Loads for Eye Bolts (Shoulder Type Only)

Stock Diameter (Inches)	Safe Working Loads (pounds) Corresponding to Angle of Pull (Fig. 5-41). Vertical	75°	60°	45°	Less than 45°
1/4	500	Reduce Vertical Loads By 45%	Reduce Vertical Loads By 65%	Reduce Vertical Loads By 75%	NOT RECOMMENDED
5/16	800				
3/8	1,200				
1/2	2,200				
5/8	3,500				
3/4	5,200				
7/8	7,200				
1	10,000				
1-1/4	15,200				
1-1/2	21,400				

Note: S.W.L. for plain (shoulderless) eye bolts are same as for shoulder bolts under vertical load. Angular loading is not recommended.

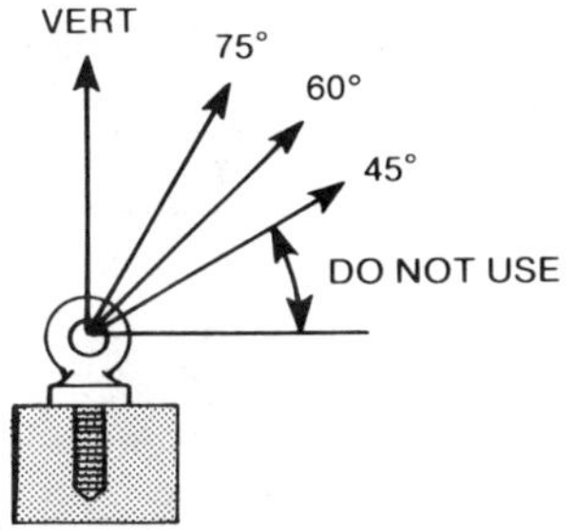

Fig. 5-41: Proper working position.

Lift Type Weld-on Lugs

Fillet or groove welded lugs used for hoisting operations should be made of either forged alloy steel or carbon steel, and their specifications are illustrated in Fig. 5-42. Check Tables 5-7 and 5-8 for their dimensions for various safe working loads.

The lift lugs are fastened in place with either a groove weld or fillet. The groove weld gives a strong joint between the lug and object to be lifted and is recommended for all heavy duty use. Even when welded with a groove joint, the safe working loads of the welded lug are reduced with angular loading. Also, when using welded lift lugs, make certain that the surface on which the lug is to be fastened is clean and free of rust or oil. Also be sure that the plate to which it is welded is heavy enough to carry the load.

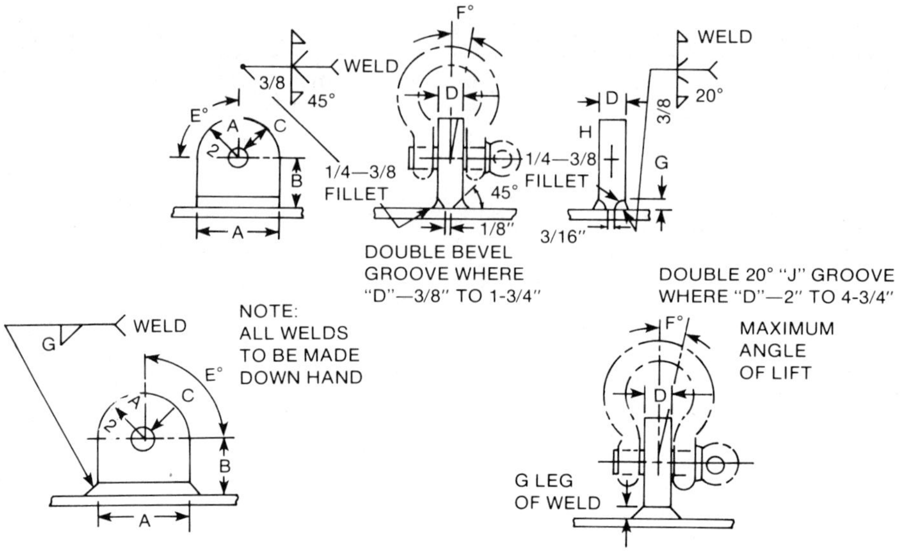

Fig. 5-42: Dimensions for typical welded lugs (see Tables 5-7 and 5-8).

Table 5-7: Standards for Fillet Welded Weld-on Lift Lug for Light Duty Use

Safe Load Tons	A	B	C	D	Load Angle E°	Load Angle F° Max.	G
1/3	1/4	7/8	3/8	3/8	0° to 90°	20°	3/8
1/2	1-3/4	1-1/8	7/16	3/8	"	"	3/8
3/4	2	1-1/4	1/2	1/2	"	"	1/2
1	2	1-1/4	5/8	5/8	"	"	1/2
1-1/2	2-1/4	1-3/8	3/4	3/4	"	"	1/2
2	2-1/2	1-1/2	7/8	7/8	"	"	1/2
3	3	1-5/8	1	1	"	"	1/2
4	3-1/2	2	1-1/8	1-1/4	"	"	1/2
5-1/2	4	2-1/8	1-1/4	1-1/2	"	"	1/2
6-1/2	4-1/4	2-1/4	1-3/8	1-1/2	"	"	3/4
8	4-3/4	2-1/2	1-1/2	1-3/4	"	"	3/4
10	5	2-3/4	1-5/8	2	"	"	3/4
12	5-1/2	2-7/8	1-3/4	2-1/4	"	"	3/4
16	6-3/4	3-1/4	2-1/4	2-1/4	"	"	7/8
21	7-1/2	3-3/4	2-1/2	3	"	"	7/8
27	8-1/4	4-1/4	2-3/4	3-1/2	"	"	1
34	9	4-1/4	3	4	"	"	1
40	9-3/4	4-1/4	3-1/4	4-1/4	"	"	1
50	10-1/2	4-3/4	3-1/2	4-3/4	"	"	1

Table 5-8: Standards for Groove Welded Weld-on Lift Lug for Heavy Duty Use

Safe Load Tons	A	B	C	D	Load Angle E°	Load Angle F° Max.	G	H Rad
1/3	1/4	7/8	3/8	3/8	0° to 90°	15°		
1/2	1-3/4	1-1/8	7/16	3/8	"	15°		
3/4	2	1-1/4	1/2	1/2	"	15°		
1	2	1-1/4	5/8	5/8	"	15°		
1-1/2	2-1/4	1-3/8	3/4	3/4	"	20°		
2	2-1/2	1-1/2	7/8	7/8	"	20°		
3	3	1-5/8	1	1	"	20°		
4	3-1/2	2	1-1/8	1-1/4	"	20°		
5-1/2	4	2-1/8	1-1/4	1-1/2	"	20°		
6-1/2	4-1/4	2-1/4	1-3/8	1-1/2	"	20°		
8	4-3/4	2-1/2	1-1/2	1-3/4	"	20°		
10	5	2-3/4	1-5/8	2	"	20°	3/4	1/2
12	5-1/2	2-7/8	1-3/4	2-1/4	"	20°	3/4	1/2
16	6-3/4	3-1/4	2-1/4	2-1/4	"	20°	3/4	1/2
21	7-1/2	3-3/4	2-1/2	3	"	20°	7/8	1/2
27	8-1/4	4-1/4	2-3/4	3-1/2	"	20°	1	1/2
34	9	4-1/4	3	4	"	20°	1-1/8	1/2
40	9-3/4	4-1/4	3-1/4	4-1/4	"	20°	1-1/8	1/2
50	10-1/2	4-3/4	3-1/2	4-3/4	"	20°	1-1/4	1/2

As is the case with eye bolts, the loads should always be applied to the plane of the lug to keep the bending of the lug to a minimum. This is particularly true when using bridle slings. Also, it is important to keep in mind that a shackle should always be used with a weld-on lug; never insert the point of a hook in a lug. Also do not use a sling reeved through a pair of lugs; one single leg only should be attached to each lug.

PLATE CLAMPS

Plate clamps are rigging devices that are capable of handling steel plate (Fig. 5-43). There are two basic types of plate clamps: serrated jaw type (Fig. 5-44A) and screw type (Fig. 5-44B).

The serrated jaw plate clamp has a serrated jaw that is designed to grip a piece of steel plate for vertical lifting. Serrated jaw plate clamps are available in locking and non-locking styles. For maximum safety, the locking grips are recommended.

Fig. 5-43: Plate clamps in use (photo courtesy of CBI Industries, Inc.).

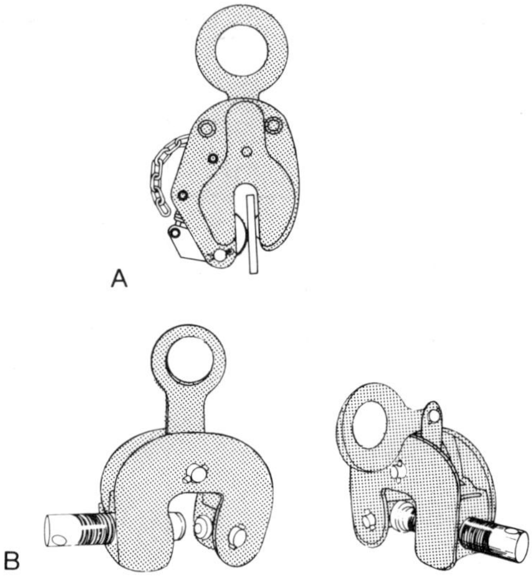

Fig. 5-44: The two basic types of plate clamps: (A) Serrated jaw type and (B) screw type.

Screw plate clamps are considered the "safest" of two types. They are capable of handling steel plate from the horizontal through a 180° arc (Fig. 5-45). Using this type, it is most important to be sure the screw clamps are securely tightened to the load. Most clamps of this type feature a spring loaded pivoting cam jaw that "cams in" (screws in tighter) when a load is applied to the lifting shackle.

When using either type of clamp, remember that it is designed to grip only **one** plate for each hoist. Also, always follow the manufacturer's recommendation as to the safe lifting load.

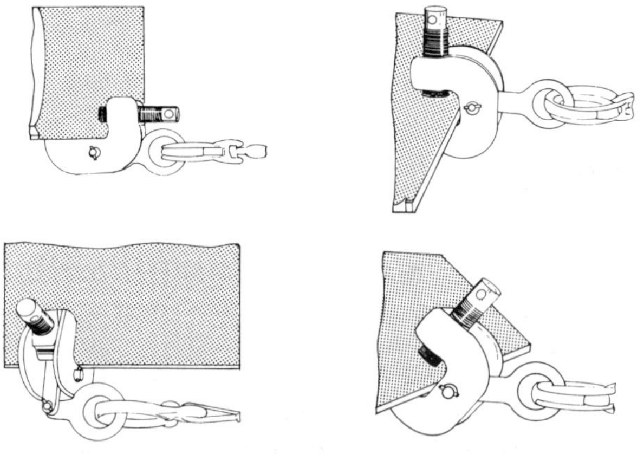

Fig. 5-45: Various methods of fastening screw plate clamps.

TURNBUCKLES

Turnbuckles can be supplied with eye end fittings, hook end fittings, jaw end fittings, stub end fittings, and any combination of these. Their rated loads depend upon the outside diameter of the threaded portion of the end fitting and on the type of end fitting. The jaw, eye, and stub types are rated equally, and the hook types have reduced ratings (Fig. 5-46).

All turnbuckles used in hoisting or rigging operations should be of weldless construction and fabricated from alloy steel. When they are supplied with hook end fittings, be sure that the hooks are fitted with safety catches. If the turnbuckle is to be used in an application where vibration is present, it is extremely important to lock the frame to the end fittings to prevent it from turning and loosening. Lock nuts or jam nuts are not effective and add greatly to the load in the screw thread (Fig. 5-47). When tightening up a turnbuckle, do not apply any more torque to it than you would to a bolt of equal size.

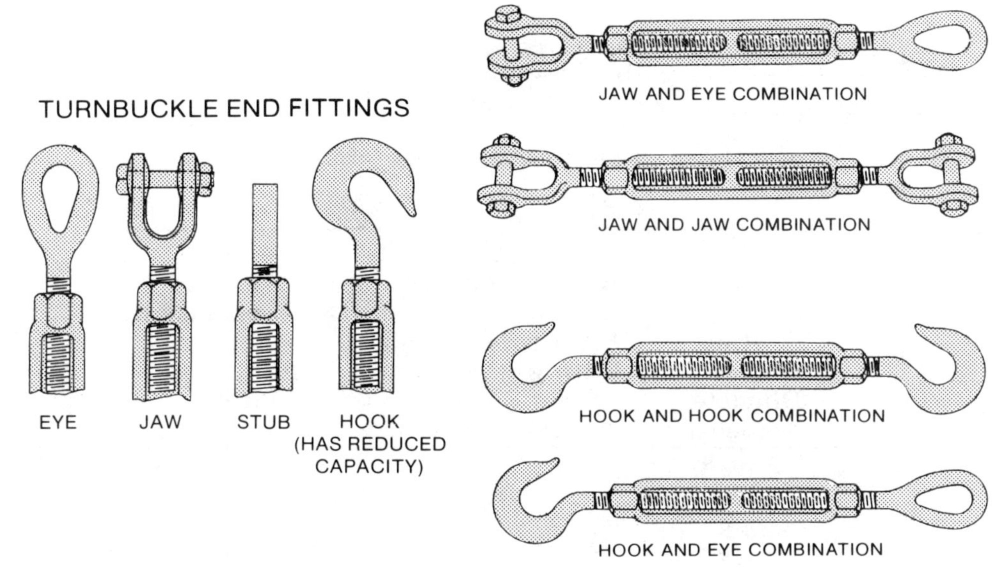

Fig. 5-46: Turnbuckles.

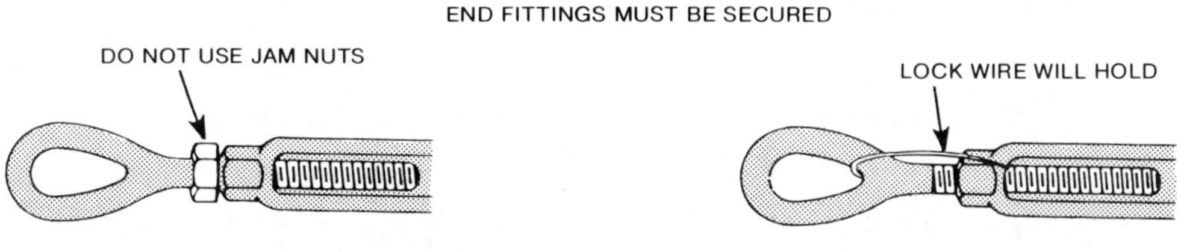

Fig. 5-47: Securing of turnbuckle end fittings.

Turnbuckles should be inspected frequently for cracks in the end fittings, especially at the neck of the shank, deformed end fittings, deformed and bent rods and bodies, cracks and bends around the internally threaded portion, and signs of thread damage (Fig. 5-48). Refer to Table 5-9 for the rated safe working loads of forged alloy steel turnbuckles of weldless construction.

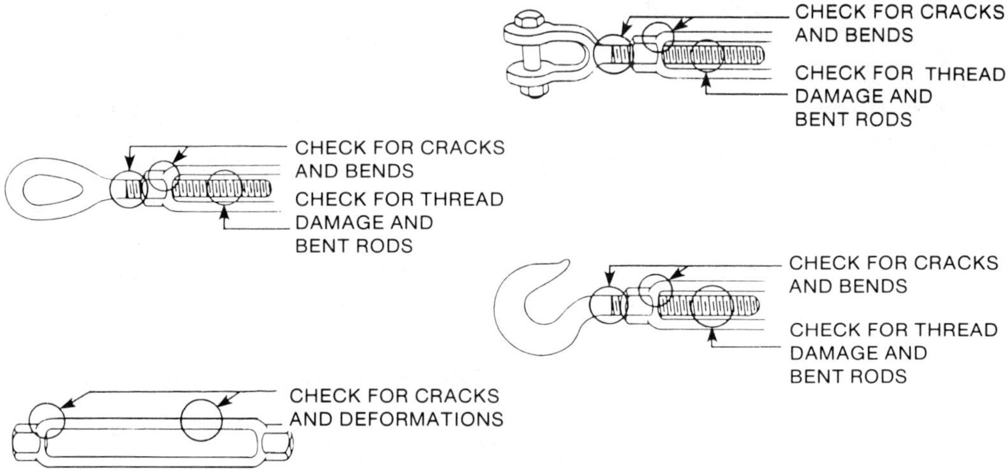

Fig. 5-48: Turnbuckle inspection areas.

Table 5-9: Turnbuckles

End Fitting, Stock Diameter (Inches)	SWL of Any Combination of Jaw End Fittings, Eye End Fittings, and Stub End Fittings (Pounds)	SWL of Any Turnbuckle Having a Hook End Fitting (Pounds)
1/4	500	400
5/16	800	700
3/8	1,200	1,000
1/2	2,200	1,500
5/8	3,500	2,250
3/4	5,200	3,000
7/8	7,200	4,000
1	10,000	5,000
1-1/4	15,200	5,000
1-1/2	21,400	7,500
1-3/4	28,000	—
2	37,000	—
2-1/2	60,000	—
2-3/4	75,000	—

Binders

Binders perform the same function as turnbuckles—increasing the tension of slings on a load—but are used almost exclusively with chains. The two most common types are shown in Fig. 5-49. The binders are operated by placing the hooks on the appropriate sling chain links to eliminate any slack. With the ratchet or steamboat binder, the amount of tension on the chain is controlled by the lever handle. In the case of the load binder, tension is increased by pushing the lever handle down. The lever handle of the load binder must be tied down with wire to prevent the release of tension. If this device has been tightened excessively, care must be taken to prevent its sudden release from injuring the person working with it. Normally, binders are used only for securing loads on trucks or other similar vehicles.

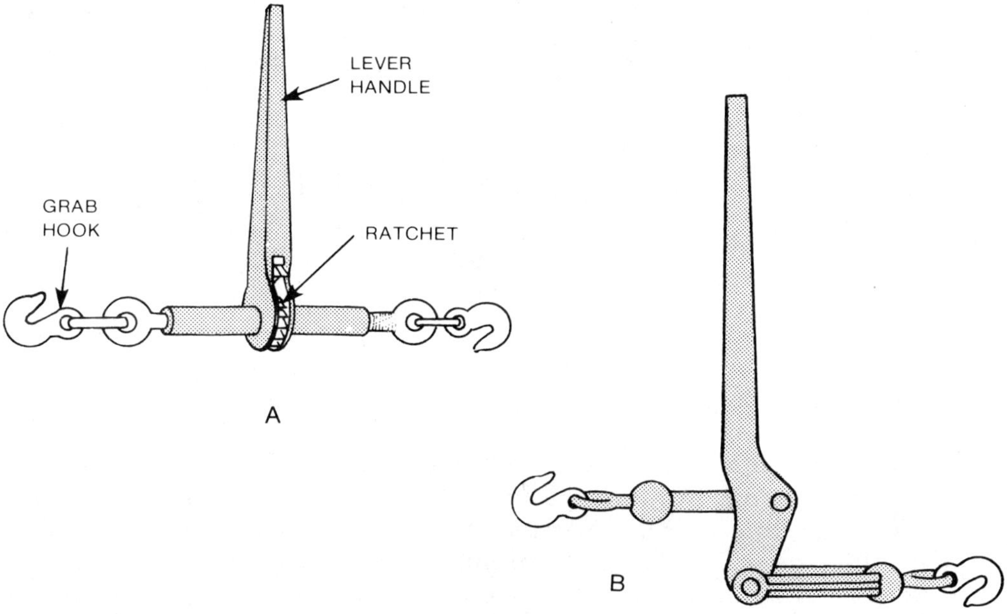

Fig. 5-49: Two common types of binders: (A) Ratchet binder; (B) load binder.

SPREADER BARS AND BEAMS

Occasionally, it is necessary to hoist loads that are not protected sufficiently to prevent crushing by the sling legs. In such cases, spreaders (Fig. 5-50) may be used with the slings. Spreaders are short bars or pipes with eyes on each end. The sling leg passes through the eye down to its connection with the load.

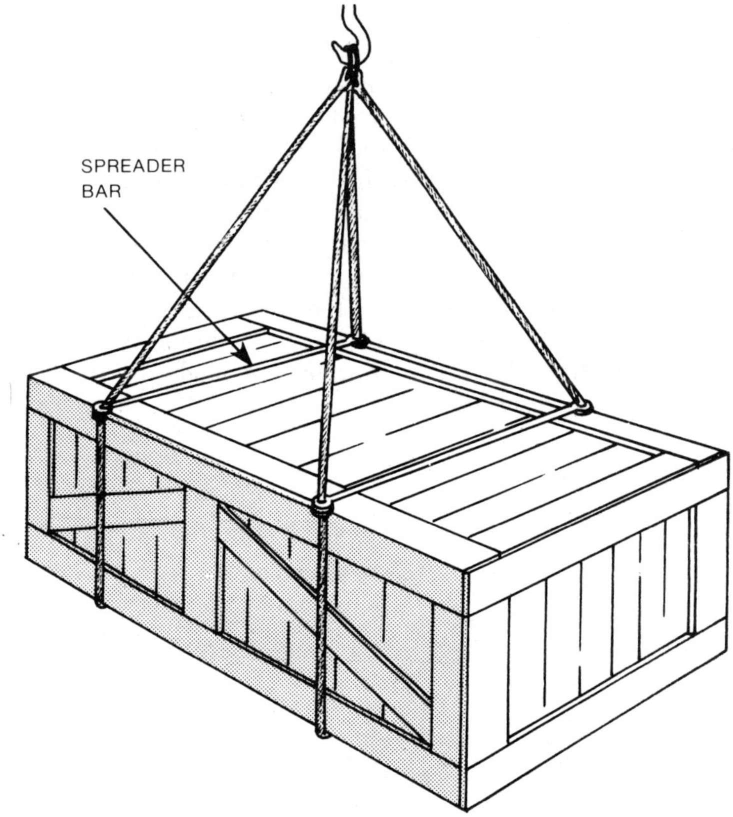

Fig. 5-50: Use of spreader bars in slings.

By setting spreaders in the sling legs above the top of the load, the angle of the sling leg is changed so that crushing of the load is prevented. Changing the angle of the sling leg may increase the stress in that portion of the sling leg above the spreaders. The determining factor in computing the safe lifting capacity of the sling is the stress (or tension) in the sling leg above the spreader.

Spreader beams are usually used to support long loads during lifts. They eliminate the hazard of the load tipping, sliding, or bending, as well as the possibility of low sling angles and the tendency of the slings to crush the load. Equalizer beams are used to equalize the load in sling legs and to keep equal loads on dual hoist lines when making tandem lifts (Fig. 5-51). Both types of beams are normally fabricated to suit a specific application. If a beam is to be used which has not been designed for the application, make sure it has adequate width, depth, length, and material. The capacity of beams with multiple attachment points depends upon the distance between the points. For example, if the distance between the attachment points is doubled, the capacity of the beam is halved.

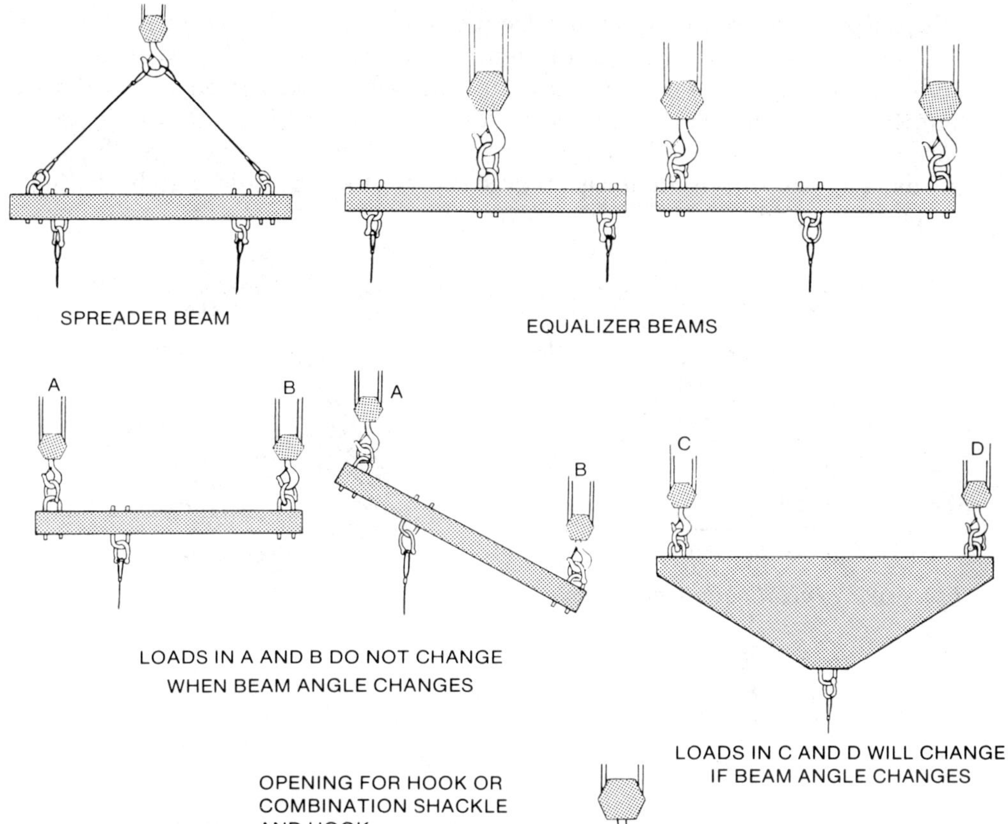

Fig. 5-51: Spreader beam design, equalization, and use.

Equalizer plates (Fig. 5-52) are frequently employed to equalize loads when the legs of a sling are unequal in length. An equalizer plate will rotate when under load conditions to distribute the load automatically on slings of unequal length. In fact, several equalizer plates can be pinned together to form a compound equalizer to equalize loads supported by four or more slings. These plates are usually custom-made for a specific job; that is, the holes can be drilled to be properly located.

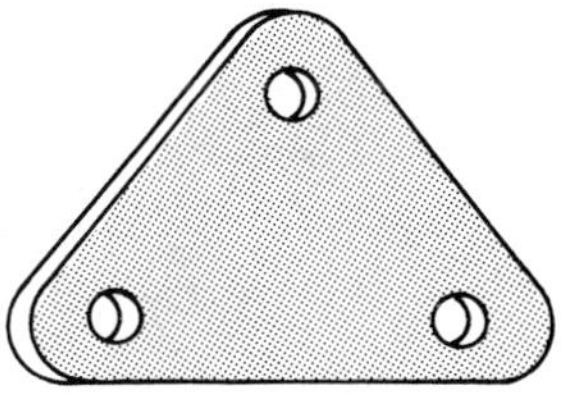

EQUALIZER PLATE

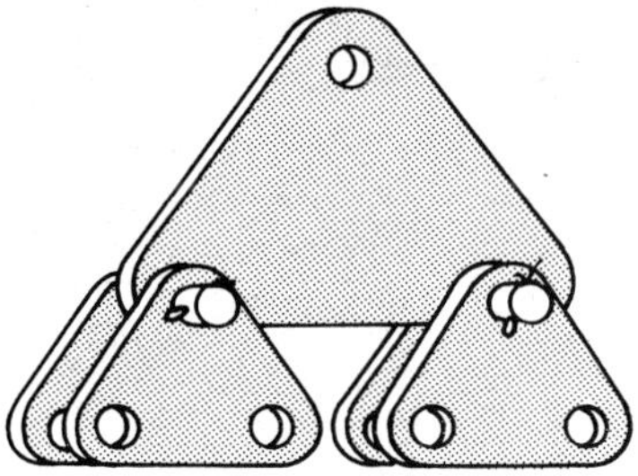

COMPOUND EQUALIZER PLATE

Fig. 5-52: Typical equalizer plates.

6

Block and Tackle Systems

INTRODUCTION

The hoisting and moving of heavy objects or objects weighing more than the safe working load of the rope being used calls for multiplication of the pulling force or changing the direction of pull, or both. Block and tackle systems make this possible.

A *block* consists of one or more sheaves fitted in a wood or metal frame supported by a hook or shackle inserted in the strap of the block. A *tackle* is an assembly of blocks and lines used to gain a mechanical advantage in lifting or pulling.

In a tackle assembly, the line is reeved over the sheaves of blocks. There are two types of tackle systems: simple and compound. A simple tackle system is an assembly of blocks in which a single line is used (Fig. 6-1A). A compound tackle system is an assembly of blocks in which more than one line is used (Fig. 6-1B).

Tackle Terms

To help avoid confusion in working with block and tackle systems, it may be wise to define some of the more common terms:

A *fall* is a line, either fiber line or wire rope, reeved through a pair of blocks to form a tackle (Fig. 6-2).

The *hauling part* is the part of the fall leading from one of the blocks upon which the power is exerted. The *standing part* is the end of the fall which is attached to one of the beckets.

The *movable* (or *running*) *block* of a tackle is the block attached to the object to be moved. The *fixed* (or *standing*) *block* is the block attached to a fixed object or support. When a tackle is being used, the movable block moves and the fixed block remains stationary.

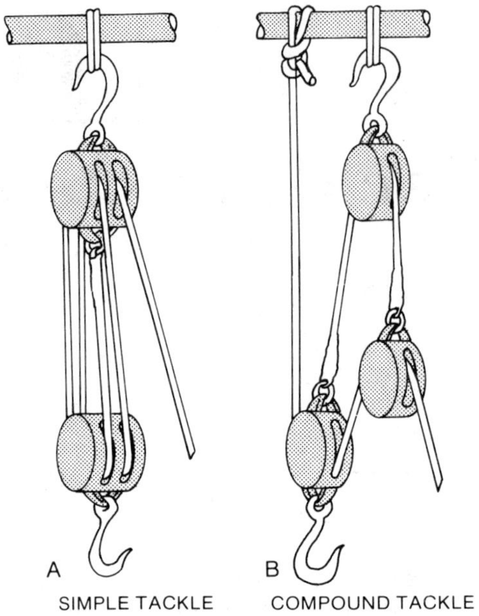

Fig. 6-1: Tackles.

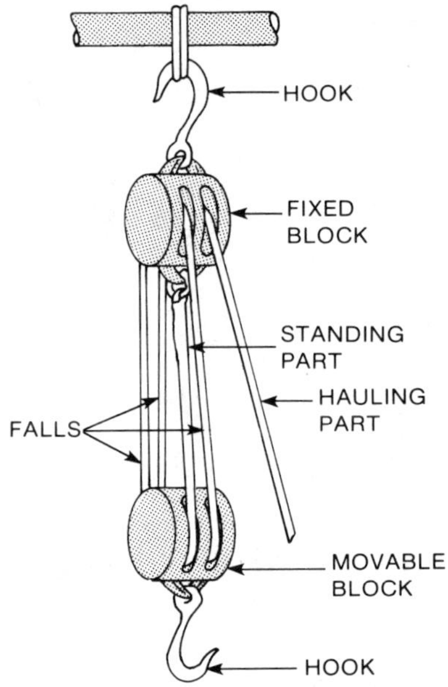

Fig. 6-2: Parts of a tackle.

The term *two-blocked* means that both blocks of a tackle are as close together as they will go. This same condition is referred to as *block-and-block.*

To *overhaul* is to lengthen a tackle by pulling the two blocks apart.

To *round in* means to bring the blocks of a tackle toward each other, usually without a load on the tackle (opposite of overhaul).

Construction of Blocks

The blocks used in construction rigging can run from the custom designed 400 ton capacity unit weighing 20 tons through all types of crane and hook blocks, wire rope blocks, and snatch blocks, down to the simplest of tackle blocks.

The essential parts of any block are the shell, the sheaves, the center pin, the straps, and the connections (Fig. 6-3). They can be equipped with a number of end fittings including hooks, wedge sockets, clevis', shackles of all types, and swivels of all types, as well as combinations of these items. The blocks may also be equipped with a becket, which is the term used for a rope end anchorage point on a block.

The shell provides protection for the sheave or sheaves and acts as a guide to keep the rope in place in the sheave groove. Wood shells add no strength to the block but merely serve as an enclosure for the sheaves. They are used for fiber ropes only. Steel shells are used on all blocks with wire rope. They add strength, rigidity, and protection for the block.

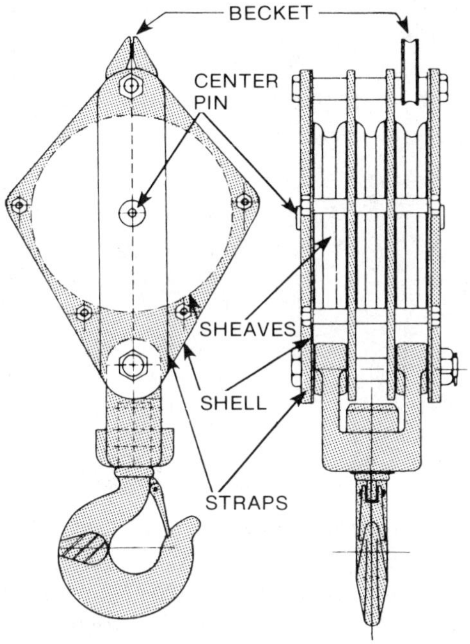

Fig. 6-3: Typical wire rope block.

The straps and side or cheek plates perform the important structural function of transmitting the sheave load to the connections and add rigidity to the block.

The sheaves transmit the load imposed by the rope to the center pin, straps, and connections. On fiber rope blocks, the sheaves are normally fabricated from cast iron; but on wire rope blocks, they should be cast steel because of its greater strength, hardness, and abrasion resistance. Never run fiber ropes over sheaves that have been used with wire ropes—rapid rope damage will result. Fiber rope sheaves must not be used with wire rope because their diameters are too small.

In *plain bore sheaves* the cast iron of the sheave provides its own bearing. They are recommended for light, intermittent service only, and they must be frequently oiled to reduce friction and avoid excessive wear (Fig. 6-4A).

Roller bushed sheaves are made with unground rollers; and without races, they are recommended for light, intermittent service only. They must be lubricated periodically with a heavy grease (Fig. 6-4B).

Self-lubricating bronze bearings are recommended where service requirements are severe or where the block cannot be frequently lubricated. They should not be subjected to high speed or continuous operation. The plug type of bearing should not be lubricated or oiled. Oil dissolves and washes away the graphite-wax mixture, destroying the bearing's self-lubricating characteristics. Oil is not harmful to the graphite impregnated type of bearing.

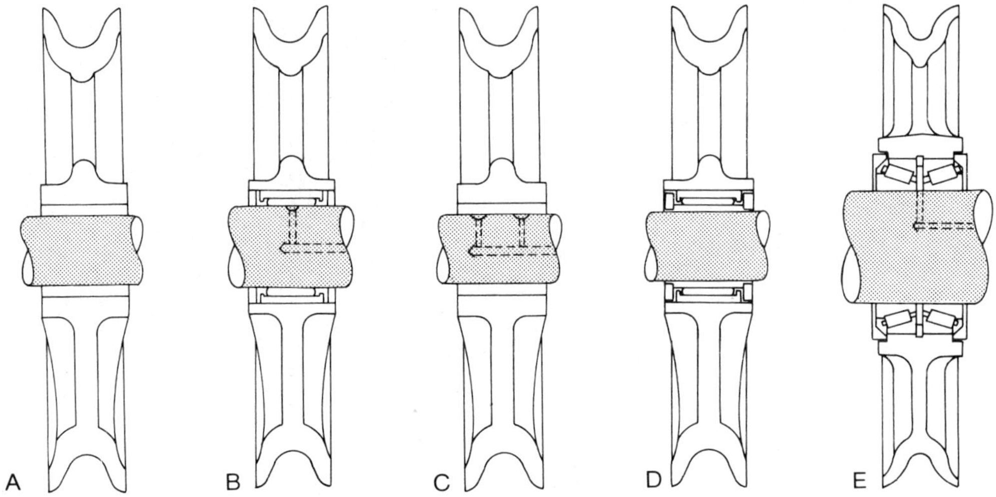

Fig. 6-4: Sheave bearing configurations.

Pressure lubricated bronze bearings are recommended where loads are extra heavy and continuous. Frequent, periodic lubrication is required (Fig. 6-4C).

Roller bearing sheaves are made with ground rollers and full races and are recommended for medium duty and high speed operation (Fig. 6-4D).

Precision anti-friction bearings capable of sustaining both radial and thrust loads are particularly suited to continuous high operating speeds and heavy loads. They combine long life with minimum maintenance (Fig. 6-4E).

The center pin, sometimes called the sheave pin, transmits the sheave load to the strap. It should be prevented or restrained from turning (Fig. 6-5).

All fittings on the blocks should be forged alloy steel. Shackles and eyes are inherently stronger than hooks; blocks equipped with them are rated at higher working loads. They are recommended for block connections where the block is

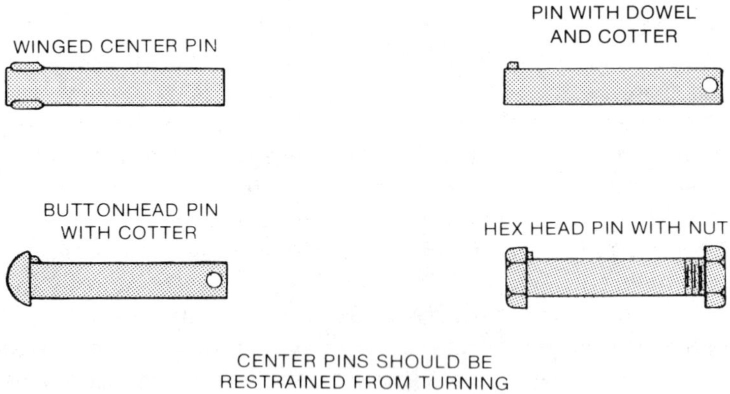

Fig. 6-5: Block center pins.

to be mounted permanently, where accidental disengagement of the hook connection would be dangerous, or where higher strength ratings are needed. Shackle connections are particularly recommended for standing blocks which must support not only the load, but also the hoisting strain of the lead line.

Types of Blocks

Blocks usually take their name from the purpose for which they are used, the position they occupy, or from a particular shape or type of construction. They can be designated according to the number of sheaves they have as single, double, triple, etc. or in accordance with the shell shape as diamond pattern, oval pattern, etc.

For the purposes of this chapter, however, the most commonly used types of blocks are designated as:

1. Crane and hook blocks (Fig. 6-6).
2. Wire rope blocks (Fig. 6-7).
3. Snatch blocks (Fig. 6-8).
4. Rope blocks (Fig. 6-9).

The basic differences between them relate to their construction and are evident from the illustrations. *Crane and hook blocks* are equipped with heavy iron cheek weights, whereas the *wire rope blocks* are normally much lighter and are equipped with cheek straps that provide strength between the end attachments and sheave center pins. Both types are well suited to high speed applications and heavy loads; however, the wire rope blocks are not intended to withstand the heavy service and abuse expected of the crane and hook blocks.

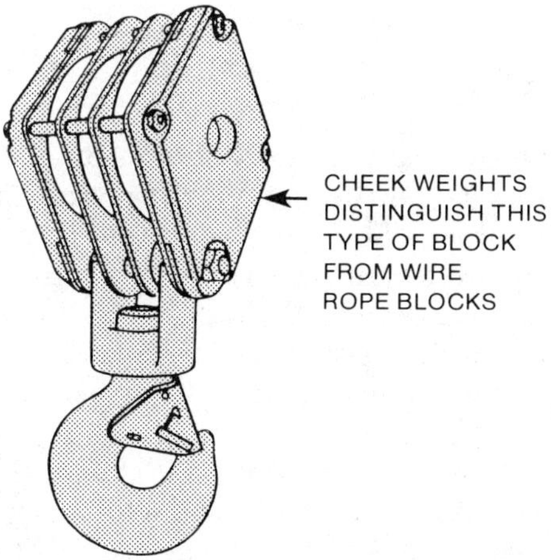

Fig. 6-6: Typical crane and hook block.

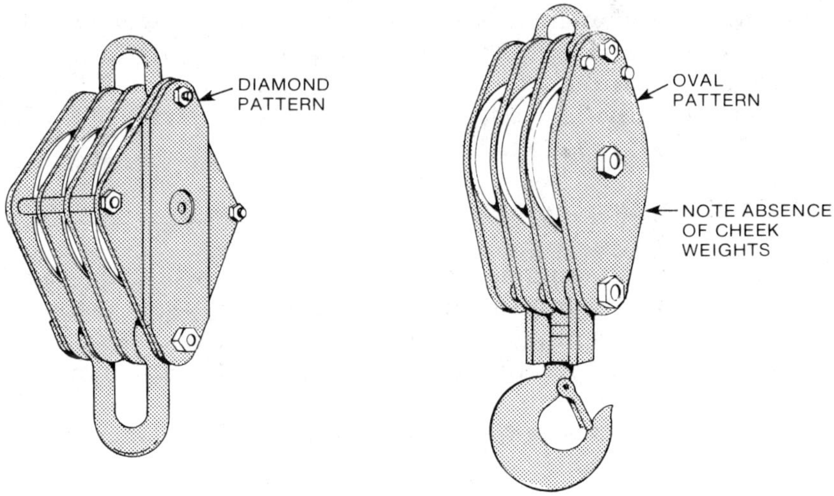

Fig. 6-7: Typical wire rope blocks.

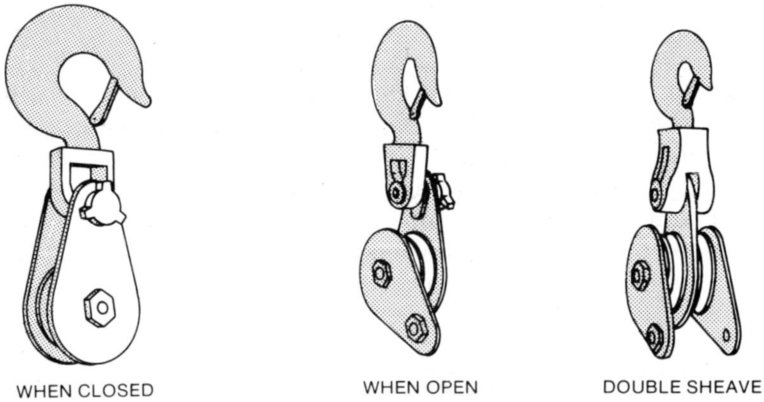

Fig. 6-8: Typical snatch blocks.

A *snatch block* can be a single or multi-sheave block that opens on one side to permit a rope to be slipped over the sheave, thus eliminating the need for it to be threaded through the block. They are available in many configurations for wire rope, manila rope, and synthetic ropes, and they can be purchased with hook, shackle, eye, and swivel end fittings.

Snatch blocks are normally used when it is necessary to change the direction of the pull on a line. In cases such as this, the stress on the snatch block varies tremendously with the angle between the lead and load lines. When the two lines are parallel, 1000 pounds (453.5 kg) on the lead line results in a load of 2000 pounds (907 kg) on the block, hook, and whatever it is hooked onto. As the angle between the lines increases, the stress on the block and hook is reduced as illustrated in Fig. 6-10. To determine the load on the block, hook, and an-

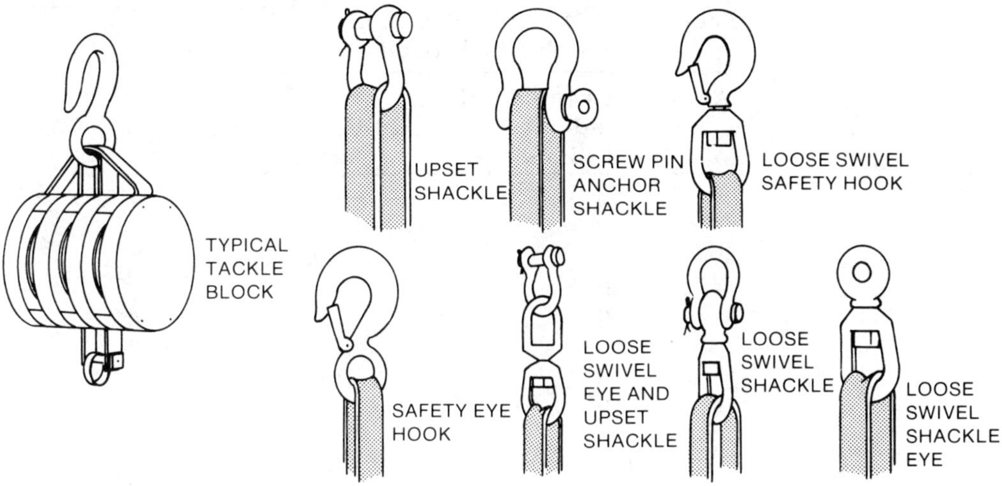

Fig. 6-9: Rope block fittings.

chorage point, multiply the pull on the lead line or the weight of the load being lifted by a suitable factor from Table 6-1 and add 10% for sheave friction.

Rope blocks are used in conjunction with fiber ropes, both natural and synthetic, and are similar to wire rope blocks except for their much lighter weight

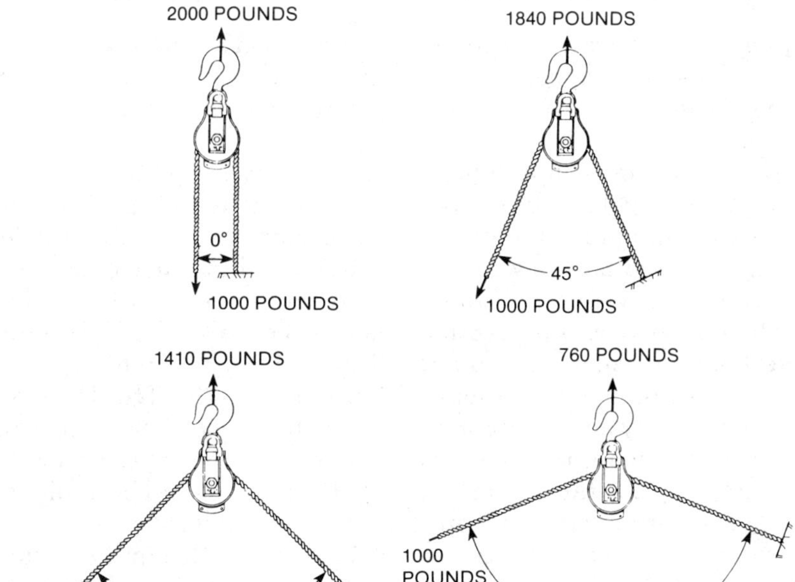

Fig. 6-10: Variation of snatch block loads with rope angle.

Table 6-1: Multiplication Factors for Snatch Block Loads

Angle Between Lead and Load Lines	Multiplication Factor
10°	1.99
20°	1.97
30°	1.93
40°	1.87
50°	1.81
60°	1.73
70°	1.64
80°	1.53
90°	1.41
100°	1.29
110°	1.15
120°	1.00
130°	.84
140°	.68
150°	.52
160°	.35
170°	.17
180°	.00

and capacity. They are available with either wood or metal shells and with or without cheek straps, depending on their capacity. The sheaves in the blocks should be of the proper size for the rope being used and must be free and well lubricated.

With the exception of snatch blocks, the other types of blocks (i.e., crane and hook, wire rope, rope) can be classified according to their position in a reeved system. A travelling or fall block is a block that is attached to the load being lifted or moved and moves with the load. A standing block is a block that is fixed to a stationary object acting as the reaction point of the loading.

Many factors govern the selection and use of rope blocks. Trouble will result from overloading, excess friction, angle of pull, condition of rope, sudden application of load, or lack of lubricant. The actual weight of the load to be moved does not necessarily determine the stress on the blocks. Obstruction to the free movement of the load, twisted ropes due to improper reeving or rigging, or improper angle of the tackle in relation to the load should be avoided. Moving heavy loads over rough ground, on an incline, without rollers or with too small rollers, can introduce severe stresses. A load suspended on two sets of blocks should be evenly distributed or one set will be subjected to more than its calculated share of the load. It is possible through careless preparation to have any or all of these conditions present to such a degree that the load on a set of blocks will greatly exceed the actual weight of the load itself.

In selecting blocks, the governing consideration should be the load to be handled, rather than diameter or strength of the rope they will carry. In multiple-sheave blocks, the load is distributed among several parts of the rope, whereas the hooks or shackles on the blocks have to carry the entire load. It is not practical to make double blocks twice as strong and triple blocks three times as strong as single blocks, since they would be too heavy and needlessly expensive for general use.

A suitable working load is not the greatest load that blocks can lift safely, but a load which blocks can handle efficiently until worn out. Increasing the number of sheaves multiplies power at the sacrifice of speed in hoisting. When blocks are used for heavy loads and fast hoisting, we strongly recommend roller or bronze bearings in the sheaves. For wire rope blocks, cast steel sheaves are recommended.

Regardless of the type or capacity of block selected, be sure that the sheaves are the maximum possible diameter for the size of rope being used and that the blocks are clearly and permanently marked with their weight, where significant, and their rated capacity. Remember also that the anchorage point for the tackle or blocks must be able to carry the total weight of the load, plus the weight of the blocks, plus the pull exerted on the lead line.

REEVING OF BLOCKS

The selection of blocks and their reeving for a particular application involves not only consideration of the working load, connections, sheaves, pins, and other structural details, but also the mechanical advantage required. Both the size of the block, the rope size, and the number of sheaves per block must be determined in relationship to the load to be lifted and the pull which can be applied to the lead line.

Mechanical Advantage

A force is a push or pull. The push or pull a human can exert depends on his/her weight and strength. In order to move any load heavier than the maximum amount a person can move, a machine must be used to multiply the force exerted into a force capable of moving the load. The machine used may be a lever, a screw, or a tackle system. The same principle applies to all of these. If a machine is used which exerts a force 10 times greater than the force applied to it, the machine has multiplied the force input by 10. The mechanical advantage of a machine is the amount by which the machine multiplies the force applied to it in order to lift or move a load. For example, if a downward push of 10 pounds on the left end of a lever will cause the right end of the lever to raise a load weighing 100 pounds, the lever is said to have a mechanical advantage of 10.

Blocks, when reeved, become a machine, which by definition is a device by which forces may be advantageously applied to do work. A multiple part reeved system is a device or machine by which the lead line pull is multiplied to lift the load, thus gaining a mechanical advantage ("purchase"). In theory, the mechanical advantage of a set of tackle blocks is determined by the number of parts of rope at the moving block (Figs. 6-11 and 6-12).

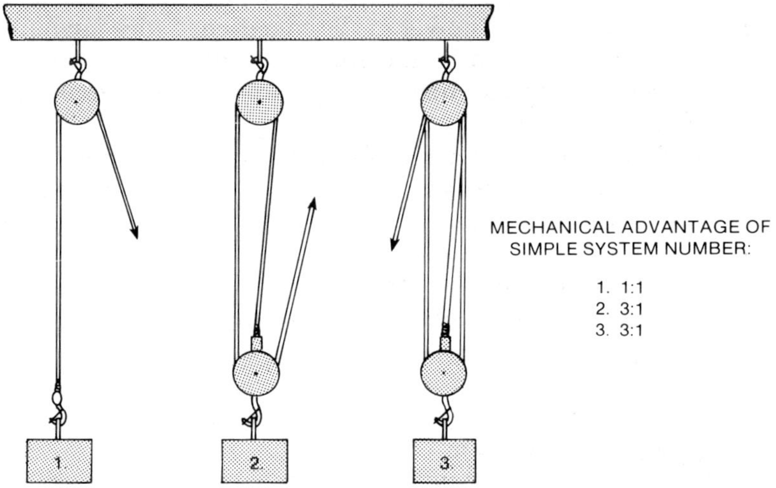

Fig. 6-11: Simple tackle systems.

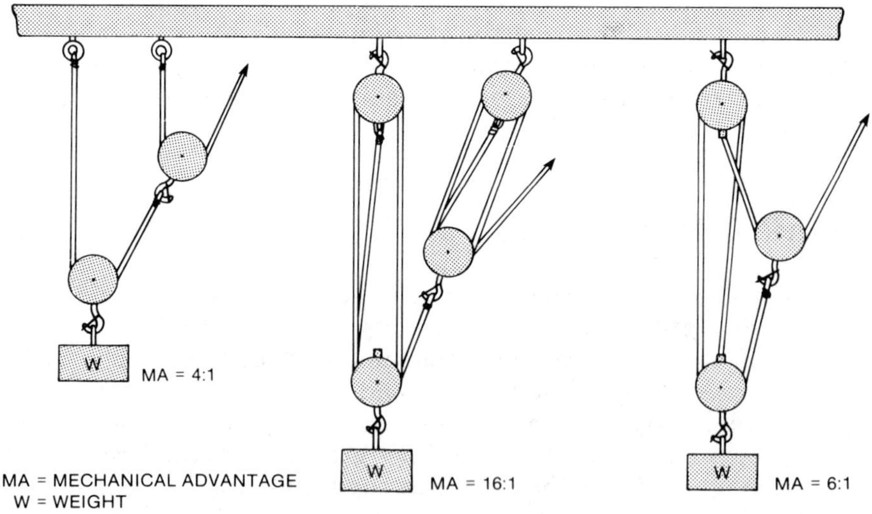

Fig. 6-12: Compound tackle systems.

Since this system is a machine, like all machines, it is not 100% efficient because of friction. As soon as the load being hoisted begins to move, part of the applied force to the lead line is lost in overcoming the friction of turning the sheave and bending the rope. Hence, the lead line pull must be increased to make up for this friction loss in the tackle (Fig. 6-13).

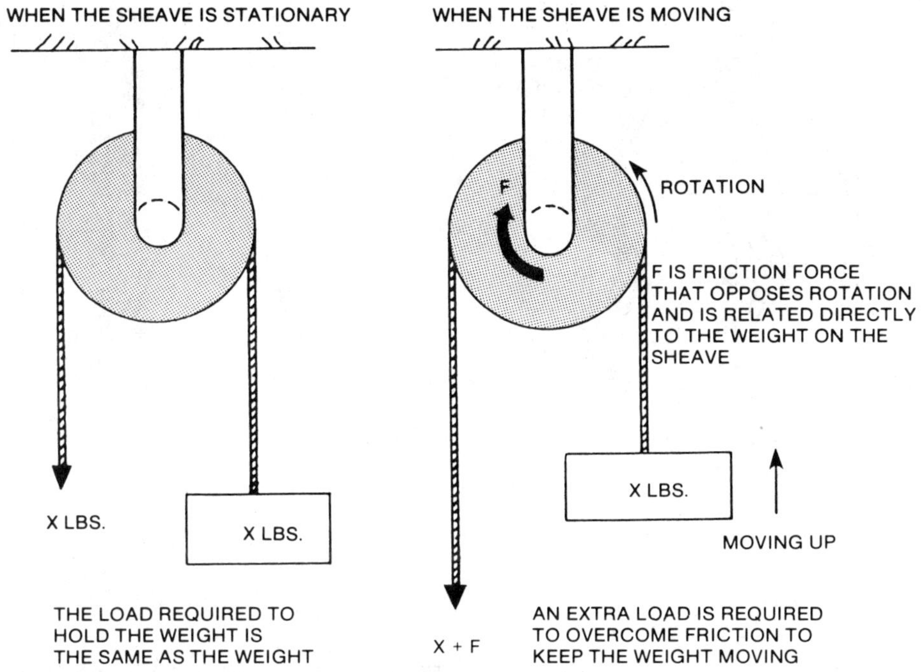

Fig. 6-13: Effect of sheave friction.

This relationship is called the efficiency and is usually expressed in terms of percent. For a greater number of sheaves and parts of line, the mechanical advantage is greater, but the friction loss is also greater and the efficiency lower. Conversely, the lesser the number of sheaves and parts of line, the lower the friction loss and the higher the efficiency, but the mechanical advantage is less.

The mechanical advantage of any multiple part reeved system is always equal to the number of parts of line supporting the running block (travelling block) and the load. The lead line should not be included (Fig. 6-14).

The number of parts of line in a reeved system is determined as follows:

- Imagine that you are going to cut all the ropes just above the load (exclude the lead line).
- Count the cut ends.
- Your answer gives you the number of parts of line which, in turn, is equal to the mechanical advantage of the system.

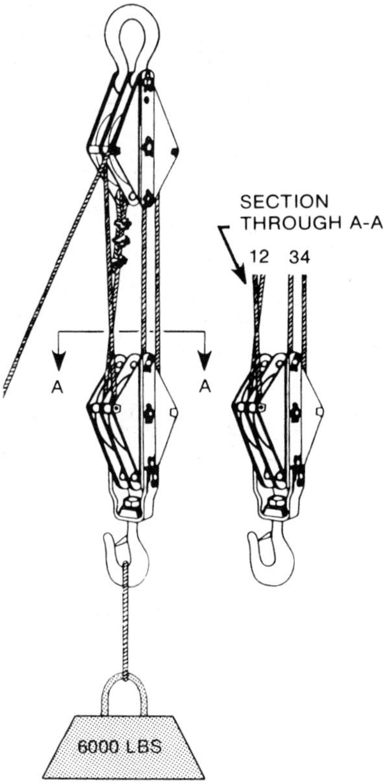

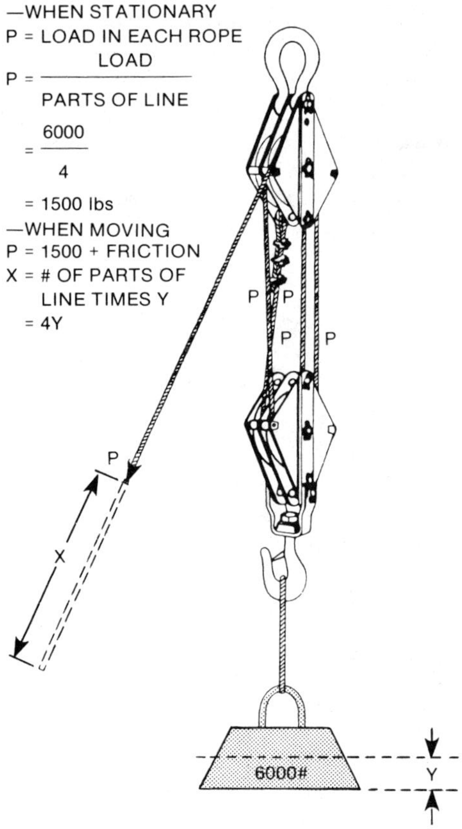

Fig. 6-14: Determination of parts of line. *Fig. 6-15: Mechanical advantage effect.*

The increase in load carrying ability, however, is accompanied by a corresponding decrease in speed of travel. The lead line must move four times as fast as the load for a mechanical advantage of 4. The travel of the lead line, divided by the travel of the load, is also a measure of the mechanical advantage (Fig. 6-15).

If this 6000 pound (2721 kg) load is not only supported but is also raised, then it is necessary to exert more than 1500 pounds (680.25 kg) on the lead line, because of the extra forces necessary to overcome friction in the sheave bearings and the bending resistance of the rope as it passes over the sheaves. The mechanical advantage is decreased by the friction.

For practical purposes, the sheave friction losses on well maintained sheaves are approximately 10% for each sheave in manila rope blocks (plain bore sheaves), about 5% for each sheave having bronze bushings, and approximately 3% for sheaves having roller bearings.

When the load is lifted, each sheave introduces a friction force equal to either 10%, 5%, or 3% *of the load it is carrying* (depending on the sheave bearings and how easily they rotate).

If the previously illustrated example was for sheaves having plain bores (10% friction), the lead line load could be calculated as follows (Fig. 6-16):

Load at A = 1500 lbs.
Load at B = 1500 lbs. + 10% of 1500 (friction @ sheave 1)
= 1650 lbs.
Load at C = 1650 lbs. + 10% of 1650 (friction @ sheave 2)
= 1815 lbs.
Load at D = 1815 lbs. + 10% of 1815 (friction @ sheave 3)
= 1996 lbs.
Load at E = 1996 lbs. + 10% of 1996 (friction @ sheave 4)
= 2196 lbs.
= Lead Line Load.

If you simply multiply the number of sheaves by 10% and then increase the static load on the lead line by that amount, the calculated lead line loads will be significantly *lower* than the actual loads.

The assessment and determination of friction loads requires a progressive type of calculation on a line-by-line basis. To simplify matters, these calcula-

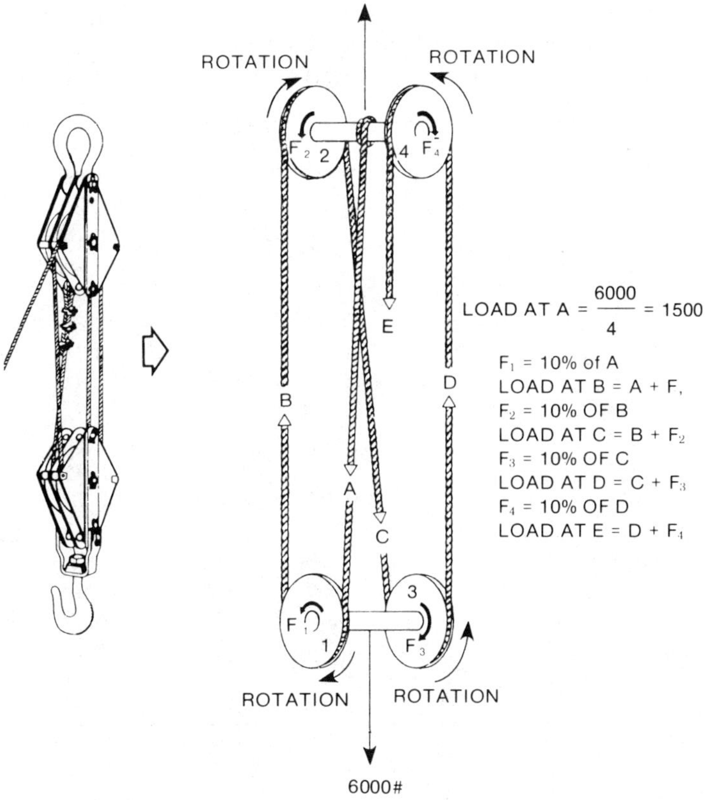

Fig. 6-16: Determination of sheave loads.

tions can be replaced with one calculation and the use of tables of multiplication factors and ratios.

$$\text{Lead Line Load} = \frac{\text{Load to be Lifted}}{\text{Parts of Line}} \times F$$

or

$$\text{Lead Line Load} = \frac{\text{Load to be Lifted}}{R}$$

Values of F and R, provided in Tables 6-2, 6-3, and 6-4, depend on the sheaves.

Table 6-2: Factors to Account for Sheave Friction Loads

Friction Force = 10% of Sheave Load (Typical for plain bore sheaves and poorly maintained bronze bushing sheaves.)

Number of Parts of Line N	Multiplication Factor F	Ratio R (R = N/F) = Actual Mechanical Advantage
1	1.10	.91
2	1.21	1.65
3	1.33	2.26
4	1.46	2.74
5	1.61	3.11
6	1.77	3.39
7	1.94	3.61
8	2.14	3.74
9	2.36	3.81
10	2.60	3.85
11	2.85	3.86
12	3.14	3.82
13	3.45	3.77
14	3.80	3.68
15	4.18	3.59
16	4.60	3.48
17	5.06	3.36
18	5.56	3.23
19	6.12	3.11
20	6.73	2.97

Table 6-3: Factors to Account for Sheave Friction Loads

Friction Force = 5% of Sheave Load (Typical for bronze bushing sheaves and stiff roller bearing sheaves.)

Number of Parts of Line N	Multiplication Factor F	Ratio (R = N/F) = Actual Mechanical Advantage
1	1.05	.95
2	1.10	1.82
3	1.16	2.59
4	1.22	3.28
5	1.28	3.91
6	1.34	4.48
7	1.41	4.96
8	1.48	5.41
9	1.55	5.81
10	1.63	6.13
11	1.71	6.43
12	1.80	6.67
13	1.89	6.88
14	1.98	7.07
15	2.08	7.21
16	2.18	7.34
17	2.29	7.42
18	2.41	7.47
19	2.53	7.51
20	2.65	7.55

Calculating Mechanical Advantage

Figure 6-17 is a graph of the R values in the tables and the number of parts of line. These graphs show the effect of the number of parts of line on the actual mechanical advantage which considers friction. For example, for a system having plain bore sheaves and four parts of line, the graph shows that the actual mechanical advantage is only 2.74, not 4. This is because the sheave friction has reduced the advantage of the extra parts of line. Increasing the number of parts of line on blocks having plain bore sheaves can work to a disadvantage beyond 10 or 11 parts of line and does not really help matters much beyond six or seven parts of line. This, of course, is due to the cumulative effect

Table 6-4: Factors to Account for Sheave Friction Loads

Friction Force = 3% of Sheave Load (Typical for good roller bearing sheaves.)

Number of Parts of Line N	Multiplication Factor F	Ratio R (R = N/F) = Actual Mechanical Advantage
1	1.03	.97
2	1.06	1.89
3	1.09	2.75
4	1.13	3.54
5	1.16	4.31
6	1.20	5.00
7	1.23	5.69
8	1.27	6.30
9	1.31	6.87
10	1.35	7.41
11	1.39	7.91
12	1.43	8.39
13	1.47	8.84
14	1.51	9.27
15	1.56	9.62
16	1.61	9.94
17	1.65	10.30
18	1.70	10.59
19	1.76	10.80
20	1.81	11.05

of sheave friction. You rapidly reach the point where the largest portion of the input load is to overcome friction rather than move the load.

The situation is considerably better on blocks that have bronze bushings and roller bearings, but they too, as shown in the graph, reach a point where the addition of more sheaves and parts of line does not help matters much. Remember, too, that a poorly maintained or lubricated sheave with either bronze bushings or roller bearings can have friction loads much higher than the values quoted.

These are three very important calculations that every rigger should know:

1. How to determine the number of parts of line required to make a lift.
2. How to determine the maximum load that can be lifted with a given reeving arrangement.
3. How to determine the line pull needed when the load weight and number of parts of line are established.

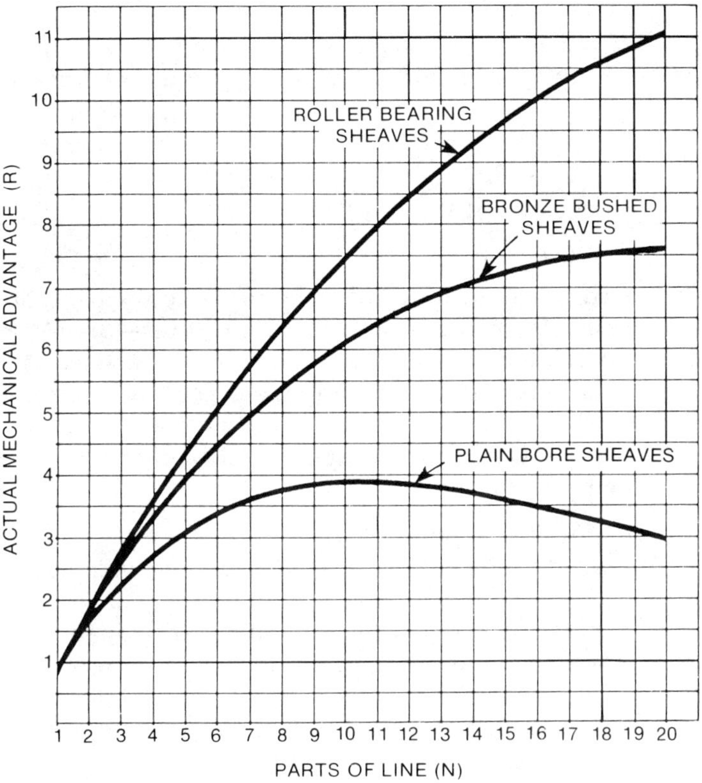

Fig. 6-17: True mechanical advantage vs. parts of line.

To determine the number of parts of line required to make a lift, proceed as follows:

- Determine the maximum safe working load of the rope and the maximum pulling load that you can or want to apply to the lead line. Use the smaller of the two values in the following calculation.
- Determine the weight of the load to be lifted.
- Calculate as follows:

$$\frac{\text{Load to be Lifted}}{\text{Rope SWL (or Desired Lead Line Pull)}} = R$$

- Opposite R in the tables read the number of parts of line or make the determination from the graphs.

Example:

Rope block sheaves have roller bearings and are well maintained. The crane is to be rigged for a capacity of 50 tons. The rope's safe working load is equal to 8.6 tons.

$$\frac{\text{Load}}{\text{SWL}} = \frac{50}{8.6} = 5.8 = \text{R}$$

By reading down the R column in Table 6-4 for roller bearing sheaves we see that our R of 5.8 falls between the 5.69 and 6.30 figures—always go to the next highest figure, which is 6.30 in this case. Opposite the 6.30 we see that this corresponds to eight parts of line. Therefore, this machine requires eight parts of line to make its lifts.

To determine the maximum load that can be lifted with a given reeving arrangement, proceed as follows:

- Determine the number of parts of line.
- Find out what kind of sheave bearings are being used and how easily they rotate.
- Determine the maximum safe working load of the rope and the maximum load that you can or want to apply to the lead line. Use the smallest of the two values in the calculation.
- Determine R from the appropriate table opposite the number of parts of line.
- Calculate as follows:

$$\text{Maximum Load} = \text{Rope SWL (or Lead Line Pull)} \times \text{R}$$

Example:

Tackle blocks have four parts of line and plain bore sheaves. Manila rope of 5/8″ diameter is used (SWL = 880 pounds), but the maximum load that can be applied to the lead line is only 400 pounds.

From the table for plain bore sheaves or from the graph, opposite four parts of line, find that R = 2.74.

$$\text{Maximum load} = 400 \text{ pounds} \times 2.74 = 1095 \text{ pounds}$$

This means that 1095 pounds is the maximum load that can be lifted.

To determine the line pull (or rope size) needed when the load weight and number of parts of lines are established, proceed as follows:

- Determine the type of sheave bearings.
- Determine the load weight.
- Determine the number of parts of line.
- Divide the load weight by the R factor that applies to the number of parts of line and the type of sheave bearings.

$$\frac{\text{Load Weight}}{\text{R}} = \text{Line Pull}$$

Example:

A hoist is rigged with six parts of line and the sheaves have bronze bushings. The load to be lifted weighs 27,000 pounds. From the table or graph for bronze bearing sheaves and opposite six parts of line, read R = 4.48. Therefore, the required line pull is:

$$\text{Line Pull} = \frac{\text{Load Weight}}{R}$$

$$= \frac{27{,}000}{4.48}$$

$$= 6{,}000 \text{ pounds}$$

This means that the hoist winch must be able to exert a line pull of 6,000 pounds if it is to lift the load. Be sure that the required line pull does not exceed the rated safe working load of the wire rope.

These recommendations and calculations apply not only to blocks but also to individual sheaves, idlers, and snatch blocks. Every sheave that a rope is in contact with adds frictional loads to that rope. As has been demonstrated, when the friction is high and there are many sheaves, these loads can combine to the point where they actually *exceed* the loading due to the weight of the object being hoisted.

TACKLE SYSTEMS

It is important that blocks and reeved systems be of sufficient size and capacity to carry the loads to which they will be subjected. If the hook on a set of blocks shows signs of opening up, then that is an indication that the blocks are being overloaded, since the hook will usually begin to bend at approximately 70% of the maximum block strength. The best method to avoid overloads is to know exactly how reeved systems work, use the previously mentioned calculations, and appreciate just how much effect friction plays on the total load.

If the blocks you are going to use have *equal* numbers of sheaves, then the dead end of the rope should be fastened to the becket of the standing (upper) block. If they have *unequal* numbers of sheaves, then the rope should be fastened to the becket of the block having the least number of sheaves (Fig. 6-18). Make sure the rope termination at the becket is secure and develops as much strength as the rope itself. Refer to Chapters 1 and 2 on wire rope and fiber rope to determine the best anchorage methods.

In reeving a pair of tackle blocks, one of which has more than two sheaves, the hoisting rope should lead from one of the center sheaves of the upper block. When reeved this way, the hoisting strain comes on the center of the blocks,

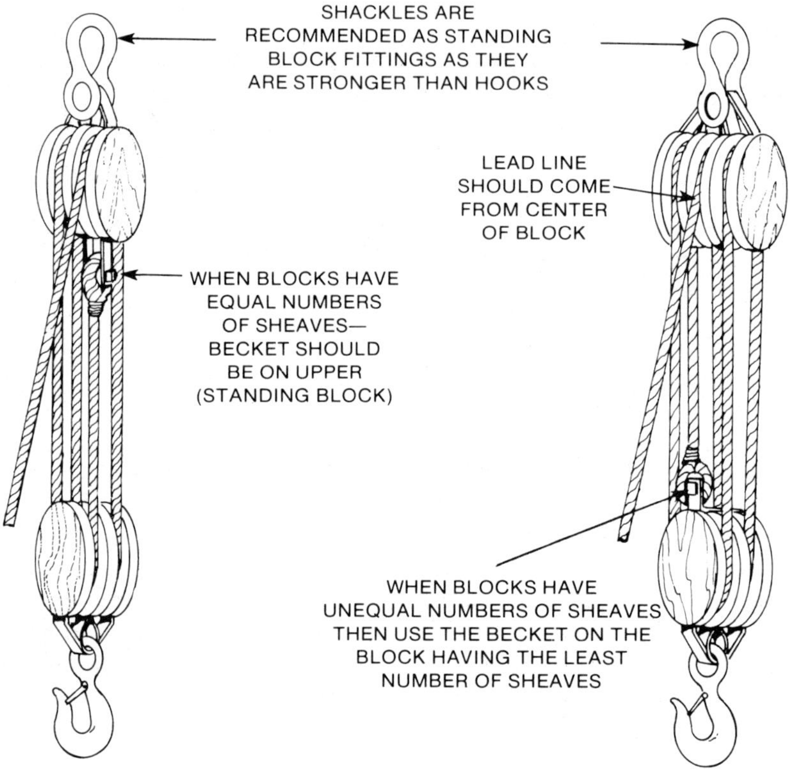

Fig. 6-18: Block orientation in a reeved system.

and they are prevented from toppling and twisting and causing injury to the rope by cutting across the edges of the block shell.

It is good practice to use a shackle block as the upper one of a pair and a hook block as the lower one. A shackle is much stronger than a hook of the same size. Also, the strain on the upper block is much greater than on the lower one. The lower block supports only the load, whereas the upper block carries the load as well as the hoisting strain. A hook is more convenient on the lower block because it is easier to attach or detach it from the load.

When making up a set of falls or reeving a system having five or more parts of line, avoid lacing the blocks because they have a tendency to tilt the travelling block when running empty (Fig. 6-19). This effect causes excessive wear and can damage the sheaves and rope. Whenever possible, *right angle reeving* should be used. The blocks should be identical and square in relation to each other.

When reeving any kind of rope block, it is extremely important that the dead end or becket connection be properly made and attached. On wire rope blocks, use either a wedge socket or cable clips for the becket. On fiber rope blocks, use the becket hitch to secure the dead end (Fig. 6-20).

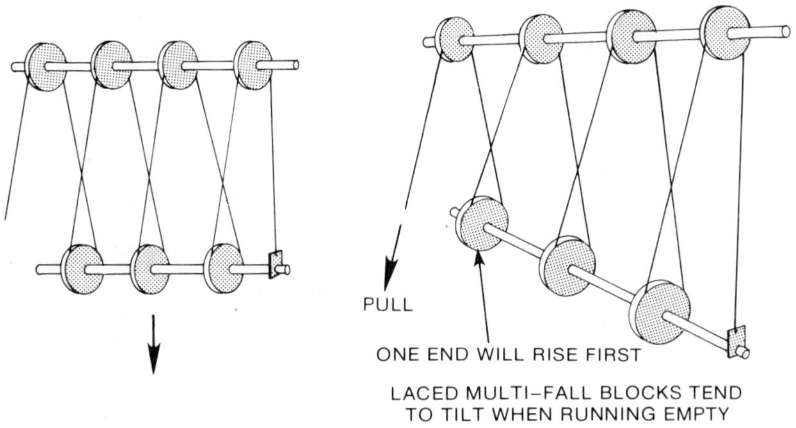

Fig. 6-19: Effect of lacing rather than reeving blocks.

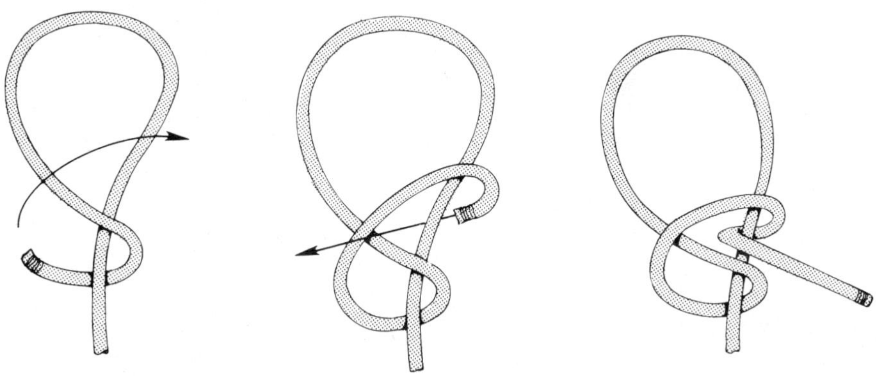

Fig. 6-20: Becket hitch.

Two Part Falls

Use two single sheave blocks. Feed the lead line over the sheave of the stationary block (Fig. 6-21), down in back of the travelling block, through the sheave, and up to the becket on the stationary block. Secure to the becket.

Three Part Falls

Use a two-sheave block and a single sheave block. Feed the lead line through the front of the stationary block at sheave "A," then feed it down in back of the travelling block (Fig. 6-22) and through sheave "C." Bring it up in front of the stationary block, and pass it through sheave "B" and down to the becket on the travelling block. Secure to the becket.

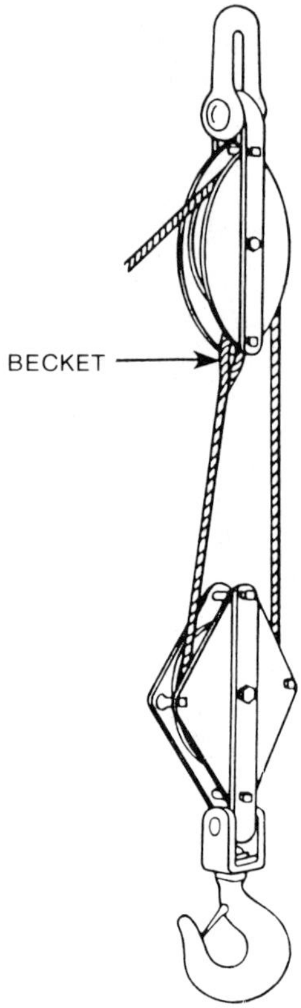

Fig. 6-21: Two part line.

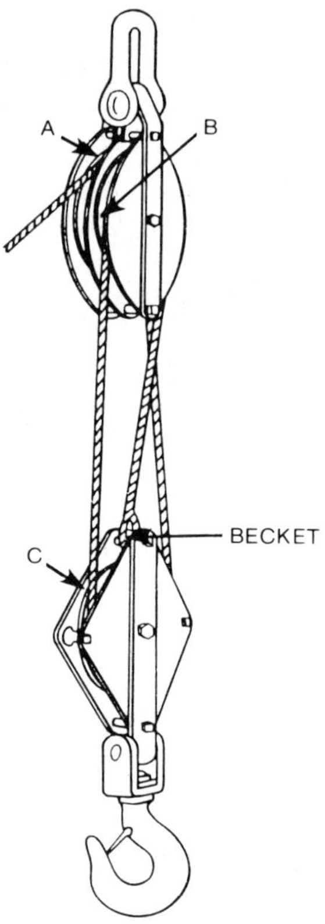

Fig. 6-22: Three part line.

Four Part Falls

Use a pair of two-sheave blocks. Arrange the blocks at right angles to each other (Fig. 6-23). Feed the lead line through the front of the stationary block at sheave "B," then feed it down in front of the travelling block and through sheave "D." Bring it up in behind the stationary block and feed it through sheave "A." Take the lead line down in behind the travelling block, pass it through sheave "C," and bring it up to the becket on the stationary block. Secure to the becket.

Five Part Falls

Use a two-sheave block and a three-sheave block. Feed the lead line through the front of the stationary block at sheave "B," then down in back of the travelling block (Fig. 6-24) and through sheave "E." Bring up behind the stationary block and through sheave "C," down in front of travelling block, and through

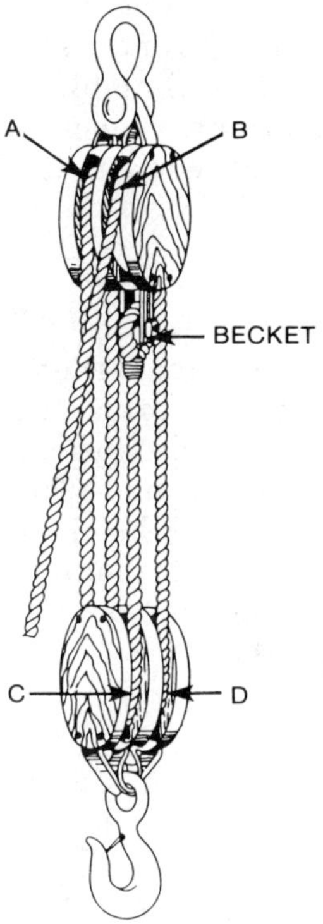

Fig. 6-23: Four part line.

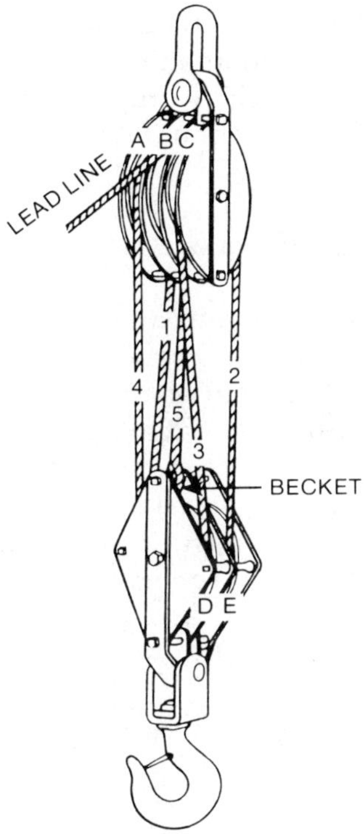

Fig. 6-24: Five part line.

at sheave "D." Then bring it up in front of stationary block and through at sheave "A," down to the travelling block, and secure to the becket.

Six Part Falls

Use a pair of three-sheave blocks. Feed the lead line through the front of the stationary block (Fig. 6-25) at sheave "B," then down in front of the travelling block and through at sheave "E." Bring it up behind the stationary block and through at sheave "A," down behind the travelling block, and through at sheave "D." Then, bring it up in front of the stationary block and through at sheave "C," down in front of the travelling block, and through at sheave "F." Finally, bring it up to the stationary block and secure to the becket.

Seven Part Falls

Use a three-sheave block and a four-sheave block. Feed the lead line through the front of the stationary block (four-sheave) at sheave "C," down in front of the travelling block, and through at sheave "F." Bring up behind the stationary block (Fig. 6-26) and through at sheave "A," down behind the travelling block through at sheave "E." Then bring it up in front of the stationary block and through at sheave "D," down in front of the travelling block and through at sheave "G." Finally, bring it up behind the stationary block and through at sheave "B," then down to the travelling block, and secure to the becket.

Eight Part Falls

Use a pair of four-sheave blocks. Feed the lead line through the front of the stationary block (Fig. 6-27) at sheave "C," down in front of the travelling block and through at sheave "G." Next bring it up behind the stationary block and through at sheave "A," down behind the travelling block and through at sheave "E." Then bring it up in front of the stationary block and through at sheave "D," down in front of the travelling block and through at sheave "H," up behind the stationary block and through at sheave "B." Finally, bring it down behind the travelling block and through at sheave "F," then up to the stationary block, and secure to the becket.

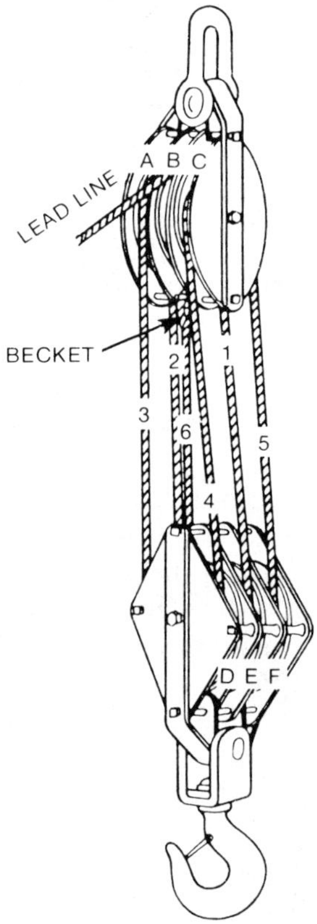

Fig. 6-25: Six part line.

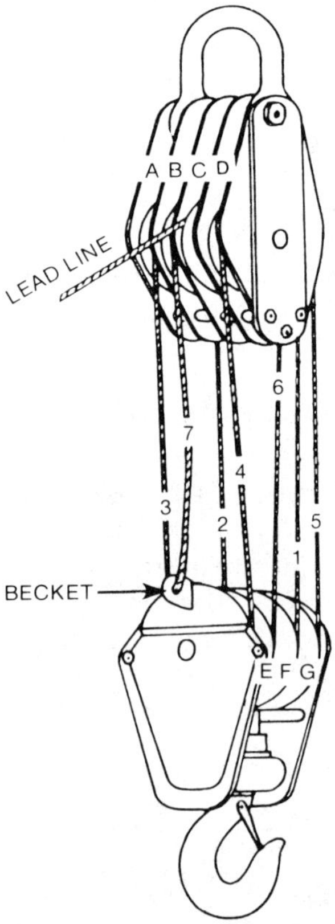

Fig. 6-26: Seven part line.

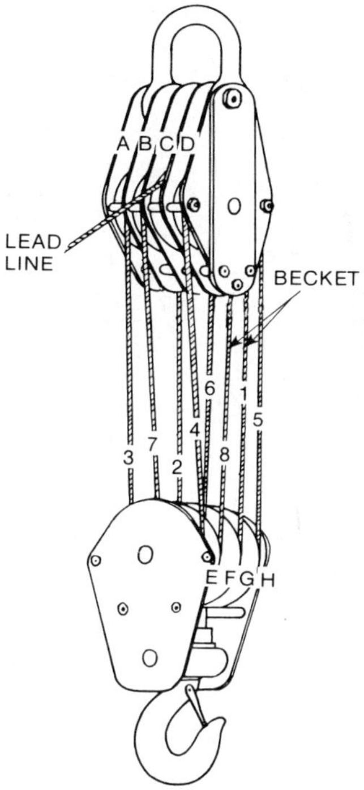

Fig. 6-27: Eight part line.

If the blocks have more sheaves than are needed, be sure that they are symmetrically reeved with the rope as equally distributed across the block as possible to equalize the loads and prevent toppling (Fig. 6-28).

If an auxiliary hook or line is run over the boom point or over a standing block, the main hook or line is generally reeved to one side of the boom point or block. This eccentric condition introduces severe torsional stresses in the boom and can cause blocks to topple. When approaching capacity loads, the blocks and boom point must be reeved in accordance with Fig. 6-29.

All tackles and reeved systems, particularly long new ones, are liable to twist primarily because of lay of the rope, not the method of reeving. If this happens, the power required to hoist increases and sheave damage and rope wear occur. Unfortunately, there is little that can be done other than using 2 in 1 braided fiber rope or nonrotating wire rope.

If the system is rigged with only a single rope fall, there is always a tendency for the rope and hook to spin after the tension is removed from the hoist rope. To avoid being hit by the hook or load, stay away from them until they are set down.

Another consideration in the use of reeved systems is that all sheaves in a set of blocks revolve at different rates of speed. The sheaves nearest the lead line rotate at the highest rates, consequently need more lubricant than the others, and wear out more rapidly.

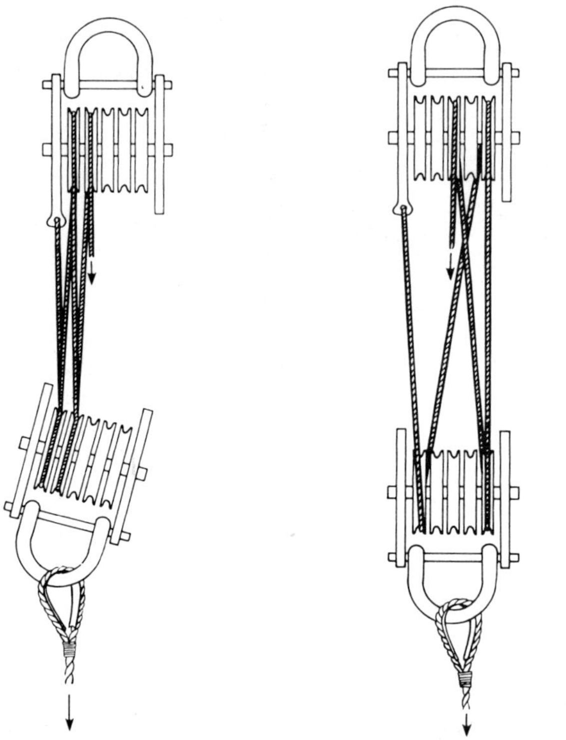

Fig. 6-28: Symmetrical reeving.

CARE OF TACKLE BLOCKS

Like any piece of equipment, tackle blocks cannot be abused or neglected without the loss of effectiveness and life. Keep them clean. Remove sheaves occasionally and clean and oil the center pins. Inspect sheaves for wear and store them in a dry place when not in use. Throwing them around carelessly will damage them and is injurious to the rope. With reasonable maintenance, replacements and repairs will be minimal.

As pointed out earlier in this chapter, many factors govern the selection and use of tackle blocks. Trouble will result from:

1. Overloading
2. Undue friction
3. Angle of pull
4. Condition of rope

5. Sudden application of load
6. Lack of lubricant.

The actual weight of the load to be moved does not necessarily determine the stress on the blocks. Avoid obstruction to the free movement of the load, twisted ropes due to improper reeving or rigging, or improper angle of the tackle in relation to the load. Moving heavy loads over rough ground or on an incline or without rollers or rollers that are too small can introduce severe stresses.

A load suspended on two sets of tackle blocks should be evenly distributed or one set will be subjected to more than its calculated share of the load. Careless preparation may result in any or all of these conditions to such a degree that the load on a set of blocks will greatly exceed the actual weight of the load itself.

When inspecting tackle blocks, the following points should be observed:

• Check the blocks for excessive wear on the beckets, end connections, sheave bearings, and center pins.

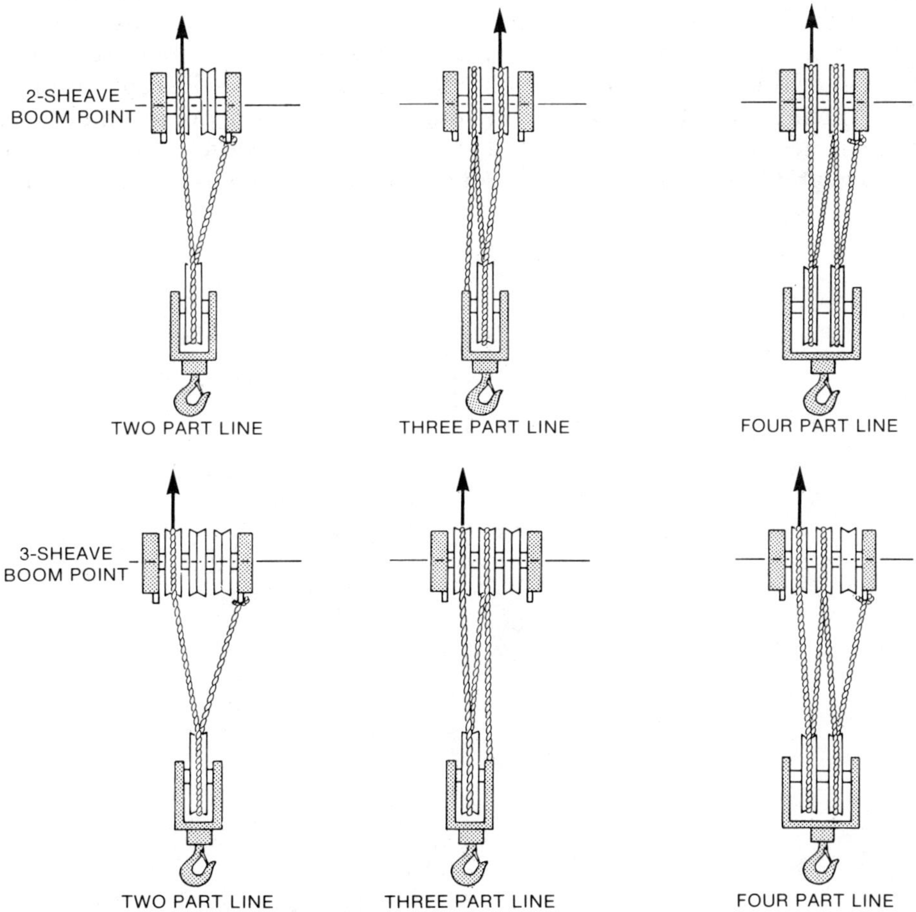

Fig. 6-29: Symmetrical boom point reeving.

- Be sure that the sheave grooves are smooth. If a wire rope sheave shows the imprint of the rope, excessive rope wear will occur.
- Look for signs of overloading such as elongated links, eyes, or shackles; bent shackle, link, or center pins; enlarged hook throats. If such conditions are found, the block will have to be replaced.
- Check the sheaves for proper rotation.
- Be sure that all cable keepers are in place.
- Check the clearance between sheaves and check and partition plates. It should be small enough that there is no danger of the rope slipping between them.

Tackle Safety Precautions

In hoisting and moving heavy objects, using blocks and tackle, keep *safety* uppermost in mind. This should include safety for personnel and materials.

Always check the condition of blocks and sheaves before using them on a job to make sure they are in safe working order. Also, make sure that the line and sheave are the right size for the job. Sheaves and blocks designed for use with fiber line must *not* be used for wire rope, since they are not strong enough for that service and the wire rope does not fit the sheave grooves. Moreover, sheaves and blocks built for wire rope should *never* be used for fiber line.

7

Skids, Rollers, Jacks, and Cribbing

INTRODUCTION

It is not always feasible to hang rigging or use cranes to move heavy objects such as vessels where, for example, overhead wiring or piping limit the work space. Under these circumstances an arrangement of skids, rollers, cribbing, and jacks can provide a safe means of transporting and/or raising large objects.

SKIDS

Timber skids may be placed longitudinally under heavy loads to distribute the weight over a greater area, to make a smooth surface for skidding equipment, or to provide a runway surface when rollers are used. It must be remembered, however, that under certain conditions, this method of movement is not completely suitable. For example, the angle of incline of the skids should be kept as low as possible to prevent the vessel from "running away" or getting out of control. If any incline is involved, it is advisable to use a tugger or a tirfor jack as a holdback. Grease may be used for skids where horizontal movement only is involved, but in most circumstances greasing is dangerous as it may cause the load to drift sideways suddenly.

Other conditions that limit the use of timber skids include the inability of the pulling power to overcome the friction of the load or an uneven or rough bottom on the object to be moved. In some instances, where the surface over which the load is to be moved is rough, wet, or soft, steel plates are used as skids. The sliding steel plate surface is frequently improved by coating it with grease or low-friction material such as bronze or Teflon. Such materials should not be used on slide plates where inclines are encountered.

ROLLERS AND OTHER MOVING DEVICES

There are various types of rollers, wheeled devices, and air and water bearings that are used to move heavy loads into position.

Wood or Pipe Rollers

Rollers made of pipe or, less frequently, hardware provide a rolling platform over the skids (Fig. 7-1), or slide plates; that is, skids are placed under the rollers to provide a smooth, continuous surface for the rollers. The rollers themselves must also be smooth and round and must extend beyond the outside dimensions of the object being moved. When using pipe rollers, keep in mind that they are not as resilient as hardwood and, therefore, cause problems. While steel pipe rollers can carry much heavier loads than wooden rollers of equal size, they can collapse or be deformed into oval shapes if the wall thickness is too thin. In other words, the carrying capacities of pipe rollers are based on their diameter and wall thickness. The larger the diameter and the thicker the wall thickness, the heavier the load the rollers can carry.

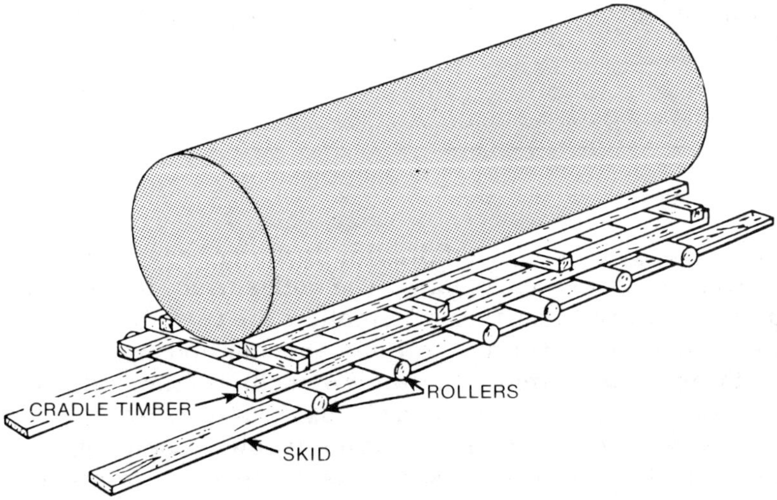

Fig. 7-1: Using skids and rollers.

To ensure continuous support for the rollers along the moving distance, the skid joints should butt securely together or the ends be arranged in a side by side overlap. Skid joints should also be staggered so that the rollers pass over only one joint at a time.

The number of rollers required will depend on the size and weight of the object being moved. For example, the general rule of thumb for hardwood maple rollers is that they will carry approximately 50 pounds per inch for every inch of diameter; that is, a 5″ diameter roller 36″ long, in contact with both the load and the skid, can safely carry:

50 pounds per inch of diameter × 5″ diameter × 36″ length = 9000 pounds.

A 36-ton load would, therefore, require:

$$\frac{36 \text{ tons} \times 2000 \text{ pounds per ton}}{9000 \text{ pounds per roller}} = 8 \text{ rollers}$$

Remember, *all* eight rollers must be in contact with the load at all times.

When moving the load, several rollers are placed in front of the load in addition to those under it. As the load rolls slowly forward, the rollers left behind are moved in front of it so that there is a continuous path of rollers (Fig. 7-2). In making a turn with a load on rollers, the front rollers must be inclined slightly in the direction of the turn and the rear rollers in the opposite direction (Fig. 7-3). This inclination of the rollers may be made by striking them sharply with a sledge.

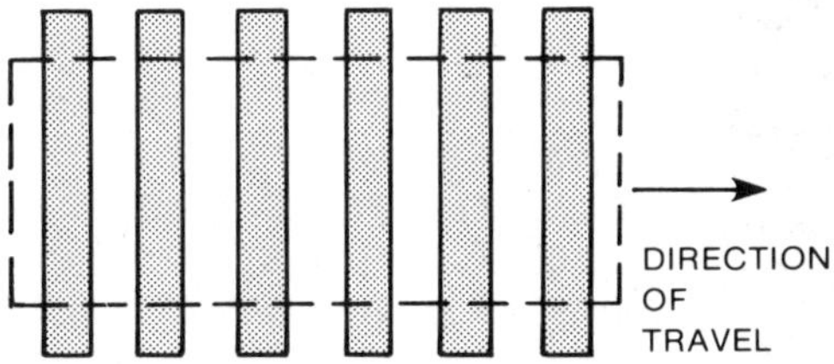

Fig. 7-2: Using rollers during a straight run.

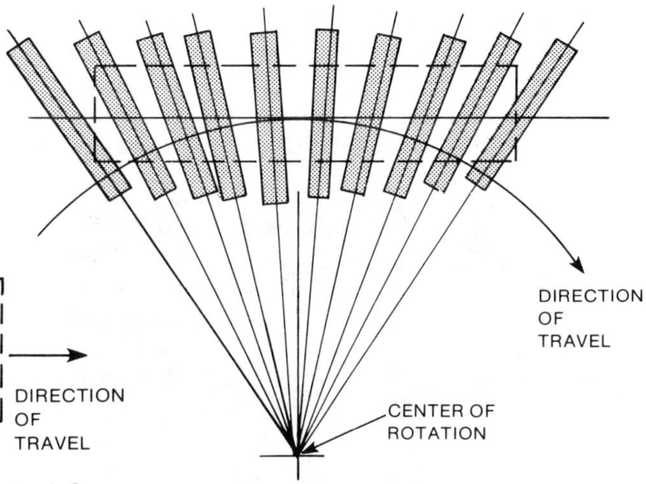

Fig. 7-3: Using rollers during a turn.

If the load to be moved does not have a smooth bottom surface, it is sometimes necessary to use steel or timber support pads or shoes (Fig. 7-4). The load each shoe is required to carry is determined by where it is located in respect to the center of gravity and whether it will support the load at all times. It must be remembered that skid runways are seldom perfectly level; therefore, sooner or later the load will not be supported by all the shoes at the same time. This must be considered when selecting the size and numbers of rollers to use with the shoe.

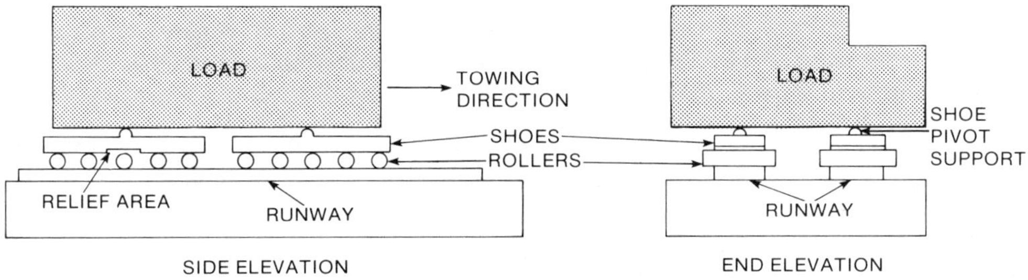

Fig. 7-4: Steel or timber support pads or shoes being used to move a load.

Express Rollers

Express rollers, also known as *endless chain rollers*, are designed to handle extremely heavy loads on relatively small rollers (Fig. 7-5). They are placed under the object to be moved (in some cases they may be welded to the bottom of the object) and they will move along with it. This makes it unnecessary to shift the rollers, thus making moving faster (express). Usually steel channels or runways are laid down on the floor or ground for the express rollers to run in. Swivel attachments are available for positioning ability of the express rollers.

When more than three express rollers are employed in combination to move a single object, it is important to equalize the load, either mechanically or hydraulically, to prevent damage to the load or the express rollers.

Fig. 7-5: Typical express rollers (photos courtesy of Hilman, Inc.).

Air and Water Bearings

Air bearings, better known as *air rollers*, "float" loads on thin cushions or films of air. This air cushion, pumped between the load and the slide plate or runway by means of an air supply system, greatly reduces the horizontal force needed to move or tow a load.

There are two basic types of air bearings. The first type consists of a synthetic, doughnut-shaped apparatus that attaches to the underside of the frame that supports the load. An air hose introduces air into the doughnut. When the pressure becomes great enough—20 to 80 pounds per square inch (psi)—the load is raised and air begins to escape between the doughnut and the runway, forming the cushion. In order for this system to operate, however, the runway must be constructed of a material such as steel that is smooth and resistant to the passage of air.

The second type consists of a rubber or plastic skirt stretched over an airtight frame. Again the escape of air between the skirt and the runway achieves the flotation effect, but this time the pressure required is somewhat lower. The main advantage of the skirt device is that it can function just as well over rugged surfaces such as railroad tracks.

Water bearings work in a fashion similar to air bearings, except that the cushioning medium is water rather than air. These devices have also been shown capable of moving large loads.

Dollies, Casters, and Conveyors

Various types of dollies are commercially available that are capable of handling loads from a few pounds up to several hundred tons and more. Also, dollies are designed and manufactured for specific horizontal movement jobs. Many different configurations of single or double rows of rollers, and rollers of various diameters, are used to make turning easy or to accommodate grade changes. The number of rollers and their size must be selected for the load weight to be handled.

Sets of individual caster wheel assemblies are sometimes employed to support a load directly or to support a platform or frame that, in turn, supports the load. The sets of wheels—generally two sets—are mounted in a diamond-shaped pattern. The two side casters are usually slightly larger in wheel diameter than the front and rear wheels. With such an arrangement it is possible for the load to ride on the side casters while the front and rear ones are used to stabilize the load.

Caster wheel diameter will vary with the weight of the load, but the minimum size is at least 6″. The wheels may be of plastic, hard rubber, forged steel, or cast iron. The latter are good for exterior use on rough surfaces.

For moving lighter loads, rollers can be made up and set on axles in side beams as a semipermanent conveyor. Permanent metal roller conveyors (Fig. 7-6) are available. They are usually made in sections, but are suitable for light loads.

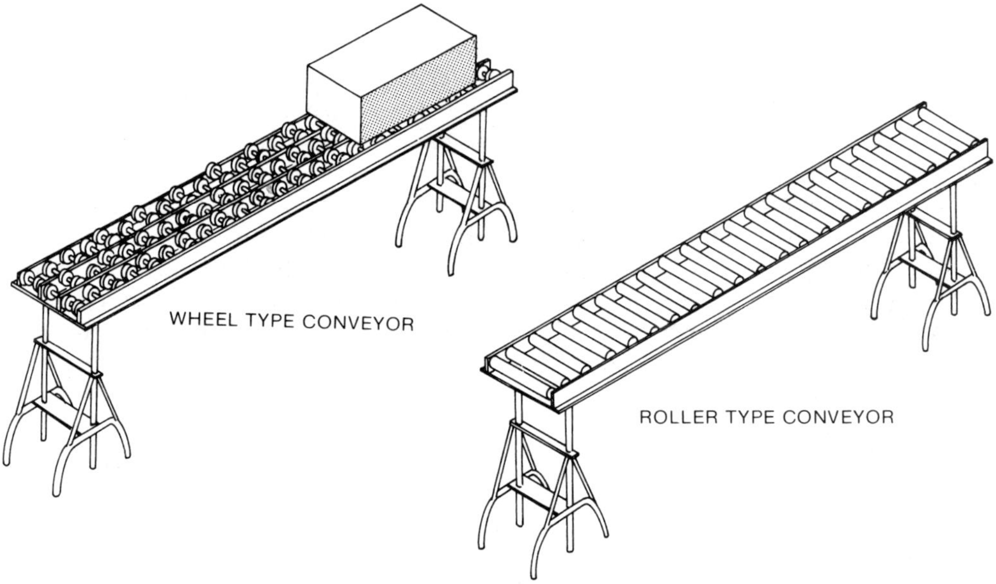

Fig. 7-6: Metal conveyors.

JACKS

In order to place cribbing, skids, or rollers, it is often necessary to lift and lower the load for a short distance. Jacks are used for this purpose. Jacks are used also for precision placement of heavy loads, such as bridge spans. A number of different styles of jacks are available, but only heavy duty hydraulic or screw type jacks should be used. The number of jacks used will depend on the weight of the load and the rated capacity of the jacks. Be certain that the jacks are provided with a solid footing, preferably wooden blocking. Cribbing is frequently used in jacking loads by stages (Fig. 7-7). The procedure requires blocking under the jacks, raising of the object to the maximum height of the jacks to permit cribbing to be put directly under the load, and the lowering of the load onto the cribbing. This process is repeated as many times as are necessary to lift the load to the desired height.

Jacks are available in capacities from 5 to 100 tons. Small capacity jacks are operated through a rack bar or screw, while those of large capacity are usually operated hydraulically.

Ratchet Lever Jacks

The ratchet lever jack (Fig. 7-8) is a rack-bar jack which has rated capacity of 15 tons. It has a foot lift by which loads close to its base can be engaged. The foot capacity is 7-1/2 tons.

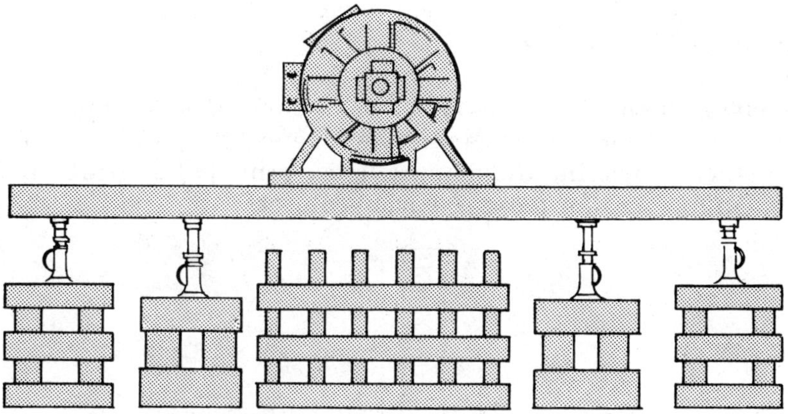

Fig. 7-7: Jacking loads by stages.

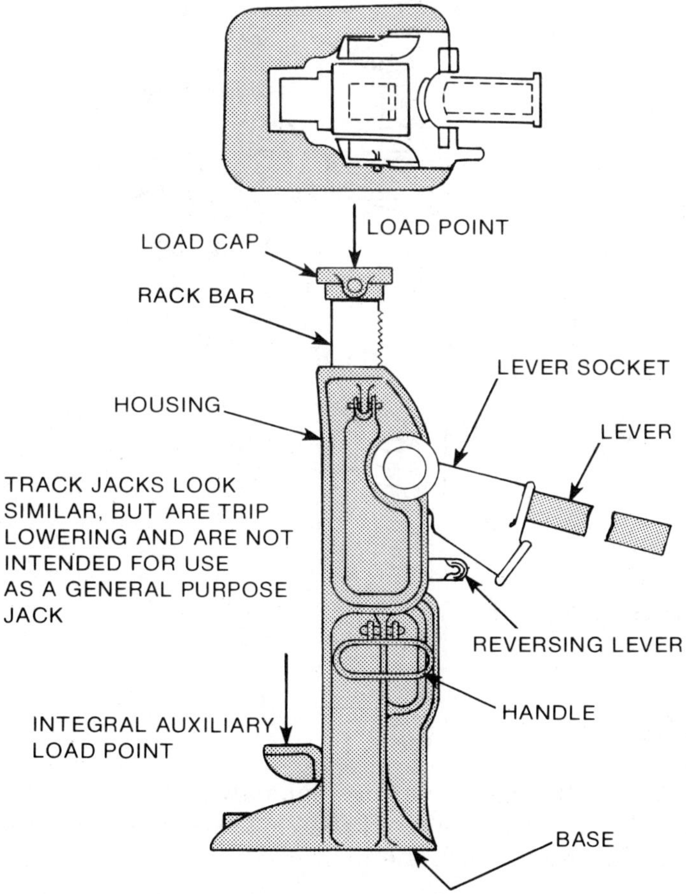

Fig. 7-8: Ratchet jack.

Steamboat Ratchets

Steamboat ratchets, sometimes called *push-and-pull jacks* (Fig. 7-9), are ratchet screw jacks of 10-ton rated capacity with end fittings which permit pulling parts together or pushing them apart. Their principal uses are for tightening lines or lashings and for moving various objects.

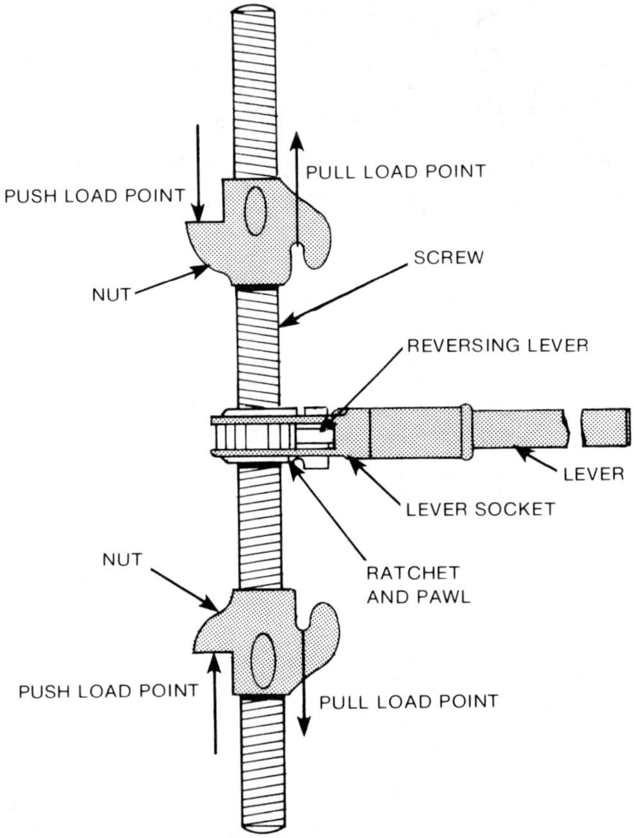

Fig. 7-9: Steamboat jack.

Screw Jacks

Screw jacks (Fig. 7-10) come in various capacities. They are about 13″ high when closed and have a safe rise of at least 7″. These jacks can be used for general purposes, including steel erection.

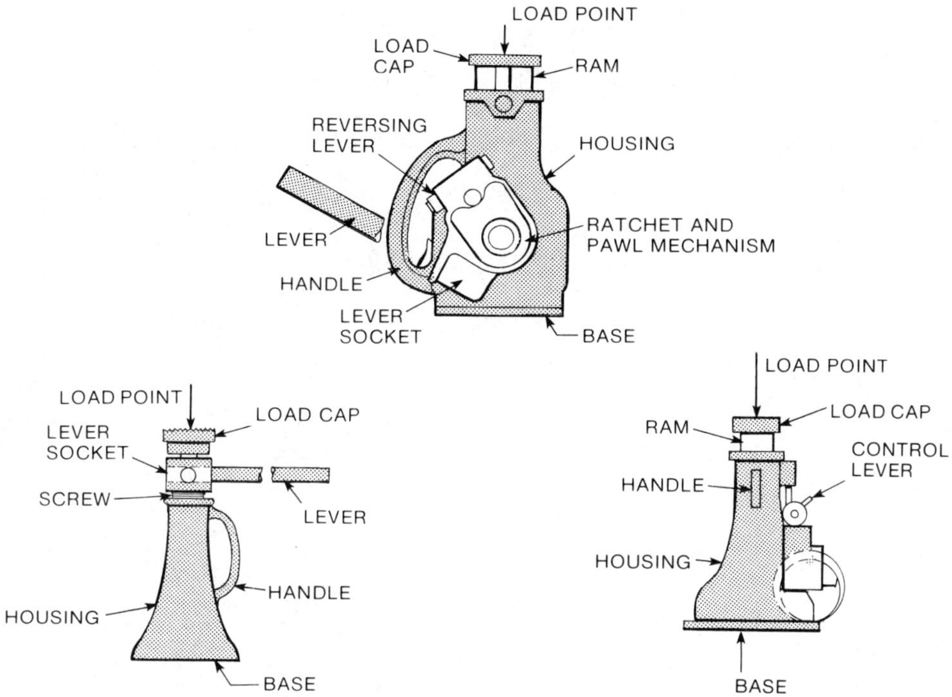

Fig. 7-10: Screw jacks.

Hydraulic Jacks

Hydraulic jacks (Fig. 7-11) are available in capacities up to 100 tons. Those supplied with the squad pioneer set are 11″ high, have a rated capacity of 12 tons, and a rise of at least 5-1/4″. They are large enough for usual construction needs. Where large weights are encountered (up to 50 tons), jacks with twin hydraulic units (often called *porta-power jacks)* are available.

Jack Operation

Regardless of the type of jack used, it should be visually inspected before use and if any deficiencies are found, it should be removed from service. Also, it is very important not to overload it. The lever or handle is made of a predetermined length by the manufacturer to ensure against overloading. Of course, with an extension pipe placed on the lever it is possible to lift the rated load more easily. Or with the same force on the extended lever, a greater load can be

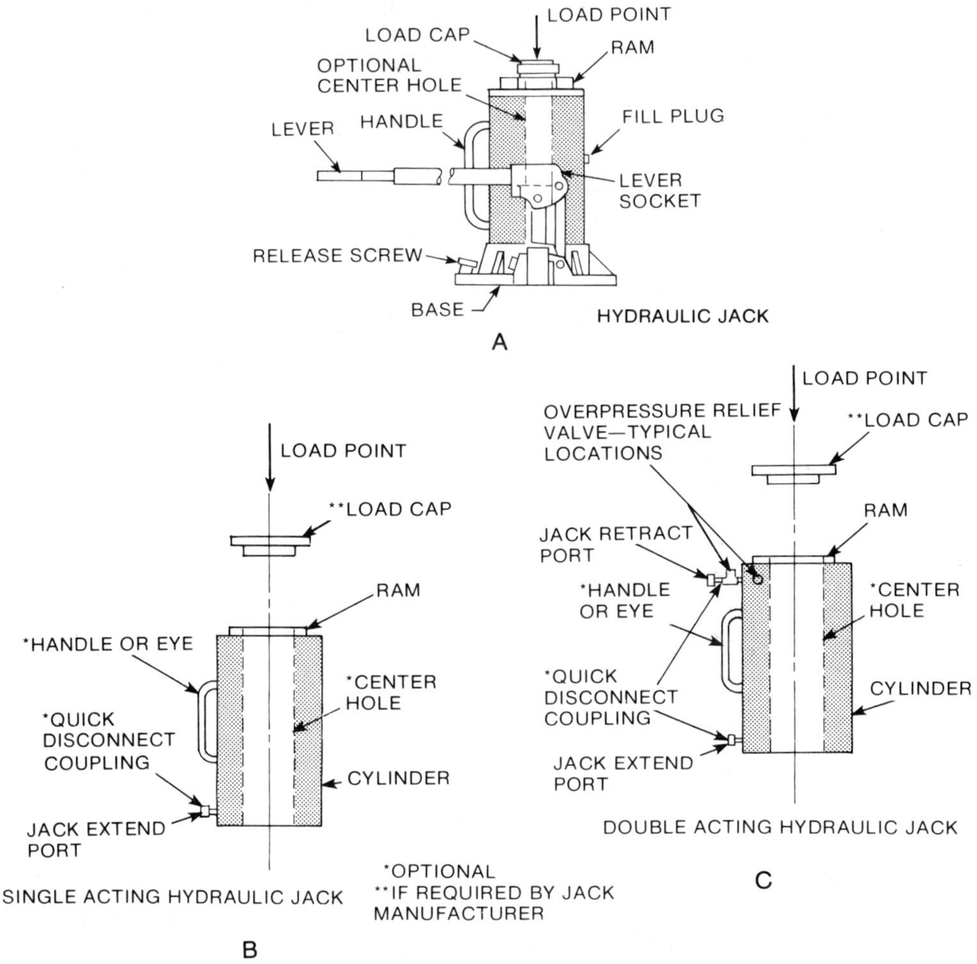

Fig. 7-11: (A) Universal hydraulic jack components, (B) specific single acting jack components, and (C) specific double acting jack components.

lifted. This load, however, may be in excess of the rated or safe capacity of the jack. So, do not use a longer lever than that furnished with the jack by the manufacturer. When placing jacks, make certain that there is sufficient swing area for the operating lever(s).

Always make certain that the jack is in a true vertical position (when lifting) and resting on a good footing. Never place a jack directly on the ground, even though the soil appears to be firm. When required, use shims or constraints to prevent slippage of the base load. Off-center jacks should be avoided. If the load is to be raised in its entirety by several jacks, it should be secured laterally by struts to prevent all the jacks from upsetting in unison.

In raising a load on a jack, it is important that it should not be raised so high as to run the ram or screw or ratchet out of the base and thereby drop the load. Most modern jacks have positive stops to prevent overtravel; however, be sure to check the jack before using it to see if it has this feature.

After the load has been raised to the required height, shoring or cribbing blocks should be placed under the load and wedged to take the load off the jacks or to be ready to take the load should a jack fail.

When operating a jack, particularly the ratchet and lever type, do not lean over the lever unless you are holding onto it. Also, do not straddle the operating lever of a mechanical jack. Remove the operating levers when not in use to avoid accidental dislocation of the jack and to reduce the tripping hazard. Jack operating lever and load bearing surfaces should be free of slippery material or fluids.

Always take measures to prevent personnel from working or passing under the load until the load is secured against accidental lowering by cribbing, blocking, or other means. In addition, be certain to take precautions to assure that all personnel are clear of the load before lowering the jack.

Jack Care and Maintenance

All moving parts of the jack requiring lubrication should be regularly lubricated. Lubricating systems should be checked for delivery of lubricant. Care should be taken to follow manufacturer's recommendations as to points and frequency of lubricant, maintenance of lubricant levels, and types of lubricants to be used.

Hydraulic jacks should be filled only with hydraulic fluid that is fully compatible with the jack manufacturer's specifications. Never use water in a jack. Keep the threads of a screw jack free from dirt and grit. Jacks exposed to rain, sand, or grit-laden air should be cleaned prior to use.

Jacks should be stored where protected from the elements, abrasive dust, and damage. Hydraulic jacks should be stored in vertical position.

CRIBBING

Cribbing or blocking is often necessary as a safety measure to keep an object in position or to prevent accidents to personnel that work under or near these heavy objects. Cribbing is formed by piling timbers in tiers, with the tiers alternating in direction (Fig. 7-12), to support a heavy weight at a height greater than blocking would provide. The timbers may either be tightly packed or spread at even intervals, depending upon the requirements of the object's weight. Where the cribbing is spaced, all layers of blocking must be in a direct

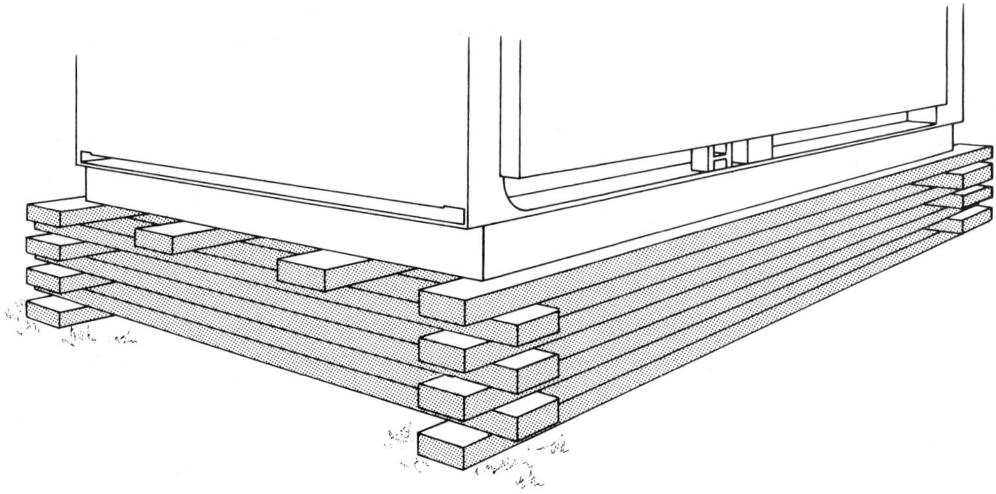

Fig. 7-12: Timber cribbing.

vertical line with other members cribbed in the same direction in the system.

A firm and level foundation for the cribbing is essential, and the bottom timbers should rest firmly and evenly on the ground. Blocking used as a foundation for jacks should be sound and large enough to carry the load. The timbers should be dry, free from grease, and placed firmly on the ground so that the pressure is evenly distributed.

Cribbing may be required to provide a level, stable platform for a load that cannot be placed directly on rollers. It is also used to raise loads in stages using jacks. The raising procedure is as follows:

1. Using jacks on a stable footing, raise the load to the maximum stable height.
2. Place one or more rows of cribbing under the load and lower the jacks to rest the load onto the crib platform.
3. Set the jacks on blocking and raise the load further, adding more rows of cribbing under the load. Repeat the stepwise jacking and cribbing until the required elevation is reached.

The weight of the load and the rated capacities of the jacks determine the number and type of jacks required for the lifting operation.

SHORING

Where construction work involves excavation, there are definite precautions that should be observed to prevent accidents. To avoid slides or cave-ins, the sides of excavations 4′ or more in depth should be supported by substantial and adequate sheathing , sheet piling, bracing, shoring, etc., or the sides sloped to the angle of repose. The angle of repose is the angle, measured from the horizontal, of the natural slope of the side of a pile of granular material formed by pouring grains or particles through a funnel, practically without impact. The angle of repose varies with the moisture content and the type of earth or other material. For ordinary earth, the angle of repose varies from about 20° to 45°, corresponding to slopes of from about 2.8:1 to 1:1. The sides of an excavation do not consist of poured particles, however; many types of earth, because of their cohesive qualities, will stand vertically without failure. But, because of the nonuniformity of most soils, the times and places of local and intermittent cave-ins and slides cannot ordinarily be predicted. Therefore, it is conservative and safe to require laying the bank back to the angle of repose, or natural slope, of the material being excavated.

It is seldom practical to slope the sides of foundation and footing excavations to the angle of repose. Therefore, any such excavation 4′ deep or more must be supported as specified. Many people wonder about this, in view of the fact that most adults are a good deal more than 4′ tall. The reason lies in the fact that a person needs to be buried only to chest level to suffocate in a cave-in. The pressure against the chest makes breathing impossible, and if the chest is not freed within a minute or two the person will suffocate, even though head and shoulders are out in the air.

Sheathing

Sheathing consists of wooden planks placed edge-to-edge either horizontally or vertically. Horizontal planking is used for excavations with plane faces, vertical planking when it is necessary to follow curved faces. Sheathing is supported by longitudinal wales or rangers which are nailed to the sheathing and which bear against transverse shores or braces.

Sheathing must be progressively installed and braced as every 4′ stage of depth is reached. Vertical and horizontal wood sheathing are illustrated in Fig. 7-13.

Sheet Piling

Sheet piling (Fig. 7-14A) is driven before the work of excavation begins. Wooden sheet piling consists of 2″, 3″, or 4″ planks, beveled at the lower end to facilitate penetration of the soil. Also, the lower ends are cut at an angle, as shown in Fig. 7-14A, so as to cause the edge of a pile being driven to bear against the edge of one previously driven. Obviously, care must be taken to place a pile with this angle inclined in the proper direction. The bevel on the lower end-edge, too, must face toward the excavation. As a pile is driven, it

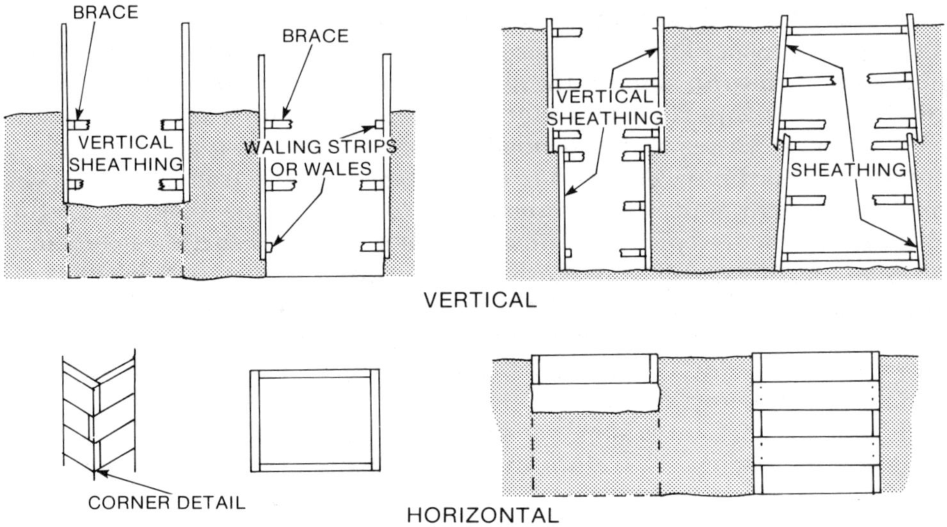

Fig. 7-13: Vertical and horizontal sheathing.

tends to slant off in the direction away from the bevel. If the bevel is incorrectly turned away from the excavation, the excavation will progressively narrow as shown in Fig. 7-14E.

In wakefield sheet piling, each pile consists of three planks, bolted together with the center plank offset for tongue-and-groove joining as shown in Fig. 7-14B.

To set wooden piles for driving, first dig a shallow notch along the excavation line, wide enough to admit the bottoms of the piles and deep enough to hold them upright (Fig. 7-14C). For additional upright support, lay pairs of wales along the notch and stake them in position. If the piles are too long to stand this way, the wales must be braced in elevated position.

In favorable soil, a 2″ sheet pile can be driven to a depth of about 16′, a 3″ sheet pile to about 24′, and a 4″ sheet pile to about 32′. When piles are too short to cover full excavation depth, they are driven in stages called *sections,* as shown in Fig. 7-14D. Table 7-1 gives the bearing capacity of soils.

Shoring

A common method of providing support is by the use of up to 12 × 12 timbers called *shores,* inclined against the wall to be supported and extending across the excavation to a temporary *footing* consisting of a framework or mat of timbers laid on the ground. The upper ends of the shores may be fitted into openings cut in the wall, or they may butt against a timber bolted to the wall. Steel

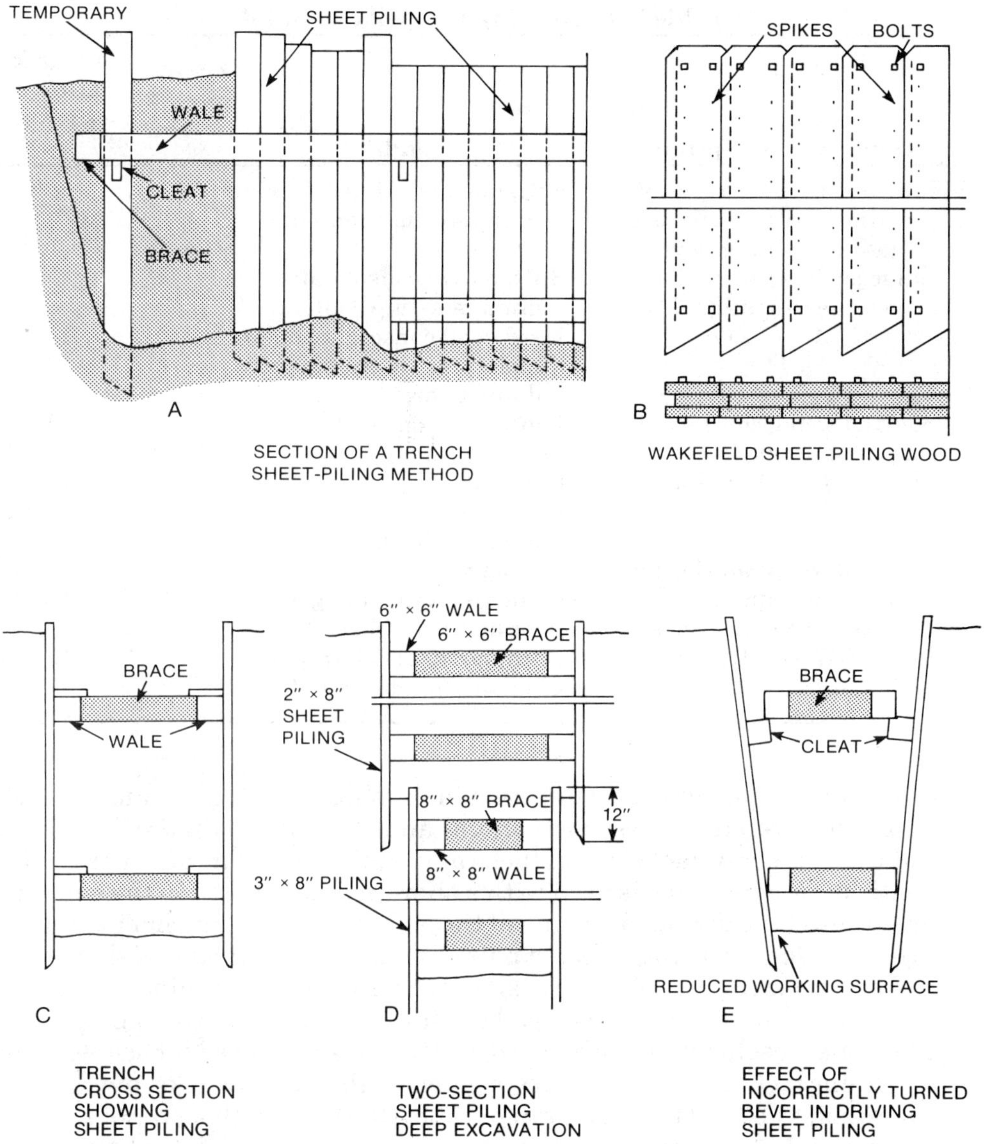

Fig. 7-14: Wooden sheet piling.

saddles may be placed in openings cut in concrete or masonry walls to support lifting or steadying shores.

It is good practice to set shores as nearly vertical as possible, to reduce lateral thrust against the wall. Heads of shores should, whenever possible, be located at floor levels, to minimize the danger of pushing the wall in.

Table 7-1: Bearing Capacity of Soils

General description	Condition	Safe allowable pressure (psi)
Fine grained soils: Clays, silts, very fine sands, or mixtures of these containing few coarse particles of sand or gravel. Classification: MH, CH, OH, ML, CL, OL.	Soft, unconsolidated, having high moisture content (mud).	1,000
	Stiff, partly consolidated, medium moisture content.	4,000
	Hard, well consolidated, low moisture content (slightly damp to dry).	8,000
Sands and well-graded sandy soils, containing some silt and clay. Classification: SW, SC, SP, SF.	Loose, not confined.	3,000
	Loose, confined.	5,000
	Compact.	10,000
	Loose, not confined.	4,000
	Loose, confined.	6,000
Gravel and well-graded gravelly soils containing some sand, silt and clay. Classification: GW, GC, GP.	Compact.	12,000
	Cemented sand and gravel.	16,000
	Poor quality rock, soft and fractured; also hardpan.	10,000
Rock	Good quality; hard and solid.	20,000

Provision for inducing a lift or thrust in the shores is usually made by inserting jacks between the bases of the shores and the footing. Figure 7-15 shows a standard steel screw jack. One of these may apply a lift of as much as 100 tons. When a single screw jack is used with a shore, a hole is bored in the base of the shore to admit the threaded portion of the jack, and the arrangement is called a *pump*. For a larger lifting effect, a pair of jacks are attached to a short timber called a *crosshead*. Pump and crosshead arrangements are illustrated in Fig. 7-16. An advantage of the crosshead arrangement is that after a lift has been applied, the crosshead can be blocked and the jacks removed for use elsewhere.

Hydraulic jacks provide a much stronger lift than screw jacks, but cannot be used to support a load over a length of time. However, with a pair of hydraulic jacks in crosshead arrangement, as many shores as desired can be set up and blocked in a short time and with a minimum of labor.

Needling

Figure 7-17 shows a project which involves a construction procedure known as *underpinning*. The feature marked W is a wall, resting on a footing BB. Excavation is to be carried down to the level indicated by the horizontal dotted line,

Fig. 7-15: Several types of standard steel screw jacks (photo courtesy of Templeton, Kenly & Co.).

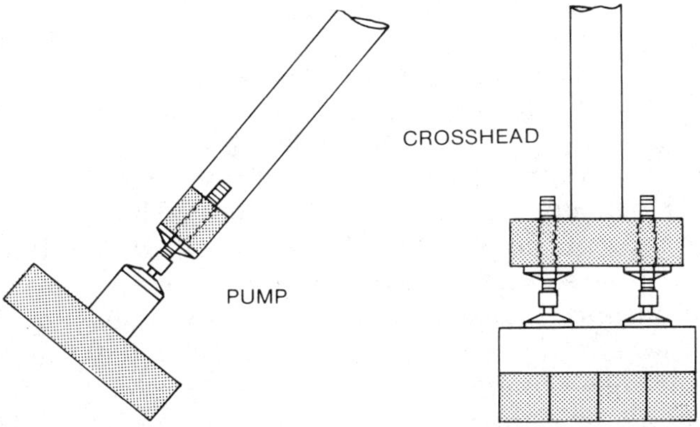

Fig. 7-16: Pump and crosshead.

which means that the earth supporting the footing BB is to be completely removed. The existing wall is to be supported during this procedure, and subsequently a new foundation wall extended upward from the footing shown below. This is the process known as underpinning.

Support for the wall will be provided by a series of needles, that is, of heavy timbers or steel beams, inserted horizontally through holes cut in the wall, and wedged or jacked upward so as to assume the weight of the wall. In Fig. 7-17, the member marked GG is a needle. The needling and underpinning procedure shown in that figure would be carried out as follows.

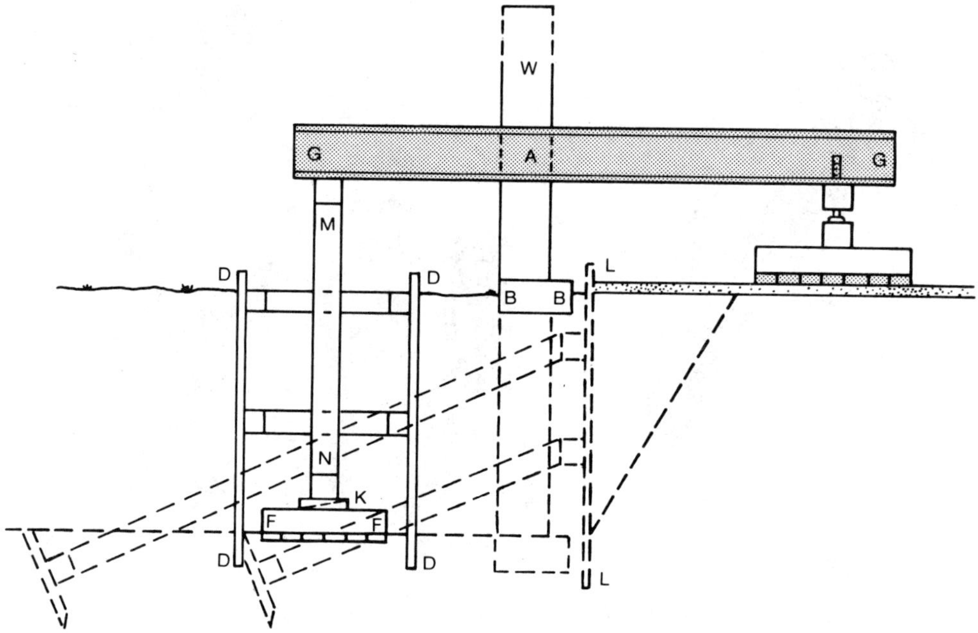

Fig. 7-17: Needling and underpinning.

Before actual underpinning operations are begun, complete preliminary investigations are made, especially with regard to soil bearing capacity. Too much emphasis cannot be placed upon this point. The neglect of adequate study and proper interpretation generally proves very costly. Soil samples should be properly taken, and the results interpreted by trained engineers. The depth and character of the soil should be determined, and it should be ascertained that a suitable stratum is not underlain by softer material.

After the preliminary investigations have been made, a pit DDDD is excavated down to the level of the top of the proposed new footing. At the bottom of this pit a layer of heavy timbers FF is laid on a layer of thick planks. At grade level on the other side of the wall a similar platform is laid. Holes are cut in the wall and the needles are inserted, each supported at the pit end by vertical timber and blocks MN and at the other end by a screw jack. The lift at the pit end is obtained by driving wedges at K, at the other end by the jack.

Before the material under the wall is excavated, sheet piles are driven at LL. As excavation proceeds downward, these are shored up as shown, to prevent a slide caused by the weight of the nearby grade-level platform.

Obviously, only that portion of the wall which is directly above a needle will receive direct support from the needle. Other portions of the wall above the needles will receive indirect support by oblique transfer of the needles' upward thrust through the material; however, some parts of the wall above the needles will receive no support at all.

SHORING AND SAFETY

Burial alive in an excavation slide or cave-in is a terrible accident which occurs in the construction industry with scandalous frequency. Application of the basic excavation safety rule—that any excavation 4′ or more in depth must be protected by one of the prescribed methods—would reduce this frequency almost to nil. There are additional safety rules, most of which have the same basic purpose; that is, to prevent slides or cave-ins. Some of these rules are as follows:

Any trees, boulders, or other surface encumbrances located close enough to create a hazard must be removed before excavation begins.

If it is necessary to bring power shovels, derricks, trucks, large quantities of supplies, or other heavy objects or materials near an excavation, the face of the excavation which is toward the object or material must be additionally shored and braced to resist the additional pressure. No object or material should be placed within 2′ of the edge of an excavation.

When soil against the face of a masonry wall is excavated, it must not be presumed that the wall will, of itself, provide sufficient support against lateral pressure from unexcavated material on the other face. The wall must be adequately shored and braced.

Temporary sheet piling installed to permit construction of a retaining wall must not be removed until the wall has developed full strength—meaning, for a concrete wall, usually 28 days.

Undercutting of earth banks should be done only when it is unavoidable, and then only as the overhang is kept progressively and adequately shored and braced.

Excavations must be inspected after every rainstorm or similar hazard-increasing occurrence, and the protections against slide or cave-in increased if necessary.

No sidewalk may be undermined without being progressively shored to carry a live load of 125 psf. Excavated material must not be piled on sidewalks or walkways.

All timber used for shoring, bracing, sheathing, and sheet piling must be sound, straight-grained timber, of strength and quality equal to that of long-leaf yellow pine or Douglas fir, and free from splits, shakes, large or loose knots, or other strength-impairing defects.

Wooden sheet piling must not be less than 2″ thick for a depth of up to 16′, not less than 3″ for a depth of up to 24′, and not less than 4″ for a depth of up to 40′.

If pedestrian and/or vehicular traffic must be maintained over or near excavations, all proper safeguards, such as bridges, walkways, guardrails, barricades, warning flags, and lights, must be installed.

Entrance to and exit from any excavation over 5′ deep must be by properly constructed ramp, ladder, stairway, or hoist. Jumping into trenches and the use of bracing and shoring members for climbing must be prohibited.

Tools, materials, and debris must not be left on bridges or walkways over excavations, on shoring and bracing members, near the edges of excavations, or in any other position from which they might fall on workers in the excavation below. Pick and shovel personnel working in excavations must keep far enough apart to avoid striking each other accidentally with tools.

8

Scaffolding and Portable Ladders

INTRODUCTION

A great deal of construction and rigging work is done on elevated levels above the ground. To provide temporary working space for such locations, scaffolding and portable ladders can be used.

SCAFFOLDING

Most construction jobs require the use of several kinds of scaffolds to permit easy working procedures. Scaffolding may range from individual planks placed on structural members of the building to involved patent scaffolding (Fig. 8-1). Scaffold planks are placed as a decking over swinging scaffolds, suspended scaffolds, needle beams, and built-up independent scaffolding. Scaffold planks are of various sizes, including 2″ by 8″, 2″ by 10″, and 2″ by 12″. Scaffold planks 3″ thick may be needed for platforms that must hold heavy loads or withstand movements. Planks with holes or splits are not suitable for scaffolding if the diameter of the hole is more than 1″ or the split extends more than 3″ in from the end. Three-inch planks should be used to build the temporary floor used for construction of steel buildings because of the possibility that a heavy steel member might be rested temporarily on the planks. Single scaffold planks may be laid across beams of upper floors (Fig. 8-2) or roofs to form working areas or runways. Each plank should run from beam to beam, with not more than a few inches of any plank projecting beyond the end of the supporting beam. Overhangs are dangerous because workers may step on them and overbalance with the scaffold plank. When the planking is laid continuously, as in a runway, the planks should be laid so that their ends overlap. Single plank runs

Fig. 8-1: Scaffolding plays a very important part in all construction operations (photo courtesy of Patent Scaffolding Co.).

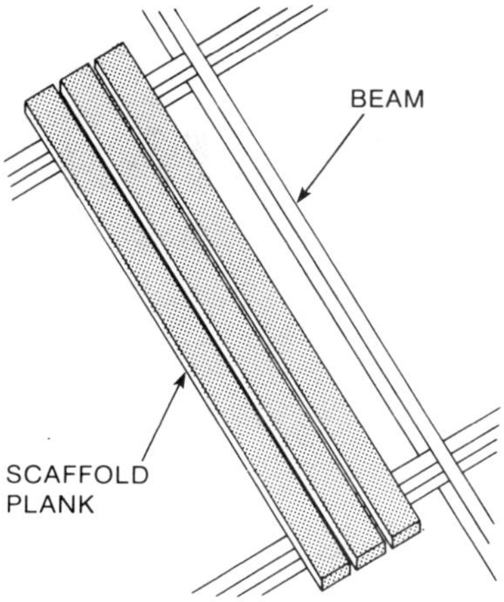

Fig. 8-2: Scaffold planks in place.

can be staggered so that each plank is offset with reference to the next plank in the run. It is advisable to use two layers of planking on large working areas to increase the freedom of movement.

Types of Scaffolding

There are several different types of scaffolds used by the rigger. They are:

Swinging Scaffolds. The swinging type of scaffold must always be secured to the building or structure to prevent it from moving away and causing the people to fall. Where swinging scaffolds are suspended adjacent to each other, planks should never be placed so as to form a bridge between them. The swinging platform scaffold (Fig. 8-3) consists of a frame similar in appearance to a ladder with a decking of wood slats. It is supported near each end by a steel

Fig. 8-3: Swinging platform scaffold (photo courtesy of Patent Scaffolding Co.).

stirrup to which the lower block of a set of wire rope falls is attached. The scaffold is supported by hooks or anchors on the roof of a structure. The fall line of the tackle must be secured to a member of the scaffold when in final position to prevent it from falling.

The swinging scaffold shown in Fig. 8-3 is raised by a pumping action of the lever handles (Fig. 8-4). To ensure that the wire ropes wind smoothly and evenly onto the drums while the scaffold is being raised, all the ratchet pawls and locking pawls must always be engaged during the raising process. The crank handle should never be wound to raise the type of scaffold shown in Figs. 8-3 and 8-4; this action is for lowering the scaffold only.

Figure 8-5 is a typical swinging scaffold safety rules sheet that is usually included with the equipment.

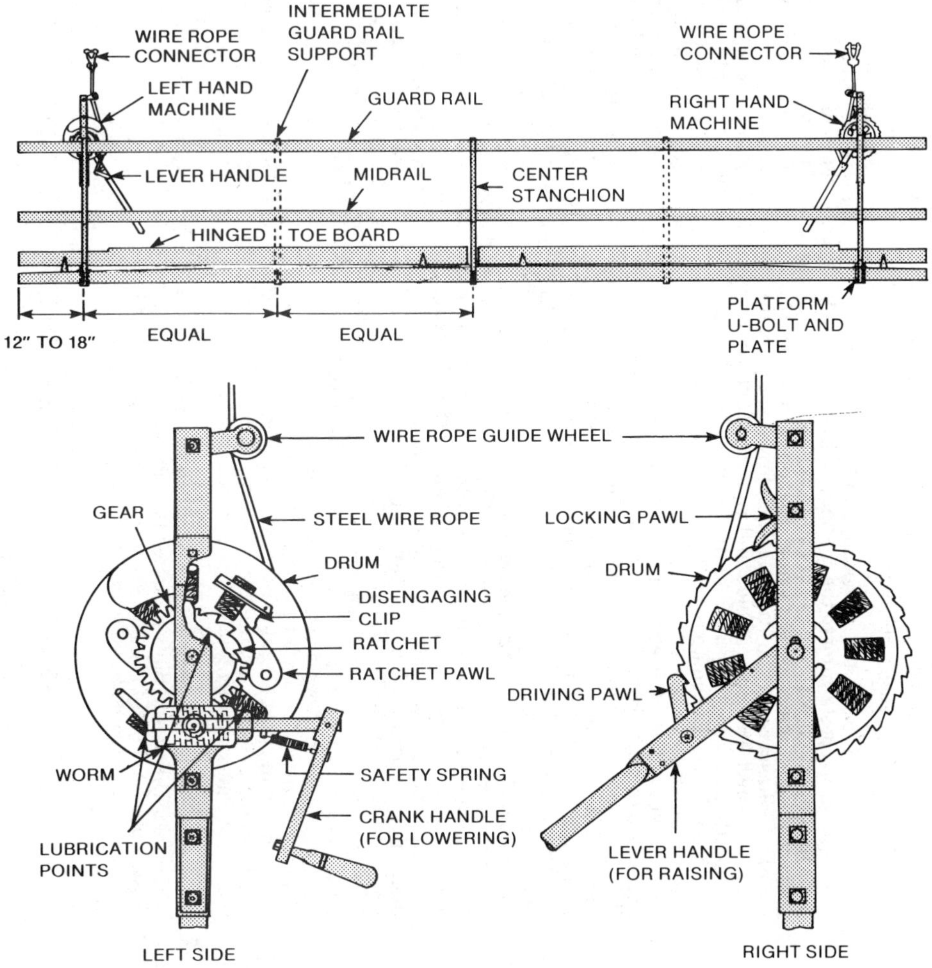

Fig. 8-4: Parts of a swinging platform scaffold.

SAFETY RULES

These rules do not purport to be all-inclusive or to supplant or replace other additional safety rules and precautionary measures. They are not intended to conflict with or supersede any state or local statute or regulation; reference to such specific provisions should be made by the user.

1. Make sure all persons who erect, dismantle, inspect, maintain or use this scaffolding are familiar with these safety rules and instructions.
2. Follow all state, local and governmental codes, ordinances and regulations pertaining to the rigging, lifelines and safety belts in connection with the use of this equipment.
3. Inspect all equipment before using. Inspect wire ropes frequently and assure they are adequately lubricated and maintained. Inspect erected scaffolds and rigging regularly and frequently to assure they are maintained in safe and proper condition. Never use equipment that is damaged or deteriorated in any way.
4. Before installation, assure roof or supporting structure is capable of safely supporting loads to be imposed by the scaffold.
5. Weight of men and materials on scaffold must not exceed five hundred (500) pounds, evenly distributed on the platform.
6. Inspect frequently the ratchet and locking pawl springs; see that they are in good working condition and pawls are properly engaged.
7. Ratchet pawls must be engaged at all times except when unreeling wire ropes before erection of scaffold.
8. Locking pawls must never be wired or otherwise held out of engagement by mechanical means. When scaffold is being lowered, locking pawls must be manually held out of engagement by the operator. Locking pawl must be engaged at all times except when lowering scaffold.
9. Two or more scaffolds shall not be combined into one by lapping platforms on one stirrup, or by bridging between them.
10. Overhang of outriggers must not exceed distance specified. Counterweights at rear of outriggers must never be less than those shown in chart herein. If other sizes or methods of overhead rigging are used, assure they are capable of safely supporting the maximum intended loads with a minimum Safety Factor of 4.
11. Platforms must be securely attached to each stirrup by means of a proper U-bolt and plate designed for this purpose.
12. Assure platform and toe-board are properly installed. Guardrail and mid-rail must be bolted to center guardrail stanchion at all times.
13. Scaffold must be tied at all times into building or structure on which it is hanging, except when it is being raised or lowered into position.
14. Handle all equipment with care, and prevent wire ropes from becoming kinked. Do not drop equipment during handling, loading or unloading from trucks. Impose loads on scaffold platform gently and without impact.
15. Take proper precautions to protect wire ropes when using welding equipment. Wire ropes and other parts of the scaffold are susceptible to serious damage if exposed to fire, undue heat, corrosive atmospheric conditions or to corrosive chemicals including hydrochloric (muriatic) or sulphuric acid, and are not suitable for use under contaminated atmospheric or exposure conditions that will affect their strength. Notify the manufacturer immediately if such exposure occurs or is suspected and discontinue use of the scaffold.
16. If extension wire ropes are used with the equipment, they must be furnished for this purpose by the manufacturer. They must be attached to the drum wire ropes with extension cable connectors supplied by the manufacturer.
17. When overhead supports or rigging methods differ from those shown, connections of wire ropes to the rigging must be made with a proper shackle designed for the purpose or with a separate, adequate wire rope sling properly protected. Never wrap drum wire ropes around overhead rigging or use makeshift devices. Assure that shackle or sling cannot slide off end of rigging beam.
18. All overhead rigging, counterweights, anchorages and connections must be secured and adequately prevented from movement in any direction.
19. Do not alter, remove or substitute parts except as directed by the manufacturer.
20. In case of apparent defects in the machines, wire ropes, platform, etc., which may develop as a result of use or abuse, notify the manufacturer and do not use the equipment until repaired or replaced.

USE OF THIS EQUIPMENT IS UNSAFE AND UNLAWFUL WITHOUT:

(a) Proper guardrails, midrails and toeboards on all open sides of the scaffold platform. It is of utmost importance that you obtain these components from us or otherwise provide for them in the erection and use of this equipment, and
(b) Proper safety belts, lifelines and lanyards being provided by you and used in accordance with the requirements of OSHA.

Fig. 8-5: Typical safety rules for suspended scaffold platforms.

Suspended Scaffolds. Suspended scaffolds, often called cable climbers, usually are supported on outriggers at the roof (Fig. 8-6) and are heavier than swinging scaffolds. From each outrigger, cables lead to hand winches on the scaffold. This type of scaffold can be raised or lowered by operating the hand winches, which must contain a locking device. The scaffold may be made up in

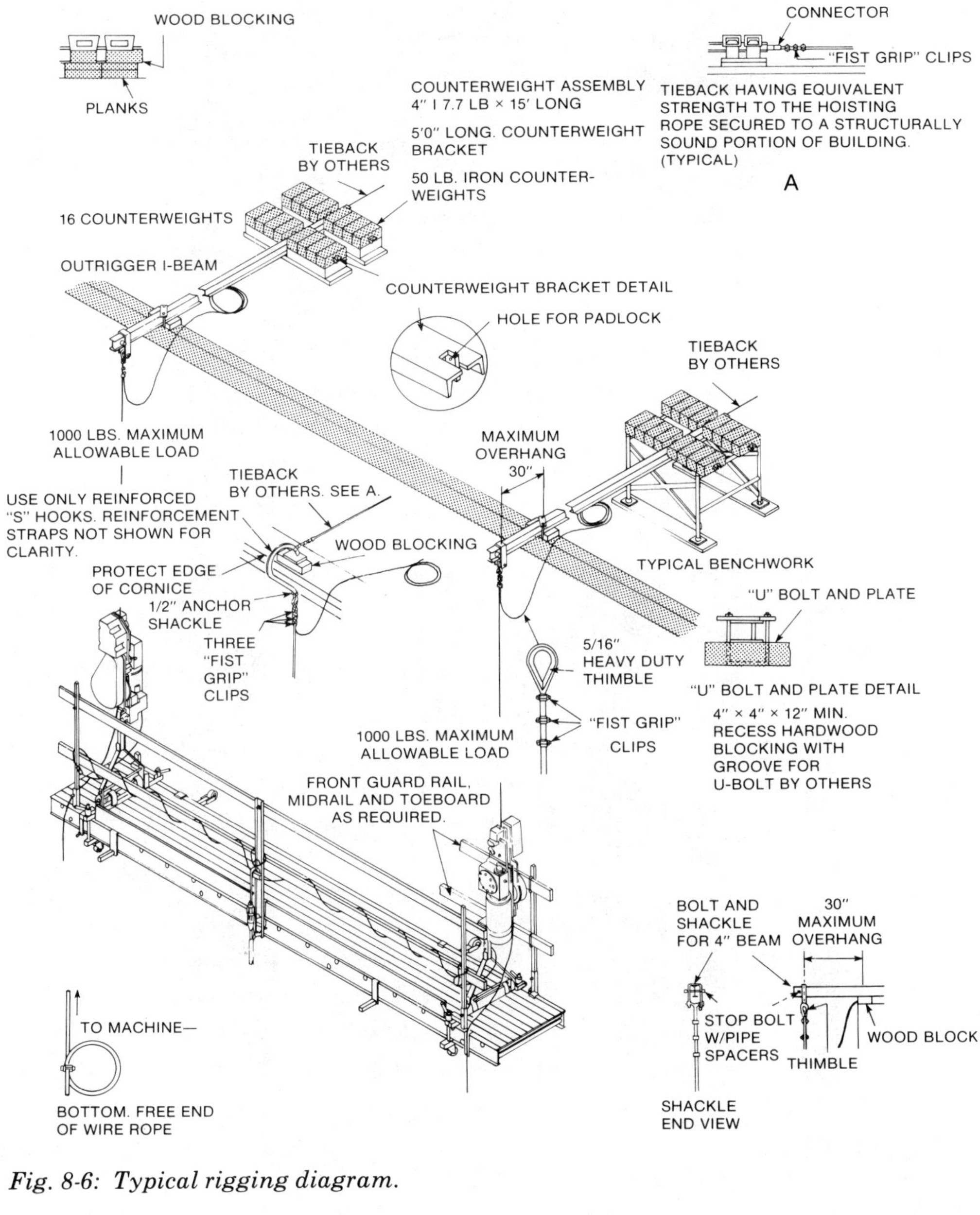

Fig. 8-6: Typical rigging diagram.

almost any width up to about 4′ to over 12′ long (Fig. 8-7), depending on the size of the putlogs, or longitudinal supports, under the scaffold. A light roof may be included on this type of scaffold to protect the workers from falling debris. The suspended scaffold can be driven by either an electric or air powered winch (Fig. 8-8).

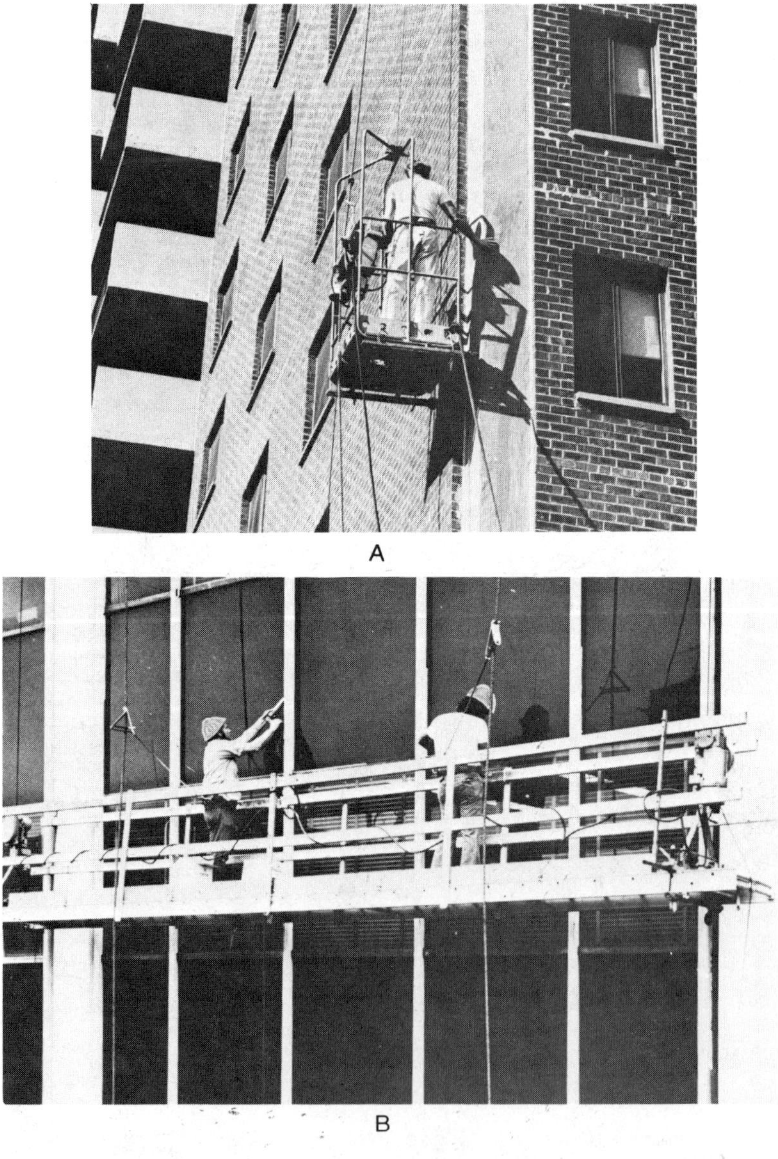

Fig. 8-7: (A) Typical single power scaffold platform and (B) large multi-men scaffold platform (photos courtesy of Patent Scaffolding Co.).

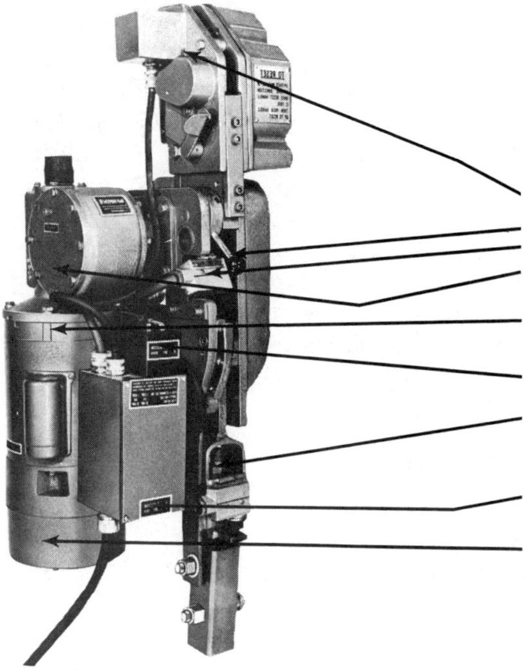

A

B

Fig. 8-8: (A) Typical electric-powered and (B) air-powered motors used with suspended scaffolds (photos courtesy of Patent Scaffolding Co.).

Needle Beam Scaffold. This type of scaffold is used only for temporary jobs. No material should be stored on this scaffold. In needle beam scaffolding, two 4″ by 6″ or similar size timbers are suspended by ropes. A decking of 2″ scaffold plank is placed across the needle beams, which should be placed about 10′ apart. Needle beam scaffolding (Fig. 8-9) is used frequently by welding gangs working on steel structures because of the frequent changes of location necessary and its adaptability to different situations. A scaffold hitch is used in the rope supporting the needle beams to prevent them from rolling or turning over. The hanging lines are usually of 1-1/4″ manila rope. The rope is hitched to the needle beam, carried up over a structural beam or other support, and then down again under the needle beam so the latter has a complete loop of rope under it. The rope is then passed over the support again and fastened around itself by two half hitches.

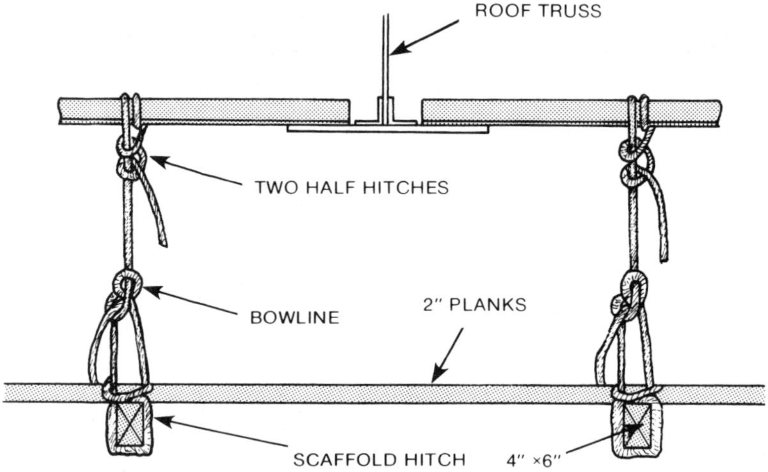

Fig. 8-9: Needle beam scaffold.

Double Pole Built-Up Scaffold. The steel double pole built-up scaffold, sometimes called the *independent scaffold,* is completely independent of the main structure. Several types of patent independent scaffolding are available for simple and rapid erection (Fig. 8-10). The scaffolding can be built up from wood members if necessary. The scaffold uprights are braced with diagonal members and the working level is covered with a platform of planks. All bracing must form triangles and the base of each column requires adequate footing plates for the bearing area on the ground. The patented steel scaffolding is usually erected by placing the two uprights on the ground and inserting the diagonal members. The diagonal members have end fittings which permit rapid locking in position. The first tier is set on steel bases on the ground. A second tier is placed in the same manner on the first tier, with the bottom of each upright locked to the top of the lower tier. A third and fourth upright can

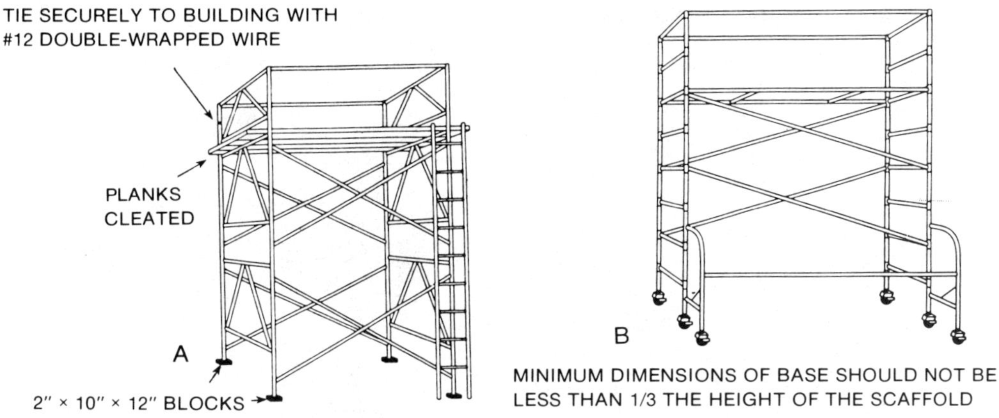

Fig. 8-10: Two types of double pole built-up scaffolds: (A) Set and (B) rolling.

be placed on the ground level and locked to the first set with diagonal bracing. The scaffolding can be built as high as desired, but high scaffolding should be tied in to the main structure.

Figure 8-11 shows the use of tubular steel clamp-type scaffolding. Clamp tubular scaffolding is found on many construction jobs because its design permits it to be shaped to any contour (Fig. 8-12) and it goes up rapidly. The safety rules given in Fig. 8-13 are recommended by the Scaffolding, Shoring and Forming Institute when installing and using steel clamp-type scaffolding.

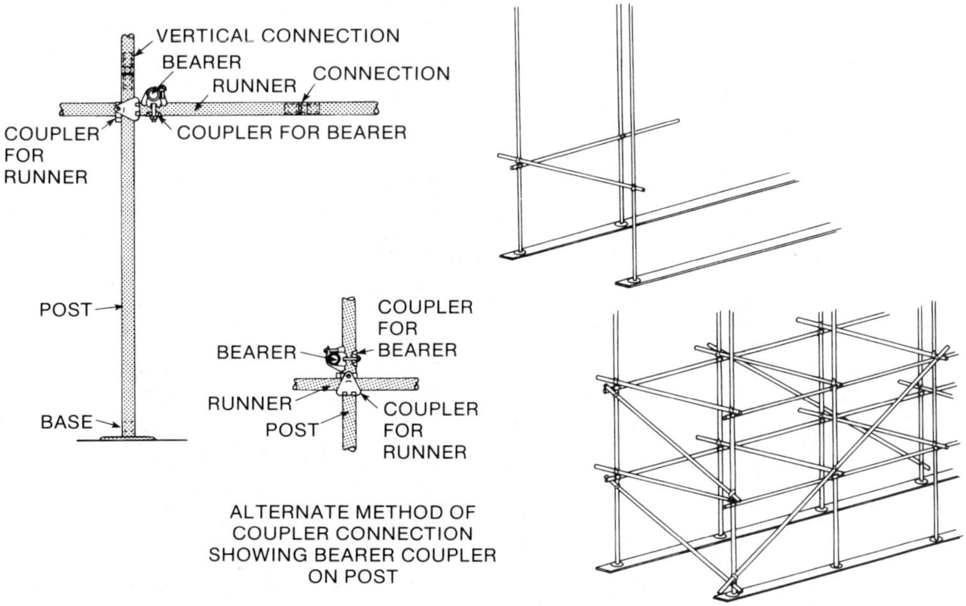

Fig. 8-11: Erecting tubular steel clamp-type scaffolding.

Fig. 8-12: Tubular steel clamp-type scaffolding is designed to permit shaping to almost any contour (photos courtesy of Patent Scaffolding Co.).

SCAFFOLDING SAFETY RULES
SCAFFOLDING SHORING AND FORMING INSTITUTE, INC.

Following are some common sense rules designed to promote safety in the use of steel scaffolding. These rules are illustrative and suggestive only, and are intended to deal only with some of the many practices and conditions encountered in the use of scaffolding. The rules do not purport to be all-inclusive or to supplant or replace other additional safety and precautionary measures to cover usual or unusual conditions. They are not intended to conflict with, or supersede, any state, local or federal statute or regulation; reference to such specific provisions should be made by the user. (See Rule II).

I. **POST THESE SCAFFOLDING SAFETY RULES** in a conspicuous place and be sure that all persons who erect, dismantle or use scaffolding are aware of them.

II. **FOLLOW ALL STATE, LOCAL AND FEDERAL CODES, ORDINANCES AND REGULATIONS** pertaining to scaffolding.

III. **INSPECT ALL EQUIPMENT BEFORE USING**—Never use any equipment that is damaged or deteriorated in any way.

IV. **KEEP ALL EQUIPMENT IN GOOD REPAIR.** Avoid using rusted equipment—the strength of rusted equipment is not known.

V. **INSPECT ERECTED SCAFFOLDS REGULARLY** to be sure that they are maintained in safe condition.

VI. **CONSULT YOUR SCAFFOLDING SUPPLIER WHEN IN DOUBT**—Scaffolding is his business, **NEVER TAKE CHANCES.**

A. **PROVIDE ADEQUATE SILLS** for scaffold posts and use base plates.

B. **USE ADJUSTING SCREWS** instead of blocking to adjust to uneven grade conditions.

C. **PLUMB AND LEVEL ALL SCAFFOLDS** as the erection proceeds. Do not force braces to fit—level the scaffold until proper fit can be made easily.

D. **FASTEN ALL BRACES SECURELY.**

E. **DO NOT CLIMB CROSS BRACES.** An access (climbing) ladder, access steps, frame designed to be climbed or equivalent safe access to the scaffold shall be used.

F. **ON WALL SCAFFOLDS PLACE AND MAINTAIN ANCHORS** securely between structure and scaffold at least every 30′ of length and 25′ of height.

G. **WHEN SCAFFOLDS ARE TO BE PARTIALLY OR FULLY ENCLOSED,** specific precautions must be taken to assure frequency and adequacy of ties attaching the scaffolding to the building due to increased load conditions resulting from effects of wind and weather. The scaffolding components to which the ties are attached must also be checked for additional loads.

H. **FREE STANDING SCAFFOLD TOWERS MUST BE RESTRAINED FROM TIPPING** by guying or other means.

I. **EQUIP ALL PLANKED OR STAGED AREAS** with proper guardrails, midrails and toeboards along all open sides and ends of scaffold platforms.

J. **POWER LINES NEAR SCAFFOLDS** are dangerous—use caution and consult the power service company for advice.

K. **DO NOT USE** ladders or makeshift devices on top of scaffolds to increase the height.

L. **DO NOT OVERLOAD SCAFFOLDS.**

M. **PLANKING:**
   1. Use only lumber that is properly inspected and graded as scaffold plank.
   2. Planking shall have at least 12″ of overlap and extend 6″ beyond center of support, or be cleated at both ends to prevent sliding off supports.
   3. Fabricated scaffold planks and platforms unless cleated or restrained by hooks shall extend over their end supports not less than 6 inches nor more than 12 inches.
   4. Secure plank to scaffold when necessary.

N. **FOR ROLLING SCAFFOLD THE FOLLOWING ADDITIONAL RULES APPLY:**
   1. **DO NOT RIDE ROLLING SCAFFOLDS.**
   2. **SECURE OR REMOVE ALL MATERIAL AND EQUIPMENT** from platform before moving scaffold.
   3. **CASTER BRAKES MUST BE APPLIED** at all times when scaffolds are not being moved.
   4. **CASTERS WITH PLAIN STEMS** shall be attached to the panel or adjustment screw by pins or other suitable means.
   5. **DO NOT ATTEMPT TO MOVE A ROLLING SCAFFOLD WITHOUT SUFFICIENT HELP**—watch out for holes in floor and overhead obstructions.
   6. **DO NOT EXTEND ADJUSTING SCREWS ON ROLLING SCAFFOLDS MORE THAN 12″.**
   7. **USE HORIZONTAL DIAGONAL BRACING** near the bottom and at 20′ intervals measured from the rolling surface.
   8. **DO NOT USE BRACKETS ON ROLLING SCAFFOLDS** without consideration of overturning effect.
   9. **THE WORKING PLATFORM HEIGHT OF A ROLLING SCAFFOLD** must not exceed four times the smallest base dimension unless guyed or otherwise stabilized.

O. For **"PUTLOGS"** and **"TRUSSES"** the following additional rules apply:
   1. **DO NOT CANTILEVER OR EXTEND PUTLOGS/TRUSSES** as side brackets without thorough consideration for loads to be applied.
   2. **PUTLOGS/TRUSSES SHOULD EXTEND AT LEAST** 6″ beyond point of support.
   3. **PLACE PROPER BRACING BETWEEN PUTLOGS/TRUSSES** when the span of putlog/truss is more than 12′.

P. **ALL BRACKETS** shall be seated correctly with side brackets parallel to the frames and end brackets at 90 degrees to the frames. Brackets shall not be bent or twisted from normal position. Brackets (except mobile brackets designed to carry materials) are to be used as work platforms only and shall not be used for storage of material or equipment.

Q. **ALL SCAFFOLDING ACCESSORIES** shall be used and installed in accordance with the manufacturers recommended procedure. Accessories shall not be altered in the field. Scaffolds, frames and their components, manufactured by different companies shall not be intermixed.

Fig. 8-13: Typical scaffold safety rules for use of steel scaffolding.

Tank Scaffold. As shown in Fig. 8-14, tank scaffold can be used for various types of installation jobs. Figure 8-15 shows how tank scaffolding is set up. The scaffold clip is first welded to the plate, then the rest of the scaffolding is erected into place. Knee braces or brackets should have a maximum spacing of 8′. All scaffold boards should be securely attached to the braces with boomers or other approved methods and have a minimum overhang of 1′ 6″ on each brace. Scaffold planks should overlap in one direction. This overlap should be a minimum of 9″. Keep scaffold braces or brackets plumb. If a brace is installed only 5″ out of plumb, it will support only half its load. Once the tank scaffold has served its purpose, it can easily be taken down and scaffold clips removed from the steel plate.

Fig. 8-14: Examples of tank scaffolding (photos courtesy of CBI Industries, Inc.).

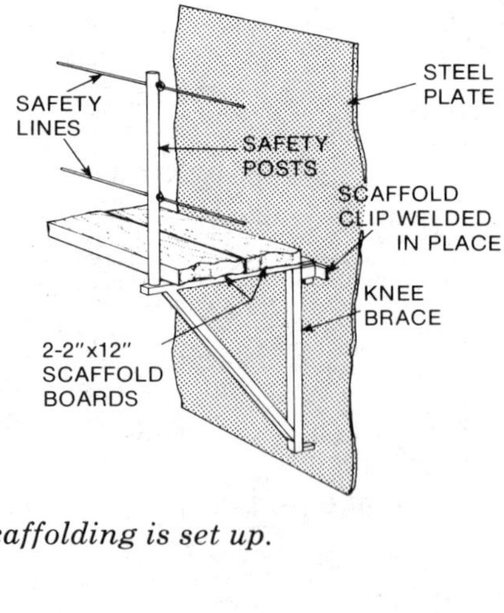

Fig. 8-15: How tank scaffolding is set up.

On any given construction job, specially designed scaffold arrangements may be needed (Fig. 8-16). But, regardless of the scaffold design or type, the lumber should be inspected before and after use in the field. The ends of each scaffold plank should be banded or bolted with S-wedges used to avoid splitting.

Fig. 8-16: Examples of special scaffolding arrangements (photos courtesy of CBI Industries, Inc.).

General Scaffolding Safety

Earlier in the chapter specific safety rules for swinging and steel scaffolding were given (Fig. 8-5). The following scaffolding safety precautions must be observed by all people working on scaffolds or tending other people who are working on scaffolds.

Standard scaffolds suitable to the work at hand must be provided and used. The use of makeshift substitutes is prohibited.

All scaffolds must be maintained in a safe condition, and a scaffold must not be altered or disturbed while in use. Personnel must not be allowed to use damaged or weakened scaffolds.

Structural members, supporting lines and tackles, and other scaffold equipment must be inspected daily before work on scaffolding is started.

When personnel working on a scaffold are directly below others working above, the personnel below must be sheltered against possible falling objects by a protective covering. Those below must wear protective headgear.

If the frequent presence of personnel directly under a scaffold is unavoidable, a protective covering must be set up under the scaffold. A passageway or thoroughfare under a scaffold must have both overhead and side protection.

Access to scaffolds must be by standard stairs or fixed ladders only.

The erection, alteration, and dismantling of scaffolds must be done under the supervision of personnel who are experienced in scaffold work. When scaffolding is being dismantled, it should be cleaned and made ready for storage or use. Scaffolding that is not ready for use should never be stored.

Work on scaffolds should be stopped during storms or high winds or when scaffolds are covered with ice or snow.

Unstable objects, such as barrels, boxes, loose brick, or building blocks, must not be used to support scaffold planking.

No scaffold may be used for the storage of materials in excess of those currently required for the job.

Tools not in immediate use on scaffolds must be stowed in containers to prevent tools left adrift from being knocked off. Tool containers must be lashed or otherwise secured to the scaffolds.

Scaffolds must be kept clear of accumulations of tools, equipment, materials, and rubbish. If part of a scaffold must be used as a loading or landing stage for materials, the scaffold must be additionally braced and reinforced at and around the landing stage area.

Throwing objects to or dropping them from scaffolds is absolutely prohibited. Hand lines must be used for raising or lowering objects which cannot be passed hand-to-hand.

A standard guardrail and toeboard should be provided on the open side of the platform on all single-pole and independent-pole scaffolds.

If the space between the scaffold and building is more than 18″, a standard guardrail should be erected on the building side.

No person should remain on the rolling scaffold while it is being moved.

When a light-duty portable scaffold is formed of planks that are supported or hitched on trestle ladders, the base of the ladder should be secured against opening up to the full spread before laying on the planks.

A scaffold must *never* be overloaded. Scaffolds are built in the following strength categories: (a) extra heavy duty, (b) heavy duty, (c) an intermediate category between light and heavy, and (d) light duty. The maximum uniform safe working load per square foot of platform for each of these categories is as follows:

Extra heavy duty—75 pounds.
Heavy duty—50 pounds.
Intermediate—35 pounds.
Light duty—25 pounds.

To get the load per square foot of platform of a pile of materials on a platform, divide the total weight of the pile by the number of square feet of platform it covers. For strength of plank, see Table 8-1.

Table 8-1: Strength of Plank

Size in Inches	Span in Feet 4	6	8	10	12	14	16	18
	Safe Load in Pounds							
2 × 8	400	260	190					
2 × 10	500	325	235	180				
2 × 12	600	390	285	220	165			
3 × 10	1245	820	600	465	370	305		
3 × 12	1495	980	720	555	445	370	300	
3 × 14	1740	1145	840	650	520	430	350	290

The loads given are in pounds concentrated at the center of the span.
The above loads are for fir or spruce planks in first class condition. For yellow pine planks in first class condition, add 10% to the above allowable loads.
Safe loads given are based on planks surfaced 1/4″ under sizes shown.

Boatswain's Chair

The boatswain's chair can be made up in several forms, but it generally consists of a sling for supporting one person.

Rope Chair. A rope boatswain's chair (Fig. 8-17A) can be made up by using a double bowline and a rolling hitch. One person can operate the rope seat to lower himself/herself by releasing the grip of the rolling hitch. A slight twist with the hand on the hitch permits the suspension line to slip through it, but when the hand pressure on the hitch is released, the hitch will hold firmly.

Rope Chair with Seat. If the rope boatswain's chair must be used to support a person at work for some time, the rope may cause considerable discomfort. A notched board (Fig. 8-17B) inserted through the two leg loops will provide a comfortable seat. The loop formed as the running end to make the double bowline will still provide a back support, and the rolling hitch can still be used to lower the boatswain's chair.

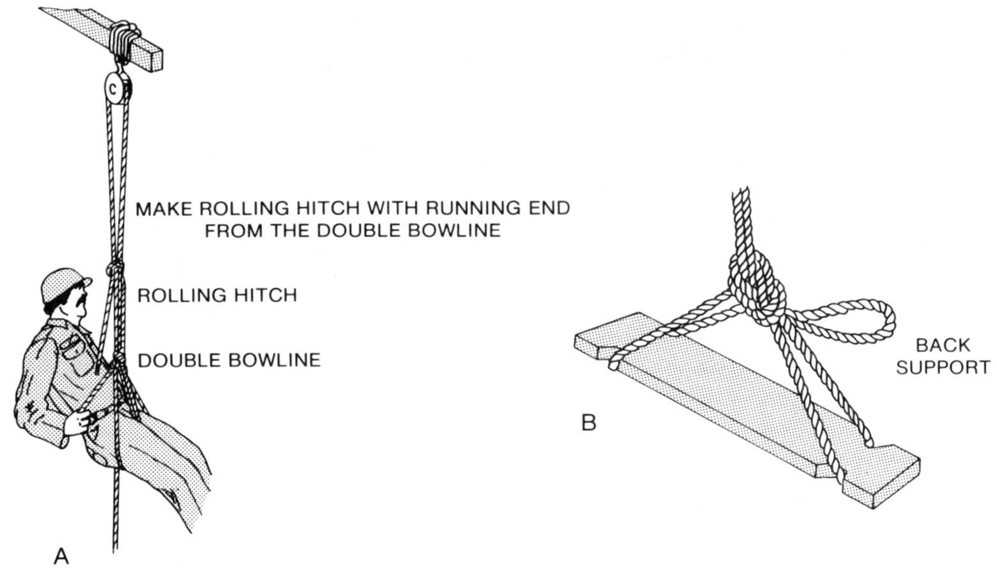

Fig. 8-17: (A) Boatswain's chair and (B) boatswain's chair with seat.

Boatswain's Chair with Tackle. The boatswain's chair is supported by a four part rope tackle (Fig. 8-18A), two double blocks. One person can raise or lower himself/herself or be assisted by a person on the ground. When working alone, the fall line is attached to the lines between the seat and the traveling block with a rolling hitch. As a safety precaution, a figure eight knot should be tied after the rolling hitch to prevent accidental untying. Figure 8-18B shows a typical powered boatswain's chair.

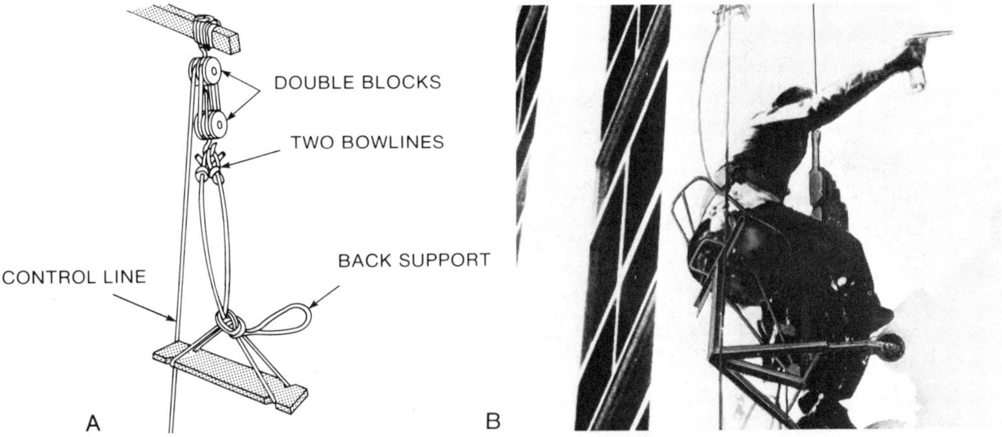

Fig. 8-18: (A) Boatswain's chair with tackle and (B) powered boatswain's chair (photo courtesy of Patent Scaffolding Co.).

LADDERS

Several types of ladders are available for construction work, including extension ladders, pushup ladders, and straight ladders (Fig. 8-19). All three types of ladders are available in both metal and wood. Ladders should always be inspected before they are used. A ladder with parts missing, with bent or cracked sides or rungs, and those made with faulty material should be condemned. Badly worn and weathered ladders and wooden ladders with rotten spots should not be used because they are subject to breaking and can cause a serious accident. Ladders with rough spots, such as protruding metal fastenings, screws, and nails, should be repaired or reconstructed to prevent injuries.

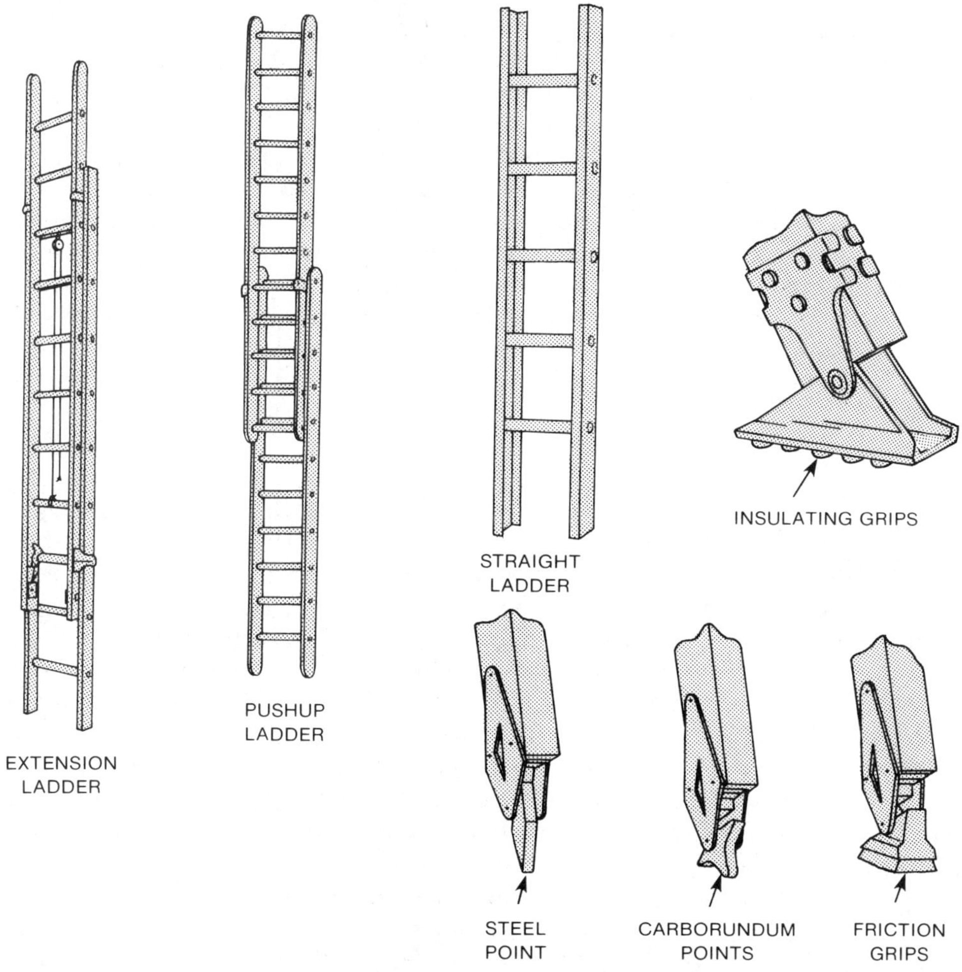

Fig. 8-19: Types of ladders and typical nonslip bases.

Extension Ladders

Occasionally sections of an extension ladder are used separately. When this is done, the upper section of the ladder must be used upside down so that the rung missing at the locks will be at the top of the ladder where it is less liable to cause an accident. In selecting an extension ladder for a particular job, it should be remembered that this type of ladder is designated by its nominal length, which is the sum of the lengths of the sections. The usable length of the ladder is 3′ to 10′ less than the nominal length due to the overlap of the sections. This overlap is 3′ on ladders up to and including 36′, 4′ on 40′ to 44′ ladders, and 5′ to 10′ on longer ones. Extension ladders and pushup ladders are placed against the wall in much the same manner with the extension, or pushup, portion lowered. When the lower section of the ladder is nearly in the final position, the upper portion is pulled out away from the building until it stands nearly vertical but leans slightly toward the building. While the ladder is held in this position, one person hauls down on the rope fastened to the extension section, pulling it upward. No attempt should be made to raise the extension section to its full extension on the first pull. It is less difficult to pull the section up in easy stages, checking the height of the ladder at intervals in order to determine the correct height. The extension section should be on the side of the ladder toward the building to lessen the danger of the user slipping at the point where the two sections join. The first time a ladder is climbed, a worker on the ground should hold the bottom, while the person climbing the ladder lashes it to the building or structure at the top.

Straight Ladders

Straight ladders are placed on the ground beside the wall which they are intended to scale, and at right angles to the wall. One person should stand at the foot of the ladder to prevent the ladder from kicking backward. A second person (or persons) grasps the ladder partway toward the top and raises it from the ground. As the ladder is raised it is "walked" toward the building, and the people keep moving toward the foot of the ladder to grasp new holds. When the ladder is in final position, the bottom of the ladder should be checked to make certain it has a firm footing. If the ground is soft or if the ladder does not rest squarely on both bottom legs, a board may be placed under the feet. If the bottom of the ladder has antislip surfaces, these surfaces should be checked before placing the ladder to be certain they are not coated with mud or debris. The ladder should be placed at a safe angle against the wall. A good rule is to place the base of the ladder about one-fourth as far out from the upper support as the length of the ladder (Fig. 8-20). The upper end of the ladder should not extend more than 2′ above the upper support, and not so far below the working area to

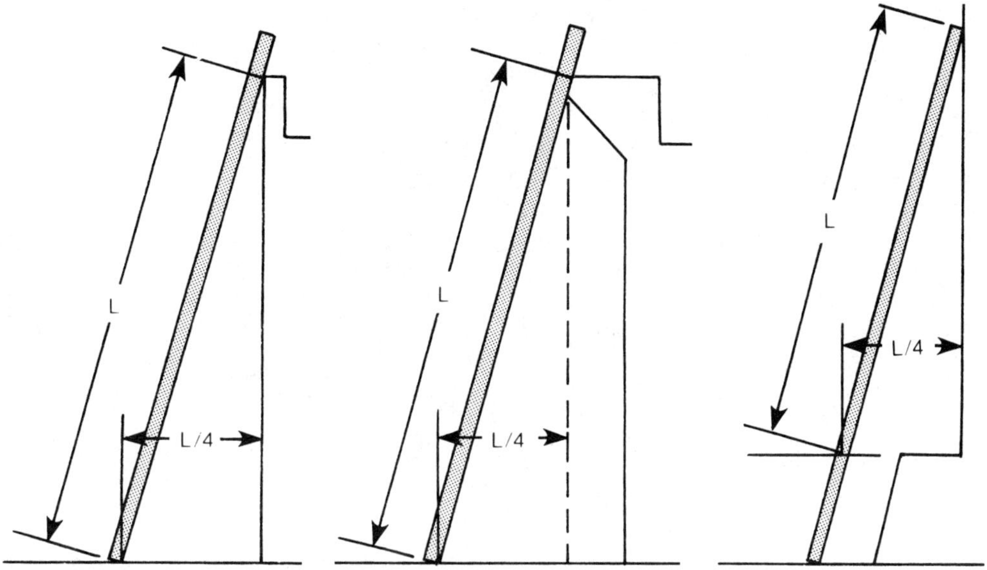

Fig. 8-20: Correct angles for ladders.

be dangerous to move from the top of the ladder to the wall. The upper end of a ladder should always be lashed to the structure with wire or fiber rope to prevent it from skidding sideways or overturning while in use.

Other Ladders

There are times when a step ladder may be used (Fig. 8-21). This type of ladder must bear evenly on all four legs on a plane surface or floor. It should have the truss rods under the treads or the step brackets in condition and drawn up tight. The diagonal bracing and the dowels on the rear section should be in good shape and not deformed by people using the rear of the ladder to climb up. The spreader bars should operate properly and should not present unnecessary finger hazard. Rope spreaders should not be used. The auxiliary step at the top of some platform ladders should be carefully examined, as the supporting members are usually very lightly constructed and are easily bent out of shape.

Trestle and extension trestle ladders (Fig. 8-22) are also used in some construction rigging work. Such ladders should not exceed 20′ in length for the base section and 20′ for the extension. A locking device must be used to hold any trestle ladder in an open position. Any trestle ladder over 12′ in length also requires a locking device to hold it in an elevated position. As with all portable ladders, trestle and extension trestle ladders must meet the minimum standards of the American National Standards Institute (ANSI).

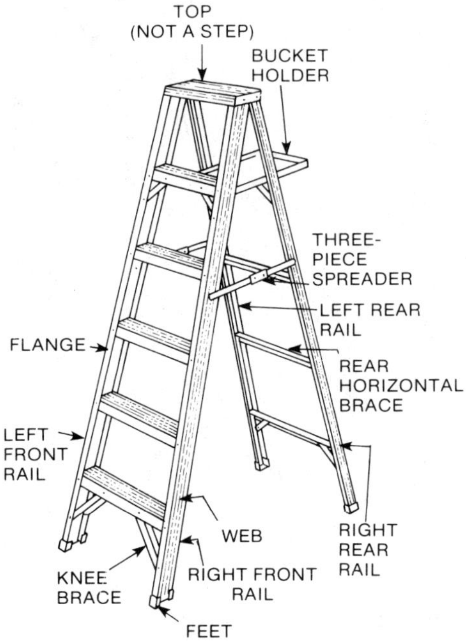

Fig. 8-21: Typical step ladder.

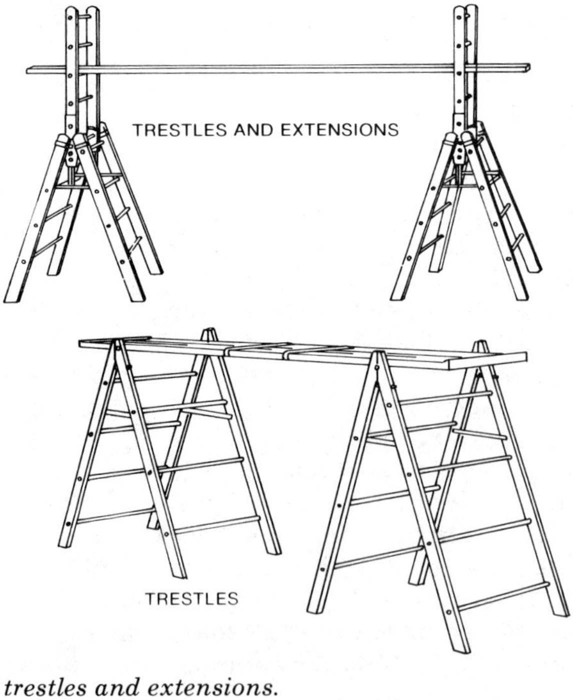

Fig. 8-22: Typical trestles and extensions.

Ladder Safety

Here are a few important safety precautions which apply to ladders in general. For example, ladders should be inspected at regular and frequent intervals. Ladders with weakened, broken, or missing treads, rungs, or cleats, or broken or "splintered" side rails should not be used.

Ladders should be kept coated with a clear shellac or other transparent material or treated with linseed oil. Painting with opaque (non-transparent) paint is forbidden.

Separate ladders for ascending and descending should be provided in a building construction of more than two stories in height or where traffic is heavy.

Where a ladder is installed wide enough to permit traffic in both directions at the same time, a center rail should be provided. One side of the ladder should be plainly marked "up" and the other side "down."

Ladders used in passageways, driveways, or thoroughfares should be guarded by barricades (guardrails). Doors which open adjacent to portable ladders should be locked or otherwise blocked or guarded while the ladder remains in use.

Ladders should be placed so that the rails have a secure footing and a substantial support at or near the top.

Ladders should not be placed against sash, window panes, or unstable supports such as loose boxes or barrels. The use of ladders during a storm or in a high wind should be avoided unless absolutely necessary, in which case the ladder should be securely lashed in position.

If a ladder is to be placed against a window frame, a board should first be spiked across the side rails at the top.

Ladders should not be placed or used in elevator shafts or hoistways. Should such a procedure be necessary, the ladders should be protected from objects from operation at higher elevations in or adjoining the shaft.

Ladders should not be left standing, especially on the outside, for long periods of time unless securely anchored at both top and bottom.

Ladders should be handled carefully when being lowered. They should not be allowed to drop on their sides or to fall heavily endwise on one rail.

Ladders constructed of metal should not be used near electricity. Metal ladders should not be used within 4′ of any electrical wiring or equipment.

When ascending or descending a ladder, you should always face the ladder.

No one should go up or down a ladder without the free use of both hands. If handling material, a rope should be used.

No one should run up or down a ladder or slide down a ladder at any time.

Before attempting to climb a ladder, you should remove oil or grease from the soles of your shoes.

When doing maintenance work above the ground, you should always use a safety belt, with a lifeline tied long enough for necessary movement, but short enough to prevent falls.

Single portable ladders over 30′ (9.1 m) in length should not be used.

Fixed ladders should be securely held in place by top, bottom, and intermediate fastenings as required.

Sloping ladders which require climbing on the under side of the ladder should not be used.

Rails of ladders fixed to top landings should extend a distance of at least 36″ (.9 m) above the landing. Rungs above the landing should be omitted when it is necessary to pass through the ladder. Landing platforms should be provided where a person must step a greater distance than 14″ (355.6 mm) from ladder to roof, tank, etc.

9

Cranes

INTRODUCTION

As the lifting needs of industry and the construction trade have increased and diversified over the years, engineers have responded by designing new, more versatile cranes to meet these needs. From giant cranes capable of lifting loads weighing hundreds of tons hundreds of feet into the air and placing them within precise tolerances, to smaller mobile cranes and stationary cranes used in manufacturing, these lifting devices have become indispensable tools in modern industrial and construction procedures.

Although cranes can differ greatly in their size and overall configuration, most operate according to the same set of general lifting principles. In this chapter, the various types of modern cranes, such as overhead cranes, tower cranes, and mobile cranes, are discussed along with their common applications.

OVERHEAD CRANES

As its name implies, an overhead crane is a lifting device which rides on an overhead runway or a pair of tracks suspended above the ground or work floor. The components which make up the overhead crane consist of a bridge that spans between the tracks and a fixed or trolley mounted hoist system (Fig. 9-1). The bridge is supported at both ends by rollers or wheels that travel along the overhead runway. The hoisting system is contained in the bridge and is either fixed or able to move along the bridge through use of its own system of rollers.

The general types of overhead cranes are classified as top running overhead cranes and gantry cranes, which include polar, semi-gantry, cantilever gantry, wall cranes, and several other crane designs having the same basic character-

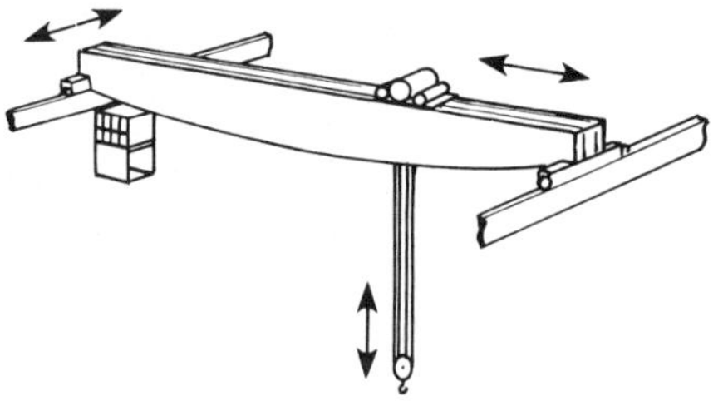

Fig. 9-1: An overhead crane.

istics. Factors which influence overhead crane design include the type of load and load weight being handled and the distance and height that loads must be lifted. Two design factors common to all overhead cranes are the use of top running trolleys and similar travel characteristics.

Overhead cranes are available in a wide variety of sizes and types designed to fit the specific needs of the user. Larger cranes, used for heavy industry and large construction jobs, are usually operated from a cab mounted at one end of the bridge (Fig. 9-2). The sizes of the loads handled by these cranes sometimes require the operator to be in radio or telephone contact with a signalperson who guides the load via either audible or visual commands.

Fig. 9-2: A large overhead crane operating in a steel warehouse (photo courtesy of Shepard Niles Crane and Hoist Corp.).

Smaller overhead cranes are sometimes operated from the work floor. They are controlled by means of a pushbutton controller which is suspended from the bridge down to the operator. The controller is arranged so that when the buttons are released, the crane ceases operation. Figure 9-3 illustrates two arrangements of this type of pushbutton operator's station. While most often employed in conjunction with overhead cranes, control units of this type can be adapted for use with other types of cranes as well. Underhung and monorail cranes are two primary examples.

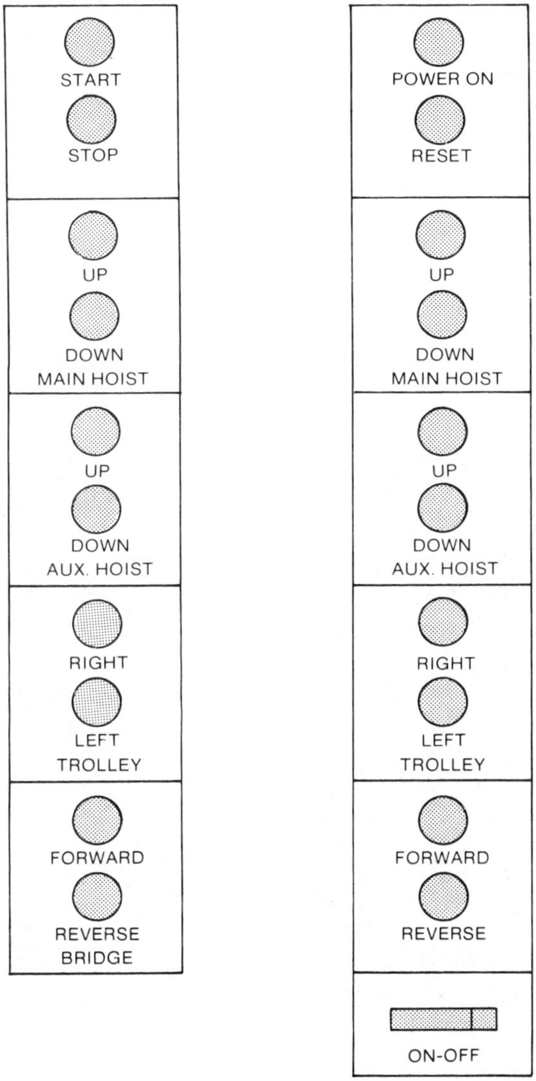

Fig. 9-3: Two arrangements of a pendant pushbutton operator's station.

Overhead cranes are generally permanent or semipermanent installations. The operating area is restricted to that area beneath the bridge and between the runways. The capacities of these cranes range from a minimum of 5 tons to in excess of 500 tons. Installations such as these are common to central station power plants, manufacturing installations, storage facilities, and plant maintenance areas where periodic heavy lifting is required. Permanent overhead cranes are usually installed during the early phases of construction. The cranes can then be used in the later stages of construction for the installation of heavy equipment.

Overhead Traveling Cranes

Overhead cranes which have the capacity to travel along fixed overhead rails are generally used for indoor work. The rails are usually built into the building's structure to provide a stable runway for the crane (Fig. 9-4). These runways can extend beyond the building's perimeter in order to bring in materials which have been stored outdoors, as in the case of the crane shown in Fig. 9-5.

Overhead traveling cranes are designed to perform specialized tasks such as stacking materials in warehouses where high density storage requirements limit the size of the equipment (Fig. 9-6) and handling large power plant components (Fig. 9-7).

Fig. 9-4: An overhead traveling crane (photo courtesy of Shepard Niles Crane and Hoist Corp.).

Fig. 9-5: Overhead crane runways extending beyond the building perimeters (photo courtesy of Shepard Niles Crane and Hoist Corp.).

Gantry Cranes

A gantry crane is similar to an overhead traveling crane due to the fact that it has a bridge for carrying the trolley system. However, the bridge frame is rigidly supported on two or more legs which run on fixed rails or other types of runway (Fig. 9-8). The operator's cab can be located at one end of the crane beneath the bridge, in between and above the bridge beams, or in an area not attached to the crane but with full view of its operation. Figure 9-9 shows a 200 ton gantry crane located on a turbine deck. The purpose of this particular gantry crane is to lift large turbines out of their housings for repairs. The crane was

Fig. 9-6: An overhead stacking crane (photo courtesy of Shepard Niles Crane and Hoist Corp.).

Fig. 9-7: The heavy lift capabilities of overhead cranes (photo courtesy of Pennsylvania Power and Light Co.).

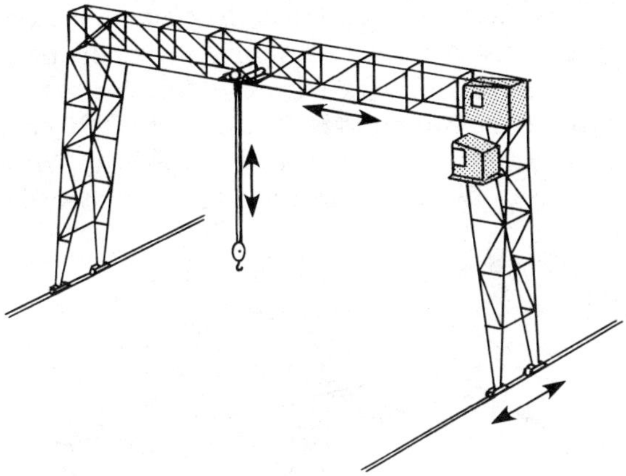

Fig. 9-8: A gantry crane.

also used for the installation of other power plant components throughout the various phases of construction.

The Semi-Gantry Crane. In many ways a semi-gantry crane resembles a gantry crane. One end of the bridge is supported by one or more legs; however, the opposite end of the bridge is supported by a truck which runs on an elevated runway or fixed rail. This truck is merely a device with rollers on which the

Fig. 9-9: A 200 ton gantry crane in operation (photo courtesy of Whiting Corp.).

bridge rests (Fig. 9-10). It is used for traveling purposes and resembles the rollers employed by overhead cranes. The semi-gantry crane is especially suited to move materials in a specific area where a top running overhead or gantry crane causes too much wasted space. Bulk materials can be stacked against the wall at the end of the trolley without hampering bridge travel or covering the track. Figure 9-11 shows a 10 ton semi-gantry crane operating in a metal center warehouse.

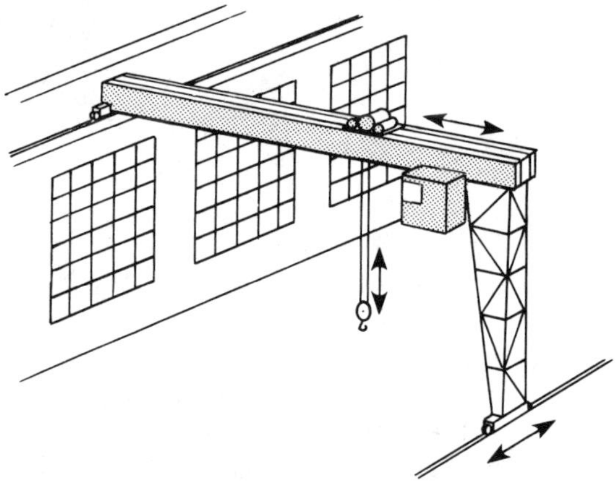

Fig. 9-10: A semi-gantry crane.

Fig. 9-11: A 10 ton semi-gantry crane in a metal center warehouse (photo courtesy of Shepard Niles Crane and Hoist Corp.).

Cantilever Gantry Cranes. A cantilever gantry crane is a gantry or semigantry crane with bridge girders or trusses extending transversely beyond the crane runway on one or both sides (Fig. 9-12). Cranes such as these are used for handling loads outside the area bounded by the gantry legs. This allows the span between the gantry legs to be lessened, which reduces the weight of the crane without reducing capacity. A 110 ton gantry crane is shown in Fig. 9-13, with the cantilever section extending in the upper right-hand corner.

Wall Cranes. A wall crane is a lifting device having a cantilever frame with or without a trolley (Fig. 9-14). It is supported from a side wall or line of

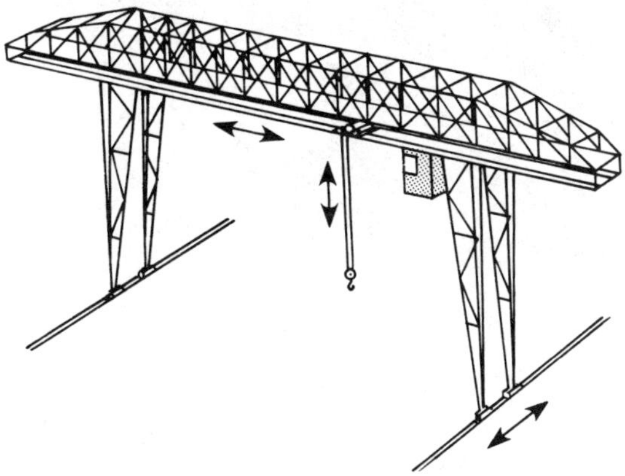

Fig. 9-12: A cantilever gantry crane.

Fig. 9-13: A 110 ton cantilever gantry crane (photo courtesy of Whiting Corp.).

Fig. 9-14: A wall crane (photo courtesy of Shepard Niles Crane and Hoist Corp.).

columns of a building. Wall cranes are capable of travel due to the fact that the runway is attached to the wall or columns. Figure 9-15 illustrates the travel and movement capabilities of a wall crane.

Polar Cranes. A polar crane is a bridge or gantry type crane which travels on a circular runway (Fig. 9-16). Cranes such as these can be found in most nuclear power plants because they are best adapted to lift the components that are housed in the circular reactor buildings. Figure 9-17 shows a polar crane in a nuclear reactor building.

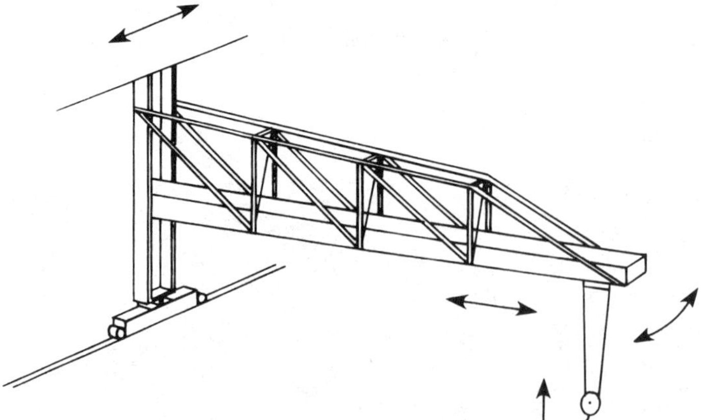

Fig. 9-15: The travel capabilities of a wall crane.

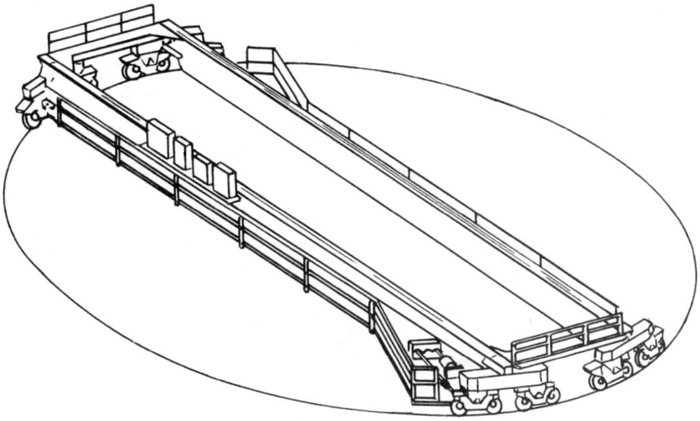

Fig. 9-16: A polar crane.

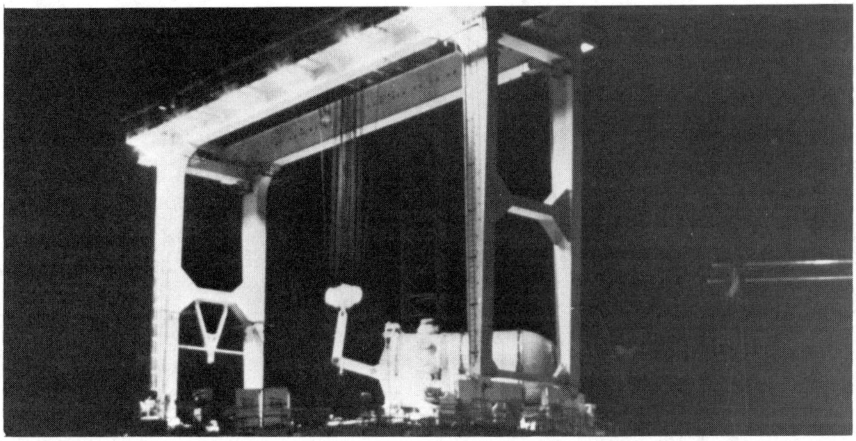

Fig. 9-17: Circular nuclear reactor building is an ideal location for polar cranes (photo courtesy of Whiting Corp.).

MONORAILS AND UNDERHUNG CRANES

Monorails and underhung cranes are quite similar in design to overhead cranes. The main difference between these two types is the fact that the end tracks on which the monorail or underhung crane travels operate on the bottom flange of the runway track section instead of on top as in overhead cranes. The wall supported jib crane illustrated in Fig. 9-18 shows the fundamental characteristics of an underhung crane. This classification applies to carriers (trolleys) operating on single track monorail systems, including curves, switches, transfer devices, lift and drop sections, jib booms and associated equipment.

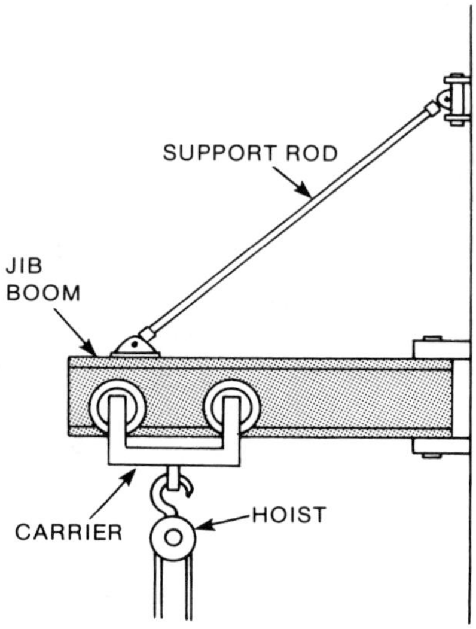

Fig. 9-18: A wall supported jib crane.

There are several types of monorails and underhung cranes which are available for industrial and construction purposes. A cab operated underhung crane, used for heavier lifts, is a device controlled from the cab situated on the same rail as the lift mechanism (Fig. 9-19). The traveling (Fig. 9-20) and semi-gantry underhung cranes (Fig. 9-21) also resemble similar overhead cranes in overall composition.

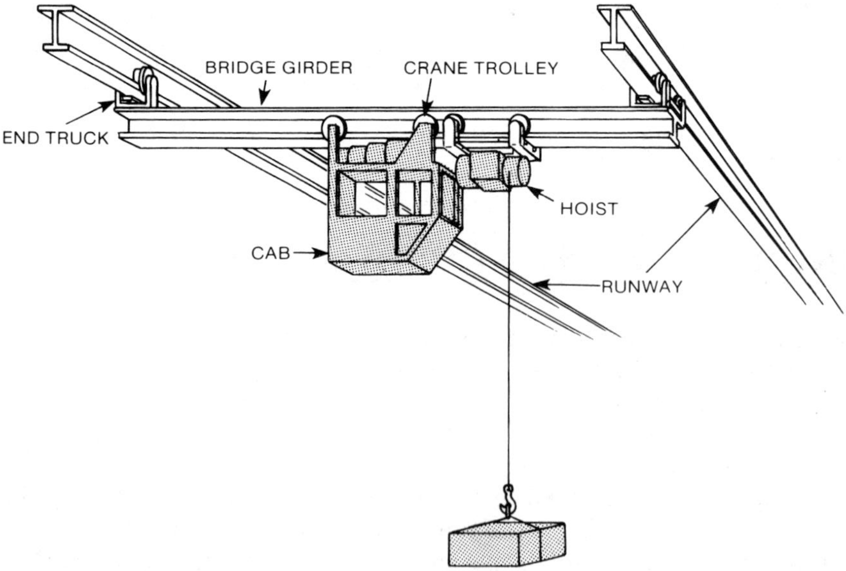

Fig. 9-19: A cab operated underhung crane.

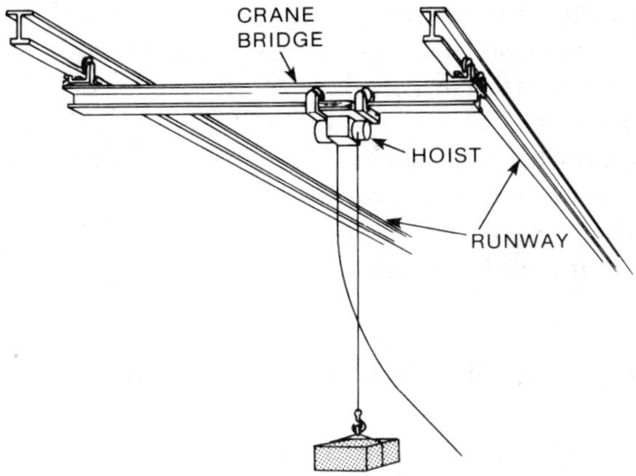

Fig. 9-20: A floor operated traveling underhung crane.

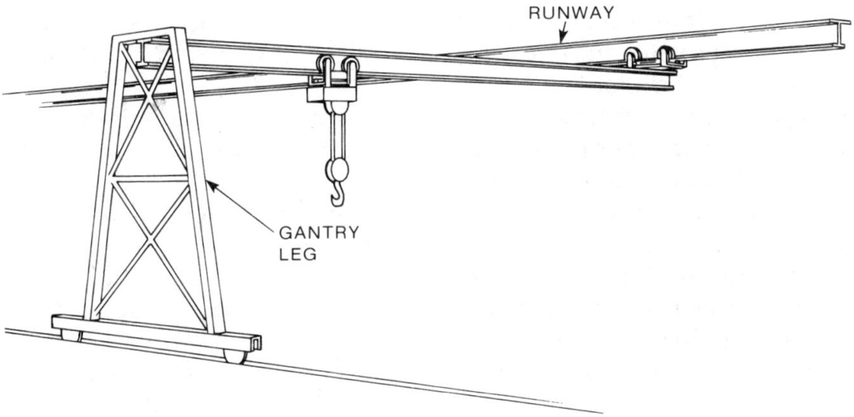

Fig. 9-21: A semi-gantry underhung crane.

MOBILE CRANES

Mobile cranes are probably the most widely used of all crane types in modern construction and erection procedures. A mobile crane is a crane capable of movement under its own power. Common examples of mobile cranes are cranes mounted on or integrated into truck or locomotive assemblies and cranes equipped with crawler units. There are two distinct types of mobile crane movement. Movements around the work area are referred to as travel movements. When a lifting unit is capable of movement from job site to job site, that movement is classified as transit movement. *Note:* All types of mobile cranes can travel. The wheeled mobile cranes are capable of transit movements, while the crawler type assemblies are not.

Any crane capable of movement has decided advantages over stationary equipment. A mobile crane's ability to move about the job site and perform lifts in different locations often minimizes the need for additional lifting equipment. In many cases, a single mobile crane can do the work of several stationary cranes. Also, a mobile crane can often move about the work site in its fully erected state. Stationary cranes, on the other hand, must be fully dismantled and re-erected if they are to be moved around the job site. This causes increased cost by delaying construction until the crane is reassembled.

Considerations for Mobile Crane Use

The weights, dimensions, and lift radii of the heaviest and largest loads along with the maximum lift height are the primary factors to be considered when mobile cranes are used. The various types of mobile crane configurations make it possible to obtain a crane that is specifically designed to handle a particular type of heavy lift or a great number of moderate lifts. Precision is also an important consideration due to the fact that some loads require placement within 1/4″ (6.25 mm) tolerances.

The type of carrier which is used depends upon the ground conditions and the capacity of the machine in its operating quadrants. The maximum capacity is normally over the rear of the carrier and decreases as the boom is swung over the side. Most mobile truck cranes are not designed to lift over the front; therefore, this practice is usually avoided unless the capacities are supplied by the manufacturer. The job site conditions, including such factors as access roads and ramps that the machine has to travel, available space for erection, operations and dismantling, and obstacles which might impede operation, all contribute to the list of variables which must be considered when using mobile cranes.

Boom Designs

Mobile crane booms can be classified as either conventional or hydraulic. A conventional type boom configuration can be lowered, raised, or swung sideways. However, its assembled length does not change (Fig. 9-22) until sections are added. The hydraulic type boom can also be raised, lowered, and swung sideways. However, the boom can be extended or retracted to a certain extent without reassembly (Fig. 9-23).

The choice of boom is based upon the functional requirements of the crane for the job. Where heavy or exceptionally high lifts are involved, the conventional boom is used. When the loads are lighter and more setups are required for shorter lifts, the hydraulic boom is more suitable.

When a hydraulic mobile crane is delivered to the job site, it is almost immediately ready to go into operation. It need only be "spotted," the outriggers extended, and the crane leveled before the first load can be lifted.

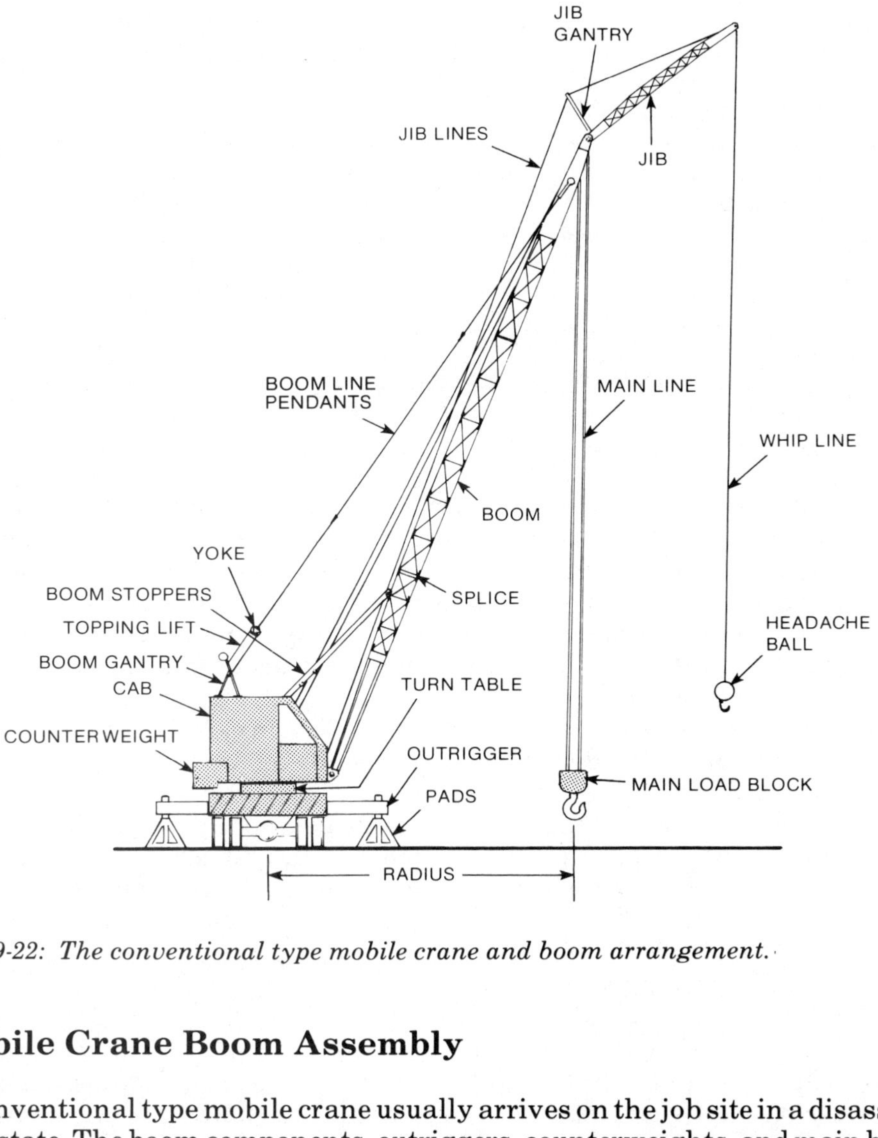

Fig. 9-22: The conventional type mobile crane and boom arrangement.

Mobile Crane Boom Assembly

A conventional type mobile crane usually arrives on the job site in a disassembled state. The boom components, outriggers, counterweights, and main block must be off-loaded (usually assisted by an on-site hydraulic crane) and assembled on the job location.

There are various precautions which must be taken in order to effectively and safely perform the task of constructing and setting up the crane. Initially, all outrigger assemblies should be checked to make certain that they extend properly, and the boom sections should be checked for damages. During assembly procedures, the boom pins should be installed from the inside of the boom to the outside and secured with cotter pins to avoid possible injury to personnel due to falling boom sections. When the boom has been assembled, it must not be raised until the machine is leveled because tipping can occur.

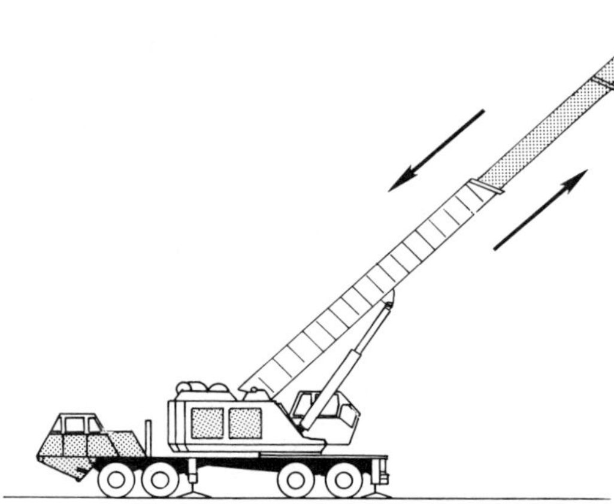

Fig. 9-23: A hydraulic type mobile crane.

On a conventional type crawler crane, the entire area that the treads occupy must be firm and leveled. In the case of a wheel mounted crane, the machine must be leveled by blocking underneath the outriggers. The bearing pressure of a modern mobile crane on outriggers can vary anywhere from 200 psi to over 1000 psi, so it is extremely important that the ground be as firm and level as possible and that large, substantial timber or metal mats be available to distribute the load as evenly as possible (Fig. 9-24). These considerations are even more important when the crane is to be set up on an existing structure such as a garage deck or bridge deck because the loads applied by the crane could be enough to collapse the whole structure.

If the ground is especially soft or waterlogged, special steps must be taken to provide a satisfactory base for the machine. Mats, steel plates, timber pads, or concrete rafts, etc. should be used to distribute the loads under the crane to ensure that the bearing strength of the ground is not exceeded (Fig. 9-25). The base must be checked often during the lifting operation as settling will occur, necessitating outrigger adjustment.

Where operation on slopes cannot be avoided, the machine can either be leveled with the outriggers or the slope should be built up locally with specially constructed foundations or cribs to give a level, firm, and stable working platform (Fig. 9-26). Remember that slope angles as small as 1° can significantly affect the crane's stability and capacity.

If the crane is to be set up in flowing water, the erosion effects of the water on the ground around the machine supports present a serious hazard unless the crane is moved frequently. When setting up in areas having high water tables or areas subject to flooding, the crane may require deep foundations or special ground consolidation.

Account must also be taken of seasonal variations in ground conditions. During the winter months, tracks, tires, and outriggers can freeze to the ground, giving a false impression of stability. In the spring, the ground bearing capac-

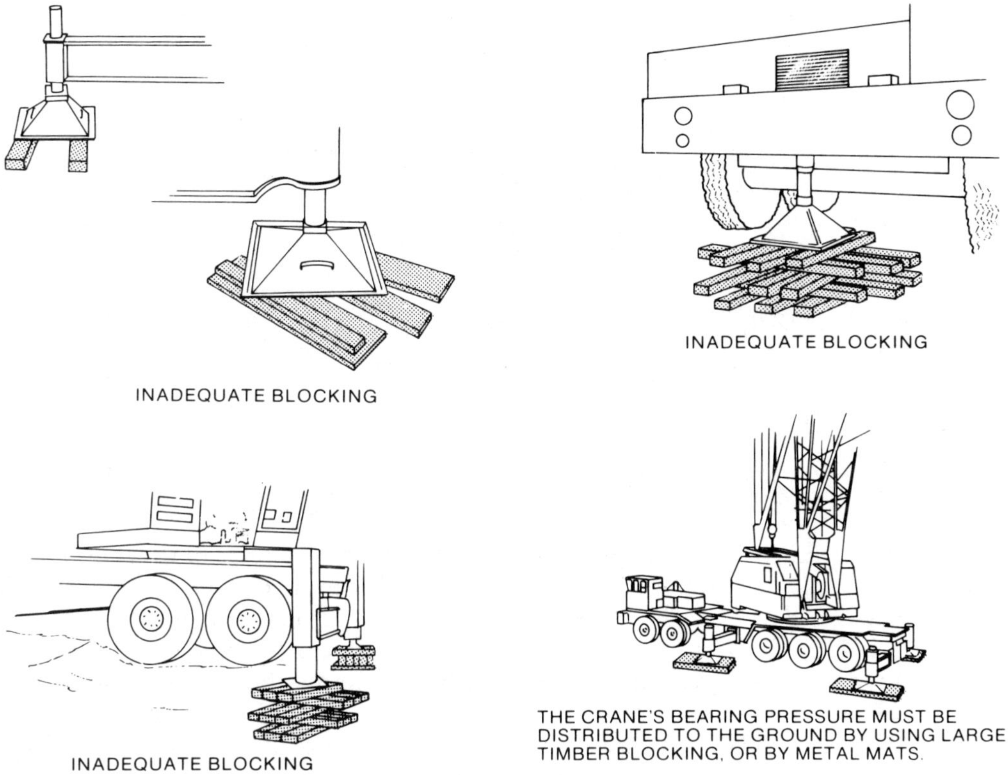

Fig. 9-24: Correct blocking is essential beneath outrigger pads.

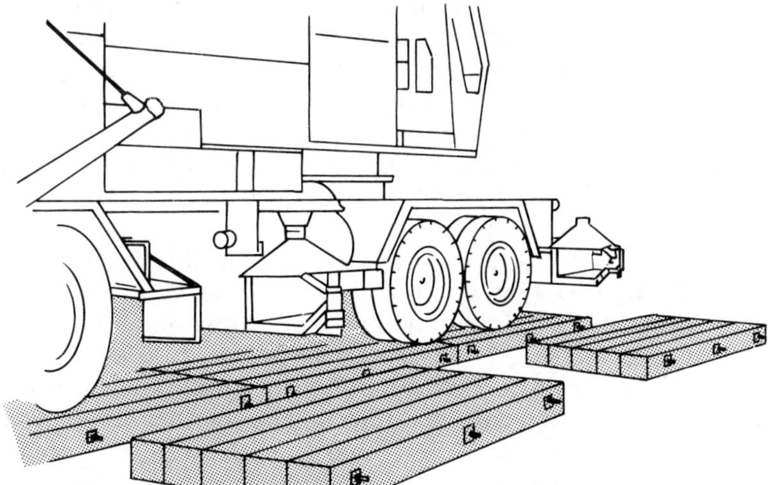

Fig. 9-25: Mats, steel plates, timber pads, or concrete rafts are often used to distribute the bearing pressure evenly below outrigger pads.

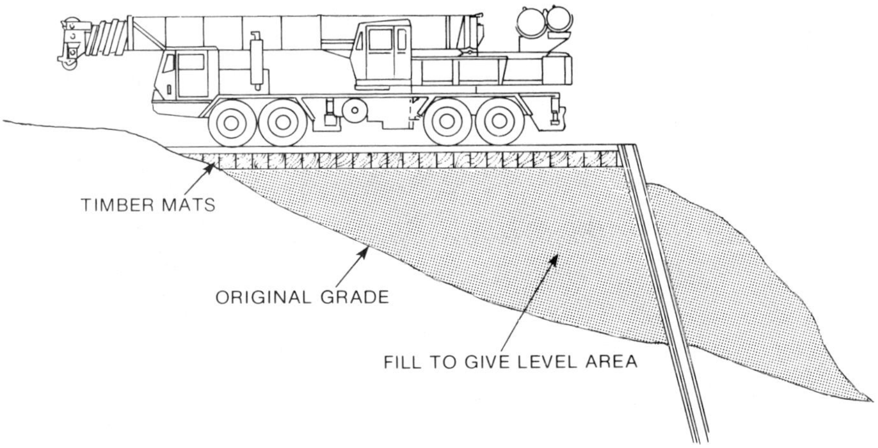

Fig. 9-26: Working platforms must sometimes be provided when rough terrain prohibits crane operation.

ity and stability are almost impossible to determine because of the frost coming out of the soil. In the late fall and early winter months, the ground bearing capacity must not be assumed to be increased because of frost. The estimate should be based on the worst possible prevailing conditions.

Erection Crew Responsibilities. The responsibilities of the operator and the erection or assembly crew begin when the crane arrives at the site. Even though the previous recommendations are to be the responsibility of whoever is going to use the crane, the operator of the machine is also responsible for ensuring that all are carried out before the machine is assembled or operated. Regardless of the type or make of crane being used, it must be set up strictly in accordance with the manufacturer's or his agent's instructions and in such a manner as to eliminate hazards to personnel, property, and the public.

The manufacturer's handbook for the particular crane being set up should be available on the site for use by the staff who must, in turn, be familiar with the procedures. Any departure from these procedures or substitution of parts or materials is permitted only if the details have been discussed with and approved by the manufacturer.

The handbook may be a large volume more suitable for retention in the site office as a reference book, in which case it is recommended that a concise summary of the assembly and dismantling procedures be set out in the form of a checklist (Fig. 9-27A).

Load Chart Usage and Safety Precautions. When assembling or dismantling the boom, some basic precautions must be observed to ensure safety.

Before putting the boom together, determine the weight of the loads and rigging plus the height of the lift and the radius of operation. Then, from the load chart, find the shortest boom length that will accomplish the job. While it is absolutely necessary to have sufficient boom for a particular lift, installing *too much* boom generates unnecessary work and may create a hazardous situation.

Note: A load chart bases its information on the center of rotation of the crane and the centerline of the boom. While the load chart may indicate that the weight of the load can be lifted, the size and shape of the load may prevent the load from achieving the proper height required to place the piece properly. If the boom is not long enough, an odd sized or shaped load could come in contact

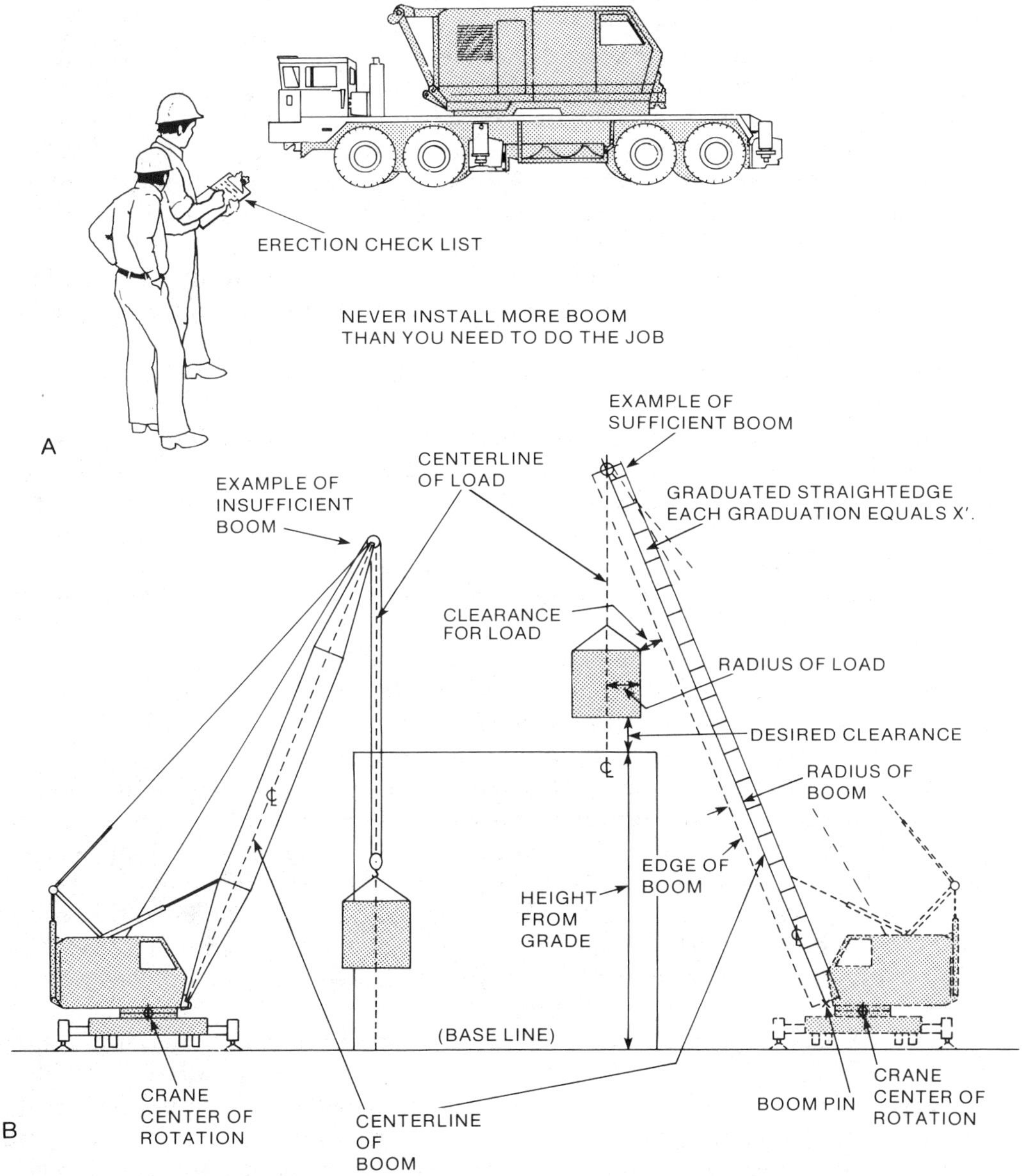

Fig. 9-27: (A) A checklist for crane assembly and dismantling procedures is essential for reference purposes. (B) Terms used on a crane layout sheet.

with the boom before it reaches the proper height. This must be taken into account when calculating boom length. It is very important to always know the radius of the center of rotation to the boom pin and the radius of the load (Fig. 9-27B). When using a crane chart, the distance must be added between the center of rotation and the boom pin, and the distance equal to the radius of the load plus the radius of the boom to the value given in the crane chart in order to find the correct length of boom.

A method of obtaining the proper boom length is to use a graduated straightedge. In this case, the graduations of the straightedge represent distances of feet scaled down to a convenient working level. The use of this method requires that the height of the lift be marked off (to scale) on the layout sheet. The distance between the crane and the building and the structure the component will be lifted onto is also laid out on the sheet. Both of these measurements are laid out in reference to a base line which represents the grade on which the crane is located. Once this is done, the straightedge can then be placed so it runs from the boom pin of the crane to a position over the building or structure. This line denotes the centerline of the boom.

The load can then be proportionately sketched in on the layout sheet at its desired height over the structure. If the clearance between the boom and the load at the desired height is greater than 1′ (304.8 mm), the lift can be made if the load is within the weight limitation of that particular crane. When this clearance is determined, the number of graduations on the straightedge can be counted and the suitable conversions made to find the corresponding correct boom length. When the distance between the load and the boom at the load's desired height is measured, the fact that the line the straightedge creates represents the centerline of the boom, not the overall boom width, should not be overlooked. The radius of the boom must always be taken into consideration when making boom length calculations. *Whatever method is used, always be sure there is enough boom to make the lift.*

Boom assembly must not begin until the counterweight is correctly attached and it is of sufficient weight to get the boom off the ground. In order to lift some long booms off the ground it may be necessary to install extra counterweights beyond that which is required for operation. On some cranes this extra weight must be removed before operations begin or when shorter booms are used, otherwise the crane's backward stability will be reduced (Fig. 9-28). Traveling a crane and carrier with front-bumper counterweights in place, however, is poor practice on some machines as damage to the unit's front end suspension and steering systems can result (Fig. 9-29).

It is never advisable to add or remove counterweight from a carrier mounted crane without first locating the machine on firm, level ground and extending all outrigger beams and jacks. The wheels must be off the ground (Fig. 9-30). If the upper works are swung over the side while on rubber, the machine may tip. It is also good practice to engage the swing lock whenever any work is being done on the counterweights and whenever the boom is being assembled, lifted,

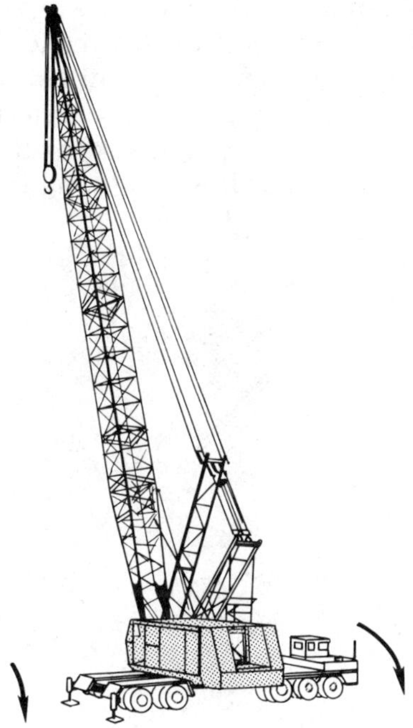

Fig. 9-28: Crane stability is reduced if too much counterweight is used.

or lowered. These procedures apply when any of the counterweights are altered (including the front bumper counterweight) or whenever the upper works are swung without boom (Fig. 9-31).

Of equal importance is the necessity of assembling, raising, lowering, and dismantling the boom with the upper works swung to face directly over the rear of the carrier, which is the area of highest capacity (Fig. 9-32).

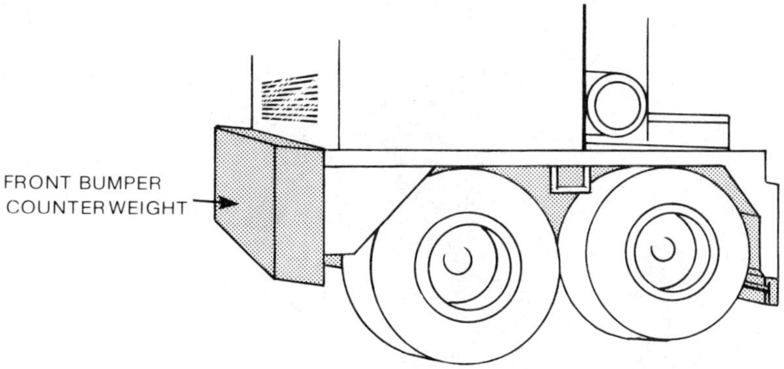

Fig. 9-29: Damage can result if a wheeled mobile crane is traveled with the front bumper counterweight in place.

If the outriggers are not set and the wheels are not off the ground, the crane will not be able to get the boom up or the front end of the carrier will raise off the ground. The boom should never be assembled, raised, or lowered with the machine on rubber. In the case of a crawler mounted crane, wooden packings placed under the ends of the crawler tracks will reduce the tendency for the machine to rock when raising or lowering long booms (Fig. 9-33).

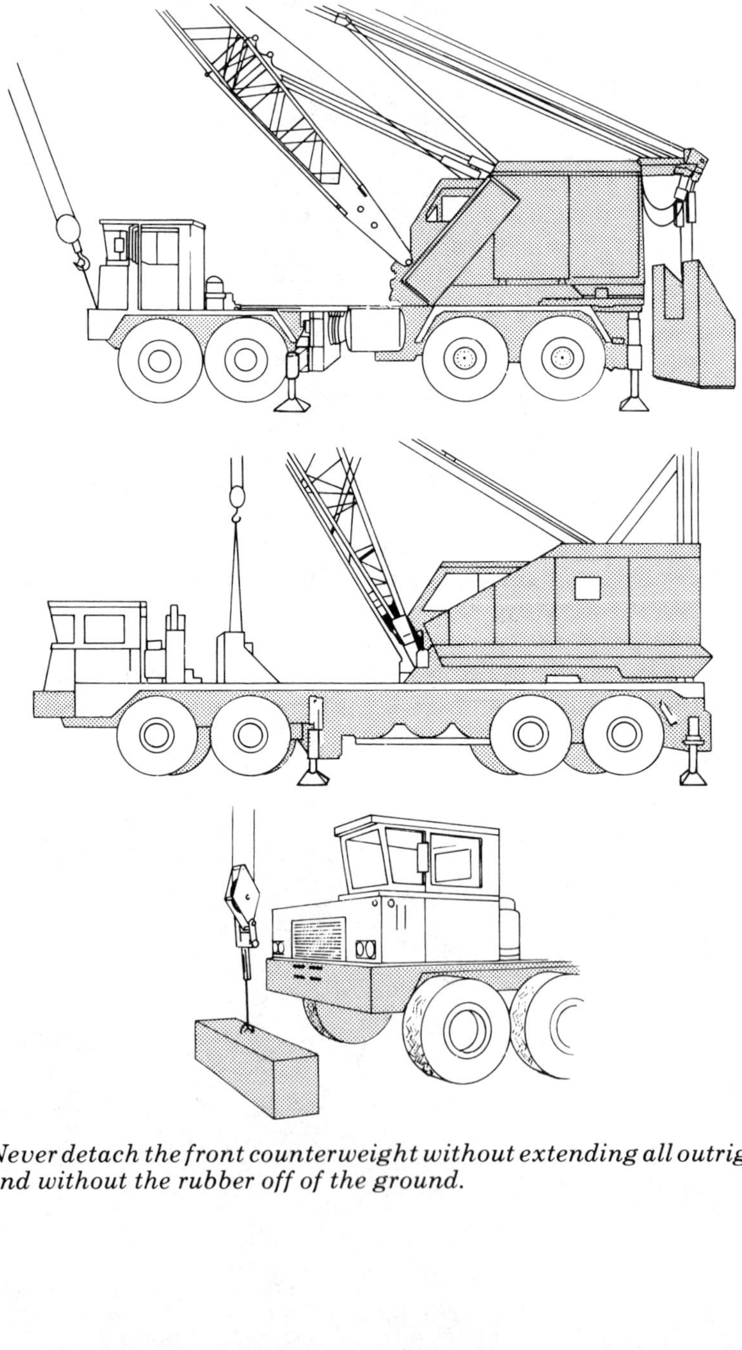

Fig. 9-30: Never detach the front counterweight without extending all outrigger beams and jacks and without the rubber off of the ground.

Fig. 9-31: Swinging the crane without the boom attached could cause tipping.

It is recommended that all components be clearly marked for proper identification during assembly and that only the correct sections for the particular crane be used. A section must not be installed if it has bent lacings, bent or damaged main chords, or any broken or cracked welds. The boom sections should go together easily without the necessity of forcing in the connecting pins. Cotter pins must also be installed in every connecting pin (Fig. 9-34).

In order to avoid any undue crane failure, use only the correct size and quality of nuts, bolts, and pins for joining the sections. Sections must be assembled in the correct manner and sequence in accordance with the manufacturer's

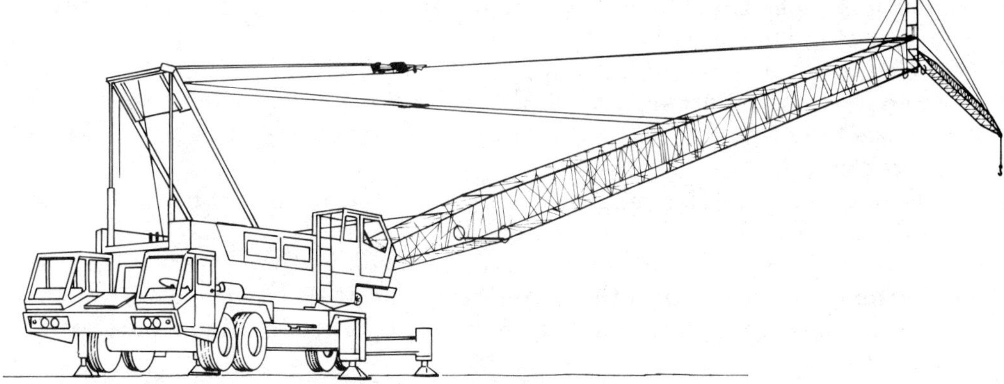

Fig. 9-32: The boom must be over the rear of the carrier and the crane up on the outriggers for assembling, raising, lowering, or dismantling.

Fig. 9-33: Timber packing under the tips of crawlers reduces the tendency to sink and rock when lifting or lowering the boom (photo courtesy of FMC Corp., Cable Crane and Excavator Div.).

instructions so that the correct lacing pattern is maintained throughout the length of the boom.

Whenever possible, the boom is assembled with short sections close to the boom foot. Two short sections will weigh more per equivalent length than one long section; therefore, keeping short sections close to the boom foot will improve machine stability (Fig. 9-35).

Boom assembly also requires that the pendants be kept behind the sections being connected. In this procedure, the two upper pin connection holes are aligned between the two sections being pinned together. The two top boom pins can then be inserted and the cotter pins installed. The rig is then boomed up until the bottom pin holes line up. When this is accomplished, the two bottom boom pins and retainers can be installed.

This procedure can be repeated by extending the pendants as needed (they should be moved out to the last completed joint) until the boom is assembled

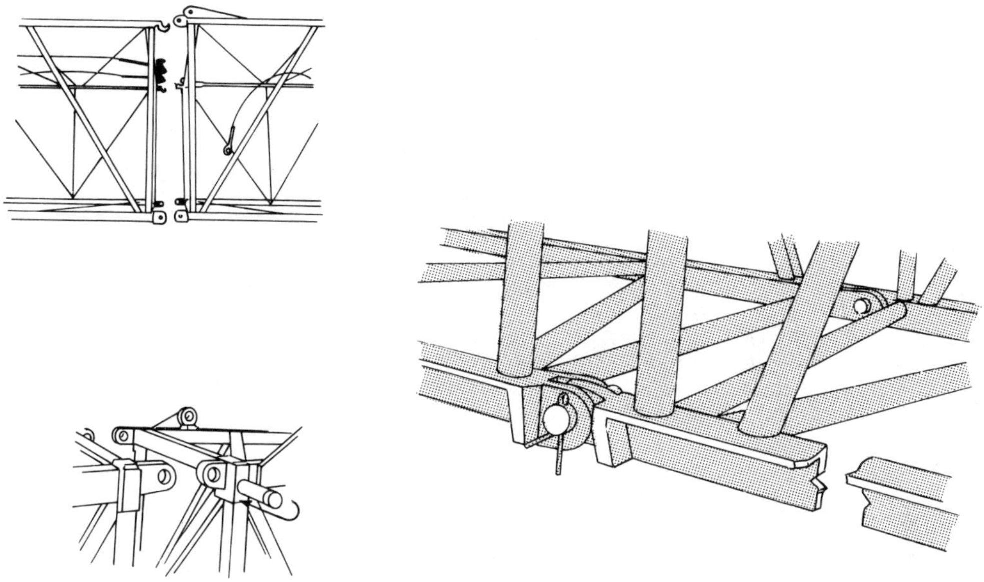

Fig. 9-34: Boom sections must fit together easily without forcing the connecting pins.

(Fig. 9-36). Figure 9-37 illustrates the assembly of a long boom. It is good policy to always block beneath each end of each section of the boom when it is being assembled. Under no circumstances during the assembly and blocking of boom sections should anyone be allowed to stand or work beneath the boom. It is also essential that the maximum allowable length of cantilevered boom for that crane not be exceeded (Fig. 9-38). Exceeding the maximum recommended cantilevered boom length could result in damage to the crane.

Note: Caution must be exercised to ensure that the suspension ropes and pendants do not foul the locking pins (cotter pins) or clips that retain the pin connections. The boom pendant pins must be checked to ascertain that they

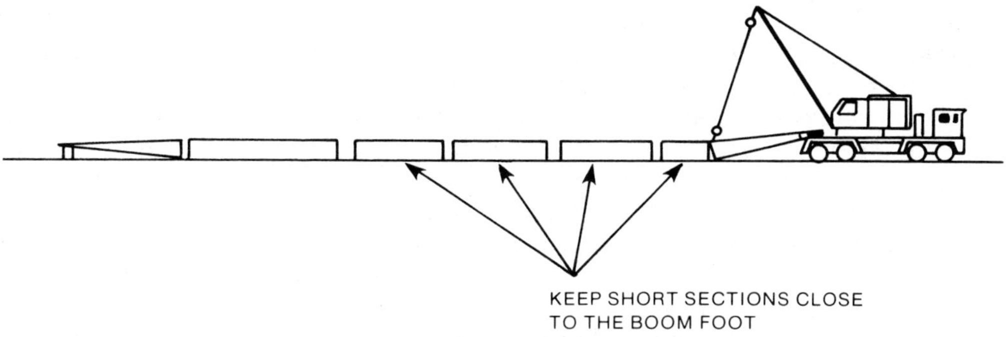

Fig. 9-35: Whenever possible, assemble the short sections close to the boom foot.

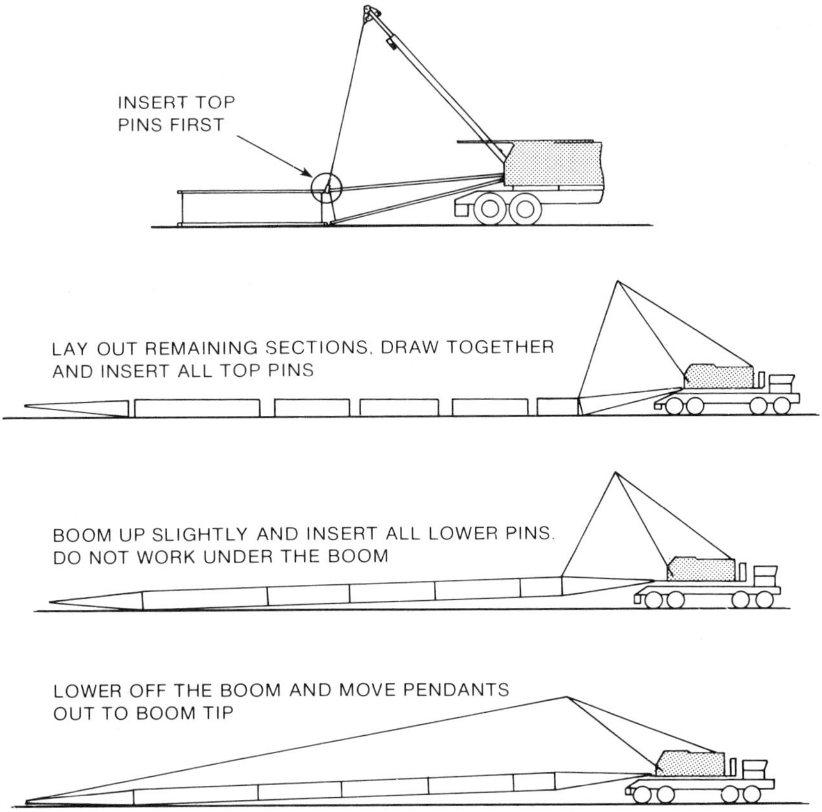

Fig. 9-36: Short boom assembly procedures.

are in good condition and are retained by proper cotter pins. Spring clips are *not* recommended as they can be forced out of position accidentally (Fig. 9-39).

Longer booms usually require special suspension equipment such as a high gantry mast, intermediate suspensions, etc. The boom should not be raised until this equipment has been fitted. Crane manufacturers specify the maximum length of boom which can be raised with standard equipment using the crane machinery, and also the lengths that require the use of special equipment or auxiliary assistance. Midpoint pendants are designed to support the mid-portion of the boom while it is being raised from or lowered to the ground (Fig. 9-40). At low boom angle working positions, some mid-portion support is provided by these pendants. The amount of support gradually reduces as the boom angle increases until, at about 45°, the midpoint supporting effect ceases.

If the crane gantry can be extended, it is good policy to use it in its highest position regardless of the boom length. Much lower stresses result in the boom hoisting equipment when this procedure is followed. This means longer life and added safety factors for the equipment (Fig. 9-41).

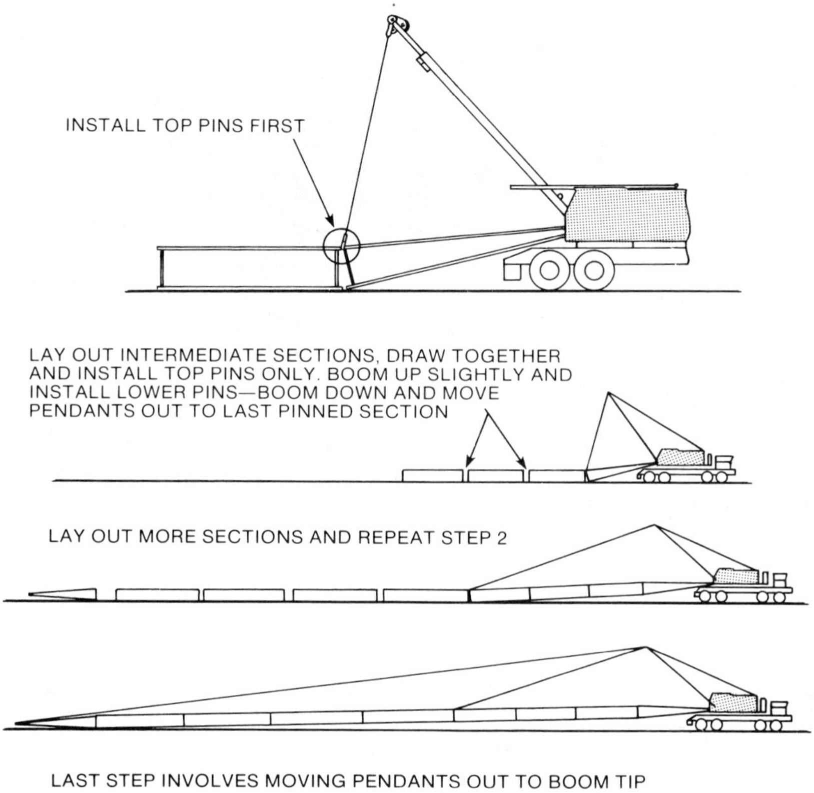

Fig. 9-37: Long boom assembly procedures.

Pendant Variables. When the boom is assembled, it is important that the pendants are of the proper length for the particular boom length. On some machines with short booms, a problem occurs if the pendants are only slightly too long. In this situation, the weight of the live mast will pull an unloaded boom into the back stops when minimum radius is approached (Fig. 9-42). This

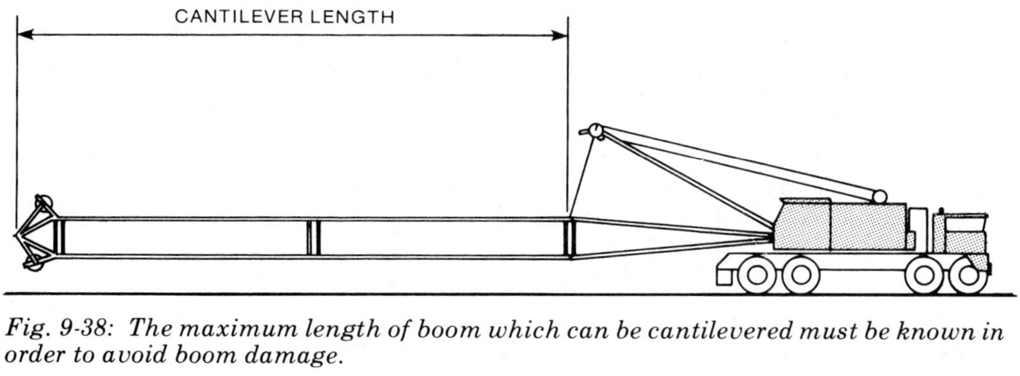

Fig. 9-38: The maximum length of boom which can be cantilevered must be known in order to avoid boom damage.

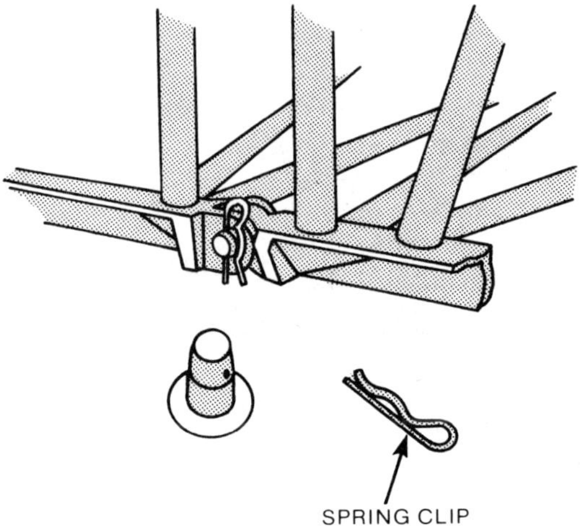

Fig. 9-39: Spring clips must not be used instead of cotter pins because they can be forced out of position accidentally.

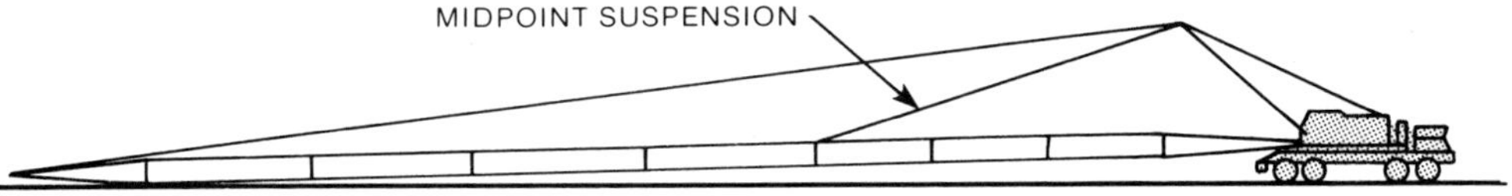

Fig. 9-40: Midpoint suspension on a long boom.

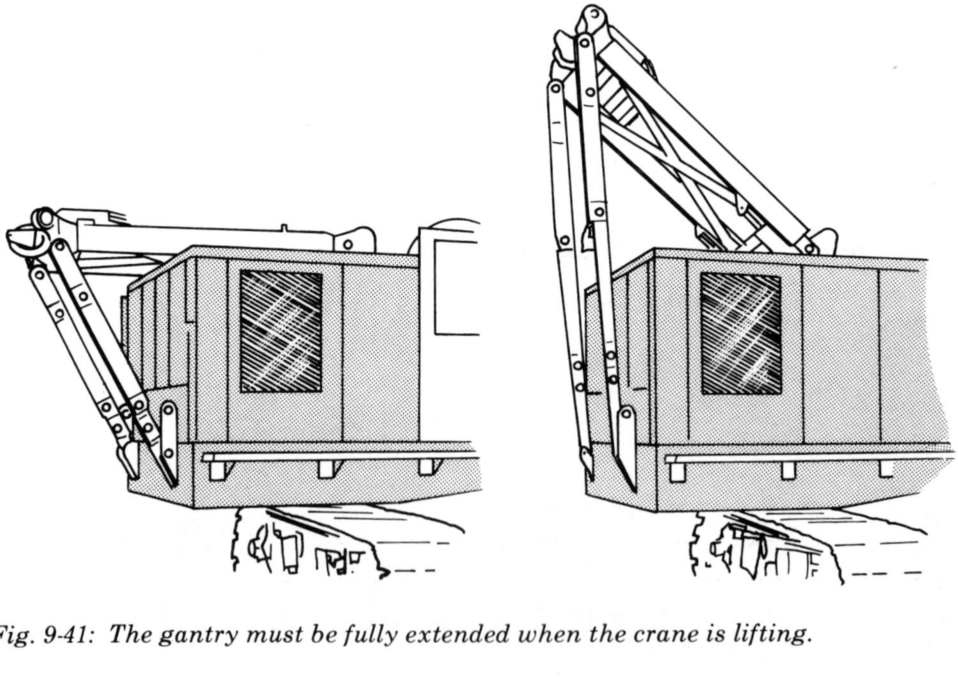

Fig. 9-41: The gantry must be fully extended when the crane is lifting.

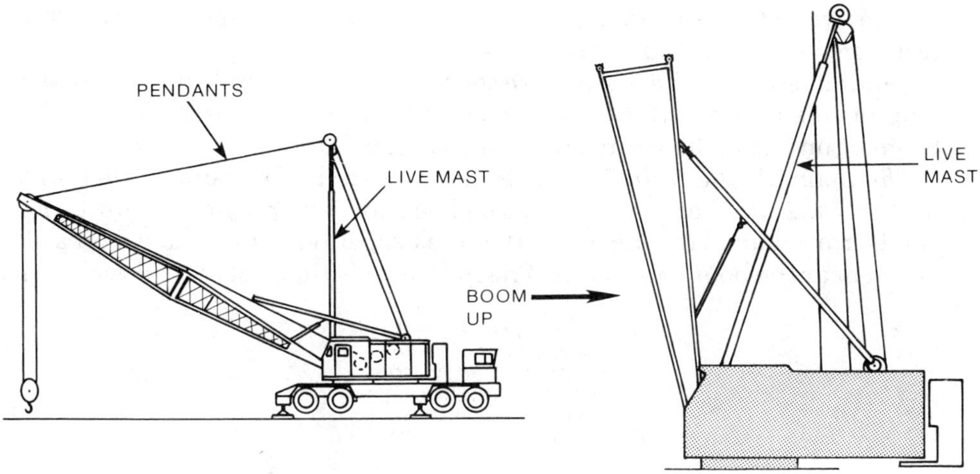

Fig. 9-42: The pendant lines must be at their proper length.

means that the boom will have to be pulled away from the back stops by hanging a load on the hook. With longer booms, the use of pendants that are too short causes overloading in the boom, mast, pendants, and boom hoist system when handling near capacity loads (Fig. 9-43). It is also very important to have the pendants exactly the same length. If they are not equal in length, the boom

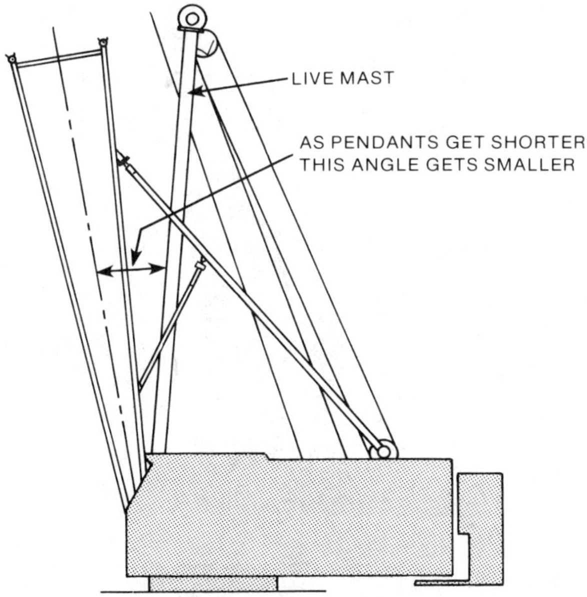

Fig. 9-43: The use of pendants which are too short causes severe overloading conditions on the boom.

can be twisted and the load distribution will be completely altered. This could result in a structural failure (Fig. 9-44).

Pendant spreader bars stabilize the pendant ropes and should be used on all long booms to prevent the main pendants from scrubbing on the side of the boom or on the boom connection pins (Fig. 9-45).

The Boom Angle Indicator. It is essential that the boom angle indicator is properly installed and adjusted. To check the accuracy of the unit, the boom can be raised until it is horizontal and a carpenter's level placed on a straight chord near the boom midpoint. The indicator is then set to read zero. After this

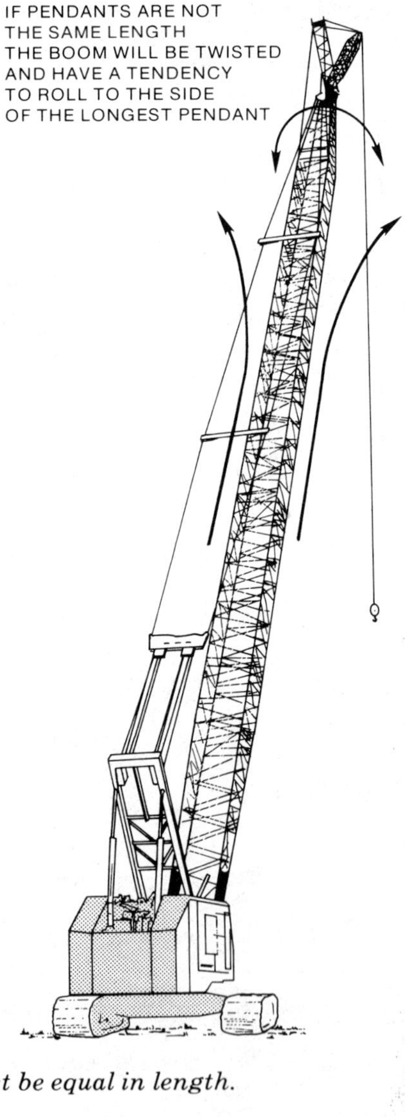

Fig. 9-44: Pendants must be equal in length.

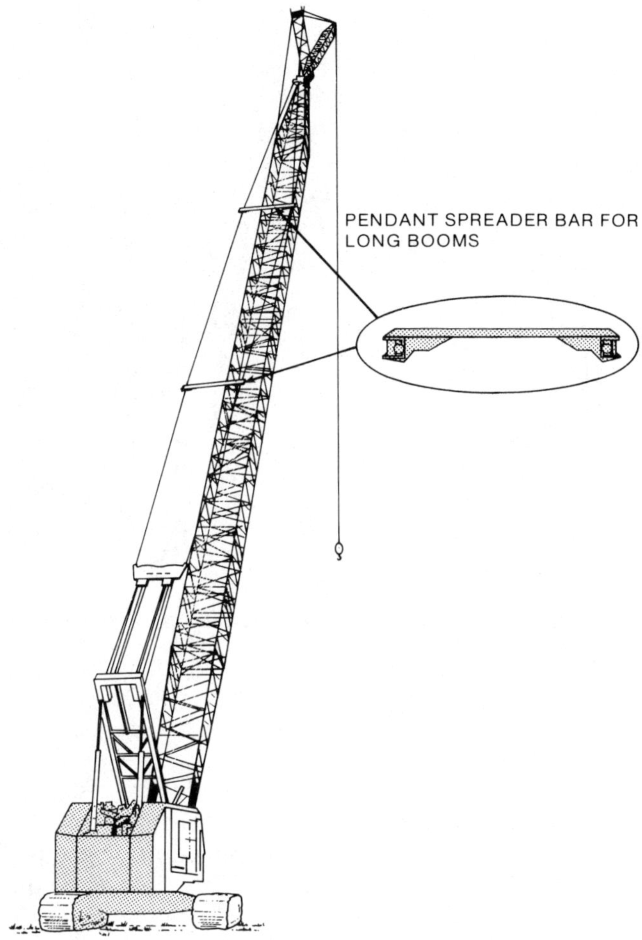

Fig. 9-45: It is good practice to use pendant spreader bars for long booms.

check, the crane can be boomed up slowly so that the motion of the pendulum in various boom positions can be observed. If the indicator sticks or drags, it should be repaired before use.

Note: Boom angle indicators are approximations only. When making heavy lifts at radii where such loads may be close to rated loads, load radius should be determined by actual measurement from centerline of rotation to the center of gravity of the load (Fig. 9-46).

Jib Boom Installation. When a jib is installed on a boom, the jib length and offset for that particular boom combination must be checked. The machine has to be level, the outriggers set, and the boom positioned over the rear of the carrier. It is also important that the shock absorbing jib back stops are functioning properly in order to avoid serious damage to the crane.

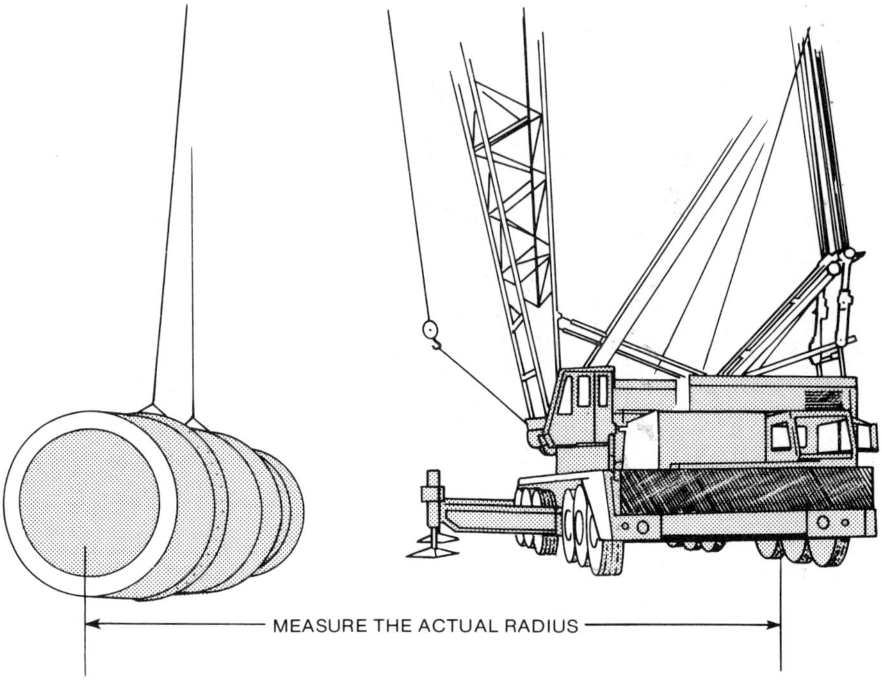

Fig. 9-46: Load radius should be determined by actual measurement from the centerline of rotation to the center of gravity of heavy loads.

All personnel not involved with the jib installation must be kept clear of the installation area and care is advised for those climbing on and off the machine. Walking surfaces must be kept clean and free of oil or grease.

On cranes equipped with swing away or folding jibs, the folded sections must be prevented from swinging uncontrollably when released from their stowed positions or when the pins are removed for folding (Fig. 9-47). There must be sufficient room for the jib to swing out from its storage position to working position and to extend the boom.

Special Boom Considerations. Some manufacturers specify limiting wind velocities for lifting long booms off the ground. That limit should be strictly observed since this operation places even higher loads on some of the components when carrying rated loads. The added wind loading could be enough to cause failure. On cranes equipped with folding booms and folding wheels, the machine must be on solid, level ground with the outriggers down while folding and unfolding the boom. All movements must be made slowly to avoid sudden shock loads being applied to the boom. Another problem is incurred when the folding wheel is hung up in a hollow in the ground. If it is obstructed and slack develops in the boom hoist line or pendants, the forward section of boom could suddenly kick loose and the rearward section of the boom drop (Fig. 9-48).

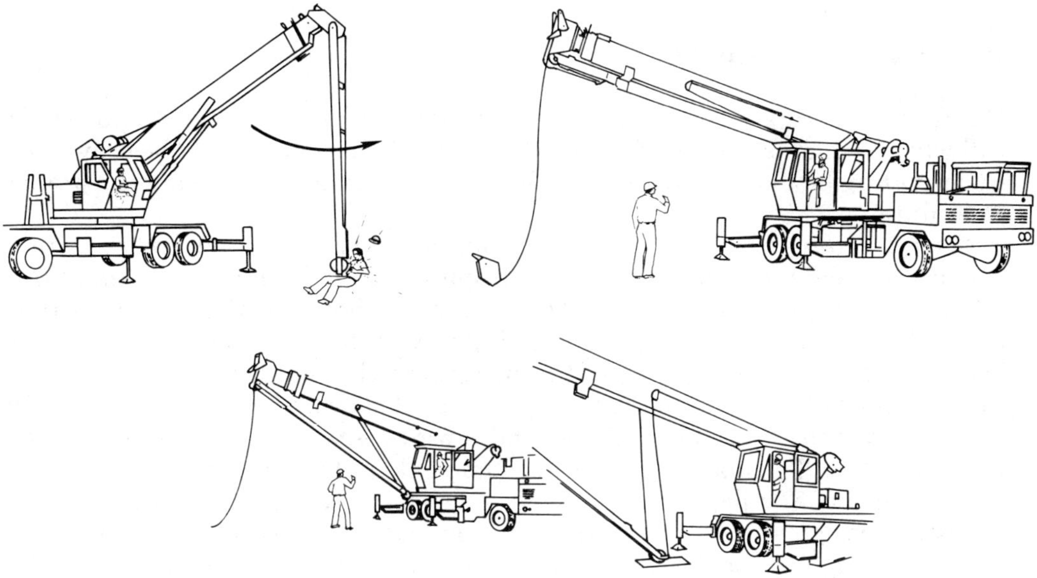

Fig. 9-47: The swing out of the jib must be controlled.

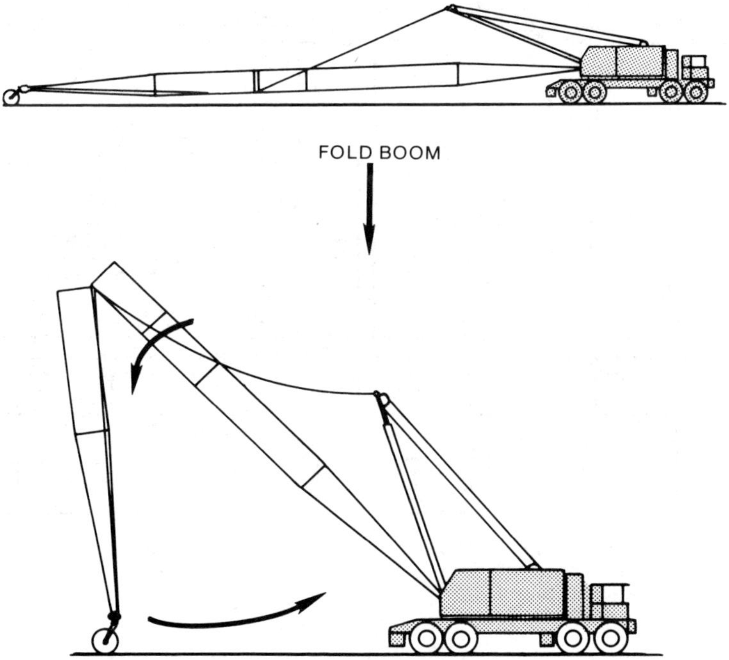

Fig. 9-48: Workers must be aware that folding booms can "kick in" suddenly if the folding wheel hangs up.

Removing Boom Sections

Operators and riggers continue to be crushed every year by boom sections that fall when being dismantled from the boom. Such accidents could be prevented by providing blocking under each side of every joint and observing the proper procedure.

When a boom section is removed, the operator or rigger must never position himself/herself under the boom in an attempt to knock out the retaining pins.

The general procedure used for the removal of boom sections requires that the rig first be moved onto a level area that is well clear of personnel. The crane is then leveled through the use of the outriggers, the counterweight is checked for correct positioning, and the swing lock is set. The boom is then rotated to a position over the rear of the carrier and lowered until it is almost horizontal. Blocking should be placed under the boom peak and under both sides of every joint on the boom if possible (Fig. 9-49). The boom is then slowly lowered until the pendants go slack. When this step has been accomplished, the pendant lines can be moved back by removing a section of boom and reanchoring the

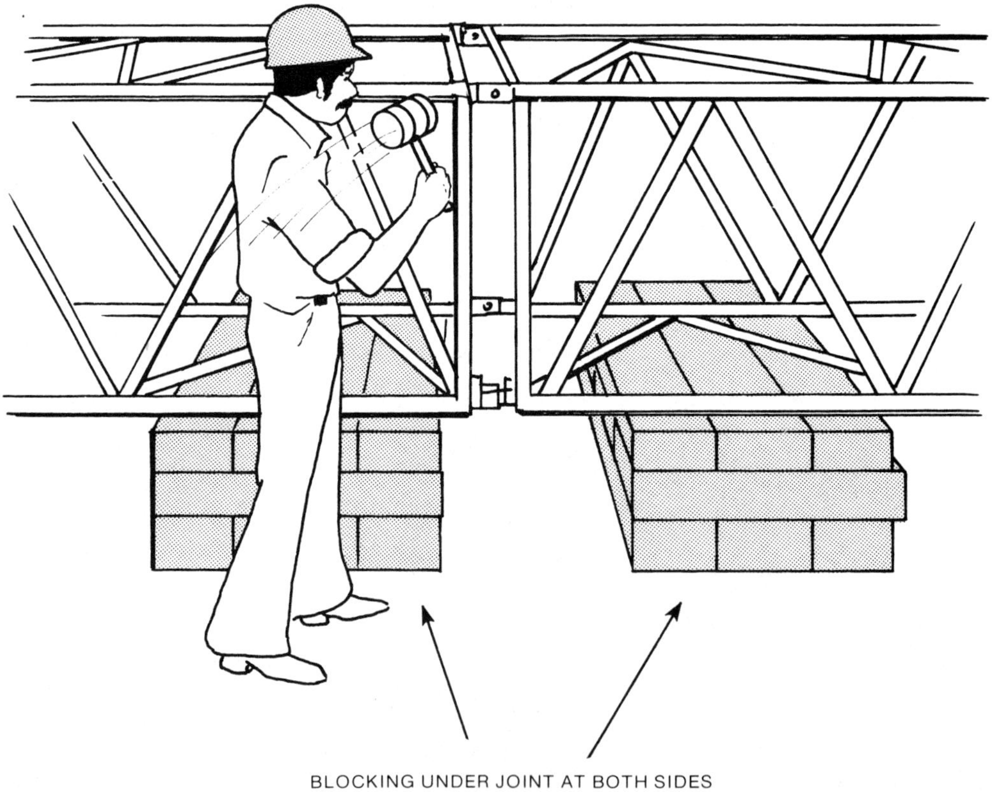

Fig. 9-49: Personnel should stay out from under the boom.

pendants to the next section toward the crane body. The boom is elevated slightly, just enough to put strain on the pendants, but not enough to lift the boom peak off its blocking. The lower pins on the sections ahead of the pendants are then driven out. (Figure 9-50 illustrates removal.) The boom is lowered back down onto the blocking so that every section is supported. The upper pins can then be driven out on the sections ahead of the pendants. This procedure can now be repeated until the boom is dismantled. Figure 9-51 illustrates the procedure for boom section removal on mobile cranes.

Safety Considerations. In some instances the float that is being used to haul the boom sections can be used as a substitute for the blocking (Fig. 9-52). However, regardless of the procedure used, no one should be allowed to walk under or work under any crane boom. Because this accident occurs so frequent-

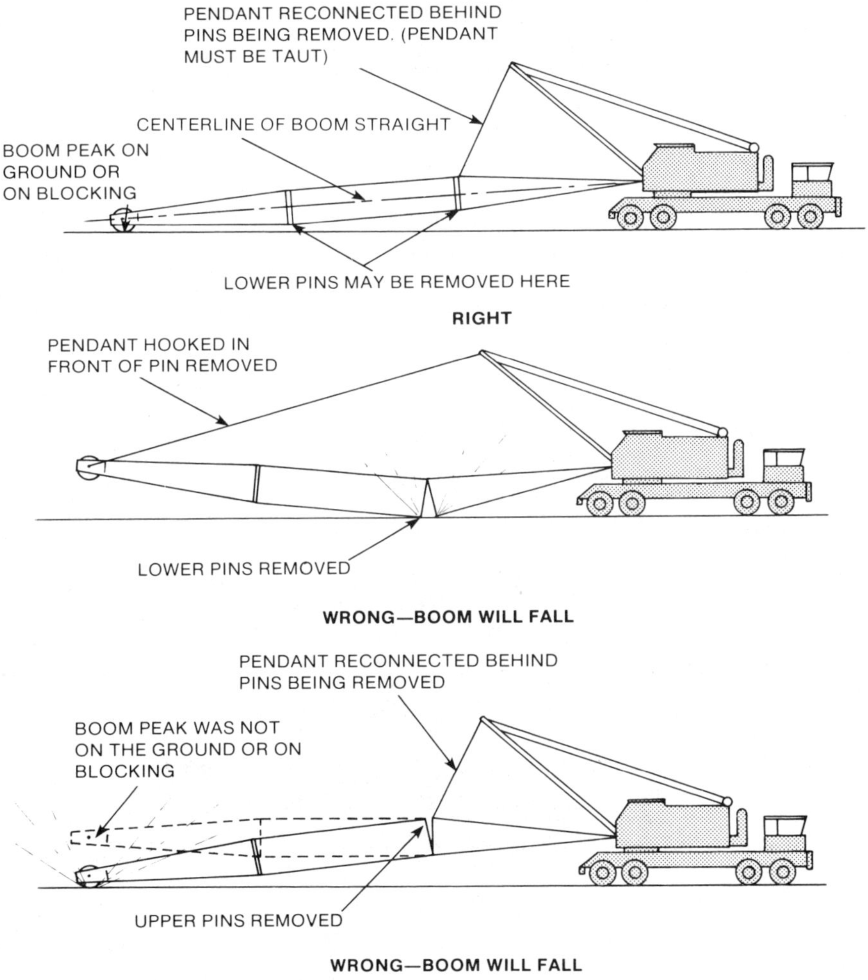

Fig. 9-50: Proper boom pin removal.

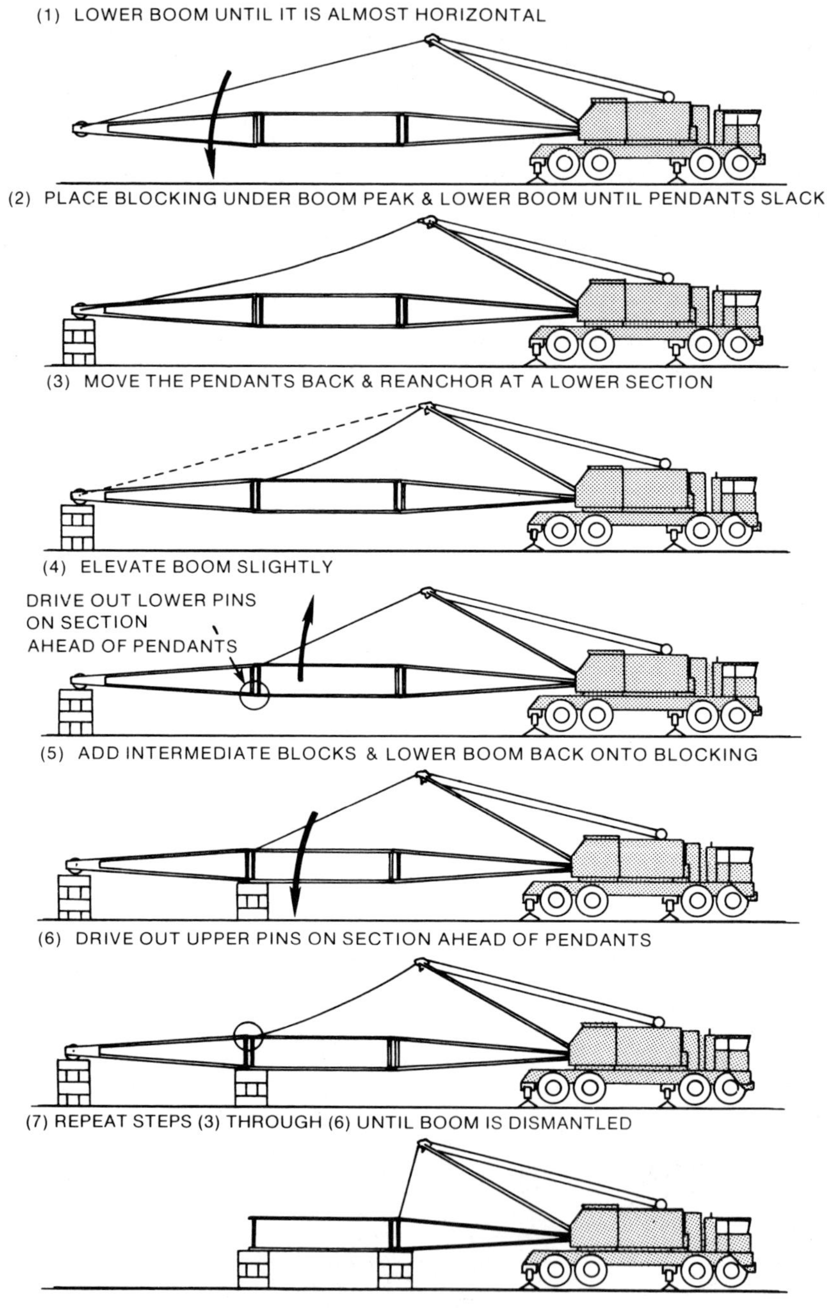

Fig. 9-51: Boom section removal.

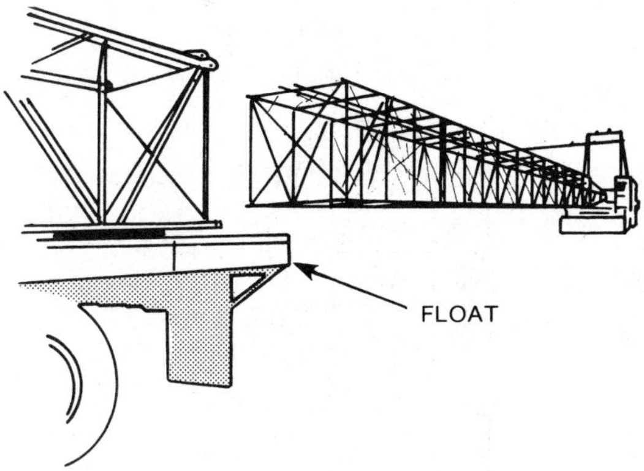

Fig. 9-52: The float can sometimes substitute for blocking.

ly to well-trained, experienced people, it is also suggested that the pin connections themselves be modified in such a way that the person removing the pin cannot stand under the boom.

Figure 9-53 shows a stepped pin which can only be placed in between the lugs from inside but must be removed by a blow outside the boom. Figure 9-54 shows a lug welded on the outside pin lug; this prevents the head of the pin from being put in from the wrong side. This type has the disadvantage of being easily circumvented by having pins without heads, by cutting off the head on one side of the pin, or by having an excessively long pin.

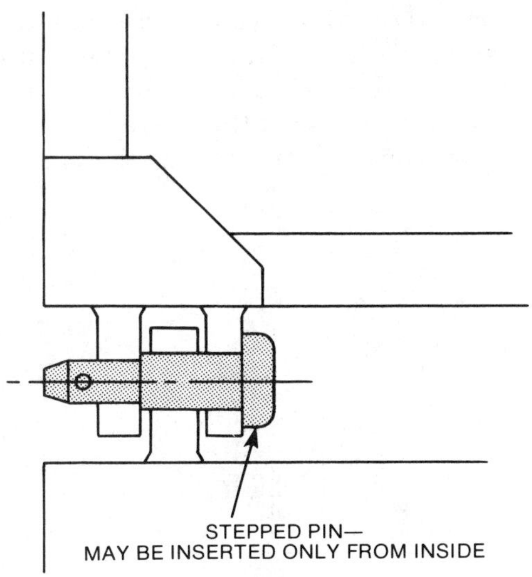

Fig. 9-53: A stepped pin.

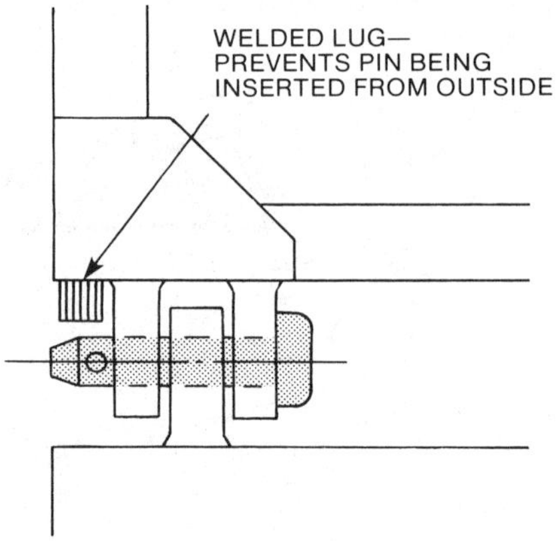

Fig. 9-54: A welded lug on the outside of the lug pin area.

Figure 9-55 shows a double-ended pin without any head at all. In this instance it is not necessary for the pin to be put in from underneath the boom. The pin can be put in from outside and also knocked out from outside.

Once the boom is erected, the machine is ready to be placed to make its lifts. The location should be selected to minimize all swinging and booming operations and to keep the machine working in its area of highest capacity (over the rear for wheeled carrier mounted cranes).

Mobile Crane Failure. Mobile crane failure on the job site can cause vast amounts of destruction and injury. Nearly all of these failures and upsets are caused by human error and could be avoided. Failure to use outriggers, failure to fully extend outriggers, failure to level the crane, and improper block beneath outrigger floats all contribute to crane accidents. Figure 9-56 illustrates several causes of crane failure. Use of the procedures detailed in the previous sections will minimize the possibility of crane failure.

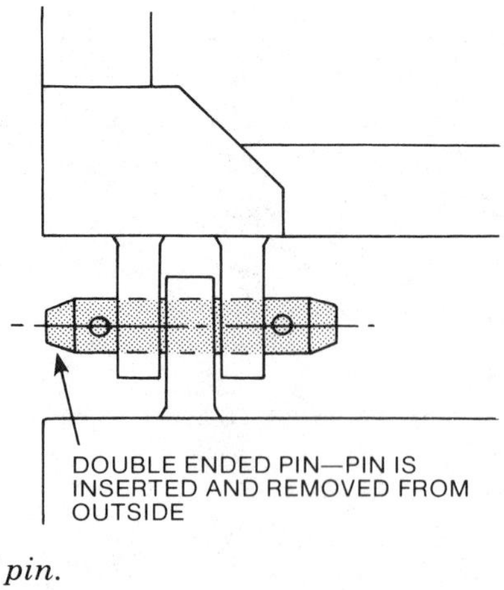

Fig. 9-55: A double ended pin.

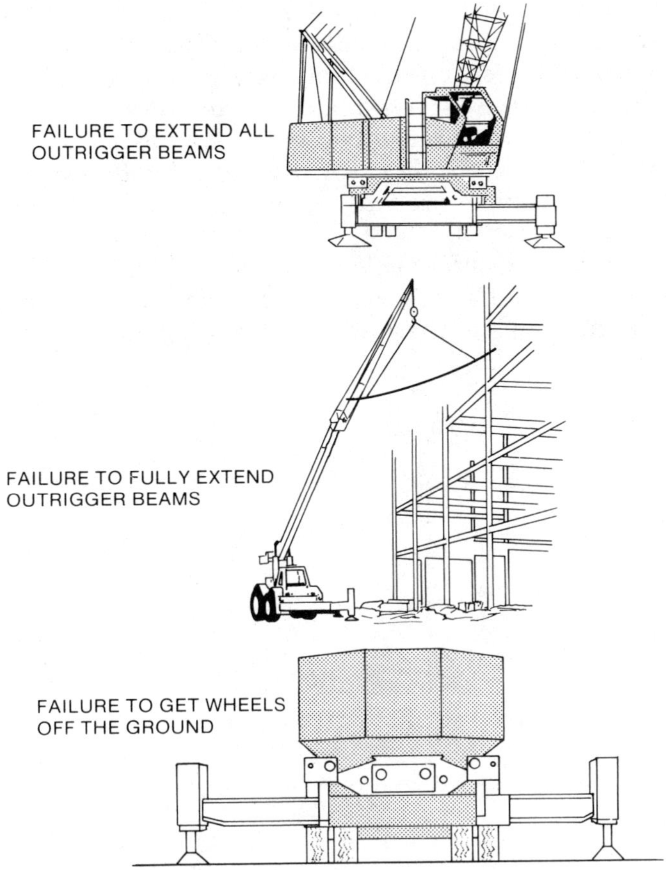

Fig. 9-56: Several causes of crane failure.

TOWER CRANES

Tower cranes are designed and used for situations which require operation in congested areas. Although these cranes are most widely used in Europe, where narrow streets and moderately tall buildings hinder the use of other types of cranes, their popularity throughout the United States is increasing. In addition to their ability to work within a limited area, tower cranes make it possible to lift moderately heavy loads to great heights while providing an extremely deep reach. Tower cranes also prove to be the most economical when situations require a large amount of lifts over a long period of time.

The use of any type of crane requires planning. Tower cranes, however, require a greater amount of planning than other lifting devices due to the fact that their structures, foundations, and presence on the job site are generally permanent as long as the heavy construction phases are going on.

The selection of the most suitable type, size, and number of tower cranes for a particular job requires that the characteristics of the various machines be considered along with the requirements imposed by the loads to be handled and the surroundings in which the crane will operate. Factors such as the weights, dimensions, and lift radii of the heaviest and largest loads are considered load requirements.

Tower crane characteristics such as the type and size of the base, the maximum freestanding height (Fig. 9-57A), the maximum braced height (Fig. 9-57B), the available jib lengths, and the area which must be covered are a few

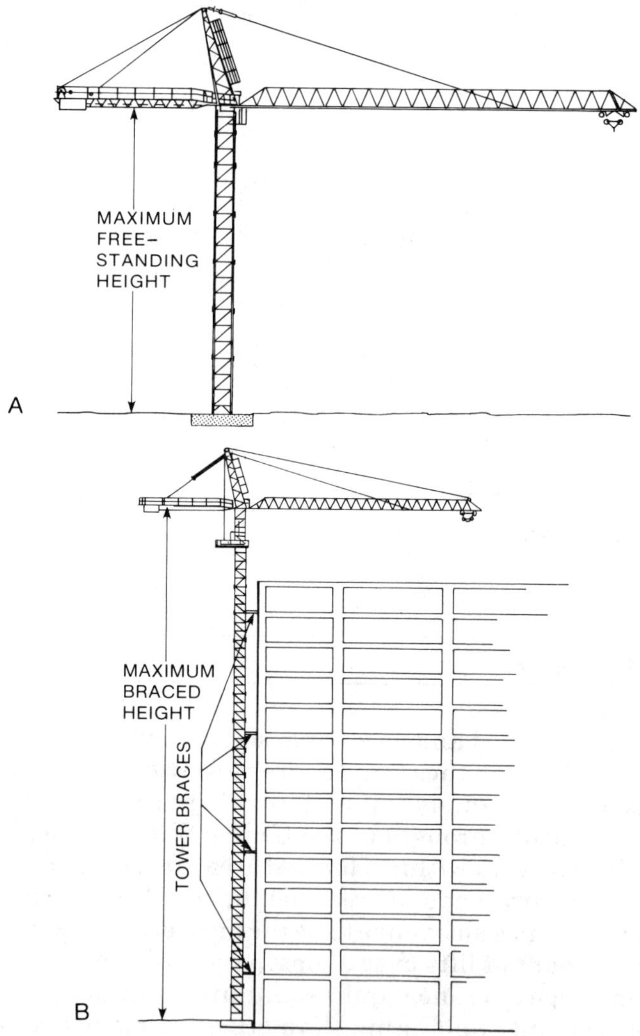

Fig. 9-57: (A) The maximum freestanding and (B) the maximum braced height of tower cranes.

factors which govern their use. Climbing arrangements, available on certain types of tower cranes, and the weight of the crane that has to be supported by the building, along with the available headroom (the space between the maximum height position of the hook in the uppermost work level [Fig. 9-58]), take on greater significance as the height of the required lifts increases.

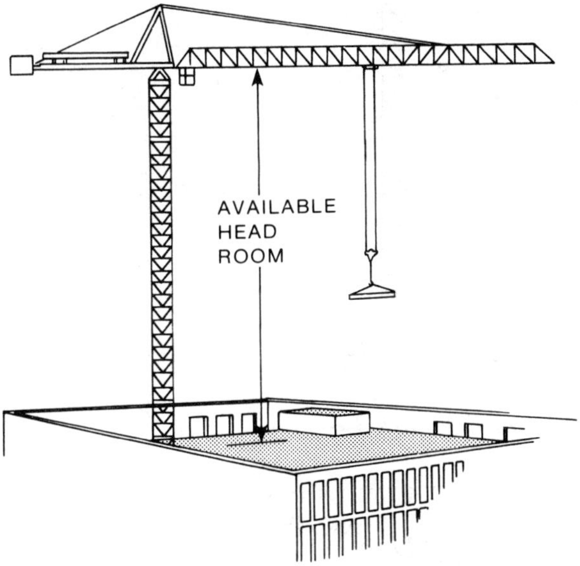

Fig. 9-58: The available headroom is the space between the maximum height position of the hook and the highest work level.

Tower Crane Configurations

There are a number of different types and sizes of tower cranes available to construction firms throughout the world. Most tower cranes are very economical to transport and set up due to the fact that the unit is made up of many components of the same design. These components can even be shared between the larger and smaller tower cranes of the same manufacturer.

The power which runs all functions of these cranes generally comes from electric motors. This type of power allows for smooth operation at relatively high (or low) speeds. The gradual acceleration provided by some models adds to the crane's efficiency by increasing lift speeds while also increasing the accuracy of load placement.

Tower cranes are available with either fixed (Fig. 9-59) or slewing (rotating) towers (Fig. 9-60). The entire tower, in some cases, is also able to travel along tracks as illustrated in Fig. 9-61.

On the fixed tower type, the slewing ring is situated at or near the top of the tower and the jib slews are located around the vertical axis of the stationary tower. The slewing ring on the slewing type tower is situated at the bottom of

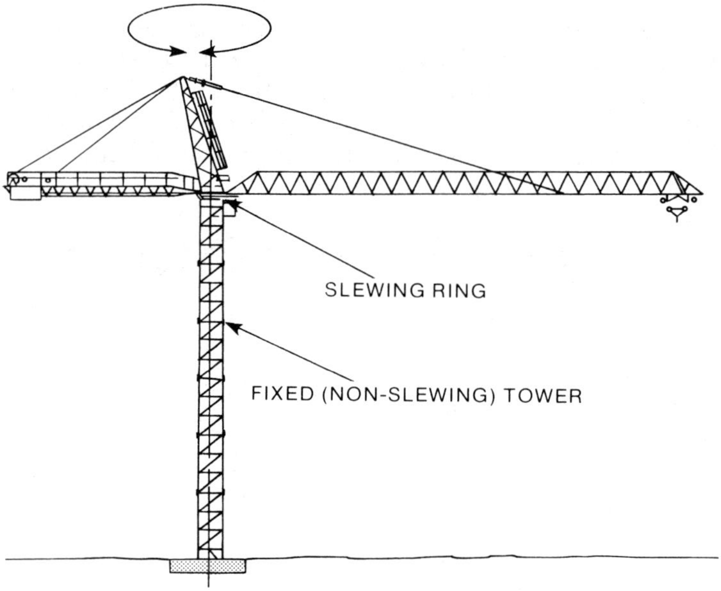

Fig. 9-59: The fixed tower configuration.

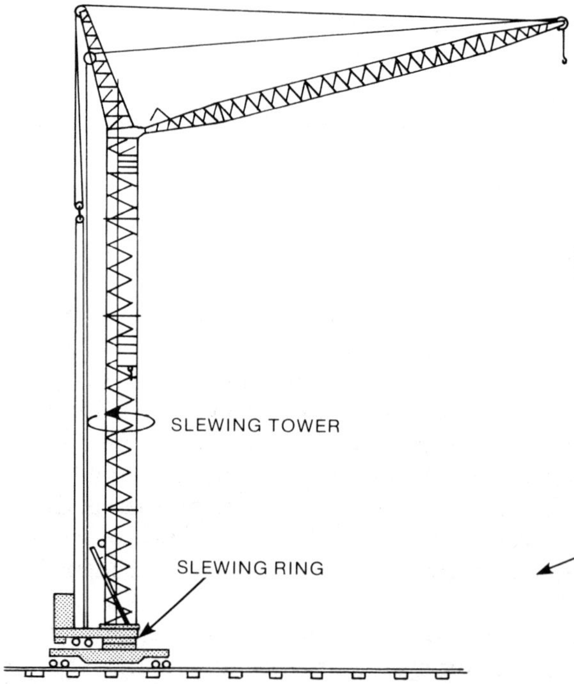

Fig. 9-60: The slewing tower configuration.

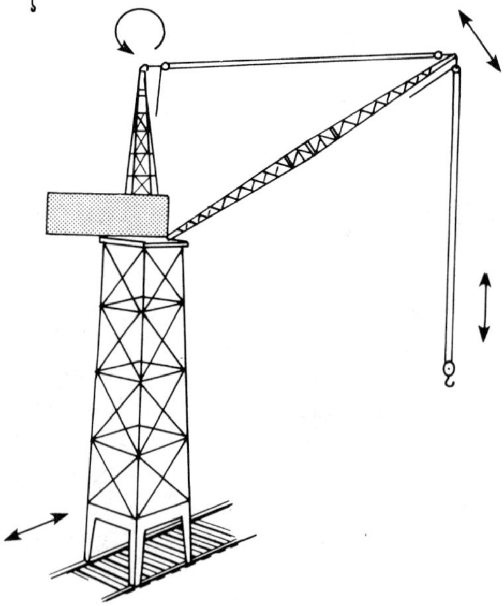

Fig. 9-61: A movable tower crane on tracks.

the tower, and the whole of the tower and jib assembly rotates relative to the base of the crane.

In addition to being either fixed or slewing, the towers are further classified as being monotowers, inner and outer towers, and telescopic towers. On the monotowers, the jib is carried by a single tower structure which can either be fixed or slewing (Fig. 9-62). The inner and outer tower types are characterized

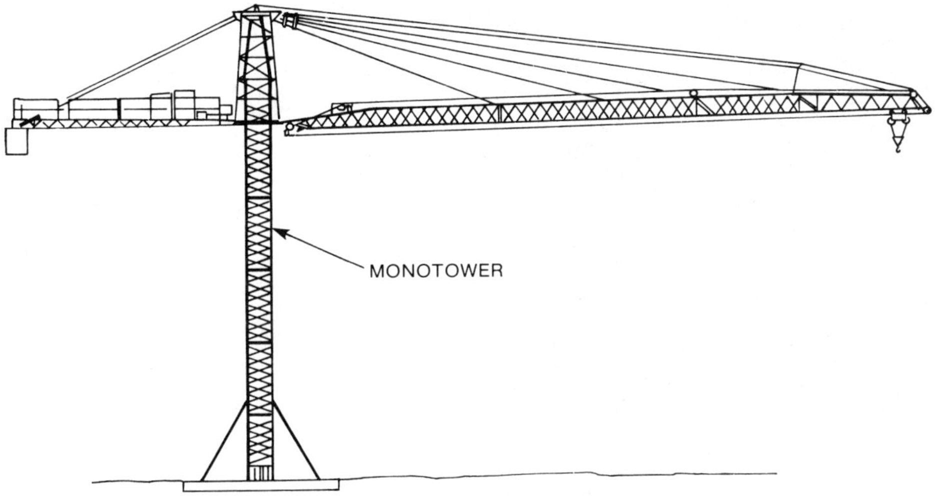

Fig. 9-62: A monotower configuration.

by the jib being carried by a slewing inner tower which is supported at the top of the fixed outer tower (Fig. 9-63). The telescopic tower structure consists of two or more main sections which nest into each other to enable the height of the crane to be altered without the need for partial dismantling and re-erection (Fig. 9-64). Telescopic towers are usually of the slewing type and are found more commonly on rail mounted and mobile tower cranes.

Jib Configuration. The main types of jibs used on tower cranes are saddle jibs, luffing jibs, fixed-luff jibs, and rear pivoted luffing jibs.

Saddle jibs are supported by pendants in a horizontal or near horizontal position, and the load hook is suspended from a trolley which moves along the jib to alter the hook radius (Fig. 9-65). Luffing jibs are pivoted at the jib foot and are supported by luffing cables much like the boom on mobile cranes (Fig. 9-66). The hoist rope which supports the load usually passes over a sheave at the jib head, and the hook radius is altered by changing the angle of inclination of the jib.

Fixed-luff jibs are also mounted on pivots at the jib foot; but unlike the luffing jibs, they are held by jib pendants at a fixed angle of inclination (Fig. 9-67). On some types the hook is suspended from the jib head and the hook radius cannot be altered, while on others the hook is suspended from a trolley which

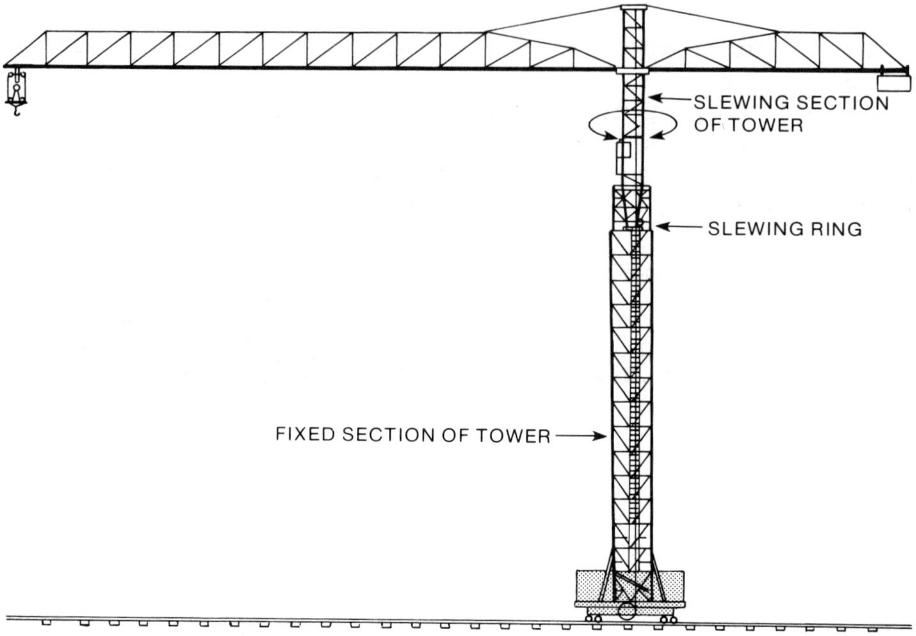

Fig. 9-63: An inner and outer type tower crane configuration.

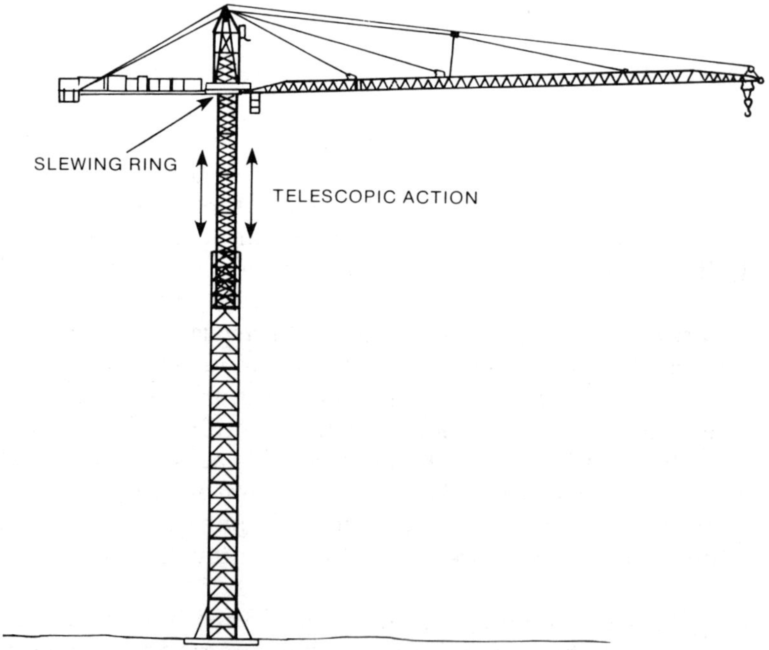

Fig. 9-64: A typical telescopic tower structure.

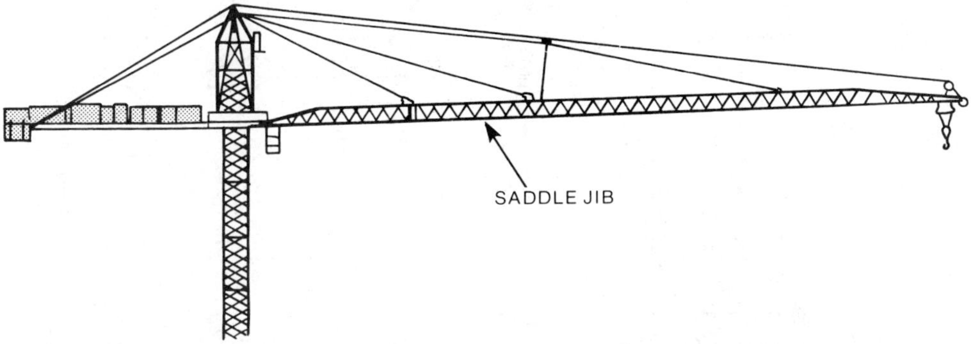

Fig. 9-65: The saddle jib configuration.

travels on the jib. On rear pivoted luffing jibs, the jib pivot is situated towards the rear of the top of the tower and the hook is supported by the hoist rope which passes over a sheave at the jib head (Fig. 9-68).

The saddle jib usually has a smaller minimum operating radius than the equivalent luffing jib and, thus, is able to handle loads closer to the tower of the crane. For a given height of tower, however, a greater height of lift is available with a luffing jib, and the jib can be raised or lowered to clear obstacles. One advantage of a fixed luff jib is that its extra height at the jib head enables it to

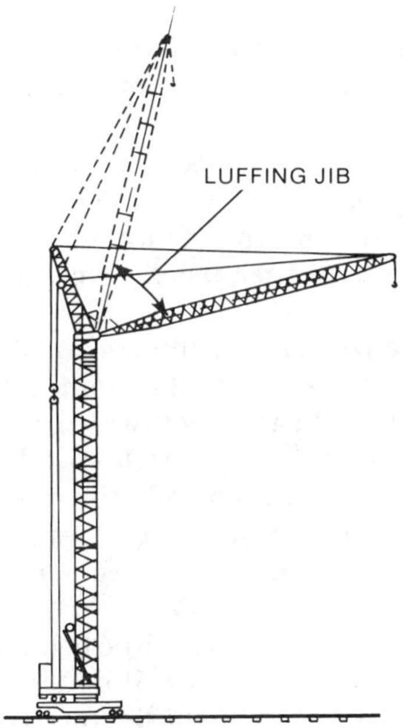

Fig. 9-66: A type of luffing jib configuration.

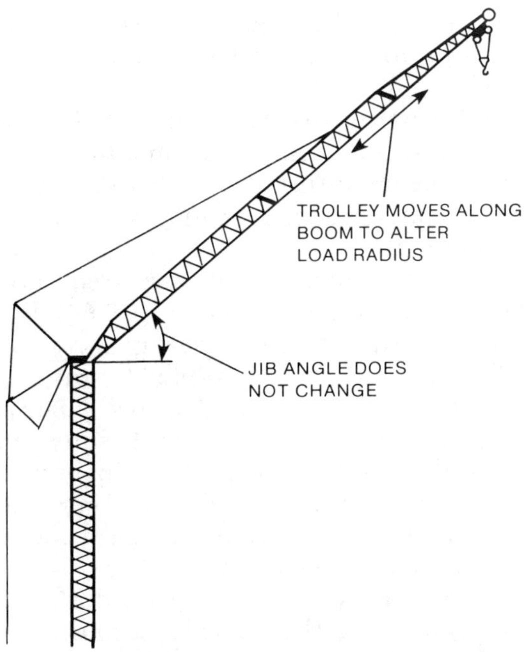

Fig. 9-67: A fixed-luff jib.

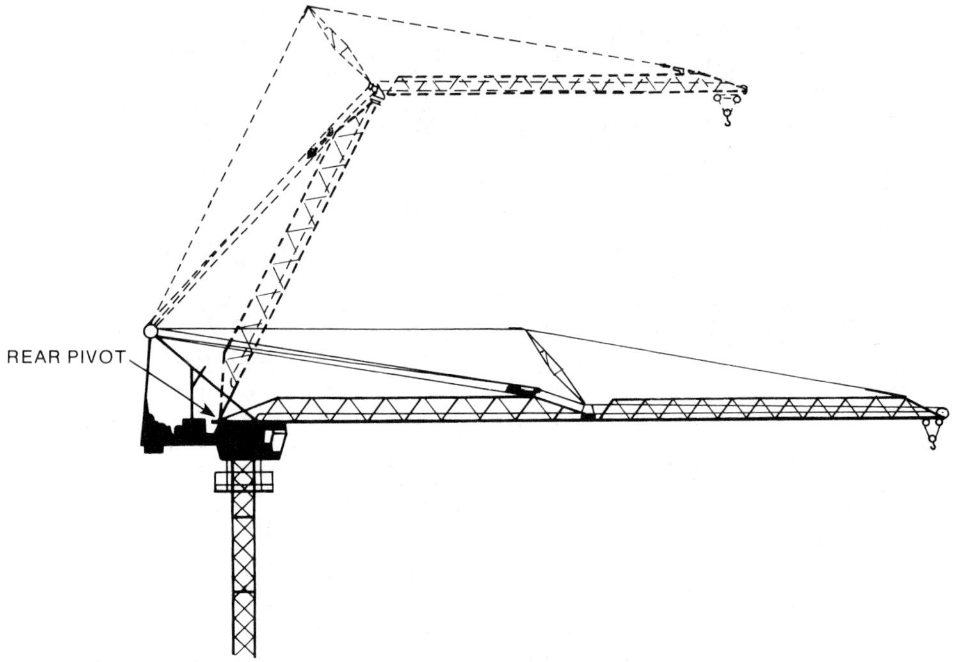

Fig. 9-68: A rear pivoted luffing jib configuration.

clear objects that would obstruct a saddle jib. The advantage of a rear-pivoted luffing jib is that it has a smaller minimum hook radius than an ordinary luffing jib.

Mounting Configurations. In addition to being classified according to tower and jib configurations, tower cranes are also characterized according to their mounting configuration. They are available as rail mounted units, stationary units, climbing units, and mobile units (either truck or crawler mounted).

The rail mounted units can be equipped with fixed or slewing towers and any of the jib configurations (Fig. 9-69). Because of their mobility, they generally have a larger area of coverage than do climbing and stationary tower cranes. They are also better adapted to traveling with loads than are mobile (truck or crawler mounted) tower cranes. Their primary advantage is maximum coverage with minimal site space. Disadvantages of the units include the expense of installing the track and the inconvenience of disrupting the crane service whenever work must be done on or near the track.

The stationary tower configurations can be used to good advantage because they occupy such a limited area (Fig. 9-70). They can be set at varying heights up to their maximum freestanding height and can be extended beyond this limit by tying the crane back to the structure. A tower crane on a static base must be able to cover from its fixed position all points at which the loads are to be handled.

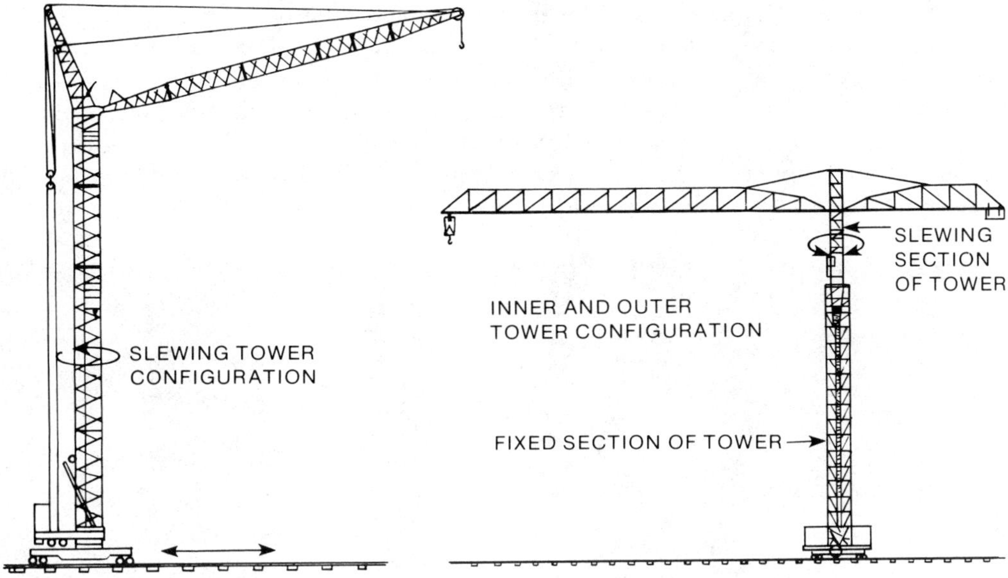

Fig. 9-69: Types of rail mounted slewing tower arrangements.

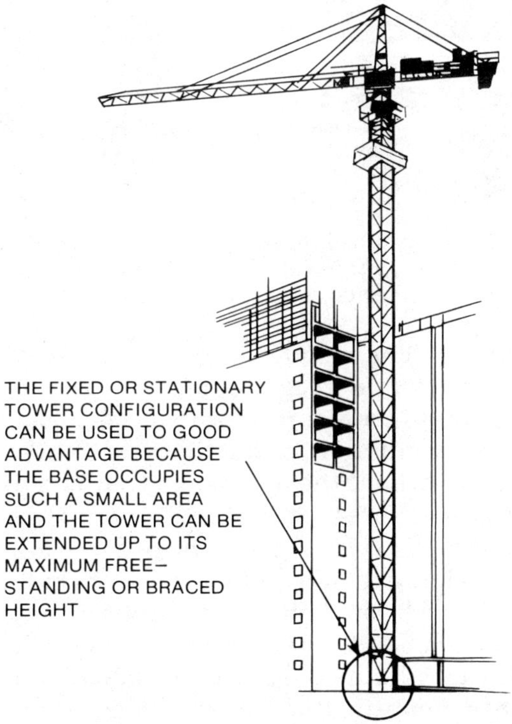

Fig. 9-70: One of the advantages of a stationary tower crane.

Static Base Mountings. Tower cranes, specifically the saddle jib or the articulated jib type, use a base for the tower which rests on top of a concrete foundation block. Various manufacturers require the use of anchor bolts for attachment, while others require that part of an expendable tower section be cast directly into the concrete foundation block (Fig. 9-71). An optional base uses steel support beams called knee braces. Both the tower and braces attach to a base frame which adds additional support to the structure (Fig. 9-72).

Fig. 9-71: An expendable tower base cast directly into concrete (photo courtesy of Tower Cranes of America, Inc.).

Fig. 9-72: A knee-braced base for a tower crane (photo courtesy of Tower Cranes of America, Inc.).

Mobile Mountings. The mobile mounting configurations for tower cranes consist of either crawler or truck mounted units, much like the mobile crane class. The crawler units are equipped with special booms that are set vertically and carry either horizontal or luffing jibs (Fig. 9-73). Their advantage lies in great inward reach without the long boom that is normally required to reach over the top of the structure.

The crawler mounted tower cranes must be set firm and level when handling their rated loads. They are able to travel over firm and level ground in their erected state, but have limited ability to handle loads while doing so. They are also able to travel in a partially erected state over unprepared ground, providing it is within certain limits of level and compaction.

The base or base mounting for the crawler crane is not truly the carrier portion of this lifting device. The crawler treads are the true carriers. The base mounting includes a central portion, called the car body, with the swing or slewing circle located on top of it. The slewing circle is the location of that point where the upper structure is joined to the base. Most crawler tower cranes are equipped with extending crawlers for greater stability. The various ways which the crawler base can be extended are shown in Fig 9-74.

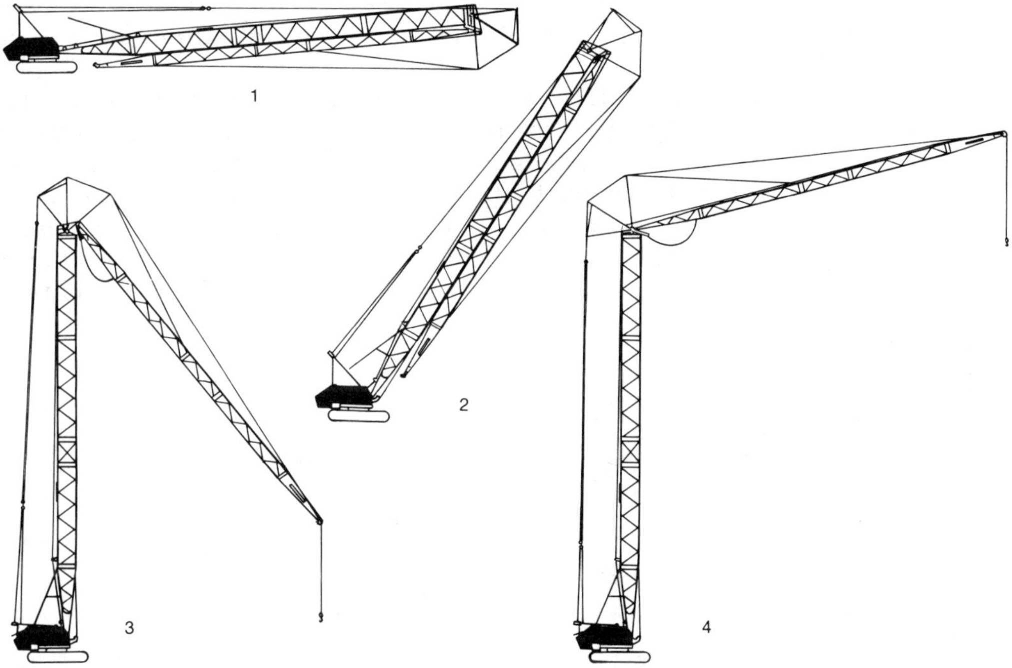

Fig. 9-73: Crawler mounted tower cranes.

Fig. 9-74: Various ways in which a crawler base can be extended (photos courtesy of Harnischfeger Corp.).

The truck mounted mobile tower crane must also have its outriggers extended and set up secure and level on its jacks when handling loads (Fig. 9-75). The majority of these machines have slewing towers and folding luffing jibs which facilitate transportation and erection. Some units are completely self-contained and self-powered and can be driven on the public highways. These machines are usually capable of comparatively rapid erection and dismantling, but are generally unable to travel in their fully erected state and cannot handle loads while traveling.

Climbing Tower Cranes. If the crane is used at a height which is to be later extended, it is advantageous if the tower has the facility for adding its own extra sections without dismantling any of the jib/counterjib assembly (Fig. 9-76). A crane such as this adds to the efficiency of the work unit because of decreased crane erection time. The climbing tower crane is generally used to good advantage when the building structure is high, the side area is limited, and the structure itself is capable of supporting the crane (Fig. 9-77). This con-

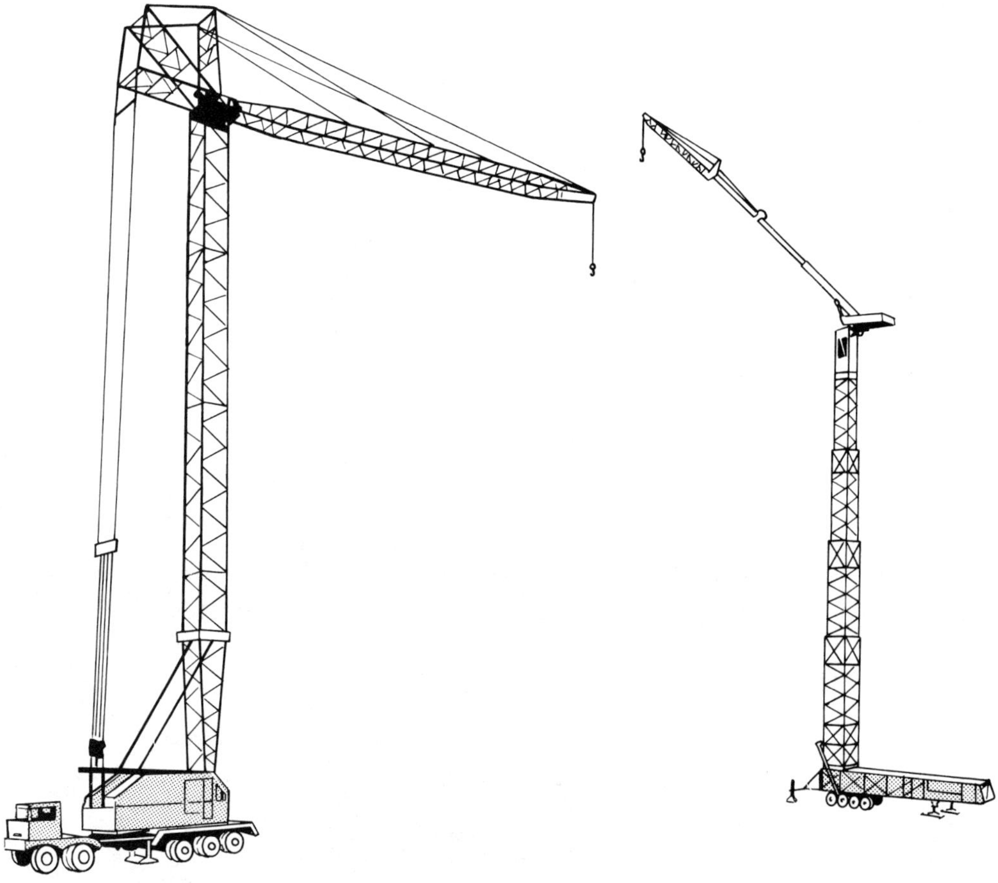

Fig. 9-75: Truck mounted and trailer mounted tower cranes.

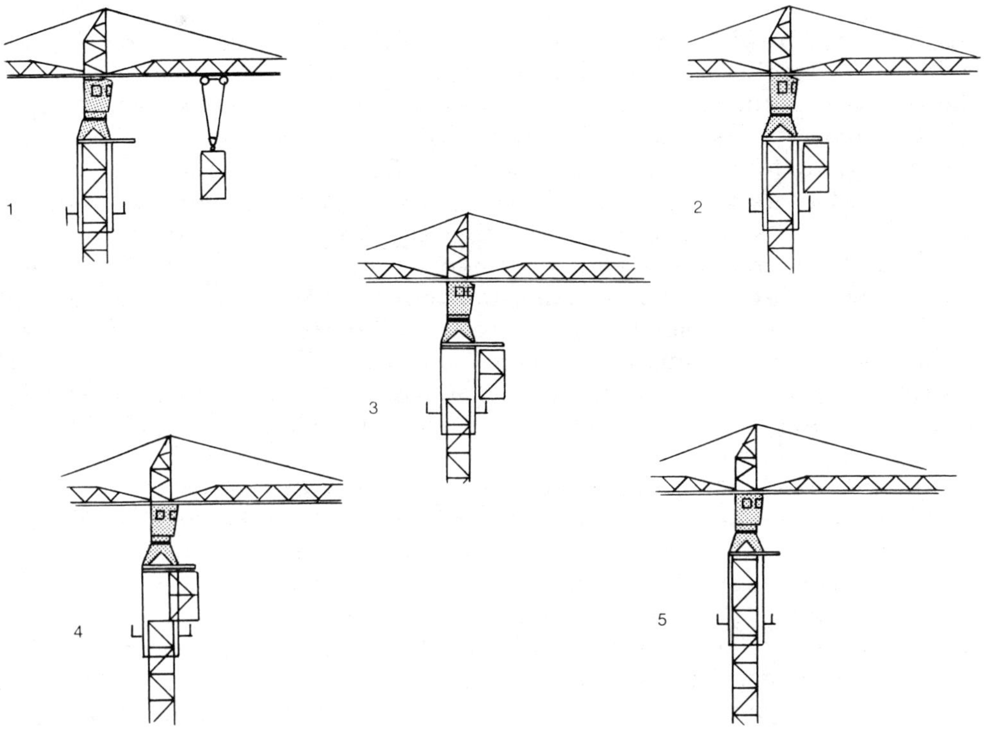

Fig. 9-76: Increasing a crane's own height by adding sections.

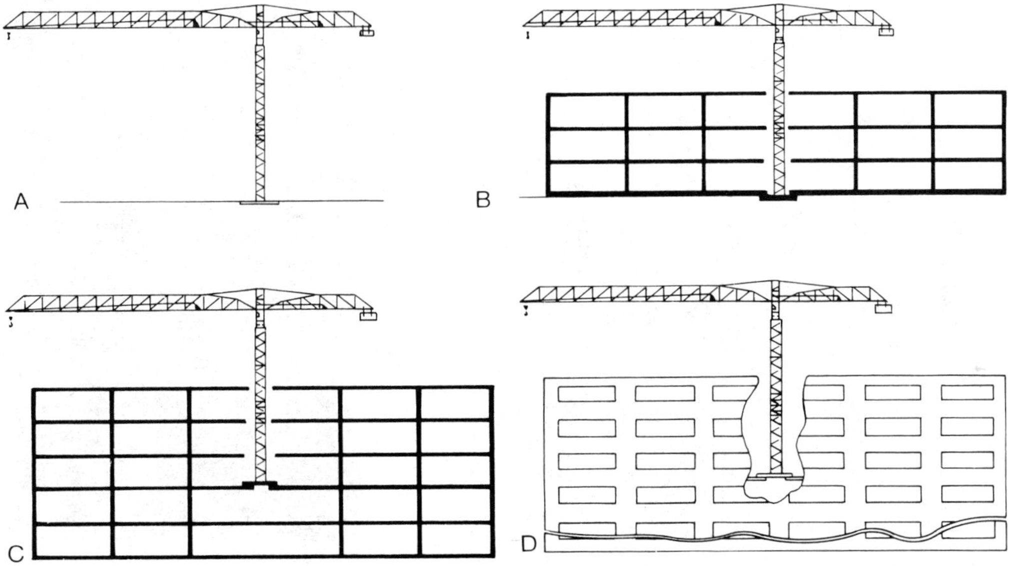

Fig. 9-77: A type of climbing tower crane.

figuration of crane is supported by the structure which it is constructing and to which it is attached by support frames and wedges. The height of the crane can be extended as the height of the structure increases by means of climbing ladders attached to the frames. The cranes are usually initially mounted on fixed bases and are later transferred to climbing frames and ladders.

Tower Crane Variations. Tower cranes have evolved into a wide variety of lifting devices, each suited to a particular purpose. If a crane's basic characteristics do not match the job requirements, however, an unsafe condition will be created before any work is done. Job personnel will be forced to "make do" and improvise in a rushed atmosphere, a combination that leads to accidents. Since there is a large amount of crane types from which to choose, it is important to properly match the equipment to the job. The following paragraphs describe a few of the variations of tower cranes which can be matched to specific lifting situations.

Pillar Cranes. This type of lifting device consists of a fixed crane with a vertical member held in position at the base to resist the overturning movement, with a constant radius revolving boom supported at the outer end by a tension member. Figure 9-78 illustrates a pillar crane.

Portal Cranes. A portal crane resembles a combination of a tower crane and a gantry crane (Fig. 9-79) without the trolley motion of the gantry crane. The boom on this type of crane is attached to a revolving crane mounted on a gantry with the boom capable of being raised or lowered at its head (outer end). Portal cranes can be either fixed or mobile.

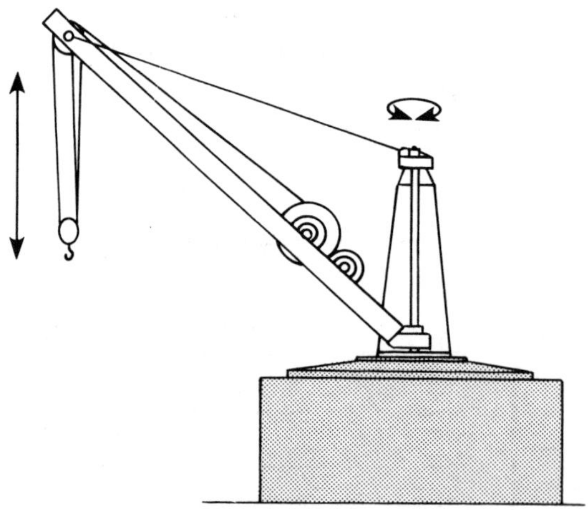

Fig. 9-78: The fundamental characteristics of a pillar crane.

Fig. 9-79: A type of portal crane (photo courtesy of F.B. Kroll A/S).

Hammerhead Tower Cranes. This lifting device consists of a mast with an upper rotating member to which is attached a jib boom extending horizontally from the rotating member, with a counterweight jib boom extending from the rotating member in the opposite direction. *Note:* Neither of the jib booms is arranged for luffing.

The main jib boom carries a trolley which travels the length of the jib and contains the necessary sheaves and associated parts that make up the upper load block and from which the lower load block is suspended. The counterweight jib boom contains the counterweight for the main jib and working load. It sometimes provides the mounting for the hoisting and travel motors and drums. Figure 9-80 illustrates a rail mounted hammerhead tower crane with a fixed tower.

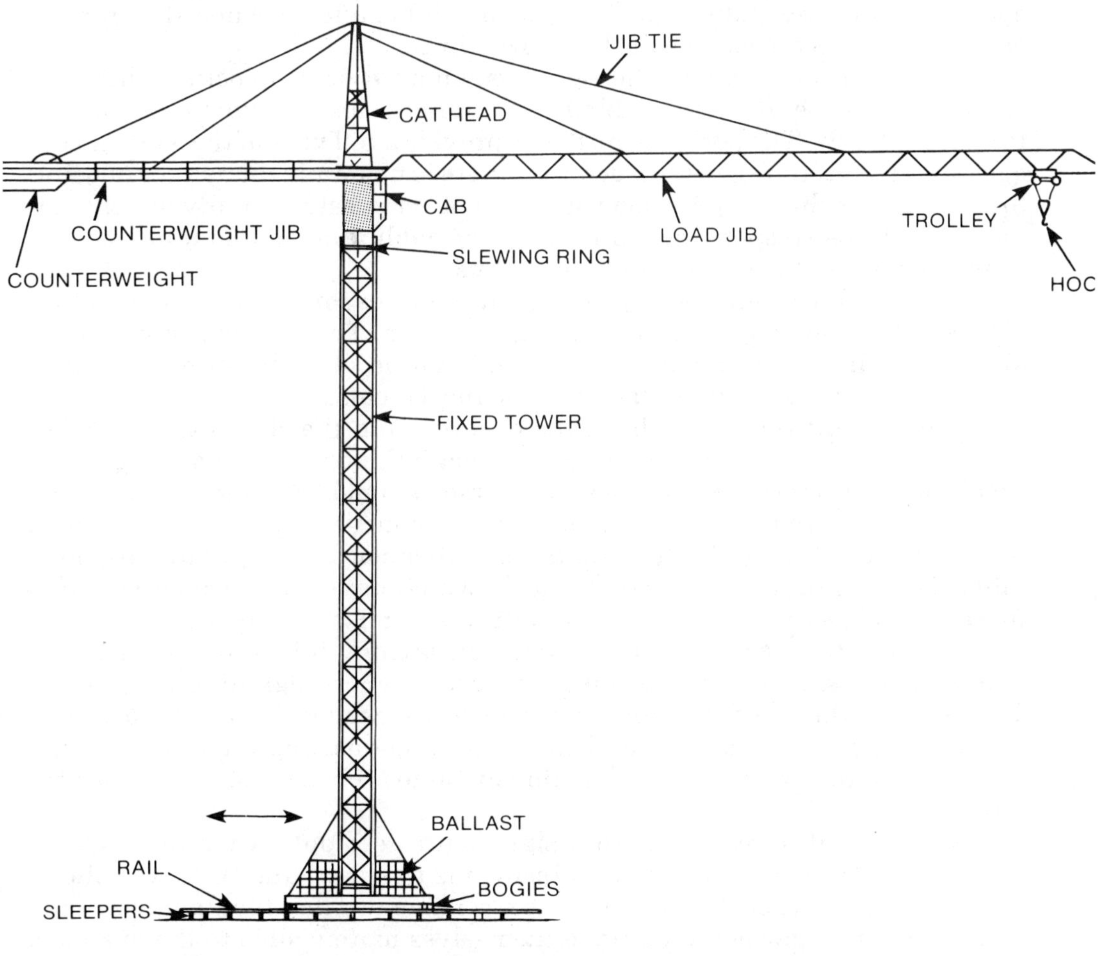

Fig. 9-80: A rail mounted hammerhead crane with a fixed tower.

Signaling

One of the most important functions performed on the construction site is that of the signalperson. It is the signalperson's responsibility not only to have the load picked and set in properly, but also to ensure the safety of those in the lift area. This individual becomes the eyes of the crane operator, and he or she must be completely familiar with the equipment being operated and the lifting procedures being employed. Whoever is in charge of any operation using any crane (contractor, foreperson, or superintendent) must ensure that a fully competent signalperson is assigned to work with the crane for as long as it is in their area of responsibility.

A signalperson is required to be present at all times when the operator cannot clearly see the hook and the load, when the crane is being moved and the operator cannot see all the parts of the machine, and/or when the crane is operating within a boom's length of power lines.

When a lift is being made, the signalperson must be in a position that is in full view of the operator and, if using hand signals, close enough for the signals to be seen clearly. The position must also provide a full view of the load and the equipment at all times, yet be such that there is no imminent danger of being hit. Along with directing the load so that it never passes over anyone, the signalperson is also responsible for keeping the public and all unauthorized personnel outside of the crane operating radius.

The safe and efficient operation of a crane requires constant communication between the crane operator and the signalperson. The standard crane hand symbols, radio communication, or a combination of both are the essential methods of communication used by these individuals.

It is important that both the crane operator and the signalperson understand the type of signals being used. Although there are standard signaling procedures, the operator and the signalperson should brief each other on the specific signals they intend to use. As today's modern lifting requirements become more exacting, the duties of the signalperson take on greater responsibility. Massive lifts in excess of 200 tons, which have now become commonplace, challenge the skills of both signalperson and crane operator.

Standard Hand Signals for Crane Operation. It is recommended that two-way radios be used for craning operations where signaling is required. This is due to the fact that many accidents have resulted from misunderstood or misinterpreted hand signals. Hand signaling does represent, however, a viable alternative when communication by radio is obstructed or in some way impeded.

It is important to remember that signals are used only when the distance between the operator and the signalperson is not great and the atmospheric conditions allow for clear visibility. It is recommended that the signalperson(s) wear distinctive garments to make themselves more visible to the operator. High visibility gloves, in this case, are a necessity. Figure 9-81 illustrates the standard hand signals for controlling crane operations.

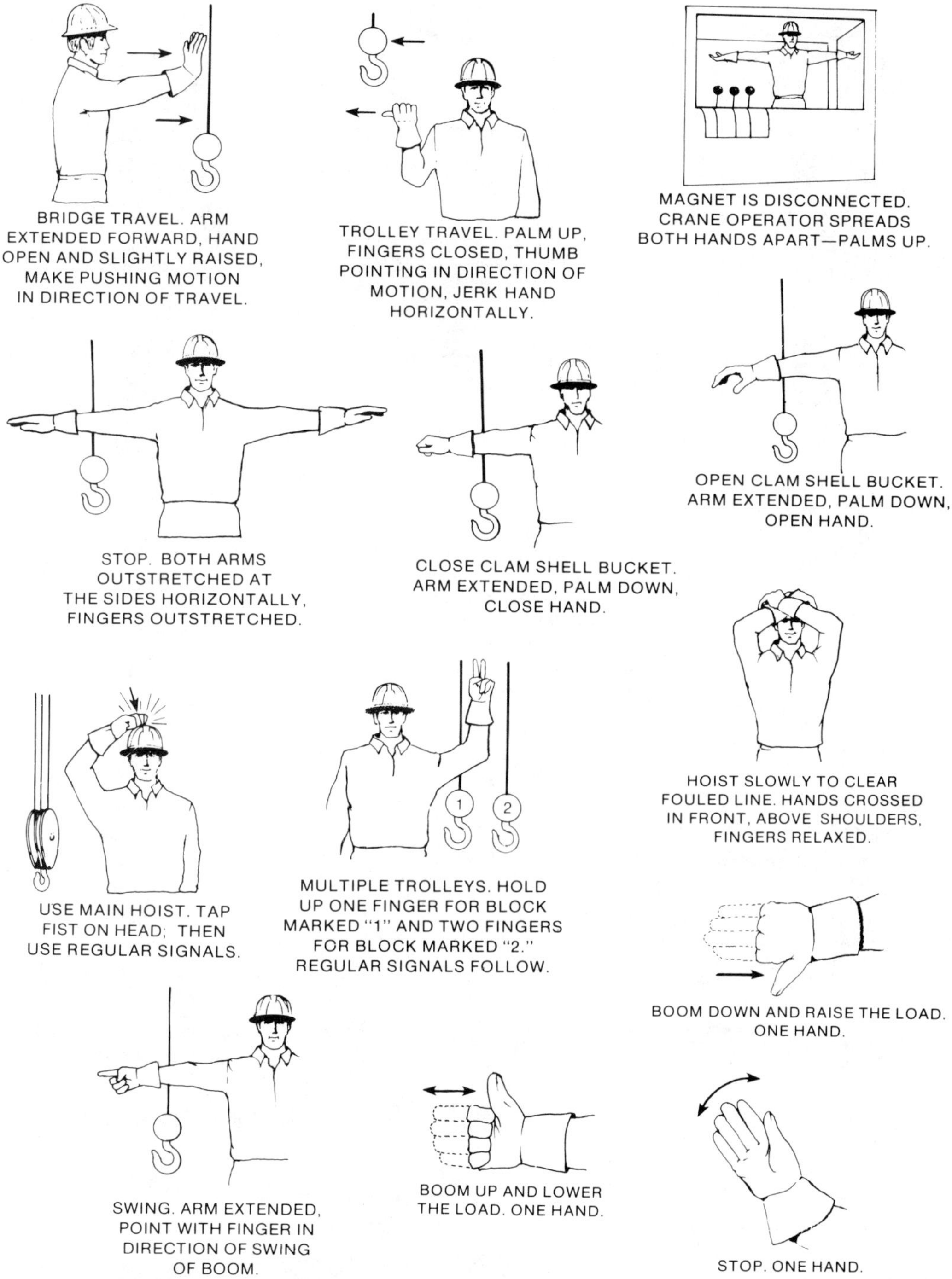

Fig. 9-81: Standard hand signals for controlling crane operations.

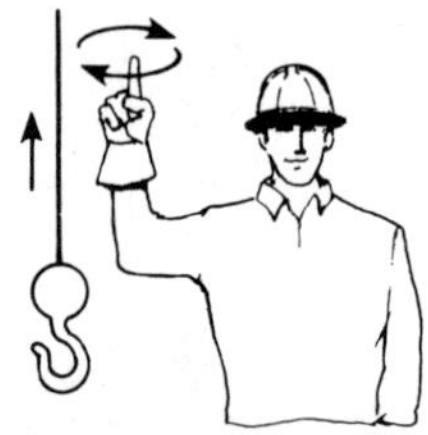

HOIST. WITH FOREARM VERTICAL, FOREFINGER POINTING UP, MOVE HAND IN SMALL HORIZONTAL CIRCLES.

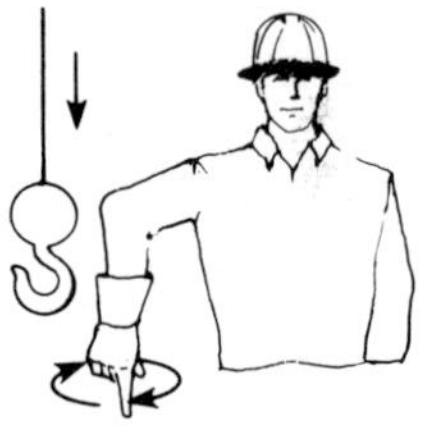

LOWER. WITH ARM EXTENDED DOWNWARD, FOREFINGER POINTING DOWN, MOVE HAND IN SMALL HORIZONTAL CIRCLES.

USE WHIPLINE. (AUXILIARY HOIST). TAP ELBOW WITH ONE HAND; THEN USE REGULAR SIGNALS.

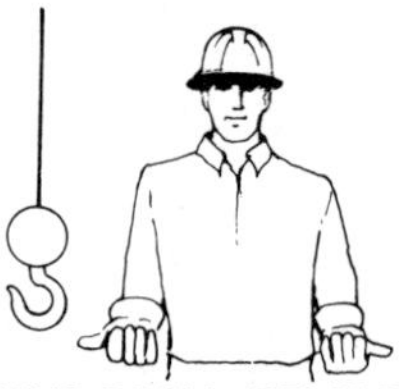

EXTEND BOOM. (TELESCOPING BOOMS). BOTH FISTS IN FRONT OF BODY WITH THUMBS POINTING OUTWARD ONE HAND SIGNAL MAY BE USED.

RETRACT BOOM. (TELESCOPING BOOMS). BOTH FISTS IN FRONT OF BODY WITH THUMBS POINTING TOWARD EACH OTHER. ONE HAND SIGNAL MAY BE USED.

LOWER THE BOOM AND RAISE THE LOAD. ARM EXTENDED, FINGERS CLOSED, THUMB POINTING DOWNWARD, OTHER ARM WITH FOREARM VERTICAL, FOREFINGER POINTING UPWARD AND ROTATE HAND IN HORIZONTAL CIRCLES.

MOVE SLOWLY. USE ONE HAND TO GIVE ANY MOTION SIGNAL AND PLACE OTHER HAND MOTIONLESS IN FRONT OF HAND GIVING THE MOTION SIGNAL. (HOIST SLOWLY SHOWN AS EXAMPLE).

LOWER BOOM. ARM EXTENDED, FINGERS CLOSED, THUMB POINTING DOWNWARD.

RAISE THE BOOM AND LOWER THE LOAD. ARM EXTENDED, FINGERS CLOSED, THUMB POINTING UPWARD, OTHER ARM BENT SLIGHTLY WITH FOREFINGER POINTING DOWN AND ROTATE HAND IN HORIZONTAL CIRCLES.

WHIP LINE. ONE HAND.

RAISE BOOM. ARM EXTENDED, FINGERS CLOSED, THUMB POINTING UPWARD.

Fig. 9-81: Standard hand signals for controlling crane operations (continued).

SPECIALLY DESIGNED CRANES

As the rigging demands of construction and industry increase, new and improved crane designs have resulted to meet these needs. Crane design engineers have become increasingly aware of the need for more versatile equipment that not only can handle heavier lifts at greater heights, but also have the capacity to literally change configurations to meet the various rigging challenges posed by each construction site. Cranes such as these have become quite popular because one crane can be used to perform many of the lifts that required several other specialized types of equipment.

Doughnut Cranes

Most cranes mounted on barges are restricted by the fact that they operate on water, and unless a complicated and expensive ballast pumping system is available, their capacities are relatively low. However, the floating doughnut crane pictured in Fig. 9-82 combines the stability of a shear leg derrick with the maneuverability and precise load spotting of a swinging crane. Because its entire self-propelled hexagonal barge turns around a fixed vertical spud during the swing, full lifting capacity and radius are available in all directions without trim compensation. Side loading on the boom and barge cannot occur, making the crane safe and efficient in both lifting and duty cycle applications. The versatility of this crane is further illustrated by the fact that, when a waterborn rig is not needed, the barge crane can be folded up for storage or transport, and the crane can be driven off for land-based lifting operations.

Fig. 9-82: A floating doughnut crane (photo courtesy of American Hoist & Derrick Co.).

Another type of water-born lifting device which is highly efficient, yet lacks the total mobility of a doughnut crane, is the barge-mounted guy derrick (Fig. 9-83).

Fig. 9-83: A barge-mounted guy derrick (photo courtesy of American Hoist & Derrick Co.).

10

Derricks

INTRODUCTION

The evolution of derricks, much like the development of cranes, has come about as a direct result of the needs of industrial and construction operations. Requirements such as these have spawned derrick designs which are capable of handling general types of lifts and highly specialized, extremely heavy lifting situations.

A derrick is defined as an apparatus consisting of a mast or equivalent member held at the head by guys or braces, with or without a boom, for use with a hoisting mechanism, such as a winch, and operating ropes. These devices are quite versatile. They can be mounted on almost any frame, such as completed buildings, roof tops, structural steel, or their own freestanding mount. The above definition most aptly applies to a lifting device known as the Chicago boom derrick, and since most other types of derricks are nothing more than variations of the Chicago boom, it will be discussed first.

CHICAGO BOOM DERRICKS

This type of lifting device is quite versatile in that it can be mounted to a structural steel building frame during or after the building's construction. These derricks can also be mounted to towers or any other type of frame. The Chicago boom derrick can range in length anywhere from 10′ (3 m) to 125′ (38 m) and is able to lift up to 35 tons. This simple apparatus can be constructed of steel pipes fitted with the proper equipment. However, most booms are latticed, trussed, or constructed of structures consisting of angle irons, tubing, or a combination of the two. Figure 10-1 illustrates the Chicago boom derrick. The topping blocks

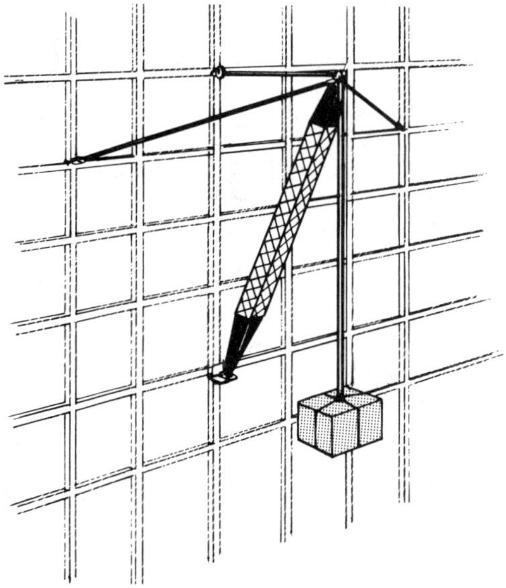

Fig. 10-1: A Chicago boom derrick.

consist of basic hoisting blocks with one block fitted at the tip of the boom and the other mounted at the pivot fitting on the support structure.

Chicago boom derricks are used to lift materials to a height somewhat above the boom foot, but their horizontal reach is limited by the length of the boom. The boom is swung by means of swing guys fitted to the tip of the boom which run laterally on each side to a point of anchorage.

During the swinging of a Chicago boom derrick, wind resistance, friction at the pivots, and resistance to the opposite swing line must be overcome. Swing of the derrick can be accomplished by manual labor or by mechanical means when the use of hand pulling is either too slow or too costly.

A winch matched to the derrick's size powers the derrick's hoisting and topping motions. When work only involves lifting objects and swinging them into position, the topping motion is not required. Locating the winch at the floor level of the boom foot often improves operational control, especially when the loads are to be hoisted to that floor, but the winch can be located on any floor. Winches can also be located at ground level.

In situations where the winch is located at ground level, the derrick operator can see both the boom and load at all times. The operator is also in direct communication with the ground crew. In cases where the winch is located on the boom foot floor, the operator has direct communication with the swing and load landing crew, and the boom (but not the load) is in view at all times. Landing the load is a critical point in the operation, so clear direct communication is highly desirable.

In theory, any column of a framed structure can be used as the mast of a Chicago boom derrick. However, corner columns should generally be avoided since arrangements to allow for swinging through a 270° arc can be quite complex.

Although a Chicago boom derrick is an efficient lifting device, it is not designed for high production usage. It can be used to great advantage when the required lifts exceed the reach or capacity of mobile cranes; however, a lower production rate can be expected.

GUY DERRICKS

Guy derricks are commonly employed in the erection of steelwork for high rise structures. These derricks are also useful in the construction of oil and chemical plants and in some types of stone quarry work.

A guy derrick resembles a Chicago boom derrick in that its boom arrangement is almost identical to that of a Chicago boom. A guy derrick, however, is a more complex and more versatile machine because of the slewing ability of its mast.

A guy derrick consists of a mast, a boom pivoted at the foot of the mast, guys, and tackle (Fig. 10-2). The mast is generally longer than the boom and is mounted on vertical pins at the foot and top so that the assembly of mast and boom may be rotated about a vertical axis. Guy ropes are attached to the top of the mast through the use of a component called a spider (Fig. 10-3). When a guy derrick is swung, both the mast and boom move and if the mast clears the guy ropes the derrick can be rotated through a full circle. A fitting called the gudgeon pin is located at the top of the mast and allows it to pivot, while a ball and socket type joint provides the pivot mechanism at the bottom. The boom swing is actuated by a large wheel called the bull wheel, which is fitted at the bottom of the mast. Wire rope is run around the bull wheel to a winch, which is the prime mover for the slewing action.

The mast and boom of a guy derrick can be made of timber or structural steel. Large derricks are made of hollow built-up sections usually consisting of four angle irons and lacing bars forming a truss. Figure 10-4 shows a guy derrick in operation.

Hoisting tackle is suspended from the end of the boom, and tackle is also provided for raising or lowering the boom in a vertical plane. Power is supplied by hand operated or engine driven hoists. The derrick can also be rotated by a bull pole located at the base of the mast. This device adds leverage so the derrick can be swung manually. The boom angle can also be changed by means of a block and tackle arrangement called the topping lift. It is attached to the mast just below the gudgeon pin and affixed to the boom tip.

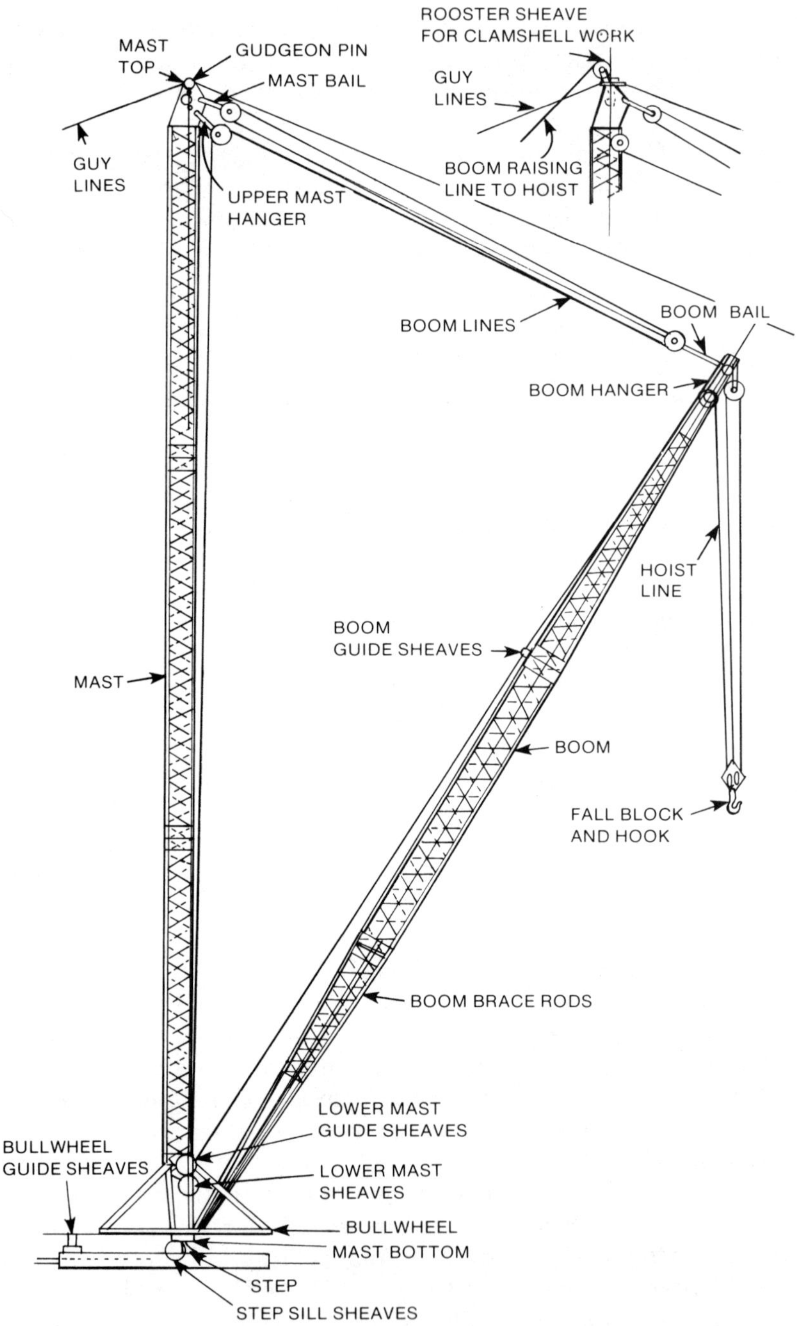

Fig. 10-2: The components of a guy derrick.

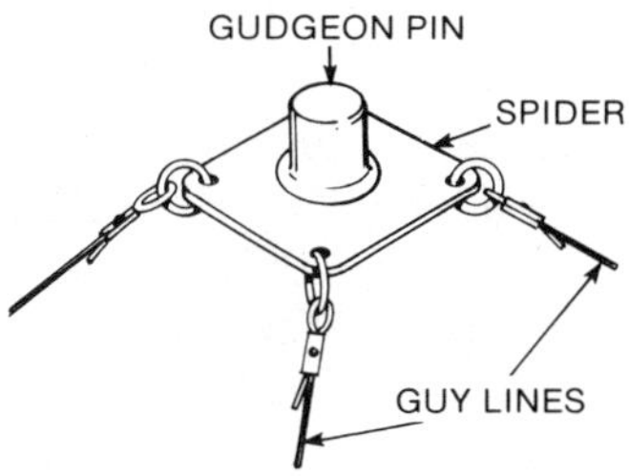

Fig. 10-3: Guy wires are attached to the spider at the top of the mast.

Fig. 10-4: A guy derrick in operation above a containment building (photo courtesy of CBI Industries, Inc.).

The mast, an integral part of any guy derrick, must be held plumb in order to maintain proper swing force levels. The function of plumbing the mast lies primarily with the guy lines. Due to their length, the guys hang in a catenary shape even when slack is taken up. When a load is applied, the guy wires tend to pull toward the load, reducing catenary sag and causing the mast to lean toward the boom. It is important, therefore, to control this type of stress. Factors such as the size, construction, tension of the guys, and load levels must be considered in order to correct mast lean.

The guyed derrick requires support at various locations and, depending upon the particular configuration, must have specific anchorage points for the guys. When a guyed derrick is mounted on the ground or has its own tower-like mounting, the guys must be properly anchored through the use of poured concrete pilings (Fig. 10-5), and the mast must be well supported. A type of steel grillage is usually placed under the mast and the guys anchored to the structural steel when the derrick is mounted on a building.

When a guyed derrick is initially installed on a job site, another lifting device such as a crane is required to place the various components. Once operational, the derrick has the advantage of "jumping" floors, or lifting itself to higher work levels. As the derrick changes levels the winch remains at the base level. The winch operator essentially works blind, relying on signals or radio communication to maneuver the rig. The derrick crew consists of the winch operator and a swing person, making this derrick a most economical lifting device when considering operational labor costs.

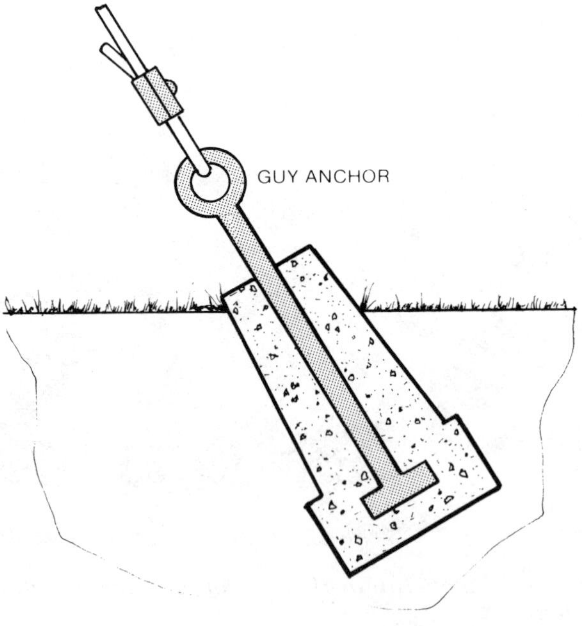

Fig. 10-5: A concrete guy rope attachment cast into the ground.

STIFF LEG DERRICKS

Stiff leg derricks are generally used in situations where the guying of other types of derricks is impossible or interferes with the placement or handling of the load.

The mast of the stiff leg derrick is held in the vertical position by two rigid inclined struts connected to the top of the mast (Fig. 10-6). The struts are spread apart 60°-90° to provide support in two directions and are attached to sills extending from the bottom of the mast. The mast is mounted on vertical pins at its foot and top as in the case of the guy derrick. The mast and boom can swing through an arc of about 270°. The tackles for hoisting loads and raising the boom are similar to those for the guy derrick. Figure 10-7 illustrates a stiff leg derrick on a platform.

A stiff leg derrick equipped with a boom is suitable for storage yard use in the unloading and transferring of material, provided the range of operations does not exceed the boom reach. Stiff leg derricks are sometimes used in conjunction with other derrick types to stage extraordinarily high lifts. For example, in the case of high rise construction, a stiff leg derrick may be used to lift

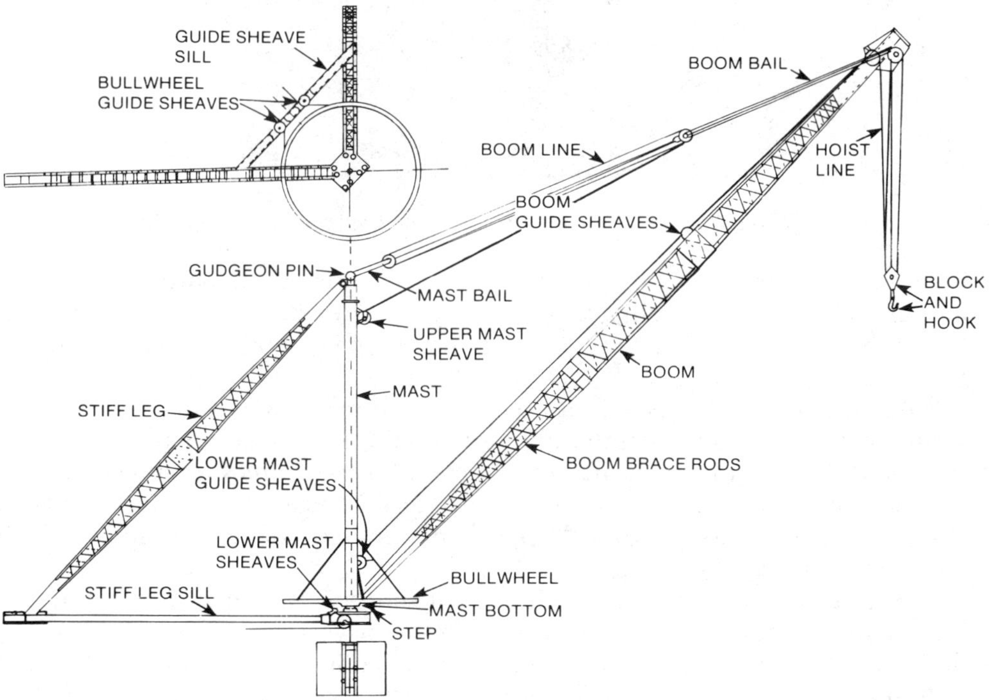

Fig. 10-6: The components of a stiff leg derrick.

Fig. 10-7: A stiff leg derrick on a platform (photo courtesy of CBI Industries, Inc.).

loads to the roof top of a building next to the construction site. This roof top actually serves as a staging platform. Guy derricks mounted on the frame of the high rise under construction pick the load up from that point and complete the lift.

When a stiff leg derrick is installed in a permanent location at ground level, structural foundations usually support the leg ends and the mast bottom. In the event that the derrick is attached to an above-ground facility, such as construction work often requires, sills or horizontal elements are added to connect the leg ends and mast bottom. These elements form vertical angles which provide the rig with added stability.

There are a wide range of stiff leg derrick sizes and capacities available for industrial and construction purposes. Light construction work may require the use of a 12 ton derrick, or lighter models with booms ranging from 30′ to 50′. For extremely high and heavy lifts, stiff leg derricks with capacities of 700 tons and booms up to 265′ are obtainable. Stiff leg derricks also have the capacity to be mounted on portal frames or towers and can either be fixed or traveling.

GIN POLES

A gin pole, like the mast of a guyed derrick, consists of an upright spar guyed at the top to maintain it in a vertical position (Fig. 10-8). It is equipped with suitable hoisting material; however, it is not intended to swing under load conditions.

The spar may be of round or square timber, heavy wall pipe, a wide flange beam section, or a built-up section consisting of angles and lacing bars. The load may be hoisted by hand tackle or engine driven hoists.

A gin pole is a very handy and versatile piece of equipment for lifting all types of comparatively light loads. To get maximum use out of a gin pole, it should be rigged with a minimum of four guys, with wire rope pendants and manila rope tackles between the pendants, and the anchorages and swivel plates top and bottom. The gin pole can then be leaned at various angles and turned to lift from all sides.

The gin pole is used mostly for erection work because of its ease of rigging, moving, and operating. It is suitable for raising loads of medium weight to heights 10′ to 50′ where a vertical lift is required.

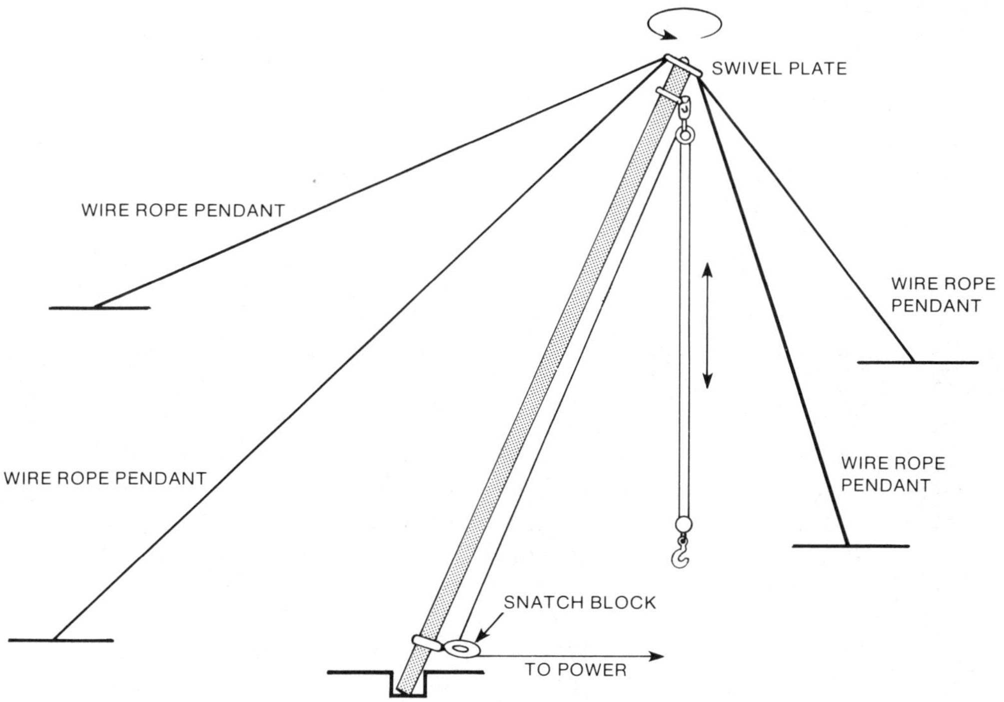

Fig. 10-8: A gin pole.

A-FRAME DERRICKS

An A-frame (or shear leg) derrick is a versatile hoisting device, requiring only two back guys and one lazy guy in front for support (Fig. 10-9). For light loads, a small A-frame derrick can be quickly constructed by drilling a hole through two pieces of square timber and bolting through the holes. A wire rope choker is put around the top of the poles with the eye hanging down to attach the block. When a load is suspended from the block, the choker tightens to hold the crossed members more securely.

A more sophisticated version of the A-frame derrick is used quite frequently on small harbor barges (Fig. 10-10). Materials are usually off-loaded from large ships onto these barges when the harbor is too shallow for docking. The barge's A-frame derrick is then used to transfer the materials from their decks to the dock.

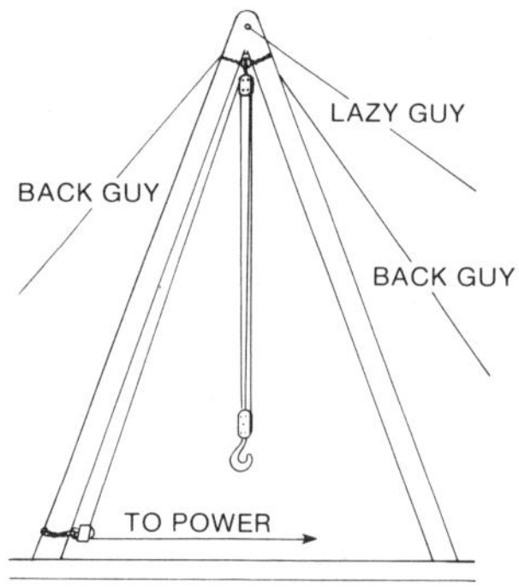

Fig. 10-9: An A-frame derrick.

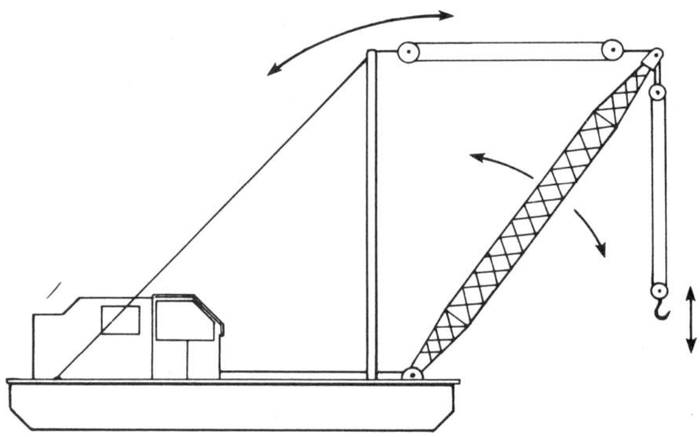

Fig. 10-10: A barge mounted A-frame derrick.

BREAST DERRICKS

This type of lifting device is a derrick without a boom. The mast consists of two side members spread farther apart at the base than at the top and tied together at top and bottom by rigid members. The mast is prevented from tipping forward by guys connected to its top. The load is lifted and lowered by ropes through a sheave or block secured to the top crosspiece. Figure 10-11 illustrates a breast derrick.

Mountings for breast derricks can include barges resembling the A-frame derrick configuration which was discussed earlier. The barge mounted breast derrick pictured in Fig. 10-12 was able to perform lifts of up to 260 tons. The barge also traveled to the shore where the loads were placed on specially designed trucks for further transport.

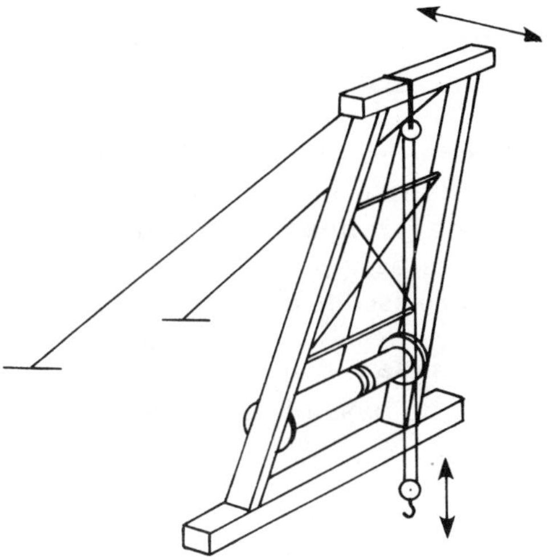

Fig. 10-11: A simple breast derrick.

Fig. 10-12: Heavy lifts can be accomplished with this barge mounted breast derrick (photo courtesy of Pennsylvania Power and Light Co.).

BASKET DERRICKS

A basket derrick resembles a derrick without a boom and is similar to a gin pole with its base supported by ropes attached to corner posts or other parts of the structure (Fig. 10-13). This type of support arrangement dictates that the base of the derrick be at a lower elevation than its supports. The top of the pole is secured with multiple reeved guys, which are used to change the position of the pole by varying the length of the upper guy lines. Loads are lifted via ropes that run through a sheave or block secured to the top of the pole. Additional ropes, called preventers, are attached from the base of the derrick down to lower sections for added stability.

GALLOWS FRAME DERRICKS

This type of lifting device consists of two gin pole assemblies with a horizontal beam across the pole heads. The upper load blocks are fitted to a boom which is arranged so that the poles are loaded concentrically. This type of arrangement provides a greater capacity than is available with similar unconnected gin poles (Fig. 10-14).

The height of a gallows frame derrick is only limited by the equipment which erects it. However, units which have hook lifts of 200′ can have capacities up to 600 tons with shorter derrick configurations capable of lifting up to 1200 tons.

The gallows beam, which connects the two poles, is a limiting factor for this type of crane, because loads which extend higher than the poles cannot be placed. Double gin poles are generally used to erect components which are taller than the poles.

CATHEAD DERRICKS

Another basic type of lifting device is the cathead derrick, which essentially consists of a cantilevered beam that supports an upper block at its outer end. When a cathead derrick is installed on a construction site, the beam is made to project well into the building to make erection and anchorage easier. The beam is locked into place either by clamping onto the floor beams below or by wedging posts between the device and the floor beams above. Figure 10-15 illustrates the lifting apparatus for a cathead derrick.

Since a typical cathead derrick is incapable of changing load radius, the load must be drifted (pulled laterally) to change its horizontal position. Loads which are suspended from relatively long ropes tend to be drifted easier and overloading on the boom is minimal. However, a shorter rope causes greater difficulty in this procedure and a greater load on the boom.

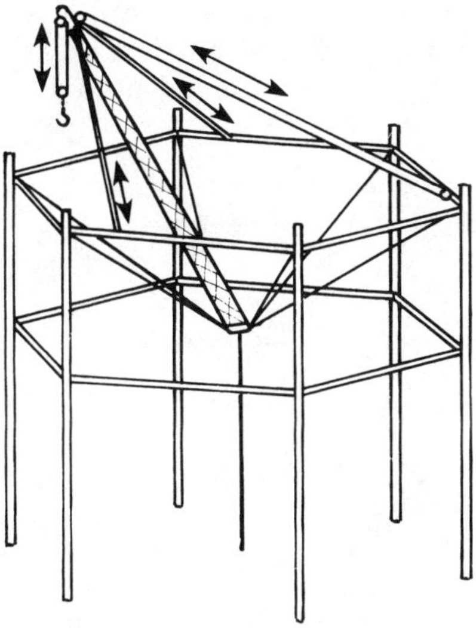

Fig. 10-13: A basket derrick.

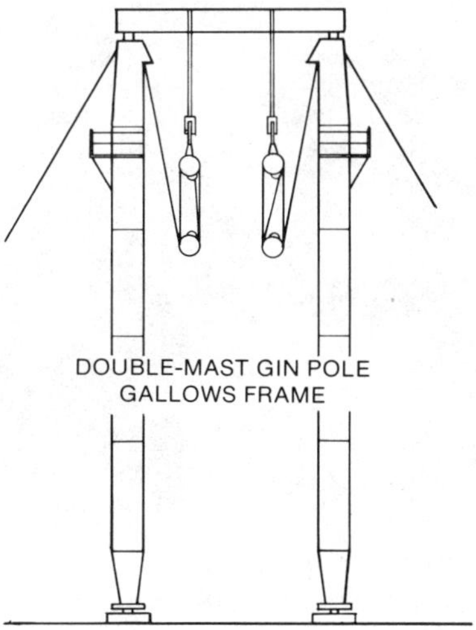

Fig. 10-14: A gallows frame derrick.

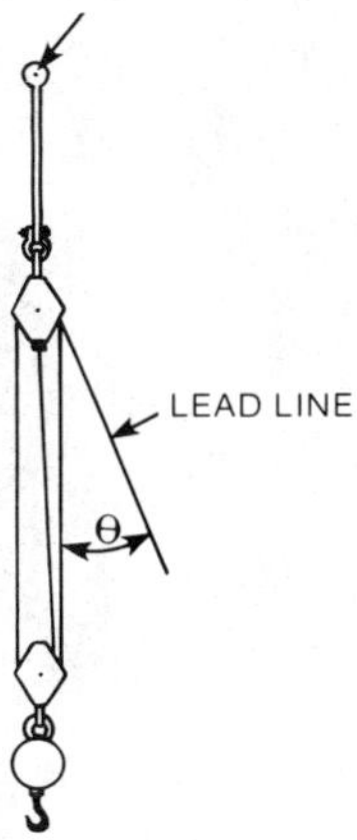

Fig. 10-15: The lifting apparatus used for a cathead derrick.

Special Application Lifting Devices

Special application derricks may be encountered when the job requires certain types of lifts. These derricks are particularly well suited to one specific type of job and are most useful if that particular lift proves too costly or dangerous for conventional type lifting devices. It is important to note, however, that these devices are generally special adaptations of the more commonly used equipment. Figure 10-16 shows a guyless derrick which is particularly suited for water tower construction while Fig. 10-17 shows a guyed pole used for special heavy lifts.

Fig. 10-16: A specialized guyless derrick for water tower construction (photo courtesy of CBI Industries, Inc.).

Fig. 10-17: A guyed pole used for special heavy lifts (photo courtesy of CBI Industries, Inc.).

11

Hoists and Winches

INTRODUCTION

Many hoists and winches are closely related in both design and function. In fact, for certain types of base mounted and power driven hoists and winches, the terms hoist and winch are often interchangeable. The definitions given in this chapter relate to the specific types of machines discussed, so a certain amount of flexibility in regard to these terms may be necessary in common practice.

HOISTS

The hoist's simple design makes it one of the most basic and easiest to use of all lifting devices. A hoist is defined as a piece of machinery, either suspended from overhead or mounted on a foundation (or other supporting structure), used for lifting or lowering freely suspended (unguided) loads. The hoist resembles the lifting components found on some cranes. In its basic configuration, a hoist can consist of a pulley and rope, or a block and tackle, operated by hand and capable of lifting up to a ton or two. Larger electrically driven chain hoists have the capabilities of lifting up to 25 tons.

A basic hoist is shown in Fig. 11-1. The load is attached to the lower block and is supported by two parts of line which are suspended from the upper block. In this case, the load is suspended from two ropes, providing the mechanism with a mechanical advantage of two. (Mechanical advantage is dictated by the number of ropes actually supporting the load.) The blocks contain sheaves so the rope can be run in one continuous piece from the load hook to the upper block to the winding drum.

Manually Operated Overhead Chain Hoists

There are several types of manually operated chain hoists, including the spur geared, the screw geared, and the differential type. These simple lifting devices are manually operated and produce differing degrees of efficiency. These three types are generally used for vertical hoisting and can be suspended from a stationary hook attached to the body of the hoist. The upper hook is always stronger than the load hook because in the event of an overload, the bottom load hook opens first, dropping the load rather than pulling the entire hoist down. The top hook must have enough support, however, to hold the load. Beam clamps are sometimes used to utilize an existing beam for hoist work. A beam clamp is a stationary device that permits safe, easy suspension of hoists from beams or girders, eliminating the use of nuts, bolts, shackles, or slings. Figure 11-2 illustrates two types of beam clamps.

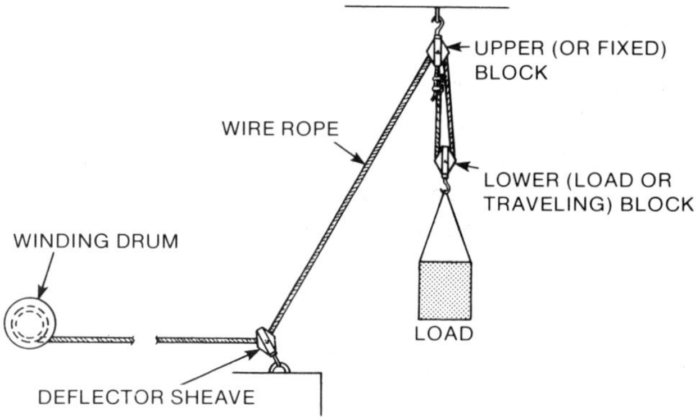

Fig. 11-1: A basic hoisting device.

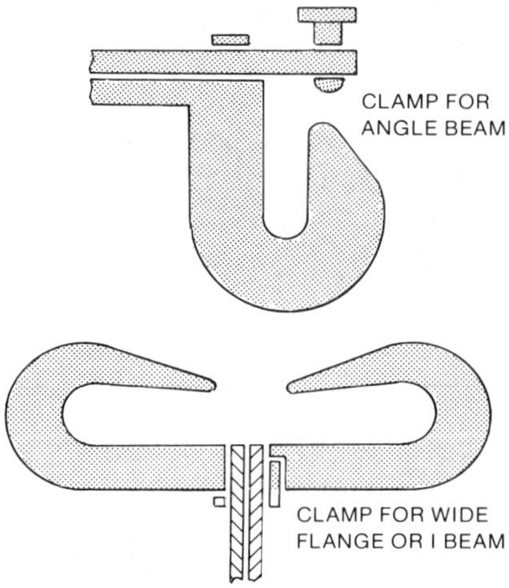

Fig. 11-2: Two types of beam clamps.

It is important to note, however, that a beam clamp must fit the beam and the beam must be sturdy enough to support the load. Also, the hoist hook should never be placed directly into the lifting eye of a beam clamp. Shackles must always be used to connect the hoist hook to the beam clamp.

Spur-Geared Chain Hoists. The spur-geared chain hoist is generally the most efficient hoist in this category. Approximately 85% of the operator's initial exerted energy is converted into useful hoisting energy, while the remaining 15% is used to overcome friction. This type of hoist requires a braking system because there is a lack of sufficient internal friction to retard the load from "running away" on the lowering motion. Figure 11-3 illustrates a manually operated spur-geared chain hoist.

Screw-Geared Chain Hoists. A screw-geared chain hoist is another type of manually operated hoist (Fig. 11-4). This lifting device, although not as efficient as the spur-geared chain hoist, still transforms from one-third to one-half of the energy exerted on the pull chain into useful hoist energy. Since this type of hoist has a relatively large amount of internal friction, there is no need for a braking system.

Differential Hoists. The differential hoist is a lifting device which utilizes two pulleys instead of a gear box for the hoisting mechanism (Fig. 11-5). This device is not as efficient as the previously discussed hoists (converts about one-third of the energy expanded); however, it is the most inexpensive to obtain. The hoist requires no braking system due to the large amount of internal friction present.

Fig. 11-3: A manually operated spur-geared chain hoist (photo courtesy of Eaton Corp., Hoisting Equipment Div.).

Fig. 11-4: A screw-geared chain hoist.

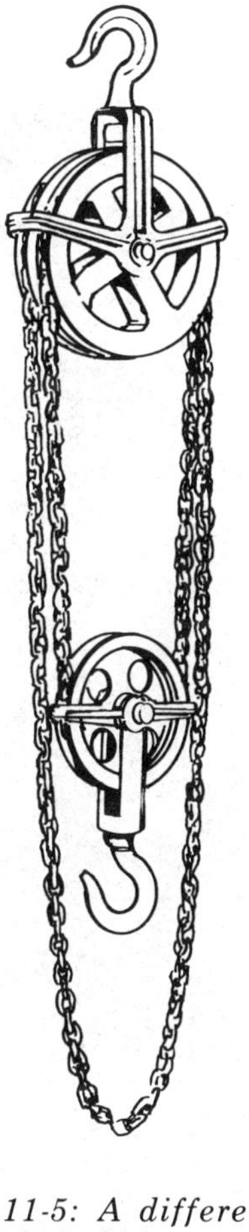

Fig. 11-5: A differential type chain hoist.

Trolley Mounted Chain Hoists. Chain hoists such as these are used when the load must be traveled. There are especially two types of trolley systems for manually operated chain hoists, including push type trolleys and hand-geared trolleys.

The push type trolley system is activated by the operator simply exerting force on the load to "push" it along the overhead system. This system usually consists of an overhead monorail type arrangement (Fig. 11-6) or a double rail system which resembles that of an overhead traveling crane (Fig. 11-7).

The hand-geared trolley system is almost identical to the push type model except that the wheel flanges on one side are geared and meshed with a pinion and are driven by a hand wheel and hand chain (Fig. 11-8).

Heavy Lift Chain Hoists. Manually operated chain hoists are available with heavy lift capabilities ranging from 10 to 40 tons. A hoist of this type achieves these capabilities by means of reduction gearing and multiple reeved chains. Figure 11-9 shows two types of manually operated heavy lift chain hoists.

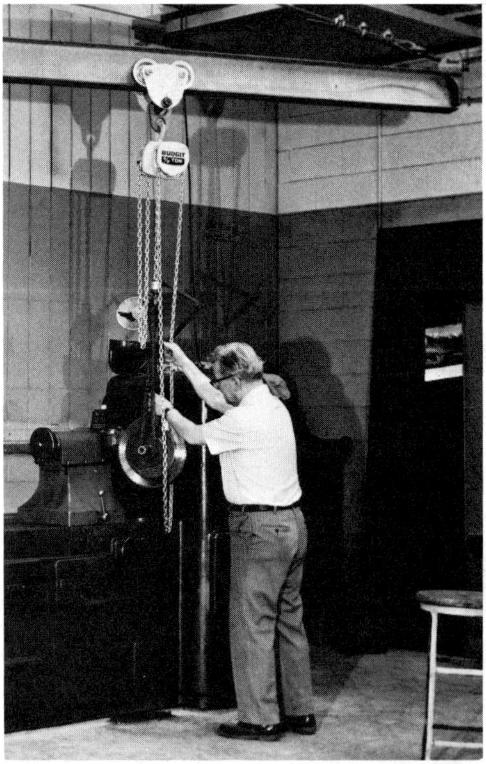

Fig. 11-6: A monorail push type trolley system (photo courtesy of Dresser Industries, Inc., Crane and Hoist Operations).

Fig. 11-7: A double rail push type trolley system (photo courtesy of Dresser Industries, Inc., Crane and Hoist Operations).

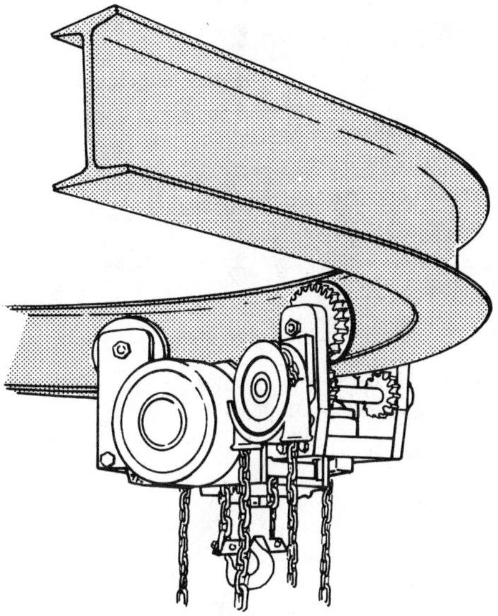

Fig. 11-8: A typical hand-geared trolley system.

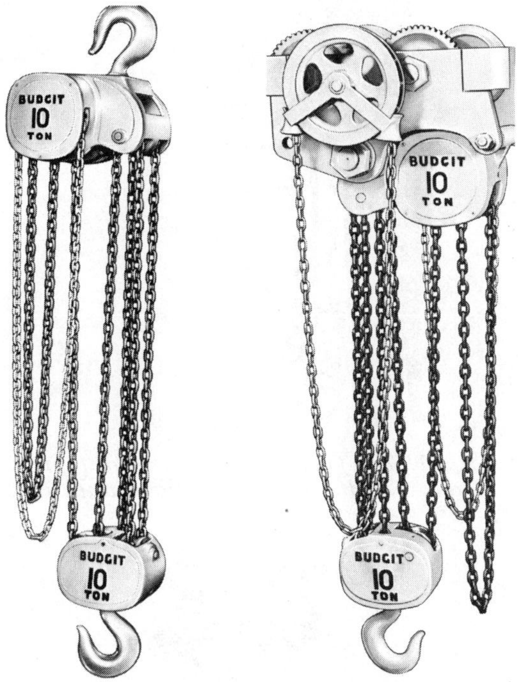

HI-CAP ALUMINUM HAND CHAIN HOIST

ALUMINUM ARMY TYPE TROLLEY HOIST

Fig. 11-9: Two types of heavy lift chain hoists (photos courtesy of Dresser Industries, Inc., Crane and Hoist Operations).

Power Operated Hoists

Hoists which utilize either air or electrical power are available with chains or wire ropes for load handling. Units such as these are controlled by pendant pushbutton stations (Fig. 11-10) or via ropes suspended from the various switches on the hoist housing (Fig. 11-11).

Since there is a wide variety of power hoist types and capacities, industry has been able to match these lifting devices to virtually any application. Equipment maintenance centers, manufacturers, and contractors, to name a few, rely on powered hoists to increase production efficiency.

Electrically Powered Hoists. Electrically powered hoists derive their lifting energy from electric motors mounted on the hoist mechanism. The load handling component can consist of chain or wire rope, depending primarily upon the user's preference.

Wire rope hoists utilize a grooved rope drum with flanges on each end to prevent the rope from overwrapping onto the hoist frame. The rope can be reeved

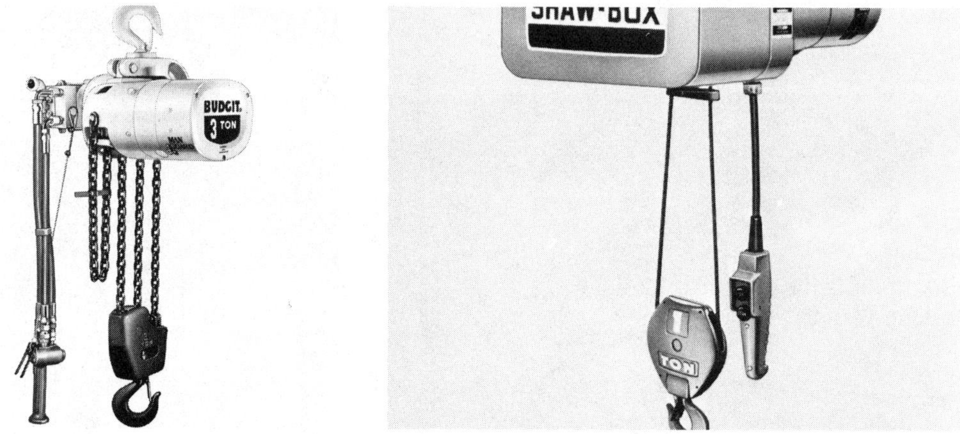

Fig. 11-10: Two types of pendant pushbutton control systems (photos courtesy of Dresser Industries, Inc., Crane and Hoist Operations).

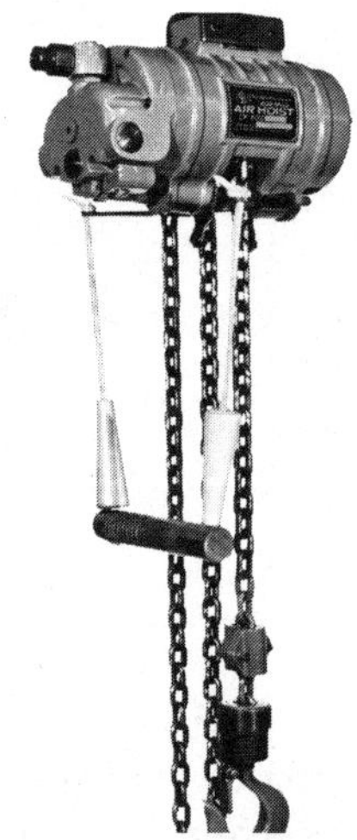

Fig. 11-11: Rope operated controls (photo courtesy of Chicago Pneumatic Tool Co.).

in various configurations according to the capacity desired. Figure 11-12 shows several reeving arrangements which increase the capacity of the hoist.

Electrically powered chain hoists are essentially the same as wire rope hoists except the chain is not wound on a drum. The chain is pulled through the hoist by means of a pocket wheel. The chain links fit over the raised portions of the wheel and its turning motion pulls the chain through. The load chain can be reeved much like wire rope to obtain a higher capacity for a particular hoist. Figure 11-13 shows a single and a double reeved load chain. Load chains are also available with either roller or coil type configurations (Fig. 11-14).

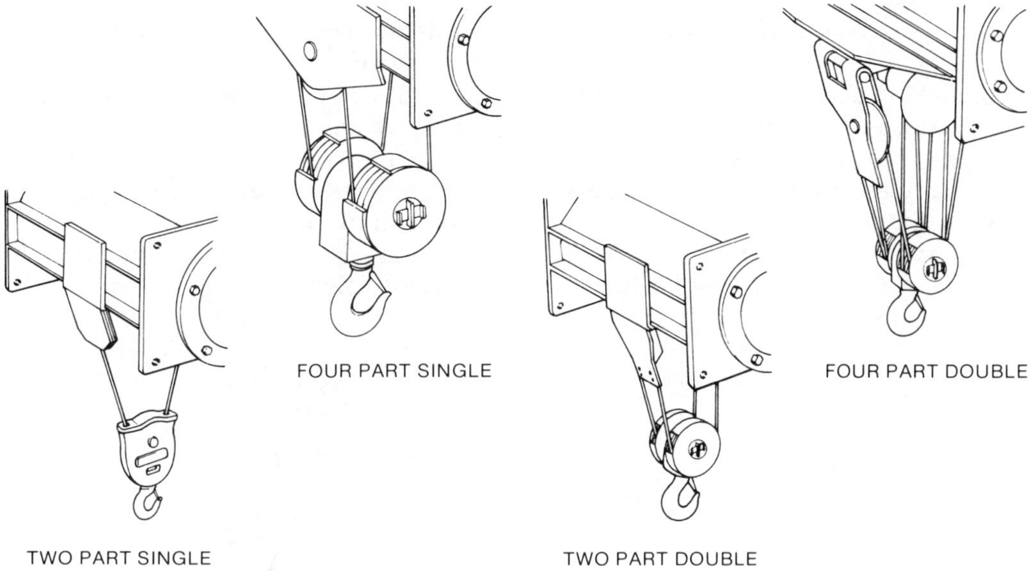

Fig. 11-12: Several types of reeving arrangements (illustration courtesy of Shepard Niles Crane and Hoist Corp.).

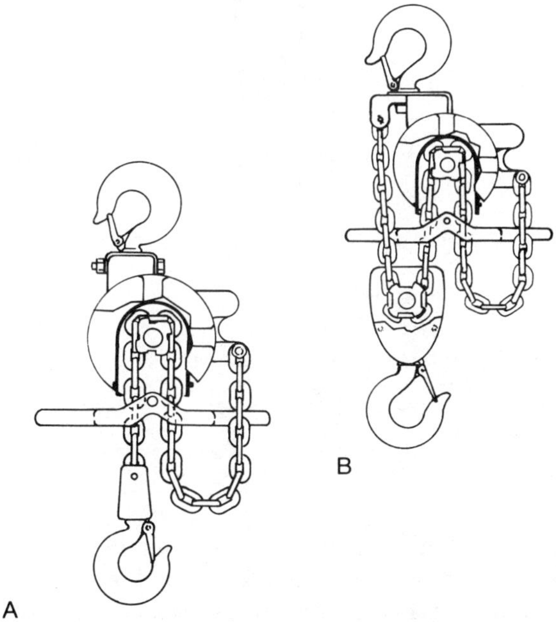

Fig. 11-13: (A) Single reeved load chain and (B) double reeved load chain (illustration courtesy of Dresser Industries, Inc., Crane and Hoist Operations).

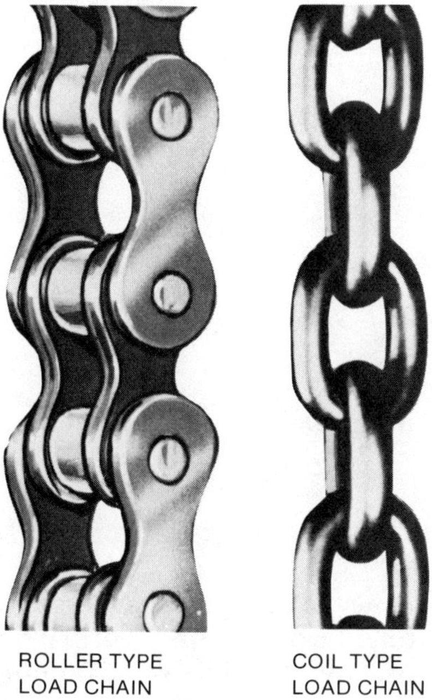

Fig. 11-14: Two types of load chains (photos courtesy of Dresser Industries, Inc., Crane and Hoist Operations).

Air Powered Hoists. The air powered hoist derives its lifting force from an air driven engine mounted on the hoist. Pressurized air must be piped to the hoist from a compressor unit usually located outside the hoist's work area. Air powered hoists are identical to electrically powered hoists in all ways except, of course, the method by which they are driven. Air hoists are capable of running either wire rope or chain load handling components and are controlled in approximately the same manner. Figure 11-15A shows a single reeved air powered hoist with a pendant operator's station, while Fig. 11-15B shows a double reeved air powered hoist with rope operated controls.

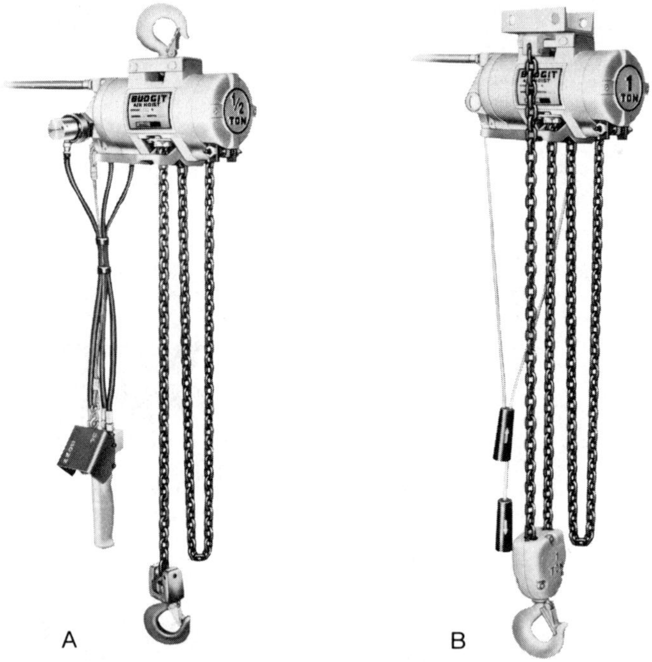

Fig. 11-15: (A) A single reeved air powered hoist with a pendant operator's station. (B) A double reeved air powered hoist with rope controls (photos courtesy of Dresser Industries, Inc., Crane and Hoist Operations).

Base Mounted Drum Hoists

This type of hoist is powered by internal combustion engines, electric motors, and hydraulics, and it utilizes drums and rope. This classification does not apply to overhead hoists, car pullers, truck body hoists, or other hoists or winches used exclusively in horizontal pulling applications. Figure 11-16 shows a three drum base mounted hoist with an attached swinger. It is powered by an internal combustion engine and consists of a gear train, drums, clutches, and brakes. A disconnect clutch is provided between the engine and gear train, with additional clutches called frictions mounted at each drum. A brake band can be provided around the periphery of one flange on each drum so as to make each drum individually controlled. For added safety a ratchet and pawl system is included on each drum which locks the drum against inadvertent spooling out of rope. When the pawl is engaged the drum is said to be "dogged" or "dogged off."

Figure 11-17 illustrates an independent direct geared swinger or single drum hoist, while Fig. 11-18 shows a single drum base mounted air hoist.

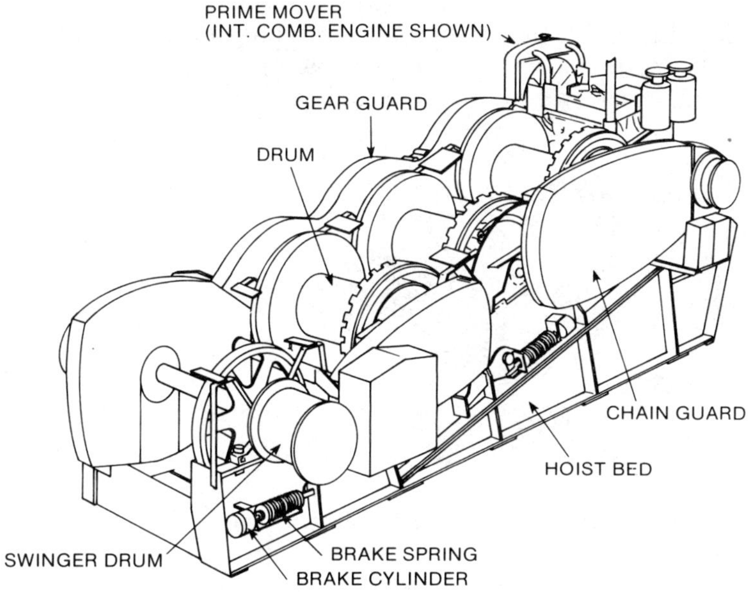

Fig. 11-16: A three drum base mounted hoist with an attached swinger.

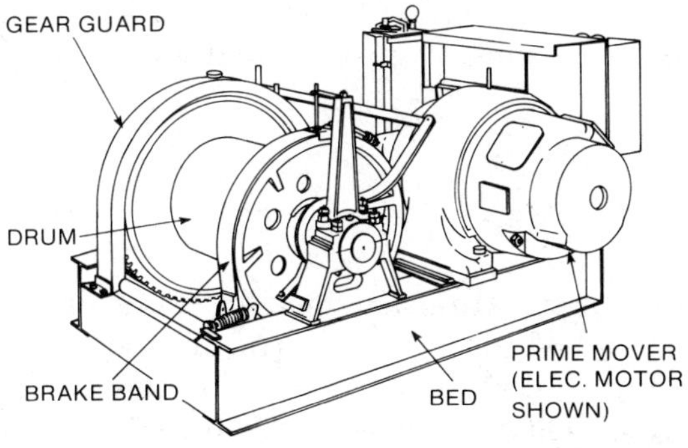

Fig. 11-17: An independent direct geared swinger.

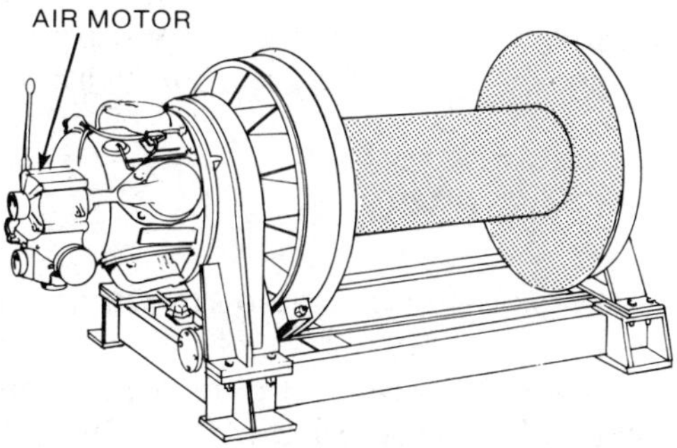

Fig. 11-18: A single drum base mounted air hoist.

WINCHES

A winch consists of a rugged bedplate and side frames upon which are mounted a horizontal drum shaft, drums or gypsy heads, reduction gearing, and usually the motor or engine that drives the winch. Figure 11-19 illustrates a typical winch.

A winch serves as a source of power for moving or lifting materials. For example, as described in the previous chapter, a winch is an integral part of any derrick configuration. Without the power of the winch, the derrick could not make the lift. Classifications for winches are based on design type and drive mechanism. There are essentially three types of winches according to design. These are drum, gypsy, and capstan winches.

Fig. 11-19: A typical winch (photo courtesy of Silent Hoist and Crane Co., Inc.).

Drum Winches

Drum winches are lifting devices that utilize a drum component(s) on which rope is wound for raising, lowering, or pulling loads. Depending upon their purpose, they can employ from one to four drums. Figure 11-20 illustrates the components of a common drum type winch.

Gypsy Winches

Gypsy winches, also known as warping or snaking winches, have one or two horizontally mounted gypsy heads around which several turns of rope can be wound in order to pull or lift a load (Fig. 11-21). Some types of gypsy winch heads are provided with raised portions called whelps, which reduce the rope's tendency to slip. *Note:* Combination winches are simply drum winches with shafts extended far enough to take gypsy heads on either or both sides.

Capstan Winches

A capstan winch utilizes a gypsy head arrangement on a vertical shaft mounted on top of the gearing apparatus (Fig. 11-22). These devices can be used to move materials in a horizontal direction. The large capstan winch shown in Fig. 11-23 is being used to move railroad cars.

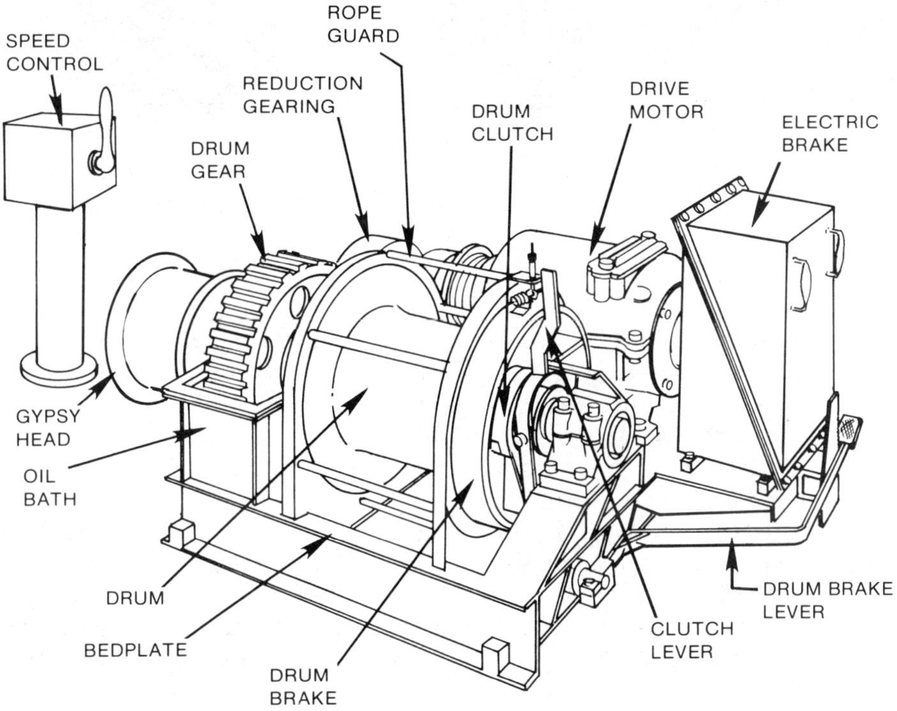

Fig. 11-20: The components of a common drum type winch.

Fig. 11-21: A double headed gypsy winch (photo courtesy of Silent Hoist and Crane Co., Inc.).

Fig. 11-22: A capstan winch (photo courtesy of Silent Hoist and Crane Co., Inc.).

Fig. 11-23: A capstan winch used to move railroad cars (photo courtesy of Silent Hoist and Crane Co., Inc.).

Fig. 11-24: A hand winch (photo courtesy of American Gage & Manufacturing Co.).

Winch Drives

Winches can receive their power from several sources and are also classified accordingly. The various types of drives consist of electric, electric-hydraulic, steam, air, gasoline or diesel engine, or hand operating.

Electric Drives. The prime mover in an electrically driven winch is the motor. Units such as these can use either alternating or direct current to actuate a drive shaft through reduction gearing.

Electric-Hydraulic Drives. An electric-hydraulic winch is a lifting device with a wide range of speeds, fine control, and smooth acceleration. The drive equipment includes a constant speed electric motor driving a variable displacement pump and a hydraulic motor that, through reduction gears, drives the winch. A manual control regulates the stroke of the pump and its output, thereby determining the speed of the motor.

Steam Drives. Prior to World War II most cargo winches were powered by steam; however, today most steam winches have been replaced by electric type drives. Steam driven winches are normally powered by double cylinder steam engines, the pistons of which drive crankshafts and, through gearing, also drive the drums and driveshaft.

Air Drives. Like steam winches, air winches provide smooth acceleration and a wide range of speeds. The drive mechanism consists of an air motor which actuates the drive shaft through reduction gearing.

Gasoline and Diesel Drives. Gasoline and diesel powered winches are used primarily when there is limited access to electricity. The prime mover, in this case, is the internal combustion engine which actuates a shaft through a torque converter and reduction gearing.

Hand Winches. A hand winch ordinarily consists of a single drum mounted horizontally on a shaft, a hand brake, reduction gearing, pawl and ratchet, and a hand wheel or crank. Figure 11-24 illustrates a hand winch used for a gin pole or tripod.

SAFETY

When working with any type of lifting device, it is extremely important that the operator and all personnel associated with the work follow safe operating procedures. All workers must be aware of the possible hazards involved and take steps to avoid these hazards. Hoists and winches are precision built tools which deliver maximum performance when operated properly and safely. They should always be handled carefully and inspected regularly to guard against unexpected mechanical failures. The following material lists some of the safe working practices used in hoist and winch operation.

Preoperational Safety Checks

The most important safety measure a hoist or winch operator can perform is to study the manufacturer's instruction manual and recommended safety guidelines for any piece of machinery he or she operates. Many accidents are caused by the operator's neglect of this procedure. Another essential function an operator must perform is to check the hoist, hooks, load rope or chain, limit switches, and brakes before the actual lifting operations begin. Failures caused by worn components could be reduced substantially by a brief test of the components at the beginning of every shift.

Safety During Operation

Hoist load blocks are designed to operate with the chain or rope run in a prescribed manner. If the lower load block is capsized or flipped over, the chain or rope twists and cannot run through the blocks correctly. This situation can cause overloading and possible failure.

It is also hazardous to side-pull or "drift" loads on the initial pull. The load must always be centered beneath the lifting device prior to the lift, because excessive side-pulling during the lift can damage the hoist and its rope or chain.

Overloading is another factor which can cause hoist or winch failure. In situations where the capacity of a lifting device is neared, the overload protection devices built into most modern equipment should be used. The operator must also know the load capacity of the beam or structure from which the hoist is suspended. It is quite possible to have cases where the hoist can handle the load weight but the overhead suspension equipment cannot.

Motorized hoists are also designed to operate at or below a specific number of lifts per hour. If a unit is run beyond its designed duty cycle, there is a chance of failure due to motor overload. Operators should always be aware of their machine's duty cycle.

Under no circumstances should a load chain or rope be used as a sling for loads or an elevator for personnel. Load handling mediums are not designed for any other uses. Shock loading is another hazard which can also damage both the hoist and load chain. Smooth, even lifts with a minimum of load jogging are necessary to prolong the usefulness of any hoist or winch.

The list given here outlines several other important safety considerations and practices:

1. Controls should be clearly marked. The operator must be totally familiar with all controls, and any setting should be checked before the work is begun. Also, a check should be made for defective electrical connections. No load should be placed on the hoist when these checks are being made.
2. Straighten any kinks in the rope, chain, or cable. Frequent kinking of the cable may cause it to weaken. A kinked chain may also fail unexpectedly when loaded.
3. All fastenings must be carefully examined to be certain that they are secure prior to lifting. Be certain that the load being lifted will not shift unexpectedly. This point is especially pertinent when slings with hooks are used for lifting. In such a case the load should be lifted slightly to take up the slack in the cable or chain. The load should then be rechecked for security and even load distribution before lifting it clear of the ground.
4. The operator should work in a clear, uncluttered area and should have a clear view of the work area. He or she must be aware of the position of other workers.
5. Avoid twists in the cable or chain because they can cause the loads to spin or twist when raised. Clear any obstructions which may snag the load, sling, or hook away from the work area.
6. If a signalperson assists in raising or lowering the load, use a set of standardized signals, and do not move the load unless each signal is clearly understood.

12

Other Lifting Devices

INTRODUCTION

Much of the lifting, moving, and handling of materials carried out in modern industrial processes does not require complex rigging equipment. Simple manual labor remains the only practical method of accomplishing many jobs. Yet the human body is probably the most misused of all lifting and moving machines. By following a number of simple lifting techniques and guidelines, it is, however, possible to substantially reduce the risk of personal injury. The tools and aids commonly associated with manual labor are described in this chapter. Sections are also devoted to the proper operation of equipment such as forklifts and highly specialized crane configurations.

Until a relatively short time ago, lifts which were unfeasible for cranes to accomplish were considered impossible and omitted from construction procedures. This factor limited the design of structures to the capabilities of ground or tower based cranes and derricks. Recently, however, an alternate means of lifting large loads has been developed which no longer limits structural designs to crane capability. Helicopters have become cost efficient and rapid means of lifting materials that are not normally hindered by congested areas or height restrictions. The important advantages and problems encountered when using helicopters as rigging tools are discussed at the close of this chapter.

MANUAL LIFTING AND HANDLING

The manual lifting and handling of objects and materials are not always as simple as they seem. Of all disabilities suffered by American workers, almost 25% are caused by improper handling of materials—more than any other single source. Strains, sprains, hernias, and other physical injuries are painful and frequently cause long layoffs from work. To reduce the chance of this type of injury, the worker handling the lift should be aware of:

- the proper way to lift.
- loads which are too heavy and ungainly to handle alone.
- the types of auxiliary tools necessary to make the lift, such as hooks, ropes, slings, or levers.
- the hazards involved in moving a particular load and the proper protective equipment required.
- the travel path through which the load is moved. (Check for obstructions, steps, slick spots, etc.)
- the appropriate areas where first aid is available.

Manual Lifting Safety

In order to lift with the least possibility of injury, certain rules should be followed as closely as possible. When a person lifts any type of load, their feet should first be situated close to the load in a position which provides a maximum amount of balance. Their back should be kept straight and as near vertical as practical while bending the knees and reaching for the load. Elbows are kept as straight as possible so that when the load is grasped the body is over the load (Fig. 12-1) and the back is as erect as possible. The load should be grasped firmly and care taken to keep the load from slipping from the grip. Once the load has been lifted, grip should not be unnecessarily readjusted or shifted during movement since this may cause the load to slip free of the worker's grasp.

After the correct lifting position has been assumed (Fig. 12-2), the load can be lifted by using the leg muscles. Lift by straightening the legs. This avoids putting stress on back muscles. Do not lift by bending your back. When changing the direction of travel, the whole body, including feet, must be turned because a sudden twist of the upper body during carrying can also cause back injuries.

If the object is to be placed on an elevation, it is not advisable to bend forward when placing the load. If possible, the edge of the load should be placed on the surface when pushed with the chest and stomach to slide it into place. It is important to free the fingers from beneath the load as it is slid because they could be pinched by the weight of the load and the sliding action.

If the elevated surface is approximately shoulder height, the object should be first lifted as far as possible by straightening the legs as described above. It should then be rested on or against something while a new grip is secured with the object higher up on the body and the legs again bent. Further lifting may then be accomplished by again straightening the legs (Fig. 12-3).

When lowering a load from an elevated surface, the above procedure should be reversed (Fig. 12-4).

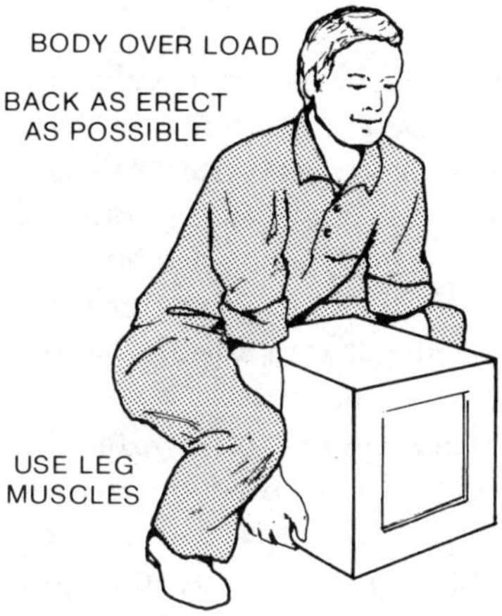

Fig. 12-1: Failure to lift a heavy load properly can result in serious back injury.

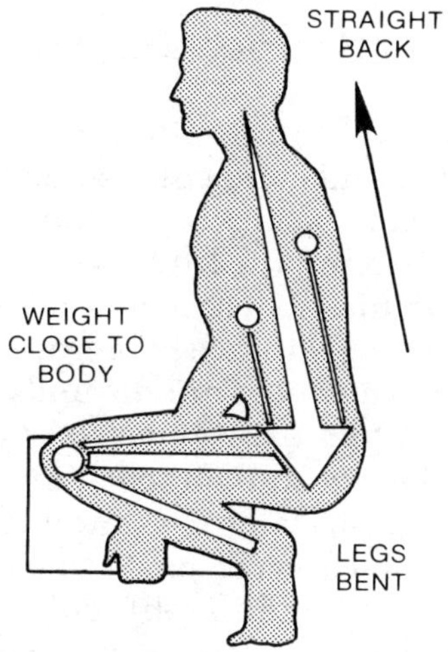

Fig. 12-2: Leg muscles must be used when lifting any size load—never the back.

Fig. 12-3: The proper way to lift a shoulder high load.

Fig. 12-4: The load must be put down as carefully as it was lifted.

Size, Shape, and Weight of Objects

The object itself often presents hazards to lifting and carrying which must be noted and compensated for if injury is to be avoided. The weight of the load or object to be moved should be considered first. It is important that the persons performing the lift know their physical limitations. Also, as a person becomes tired, his/her capacity for safe lifting is lowered and the possibility of tripping or losing balance is increased. If there is any doubt as to whether a load may be too heavy to handle under existing conditions, workers should be instructed to contact their supervisor to secure additional help.

The size and shape of an object can make it awkward to handle (Fig. 12-5). Sometimes a proper hand grip is not possible, visibility may be limited, clearance in carrying past other objects may be small, or other conditions may exist making handling difficult and increasing the possibility of dropping the load accidentally. Under such conditions, the help of others may be desirable.

When an object is heavy or the carry is long, it should be ascertained that the grip is secure and comfortable so that a person will not tire unduly before he/she reaches the destination. Interim places of rest should be planned when necessary.

Always keep hands, feet, and body in the clear of all obstructions because wide objects or narrow spaces often expose the hands and knuckles to injury. Blocks placed under loads often create the necessary clearance to prevent

Fig. 12-5: Some objects, while not heavy, are awkward and therefore dangerous to handle.

Fig. 12-6: Blocks should be used to protect fingers when lowering or lifting heavy objects.

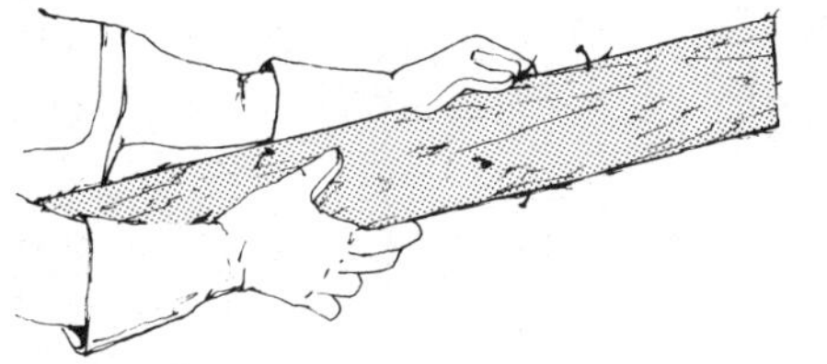

Fig. 12-7: Always be alert for protruding nails, splinters, or other sharp objects.

Fig. 12-8: Be sure to clear all surfaces of oil, water, or dirt before lifting.

pinching of the hands as the load is placed (Fig. 12-6). The following sections list some other potentially dangerous conditions which should be avoided or corrected:

Sharp Corners or Edges. Avoid taking hold of cutting edges or having the body come into contact with them. This hazard may be controlled by the way the object is gripped or by using protective clothing. Be on the alert for protruding nails, splinters, and other sharp objects (Fig. 12-7).

Abrasive Surfaces. Rough, burred, or jagged objects may not cut immediately upon being gripped, but will penetrate or abrade in a heavy lift or long carry. The person making the lift should be sure that his/her grip will hold until the object can be set down. Hard hats should be worn when there is any danger from falling objects. Goggles should also be worn when there is exposure to flying particles or to chemical splashes. Leather gloves or hand pads, leather aprons, and shin guards often are effective protection against rough or sharp materials. It is also advisable to wear safety shoes at all times when handling materials.

Slippery Surfaces. Objects that are wet, dirty, or covered with oil or grease are especially difficult to handle. If possible, they should be cleaned before lifting or placed in containers which will not slide from the hands. Slick or dirty floors can also be dangerous to maneuver about (Fig. 12-8), and such situations should be remedied.

Hot Substances. Items that are dangerously hot (or cold) should, when practical, be permitted to return to normal temperature before handling. Otherwise, the hands and body should be protected by wearing the proper clothing, shielding the hot item, or using special tools for handling hot substances (Fig. 12-9).

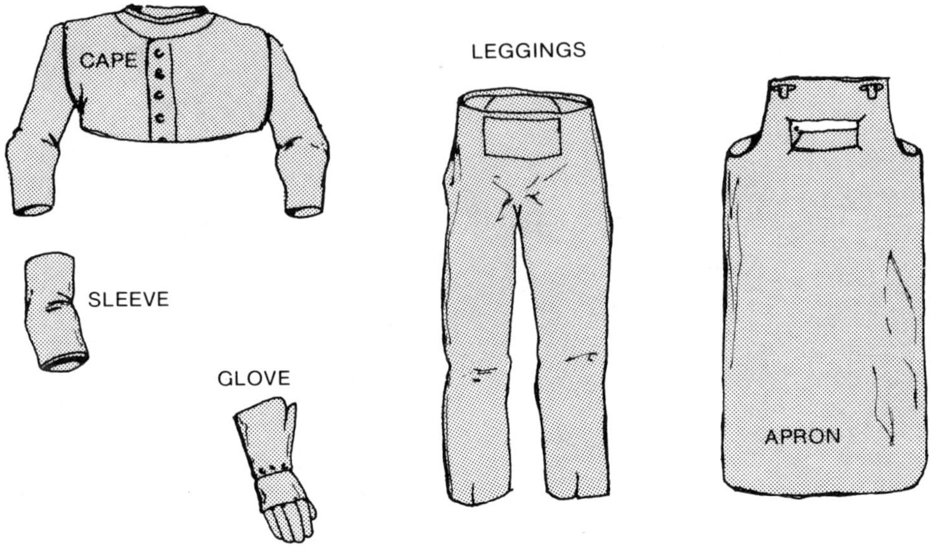

Fig. 12-9: If an item is too hot or too cold, be sure to have the proper protective clothing before lifting it.

Acids and Alkalies. Corrosive substances are particularly dangerous to everyone exposed and should be moved only in suitable containers. Additional body protection, such as goggles, hoods, aprons, etc., should be worn wherever necessary and the location and proper use of first aid equipment, such as deluge showers, should be known so that immediate treatment may be obtained in case of accident. Some of the proper protective equipment that can be worn consists of rubber or leather gloves, suits, aprons, leggings, boots, cut-type goggles, respirators, and face shields.

Flammable Substances and Explosives. As with corrosives, these substances are hazardous to everyone in the exposed area, and in some instances personnel should be evacuated along the path of movement. Approved containers, properly marked, should be used and the path of transportation should be checked and prepared. When moving such substances, proper protective devices should be worn and all necessary precautions taken.

Loose Materials. Material that is not packaged (or which has broken from containers) presents special hazards. Most finely divided materials, whether vegetable, mineral, or metallic, are explosive when suspended in air in proper concentrations. Releasing such dusts in the presence of ignition sources may be dangerous. Many dusts also are irritating to the eyes and to the skin, and protective clothing and goggles are desirable. *Note:* All loose materials should be handled carefully and masks or respirators worn.

Handling Aids

There are many tools or lifting aids which, when properly selected, maintained, and used, make the job easier, faster, and safer. Any type of handling aid, when misused, can pose a serious threat to the safety of those persons working in the area. It is necessary, therefore, to discuss the safety elements associated with this type of equipment.

Hooks. A hand hook has a hardwood handle fastened at a right angle to a highly tempered steel shaft. The shaft is curved at the end to form a hook which is tapered to a sharp point (Fig. 12-10). The primary purpose of the hand hook is to move or turn cases and bales.

A person using a hand hook should hold it in a position which is most comfortable. When breaking out or turning a box over, it is important to keep the knuckles turned away from the case (Fig. 12-11A). This gives better control over the hook when securing a "bite" with the point. To lift a box, however, the back of the hand should be turned toward the case in order to prevent pinching the fingers between the handle of the hook and the box (Fig. 12-11B). When rolling a case, the point of the hook is sunk lightly into the top edge farthest away from the person moving the case and the heel of the other hand is placed on the nearest edge. By pressing down with the heel of the nearest hand and lifting up and pulling with the hook, the box can be started to roll.

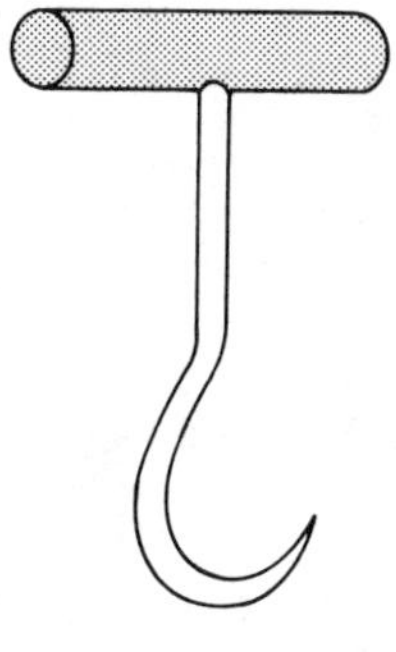

Fig. 12-10: A hand hook.

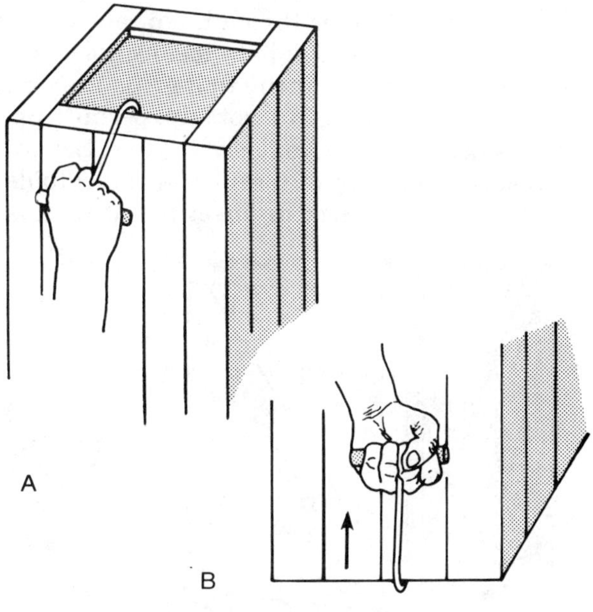

Fig. 12-11: The use of a hand hook.

Injuries happen when hand hooks glance off hard objects and strike the lifter or other workers. Hooks should always be carried in a carrier which guards the point. They should not be used where they will damage objects being handled or make holes permitting loose materials to escape. Points should be kept sharp. Hand hooks should never be used on barrels or metal unless there are holes in which the point of the hook can be inserted.

Crowbars and Car Movers. The principal hazard of a crowbar occurs when a worker does not get a good "bite" and the bar slips and either strikes the person or throws them off balance. Using the crowbar in positions between the worker's legs invites serious injury, as does using a crowbar in place of a car mover. Types of car movers which do not slip on steel rails are available for use and will save many painful injuries. When bars are not in use they should be stored where they will not fall or create tripping hazards.

Rollers and Skids. Heavy objects are frequently moved on rollers (Fig. 12-12). The principal hazard is pinching fingers or toes either between the load and the roller or between the roller and floor, although injuries are sometimes sustained when the load drops off a roller. Parts of the body should not be exposed to the roller; if the roller must be adjusted under the load, a bar or sledge should be employed. When items are dragged on skids, the skids should be soaped or greased to avoid strain on the gear.

Tongs, Pliers, and Other Hand Tools. When objects must be gripped by hand tools, because of hot, slippery, or otherwise hazardous surfaces or because of danger through contact with any part of the body (such as corrosives or poisonous substances), care should be taken that the right tool is used. It is important that the tool grips the object firmly and that the grip is firmly applied so that it will not slip, permitting the object to fall or pinching the fingers between the handles of the tool. Beware of splintered handles, mushroomed heads on chisels, etc., and makeshift tools. Figure 12-13 illustrates the method of carrying a piece of sheet metal with a plate clamp.

Fig. 12-12: Heavy objects are frequently moved on rollers.

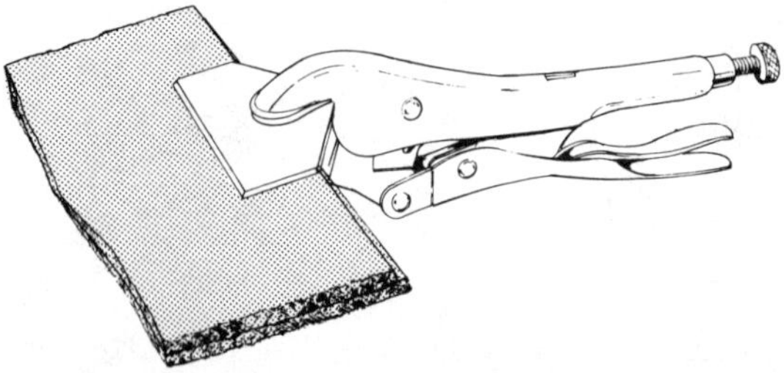

Fig. 12-13: The proper plate clamps should be used when carrying a piece of sheet metal.

Path of Travel

The immediate surroundings and the path of travel over which materials are to be moved are of considerable importance when moving materials. Look for electric wires and equipment with which parts of the body or items being handled may contact—especially if the object is long and a good conductor, such as metal pipe. Also be aware of electric conductors on overhead traveling cranes. Be careful of hot surfaces, such as steam pipes and unit heaters, not only because of possible burns, but also because contact may startle one and cause the load to be dropped.

Stumbling, tripping, and slipping hazards are present frequently and must be avoided with careful deliberation. Loose shoe laces and pants cuffs are included in this hazard. Be careful going up or down stairs and ramps, especially if there are defective treads or a lack of guard rails.

Illumination is important. When stepping into the dark from a brightly lighted area, temporary blindness and confusion can result. Be sure that necessary lights are turned on along the path of carry and that all stumbling and slipping conditions can be seen and avoided.

Dusts, vapors, and other atmospheric conditions are involved frequently in material handling accidents. Some materials create their own dusts and seriously limit visibility or otherwise impair a worker's sense of judgment. In going from a warm to a cold room, goggles or glasses may become fogged. These conditions can be anticipated and steps taken to avoid existing dangers.

Materials must at times be moved through hazardous areas; that is, places where the concentration of gases, vapors, or dusts may be ignitable, explosive, poisonous, suffocating, or irritating. Such areas should be known and special instructions received from the supervisor before entering. Rules which call for avoiding ignition sources, such as not lighting matches in dark and dusty atmospheres, not bringing dangerous objects or materials into dark areas, and the wearing of proper protective clothing, should, of course, be followed.

Traffic hazards are often encountered by persons handling materials. Foot traffic, plant trucks, cranes, conveyors, and other material handlers may all be operating in the same area and should be avoided. While personal injuries may be caused by collisions, there is also the danger of the load slipping or dropping. Not only should the path or route be picked carefully, but sometimes a helper should precede those carrying objects to assure traffic safety.

Additional Hints

There are many "tricks" or "kinks" in lifting and carrying which make the job easier, more efficient, and safer. Workers learn these through experience and sometimes, unfortunately, through accidents. A few suggestions are included here:

1. Team lifting—When two or more persons are required to handle an object, coordination of effort is essential.
2. One person should be in charge and give signals for lifting, lowering, and dropping in unison.
3. Long objects should be carried on the same shoulders (left or right) of all persons making the carry.
4. Persons carrying should keep in step.
5. Long objects, such as planks and pipes, when carried on the shoulder of one person, should be transported with the front end low (Fig. 12-14).
6. Be aware of electric wires.
7. When placing materials on supports, be sure they are strong enough to carry the load. Shore and lash piles of objects that are dangerously tiered and that cannot be broken down.
8. The use of blocks placed under objects to keep them from resting directly on the floor is an acceptable practice. Objects should be placed so that blocks will not tip. If the object is round, it should be braced or "checked" to prevent rolling.
9. When placing an object on a floor or platform, put one corner down first to avoid pinching fingers.
10. When stacking or piling, be sure objects or materials are locked, supported, and trimmed so that the pile will not topple or give way. Never stack objects so that the field of vision is obstructed.
11. Pull, do not push, a heavy hand truck or dolly up an incline. Do not hesitate to ask the supervisor to provide assistance when in doubt about moving a heavy load.
12. Objects or materials should not be thrown. If throwing cannot be avoided, keep parts of the body out of range and be careful of other workers.
13. Bags or sacks should be lifted by stooping down as far as possible and "rolling" the bag up to the knees, chest, and shoulder before straightening the legs, and carrying with the back vertical. Place hand on hip so that sack rests partly on the shoulder, arm, and back. Be sure of your grip and footing.
14. Upending of barrels and drums is difficult (Fig. 12-15). Grasp both ends, press down with one hand, and lift the other in a rocking fashion until the drum is balanced on the bottom chime. The lower grip may then be released and the drum placed on end. Keep feet in the clear.
15. When rolling drums or barrels, do not place hands on ends. Move slowly; maintain control.
16. If drums or barrels are being lowered or raised on skids, do not stand between skids. It is better to use ropes for such work, snubbing the rope on some rigid object to keep the drum in safe control. The ropes should be long enough to permit three full turns to remain on the drum at the farthest point of the line's travel; otherwise, slide the drums down the skids endwise.
17. Be sure bridge plates are secured properly before carrying a load over them.
18. Use only proper ladders and gang planks when going over the side of tanks or to other elevations. Avoid short cuts over the side to the dock or deck. When

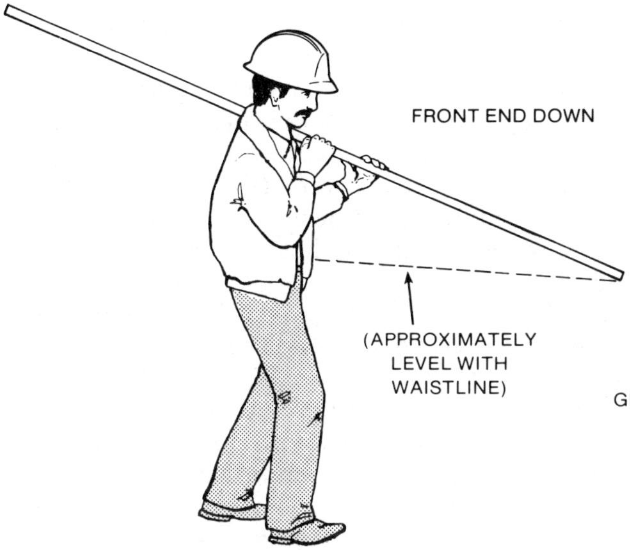

Fig. 12-14: Long objects should be transported with the front end low.

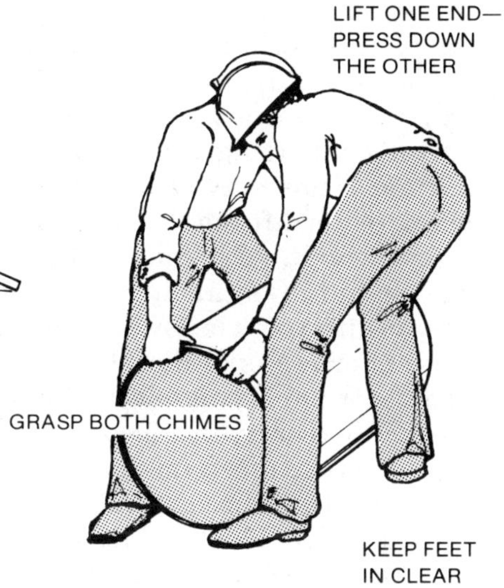

Fig. 12-15: Upending a barrel.

using a ladder, grasp the side rails, not the rungs. Carrying materials by hand when climbing up or down ladders should be avoided whenever possible. It is preferable to lower or lift materials by rope and boxes or buckets.

19. Piling objects higher than the height of an average person or tunneling should be avoided. Keep materials leveled off as much as possible.
20. Wait until drafts are lowered to proper handling level before taking hold—never reach up for a draft.
21. Stand clear of drafts being raised or lowered.
22. When stacking materials, plan an emergency exit in case the pile collapses unexpectedly.
23. Stand well away from any line under strain and face in the direction of the strain; plan an escape route in case the hoisting rope or slings fail.
24. Always stand outside of the loop of a line.
25. Injuries may occur even to the safest worker. Know where the nearest first aid room is and report for immediate treatment, even when the injury is minor.

FORKLIFT TRUCKS

Forklift trucks are mobile four-wheel or three-wheel automotive units which enable a person to pick up a unit load, transport it, and lift it to different heights when necessary. It is designed on the principle of the cantilever (a projecting beam or member supported only at one end), with the load counterbalanced by the weight of the truck in back of the center of the front wheels, which act as the fulcrum or center of balance.

The forklift truck is equipped with a two-tined fork secured to the supporting frame. These tines can be adjusted for separation. The fork is moved vertically on the supporting frame by either a mechanical or hydraulic lift. This supporting frame may be either nontelescoping or telescoping (Fig. 12-16) to provide greater lift height. The fork and supporting frame can be tilted forward from the vertical to pick up the load and can be tilted backward from the vertical to aid in balancing the lifted load.

A forklift truck is powered by a gasoline engine, battery-electric motor, diesel engine, or a gas or diesel-electric unit that utilizes a small engine to drive a generator supplying electric power. Forklift trucks are equipped with either mechanical or hydraulic braking systems and with either automobile or lever-type steering mechanisms. Some forklifts are provided with pneumatic tires, whereas others have solid rubber, cushion-type tires, depending on where they are to be used. Most forklifts have a place for the operator to ride, either standing or sitting, while others require that the operator walk alongside.

Fig. 12-16: A forklift truck with telescoping support frame (photo courtesy of Clark Equipment Co.).

Forklift trucks are designed primarily for handling palletized unit loads, although they can be used for loads assembled on skids or loads in boxes. In addition, special attachments or devices (such as a tilt box, ram, crane, and rotating clamp) can be assembled on the forklift truck in lieu of the fork, to increase the truck's usefulness. Components for drum handling (Fig. 12-17), scooping (Fig. 12-18), and even scrap handling (Fig. 12-19) are a few of the specialized attachments available for forklift trucks.

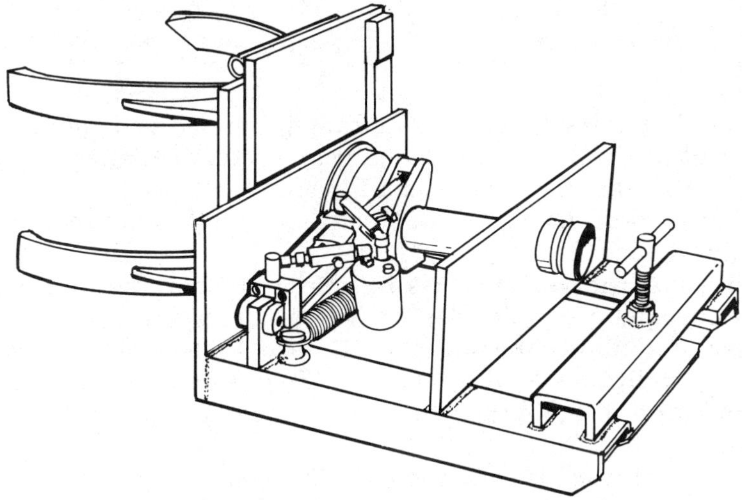

Fig. 12-17: An attachment for moving steel drums.

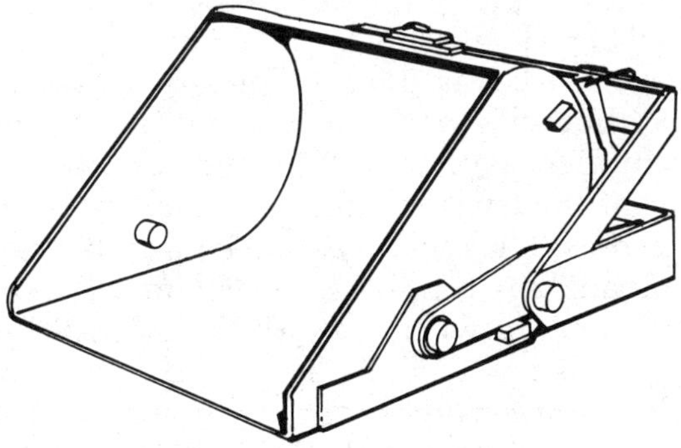

Fig. 12-18: A scooping device for a forklift truck.

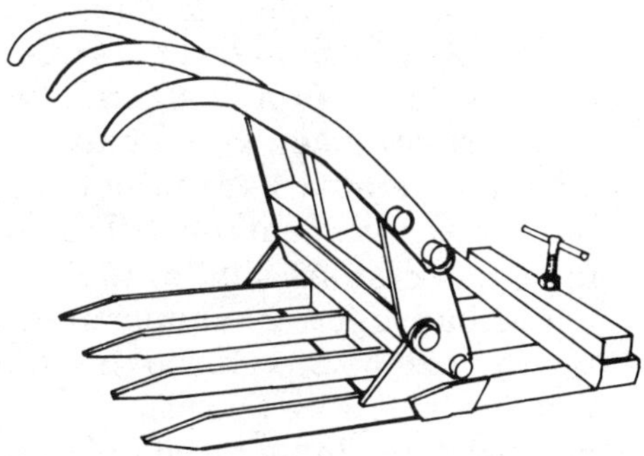

Fig. 12-19: A scrap clamp.

A forklift truck's greatest efficiency is obtained in its utilization for stacking and removing palletized loads in storage and for loading and unloading palletized loads on trailers, trucks, railroad cars, and warehouse platforms.

Forklifts should not travel with individual loads for excessive distances. General practice has shown that when the distance the load is to be carried exceeds 200′, warehouse tractors and trailers should be used to handle the hauling function for which they were specifically designed, and forklifts should be utilized for loading and unloading.

Pallet Trucks

Pallet trucks are closely related to forklifts in design, control, and operation. These machines are utilized for picking up and transporting palletized loads. They are primarily for situations where space is limited and handling operations are so small that forklift trucks are not justified from the standpoints of efficiency and economy.

Unlike the forklift truck, the pallet truck is designed to support the load on its entire wheelbase. Consequently, the pallet truck is smaller because it does not need its weight to counterbalance the weight of the load. Pallet trucks are classified as high-lift or low-lift, depending upon the height to which the truck can elevate the load, and as self-propelled or hand-operated, depending on whether motive power is provided.

The frame of the low-lift truck is constructed with a fork so arranged that it can enter between the top and bottom decks of a pallet (Fig. 12-20). A hydraulic lift mechanism, which may be operated manually or electrically, is provided to sufficiently raise the forked frame for the load to clear the floor and be moved. The low-lift pallet truck may be powered by a battery-electric motor or moved manually. It usually is steered by a tow handle which also operates the mechanical brakes. However, some self-propelled low-lift pallet trucks have automobile or lever-type steering. The usual capacity is 2 to 3 tons.

The *high-lift pallet truck* is designed to pick up, transport, and lift to different heights a load supported on a two-tined fork (Fig. 12-21). The fork is attached to a vertical mast which may or may not telescope and does not tilt. An electrically operated hydraulic lift mechanism is provided to raise or lower the fork on the vertical mast. The frame of the truck extends beyond the vertical mast in a manner similar to the fork tines. The two leg extensions of the frame act as outriggers in supporting the load. One type of frame construction has the leg extensions approximately coinciding with the fork tines so that the fork in the lowered position rests on the leg extensions. This type of truck can be used with single-faced pallets only. Another type of frame construction has the leg extensions protruding outward so as to straddle the fork tines; that is, the leg extensions are wider apart than the fork tines. This truck may be used with single- or double-faced pallets because the frame of the truck straddles the

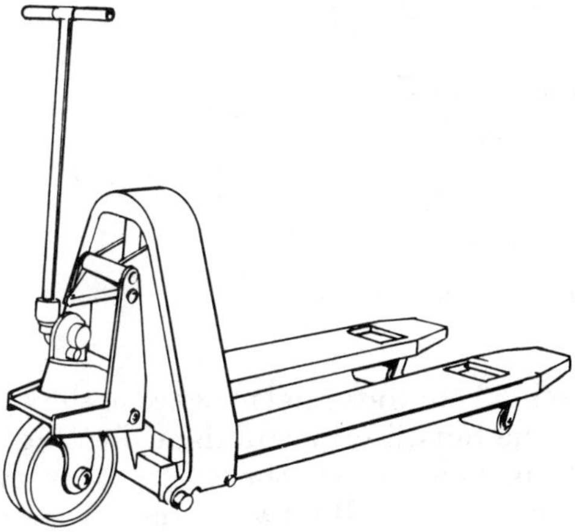

Fig. 12-20: A low-lift pallet truck.

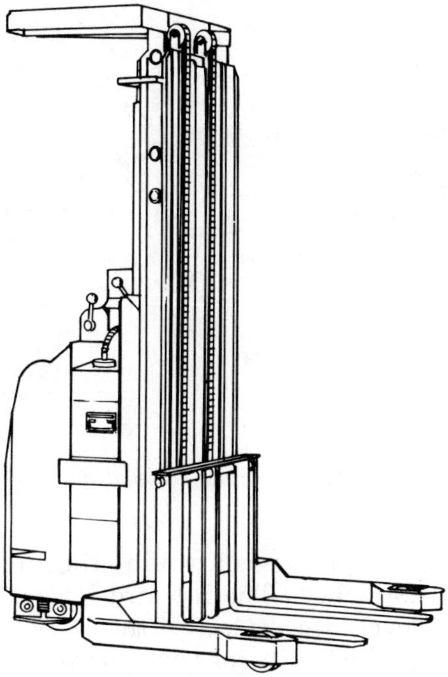

Fig. 12-21: A high-lift pallet truck.

load. The drive wheels are positioned under the steering mechanism at the other end of the frame.

Operating Controls of Forklift Trucks. There are so many different makes and models of forklift trucks that it is difficult to generalize on their controls. All, however, must have one set for the drive of truck and one set for the fork.

Figure 12-22 shows an arrangement of control levers on a typical gasoline-powered forklift truck. Under the operator's seat, and protected by a curved rod, is a panel on which are located the ampere and oil gauges and the ignition switch. To the right of the panel is the starter button. The two levers extending from the deck of the machine control the drive. This particular type of truck is geared for both low and high speeds. The clutch, brake, and accelerator follow the usual automobile arrangement—clutch on the left and brake on the right of the steering column. The accelerator is to the right of the brake pedal. There is also a parking brake near the drive levers. This brake should always be set when the operator leaves the truck. The truck is put into motion the same as an automobile, except that when either gearshift lever is in its forward position, the truck's motion is forward; and when either lever is in its rearmost position, the truck's motion is in reverse. The truck may be operated in either the high- or low-speed range.

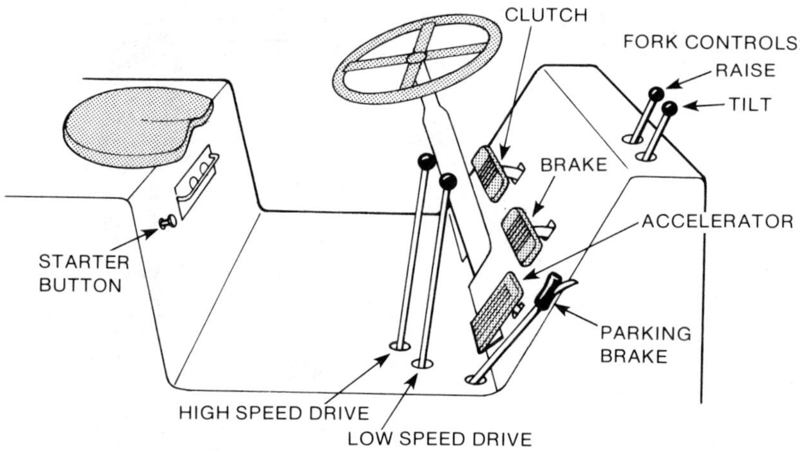

Fig. 12-22: The controls of a gasoline-powered forklift truck.

The levers controlling the forklift extend from the panel forward of the steering wheel. The inboard lever controls the hoisting mechanism. Putting the lever forward lowers the forks, and pulling it back raises them. The other lever, pushed to its forward position, causes the forks to tilt forward and away from the operator; pulled back, it causes the forks to tilt backward.

Figure 12-23 shows an arrangement of controls typical of an electric-powered forklift truck. The fork control lever, the drive lever, and the switch are located on the panel forward of the steering column. When the drive lever is in the forward position, the truck is in forward gear; when the lever is pulled back, the truck is in reverse. Speed is regulated solely by the accelerator.

A brake pedal is to the right of the steering column, and the parking brake is attached to the operator's seat. When the operator rises, the brake is applied automatically.

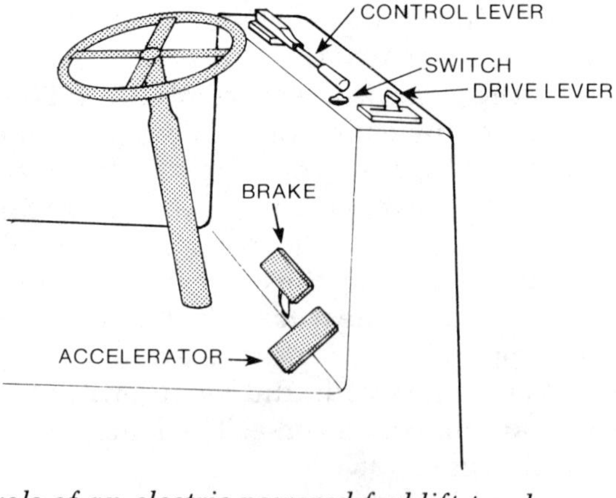

Fig. 12-23: The controls of an electric-powered forklift truck.

Lifting or depressing the fork control lever raises or lowers the forks. Pushing the lever ahead tilts the forks forward; pulling the lever rearward tilts the forks back.

Forklift Safety. Forklift trucks enable the operator to load and unload materials from other carriers at a relatively fast rate. It is important, therefore, that safety be a prime concern in the operation of this equipment. The following list provides several basic rules for the safe and efficient operation of forklift trucks:

1. All starts and stops should be gradual to prevent the load from shifting. Turns should be made at a safe speed, smoothly and gradually.
2. Never carry loads in an elevated position. When in travel, the forks should be held at a level approximately 4″ (101.6 mm) above the floor.
3. Do not permit helpers or anyone else to ride on the load or any part of the truck.
4. Do not permit any part of the load to obstruct vision while driving. The truck should be operated in reverse if the load obstructs the front view.
5. Be sure the path is clear when raising or lowering a load or moving with a load. Watch for overhead obstructions.
6. Fork trucks operated by gasoline or liquefied petroleum gas should be fueled only in areas specifically designed for that purpose. Watch for signs of leaks, overheating, etc., and report any such conditions to your supervisor.
7. Power trucks should never be left unattended without first shutting off power, neutralizing controls, setting brakes, removing ignition key, and lowering platform or fork.
8. Check all controls at the beginning of each shift. Report unsafe conditions to a supervisor and do not use equipment when there is any question as to whether or not it is safe.

DRAGLINE AND CLAMSHELL

Dragline and clamshell equipped cranes are two mobile lifting devices which have the capabilities of scooping, razing, and moving earth. These cranes resemble the mobile crane configurations explained in Chapter 9, except each is equipped with a specialized earth moving device or bucket. Boom configurations for the dragline and clamshell cranes are generally latticed or trussed with gantry arrangements similar to those found on construction cranes.

Dragline Components

The dragline component, from which this crane derives its name, is a length of wire rope run from a winch through the fairlead to the drag bucket. Figure

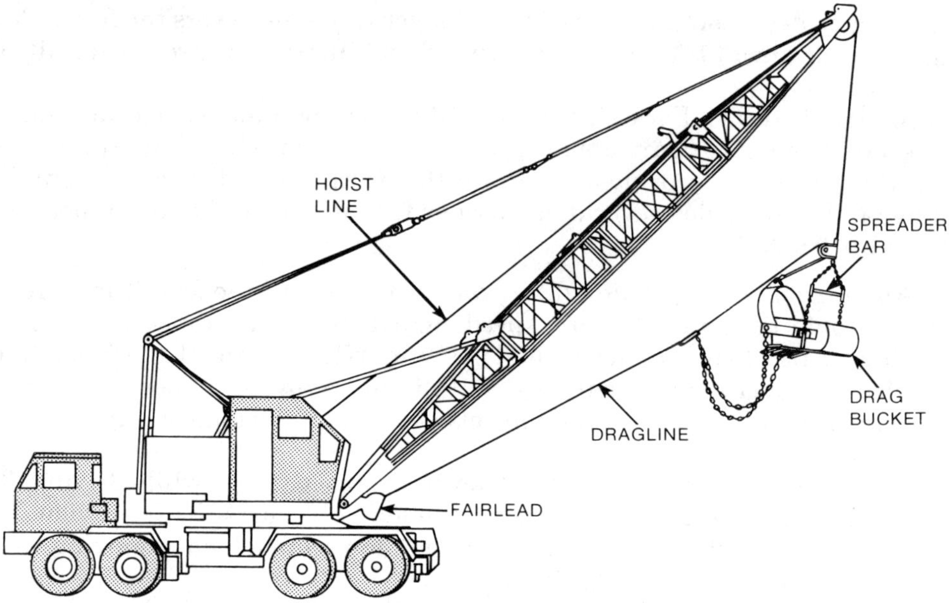

Fig. 12-24: A dragline.

12-24 illustrates the components of a dragline crane. The fairlead is a component which is mounted on the boom or boom foot that lines up the dragline so it will spool properly onto the winch drum.

Dragline Buckets. There are several types of dragline buckets which are engineered for various applications. Lighter buckets are used for digging soft earth such as top soil, medium buckets are usually used for general all-purpose work, and heavy buckets are designed for deep lifts and rock digging. Figure 12-25 illustrates several types of dragline buckets including a sieve type bucket which is useful for digging in wet areas.

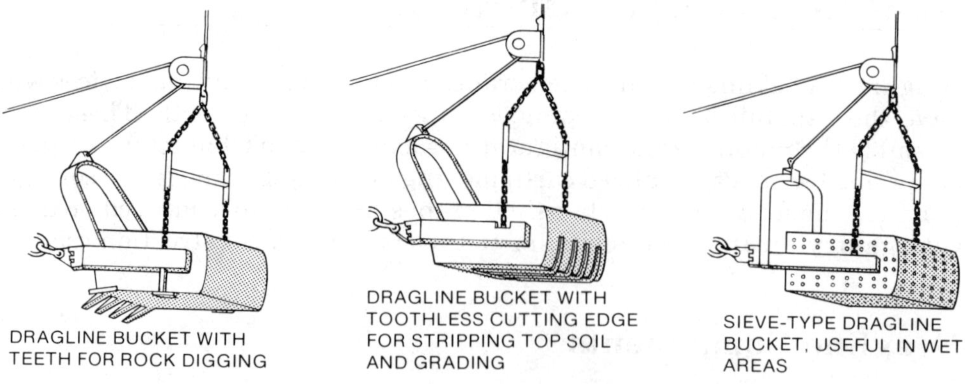

Fig. 12-25: Dragline bucket types.

Dragline Rope Arrangements. There are several ropes which actuate the drag bucket (Fig. 12-26). The boom line consists of a standard four part rigging arrangement, while the hoist line runs from the main winch over the boom point sheave to the bucket. The dump cable runs from the dragline winch to the forward drag yoke, over the dump sheave to the top of the bucket arch.

Dragline Applications. Draglines are extremely useful pieces of excavating equipment. The coal industry (both anthracite and bituminous) relies heavily upon the high capacity and accurate digging capabilities of the dragline bucket. The construction industry also employs draglines to move earth in the locations which prohibit the use of conventional bulldozers (Fig. 12-27).

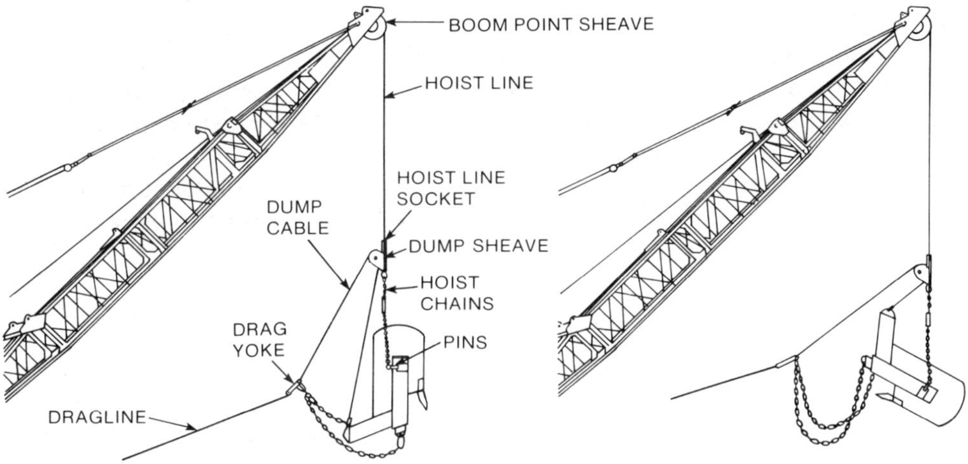

Fig. 12-26: Dragline rope attachments.

Fig. 12-27: Draglines are used in wet areas where soil conditions prohibit the use of bulldozers (photo courtesy of American Hoist & Derrick Co.).

Clamshell Cranes

A crane which has the special capabilities of digging deep, narrow, straight sided excavations is known as the clamshell (Fig. 12-28). The basic configuration of the clamshell, like the dragline, resembles that of a mobile crane. The primary difference in this case is the bucket. The clamshell bucket is suspended from the boom point sheaves by two ropes. The tagline winder keeps tension on the rope to prevent the clamshell bucket from turning.

Clamshell Bucket Applications. This type of lifting device is best suited to handling bulky objects such as logs, stumps, or boulders. Loose brush and trees can also be piled neatly, but it does not move earth to unwanted locations in its digging motion.

The clamshell bucket is an excellent utility bucket which can handle almost any lifting or excavating job; however, it is relatively slow in operation. Since the bucket is always located under the boom, the rig must be moved when the materials under the boom radius have been cleared. The time consumed when closing the bucket and also the lack of precise control of bucket motions add to its inefficiency. Careful operation is needed to position the bucket precisely.

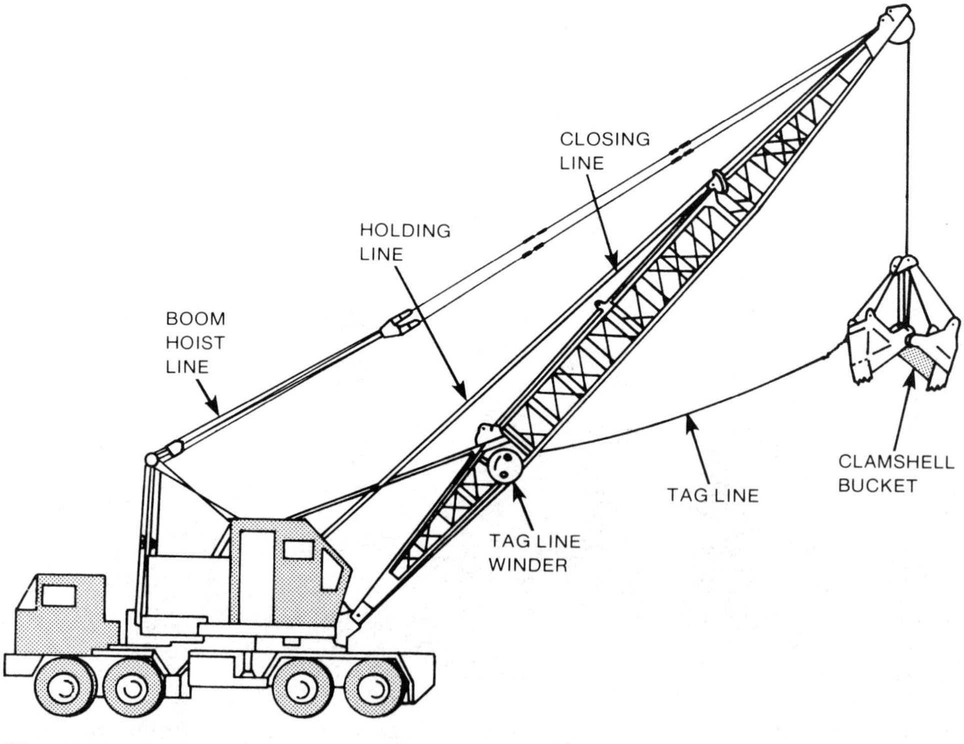

Fig. 12-28: A clamshell.

HELICOPTER LIFTING

When special rigging problems arise, it sometimes becomes necessary to use helicopters to perform the task. These lifting devices are especially suited to unique lifts which conventional cranes or derricks cannot handle. Areas of limited access or areas in which conventional cranes are not able to operate require the services of helicopters. Rooftops which are not able to support cranes and areas in which the terrain does not permit crane operation are some examples of areas in which helicopters are used.

Some of the major types of heavy lift helicopters presently in use are the Sikorsky S-55, capable of lifting 2000 pounds (907.2 kg); the Sikorsky S-58, capable of lifting up to 4500 pounds (2041.2 kg); the Sikorsky S-61, capable of lifting up to 8500 pounds (3855.6 kg); and the Sikorsky Skycrane, capable of lifting up to 18,000 pounds (8164.7 kg). These capacities are determined on the operation of the helicopter at sea level on a 60°F (15.6°C) day. Variations in capacity are due to temperature, humidity, altitude, and wind conditions.

Cargo Hooks

There are several types of cargo hooks which can be used for helicopters. Each type is designed primarily to handle specific load weights. They are also designed with hook release mechanisms which can be operated from the helicopter or by a member of the ground crew. A typical helicopter cargo hook is illustrated in Fig. 12-29.

The operating position of cargo hooks is situated in a manner which suspends the sling load beneath the center of gravity of the helicopter. It is very important that loads are lifted and transported in this position because loads which are improperly hooked threaten the in-flight stability of the aircraft.

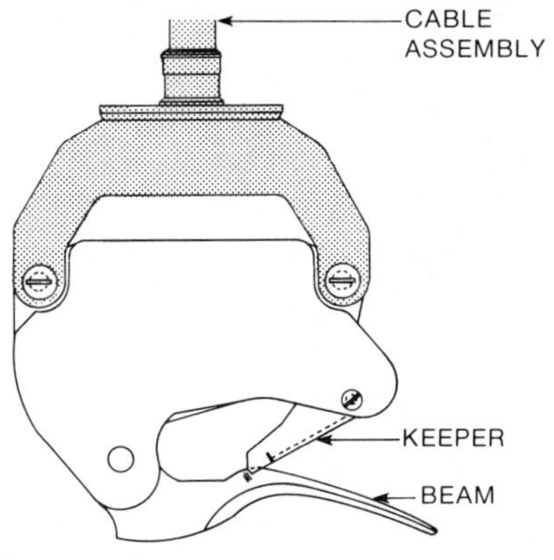

Fig. 12-29: A helicopter cargo hook.

Slings

The slings which support helicopter loads are generally the same type as used for general rigging purposes. However, it must be recognized that rigging for helicopters requires that the load be secured not only for the lift, but also for high speed travel. Cargo must be secured well enough to withstand air speeds that sometimes exceed 60 miles per hour. Figure 12-30 illustrates an eight-legged sling consisting of chains and nylon slings connected by a clevis and attached to a nylon ring.

Cargo Nets. Cargo nets can be used to haul composite loads which could not be safely or efficiently lifted through the use of conventional slings. This type of sling usually is designed in a web-type pattern and can be manufactured of either nylon or metal. Figure 12-31 illustrates a nylon cargo net.

Load Stabilization

The aerodynamic phenomena of side to side movement and forward-aft oscillation of cargo suspended beneath a helicopter show that the pilot is actually

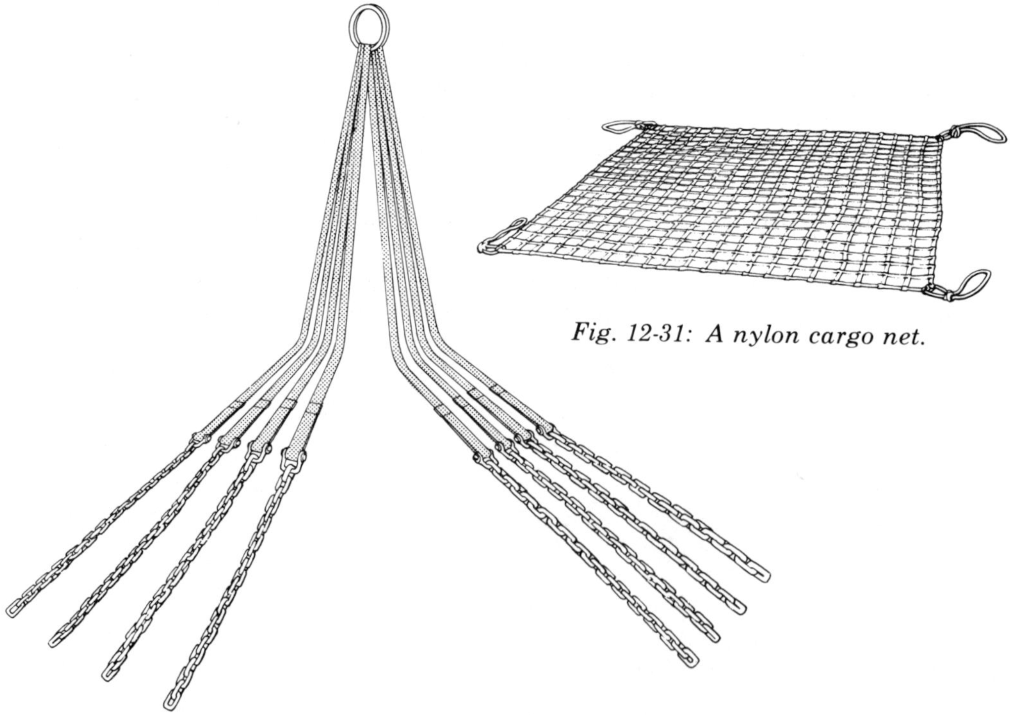

Fig. 12-31: A nylon cargo net.

Fig. 12-30: An eight-legged sling.

flying two different bodies. One is the helicopter, which is aerodynamically stable; the other is the sling load, which is seldom aerodynamically stable. Therefore, most of the problems encountered in helicopter lifts concern the instability of loads in flight.

Load instability will occur when the weight of the suspended load is not sufficient to hold it down against the drag of the air through which it moves. It is commonly experienced with elongated loads that are symmetrical about their centers of gravity. Such loads will always turn broadside to the direction of flight, thus exposing maximum drag surface (Fig. 12-32). The lighter the load in proportion to the exposed drag surface, the lower is the airspeed at which instability will occur. Stabilization of such loads may be assured by one or more of the following means:

1. Reducing the airspeed of the helicopter. *Note:* This also reduces the craft's flying performance and is not always advisable.
2. Increasing the weight of the load, which also has certain limitations.
3. Reducing the drag surface by altering the relationship between the center of gravity and the center of pressure of the suspended load in such a way as to assure that the narrowest surface faces in the direction of flight. This is often the best stabilization method.

Examples of Load Stabilization. There are two expedient means of stabilizing sling loads by alternating drag surfaces. One is by adding surface to the rear of the load (Fig. 12-33). In this concept, the added surface consists of

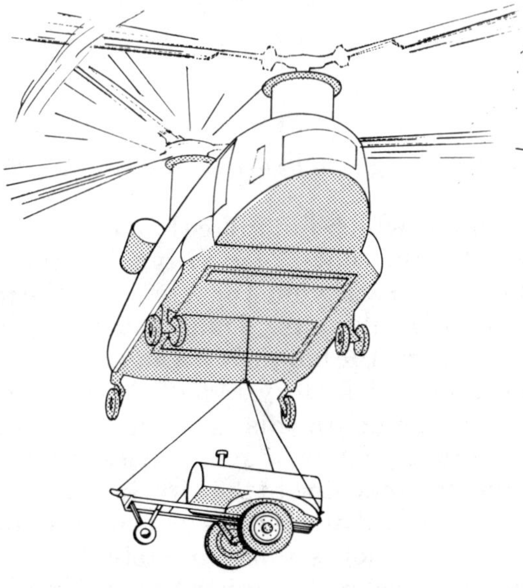

Fig. 12-32: A load having a symmetrically located center of gravity will fly crosswise when rigged beneath a helicopter.

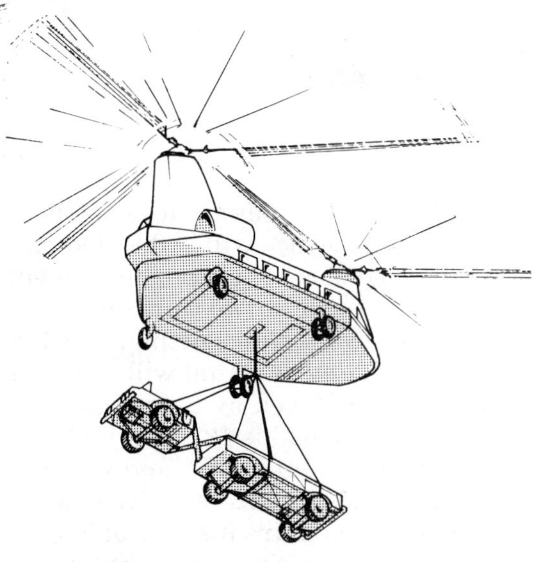

Fig. 12-33: When a truck and trailer are coupled together, the load flies lengthwise and tends to maintain its stability at high speeds.

related equipment of light density, such as connecting a trailer to the vehicle to be lifted. Since the amount of additional surface is out of proportion to the weight that has been added, the relationship between the center of gravity and center of pressure of the total load has been altered and the load tends to stabilize in flight. The addition of related equipment is a more practical technique than that of installing rigid fins or similar airfoil spoilers onto the rear of the load.

Another means of stabilizing a sling load is by adding weight to the front of the load (Fig. 12-34). In this concept, supplies of high density are secured to the front end of the load. Since the amount of additional weight is out of proportion to the amount of surface that has been added, the relationship between the center of gravity and center of pressure of the total load has been altered and the load tends to stabilize in flight. While this basic concept is simple, its mathematical application under field conditions is complex because the amount of weight to be added must be calculated in moments rather than raw weight.

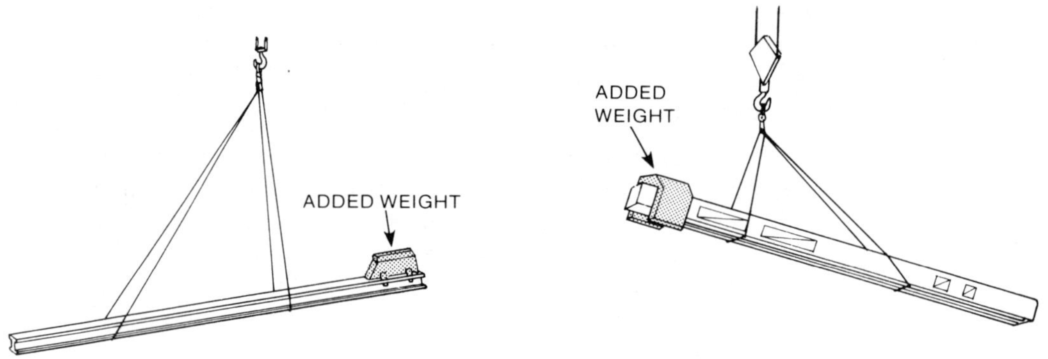

Fig. 12-34: Maintaining in-flight stability by adding weight to the front of a sling load.

The desired balance location is a point at which not more than one-third of the total surface will be located forward of the center of gravity. Weight must be added so as to cause the moment of the forward one-third to equal the moment of the rear two-thirds of the load. This value can be mathematically calculated by the same methods used to determine the balance of a loaded aircraft. A suitable system that will usually be more satisfactory under field conditions can be employed by using Fig. 12-35 as a guide. In this example, the typical configured load is 12′ by 3′ by 1′, weighs 600 pounds, is symmetrical about its center of gravity (CG), and will always fly broadside to the direction of flight.

The center of gravity (CG) of a suspended load always lies beneath the point of suspension. In this case of Fig. 12-35A, each leg is approximately 8′ long. As stated earlier, the desired balance location lies in the forward one-third of the load. In order to find the correct sling length for this rigging arrangement, an imaginary line must be projected perpendicular to the load and passing through the desired CG. The correct sling length will complete that length, which passes from the point of suspension to the suspension points on the load. The front sling legs in Fig. 12-35B will each be approximately 6′ long, while each of the rear sling legs will be about 10′ long.

Once the slings are properly set, the load should be lifted with a hoisting device to allow the actual CG to fall below the suspension point (Fig. 12-35C). If the load were flown when rigged in this manner, it would fly broadside to the direction of flight.

In order to ready the cargo for flight, a high density weight must be placed on the front end of the load to cause it to be suspended on a level plane. As shown in Fig. 12-35D, the actual CG of the load is now at the desired point, and the load will fly in the manner needed to ensure stability.

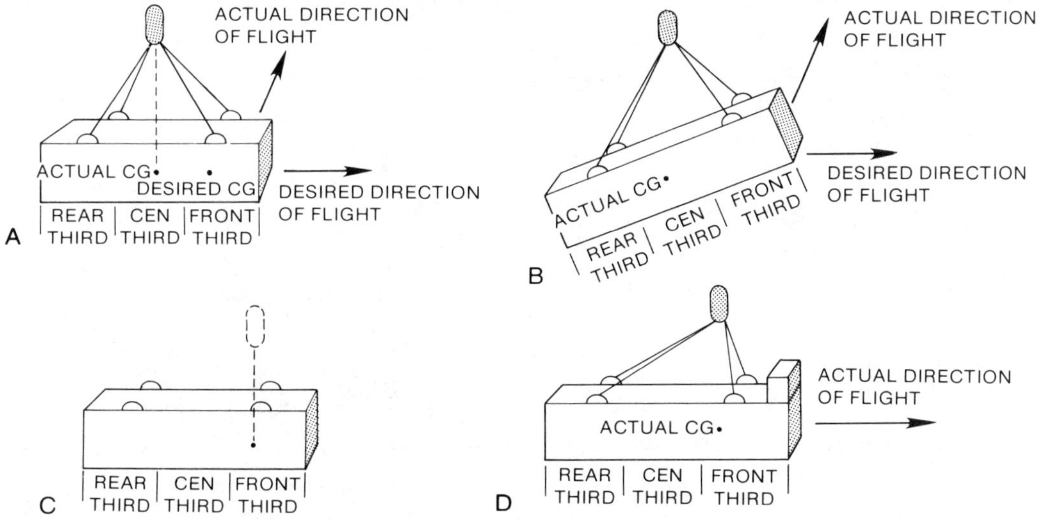

Fig. 12-35: Load stabilizing considerations.

Helicopter Hand Signals

There must be reliable and constant communication between the pilot-in-command and a designated member of the ground crew during lifting operations. Signal systems must be understood and checked prior to the beginning of each new operation. The hand signals given in Fig. 12-36 are accepted as the proper type of guidance to be used. *Note:* The speed at which these signals are given determines the rate of compliance requested on the pilot-in-command.

Radio communications are the primary means by which the pilot and ground crew are connected, along with hand signals. In the event of radio communication failure, an effective backup system of communications must be immediately available.

Figure 12-37 shows the type of rigging used to lift an air conditioning unit to the top of a building. Conventional cranes proved too costly for this operation; however, a helicopter made the lifts with ease.

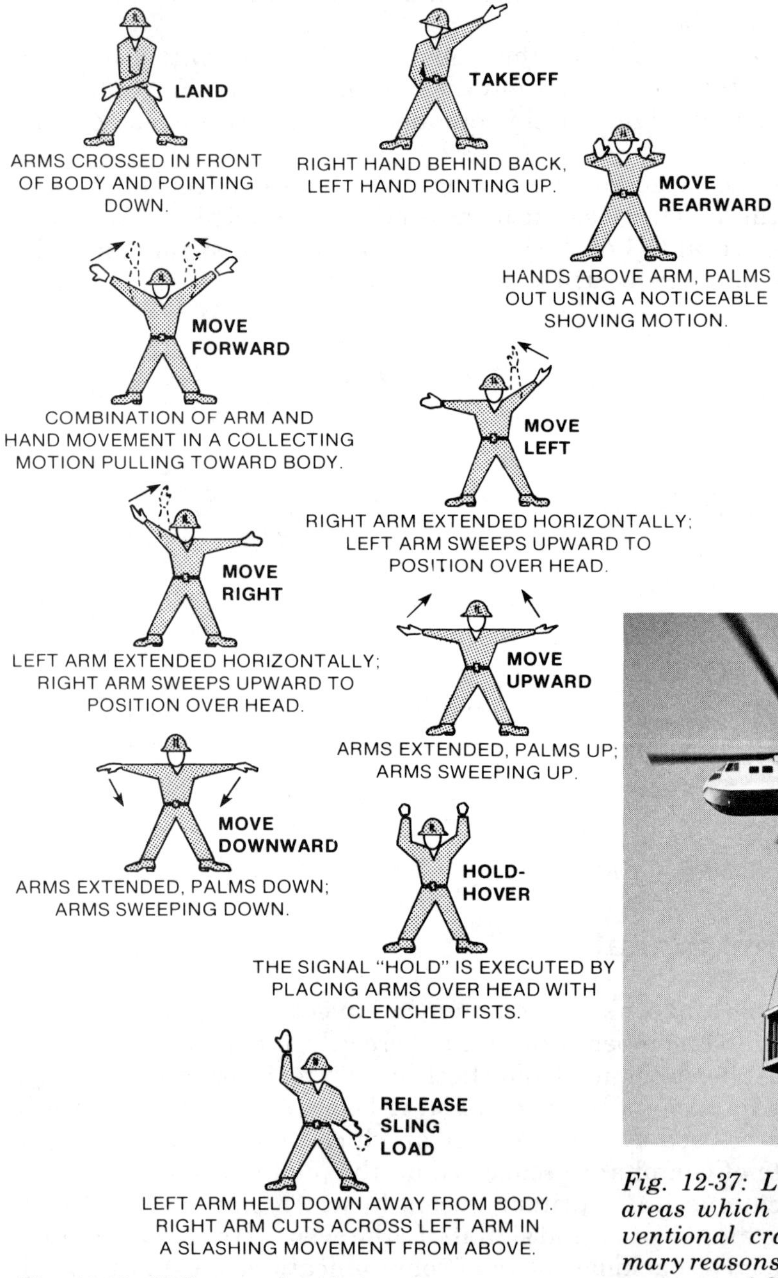

Fig. 12-36: Helicopter hand signals.

Fig. 12-37: Lifting heavy loads into areas which are inaccessible to conventional cranes is one of the primary reasons for helicopter use (photo courtesy of Carson Helicopters, Inc.).

13

Distributing Rigging Weights

INTRODUCTION

The secret of a stable lift is distributing the weight so that every force that would cause the lift to fail is counteracted by sufficient rigging. The weight must be distributed in the legs of the sling through the crane to the outriggers, tracks, or tires on the ground. The importance of weight distribution lies in the serious consequences of failure in the rigging system. Failure can occur at any point: pins may shear, ropes may snap, lugs may pull away, the boom may buckle, the crane may tip, and the list could go on and on. To avoid failure, every precaution must be taken to ensure that no element of either the crane or the rigging is subjected to stresses or moments beyond its capacity.

DISTRIBUTING WEIGHT TO THE RIGGING

The strength of a load's rigging must be sufficient to withstand all forces brought to bear on it. The ideal conditions are seldom if ever encountered in any lifting task. In addition to the static weight of the load, the rigging must withstand forces contributed by the operating environment. These are associated with wind, snow, ice, cold, and heat in addition to earthquake effects. Also in this category are the forces induced by an out-of-level state or misalignment of the mountings, tracks, or support surfaces and by increases in friction brought about by dust, mud, sand, or other particulate matter entering the interfaces of the moving parts.

From the work being done, there are dynamic forces due to the lifted load and redistributions of forces accompanying swing, trolleying, luffing, or travel motions. Normal frictional forces can also be included under this designation.

Finally, there are those forces which result from overcoming the inertia of the load and component masses in acceleration and deceleration processes for all functional motions. Several motions also introduce pendulum or centrifugal action in the load, which is accompanied by load radius change and/or lateral force.

Because of the dynamic forces which affect any lift, the rated working capacity of rope used in rigging is 1/5 of its designed capacity. This is referred to as a "safety factor of 5." The dynamic forces are too complex and variable to calculate, so they are accounted for in the safety factor.

DETERMINING LOAD WEIGHT

Before the capacity of slings, blocks, etc., can be determined, the weight of the load must be known. If this information cannot be obtained from shipping papers, design plans, catalogue data, or from other dependable sources, it may be necessary to calculate the weight. Weight can be determined on the basis of either area or volume. Appendices I and J give the weight of certain materials in area and volume.

Determining Weight by Area

On erection plans, the size of steel beams is usually supplied together with their weight per foot of length and the length of the member, so it is easy to compute the weight of any member to be lifted. For example, a 25′ S-shape beam weighing 33.5 pounds per foot has an overall weight of 25 × 33.5 = 837.5 (Fig. 13-1). Where angle, plate, or built up members are involved, however, the calculation is more involved, and the memorization of one basic weight and two formulas will give a reasonably accurate estimation of the weight.

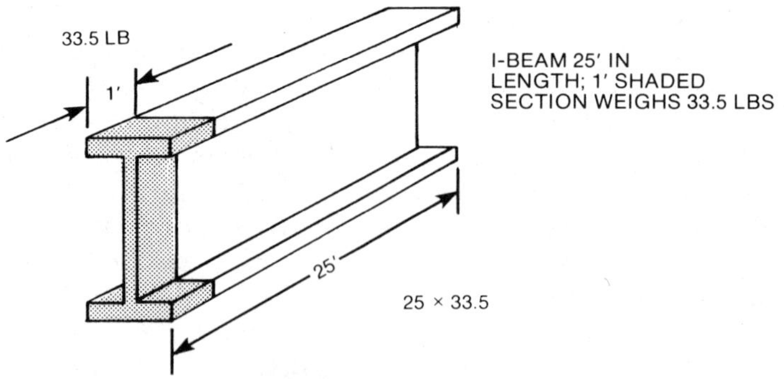

Fig. 13-1: Many structural shapes are specified as weighing so much per foot.

One square foot of steel an inch thick is approximately 40 pounds. Thus, two plates of steel each measuring 1-1/2″ × 3′ × 6′ (Fig. 13-2) would weigh a total of:

$$2 \times 1.5 \times 3 \times 6 \times 40 = 2{,}160 \text{ lbs.}$$

The weights of angles also can be approximated with results exact enough for safe job use. An angle is a structural shape which can be considered as a bent plate with some additional metal at the center for strength and a lesser amount of metal at the tips for ease in rolling. If the angle is flattened out, a plate results (Fig. 13-3). For example, a 5″ by 3″ by 1/4″ angle would flatten out to approximately an 8″ by 1/4″ plate. This should weigh 40 lbs × 8/12″ × 1/4″, or 6.65 pounds per foot.

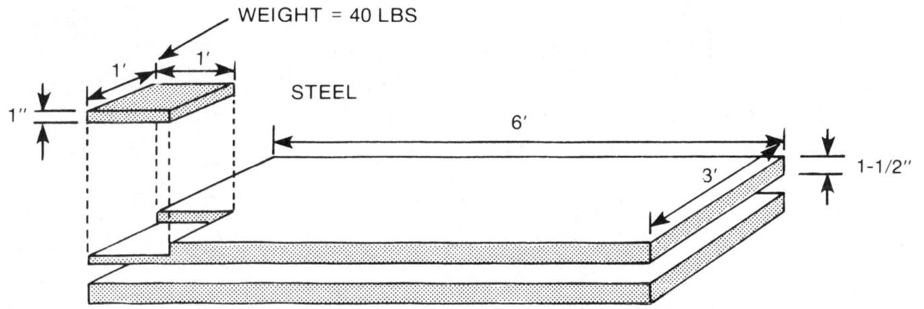

Fig. 13-2: Load weight determination of steel plate.

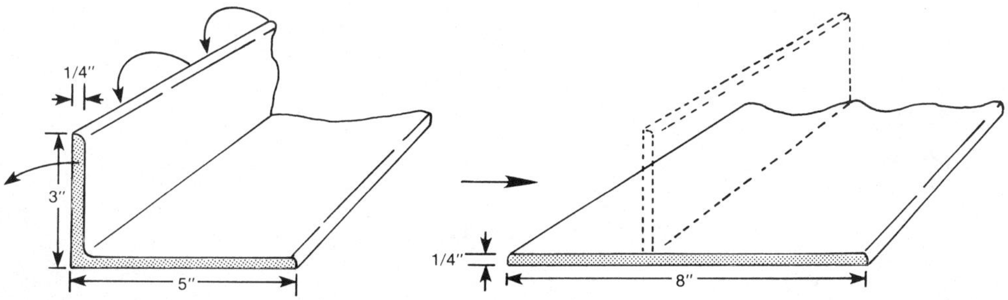

Fig. 13-3: Load weight determination of angles.

Weights of any structural shape can be computed in this manner by separating the parts or flattening them into rectangles which, in turn, become parts or multiples of a square foot of steel an inch thick.

Plates, however, are often rolled into tanks or other shapes which must be handled. It is necessary to determine the square foot area of such parts before the weight can be calculated. This necessitates learning the two simple formulas for computing the circumference and the area of a circle. The circumference (Fig. 13-4), or distance around the edge of a circle, is found by multiplying the diameter by the number 3.14 (referred to as "pi" and given the symbol π).

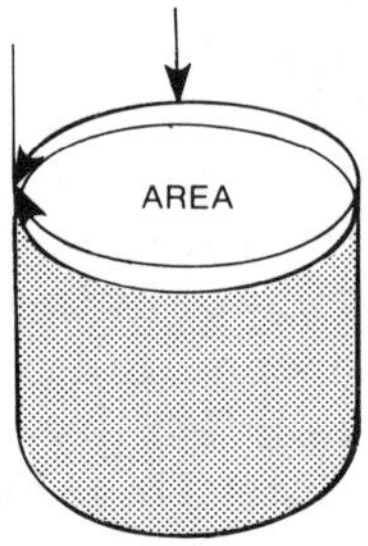

Fig. 13-4: Circumference and area of a circle.

A stack 6′ in diameter would have a circumference of 6′ × 3.14, or 18.84′. To compute the weight of this stack, if it were 30′ high and made of 3/8″ plate, mentally unroll it and flatten it out (Fig. 13-5). This gives a plate 18.84′ wide by 30′ long by 3/8″ thick. The weight is:

$$18.84 \times 30 \times 3/8 \times 40 = 8{,}478 \text{ lbs.}$$

The second formula gives the area of circular objects.

$$\text{Area} = 3.14 \times \frac{\text{diameter}}{2} \times \frac{\text{diameter}}{2}$$

Thus, if the stack had an end cap 3/8″ thick and 6′ in diameter, it would have a surface area of:

$$3.14 \times \frac{6}{2} \times \frac{6}{2} = 28.3 \text{ sq ft}$$

and would weigh:

$$28.3 \times \frac{3}{8} \times 40 = 425 \text{ lbs (Fig. 13-6).}$$

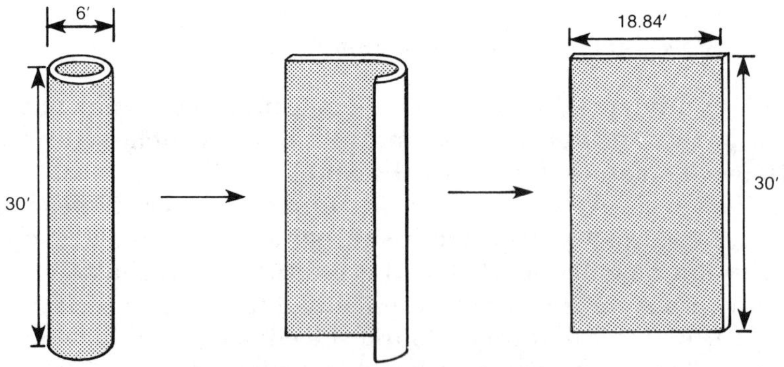

Fig. 13-5: Load weight determination of a section of stack.

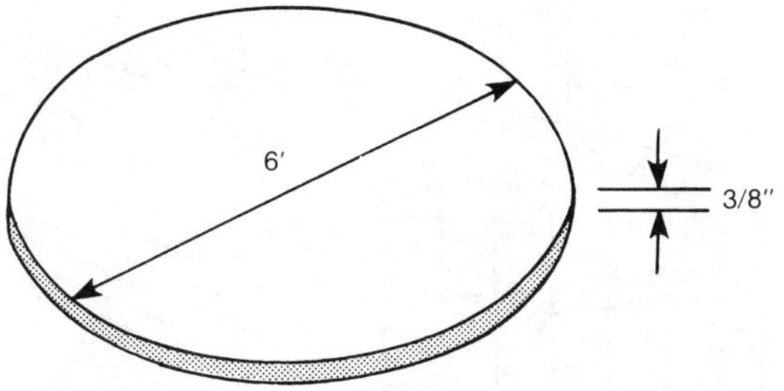

Fig. 13-6: Load weight determination of a circular disk.

Determining Weight by Volume

For other materials the weights are normally based on their weight per cubic foot, so the cubic feet of material (the volume) being hoisted would have to be determined in order to estimate the weight.

For example, suppose a bundle of spruce lumber is to be hoisted and the bundle is 12′ long, 3′ high, and 4′ wide (Fig. 13-7). The weight per cubic foot from Appendix I is 28 pounds, so the weight of this bundle is 12 × 3 × 4 × 28 = 4,032 lbs.

Figuring volume is not always easy. Odd shaped objects often require lengthy equations. The half hollow cylinder in Fig. 13-8 is a good example. The volume formula for this object is:

$$V = \frac{\pi L}{8}(D^2 - d^2)$$

Using values from the dimensions in Fig. 13-9, the equation should be worked as follows:

$$\begin{aligned} V = \frac{\pi L}{8}(D^2 - d^2) &= \frac{\pi 12}{8}(10^2 - 9.5^2) \\ &= \frac{37.68}{8}(100 - 90.25) \\ &= 4.71\ (9.75) \\ &= 45.92\ \text{ft}^3 \end{aligned}$$

By multiplying 45.92 cubic feet by the weight of steel (490 lb per cubic foot), the approximate weight of the object can be obtained:

$$45.92\ \text{ft}^3 \times 490\ \text{lb} = 22{,}500.8\ \text{lb}.$$

Volume formulas for odd shapes you may encounter are given in Appendix K.

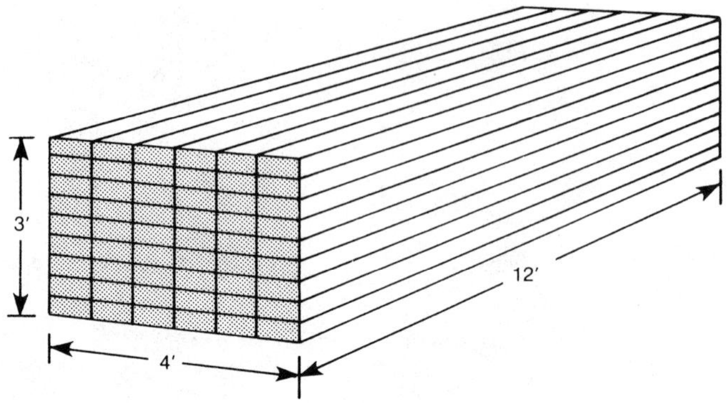

Fig. 13-7: Load weight of an object by volume.

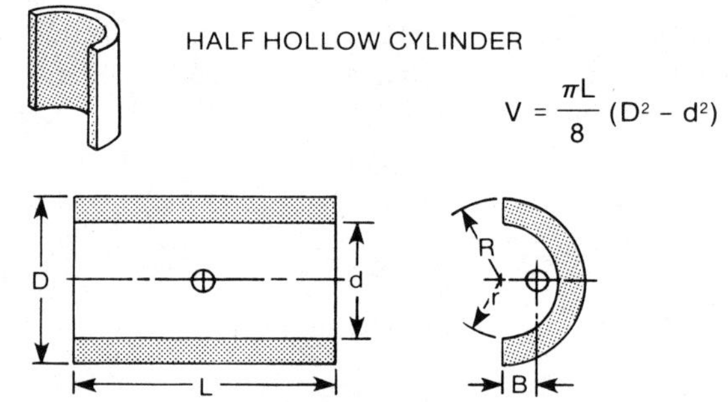

Fig. 13-8: Formula for finding volume of an asymmetrical object.

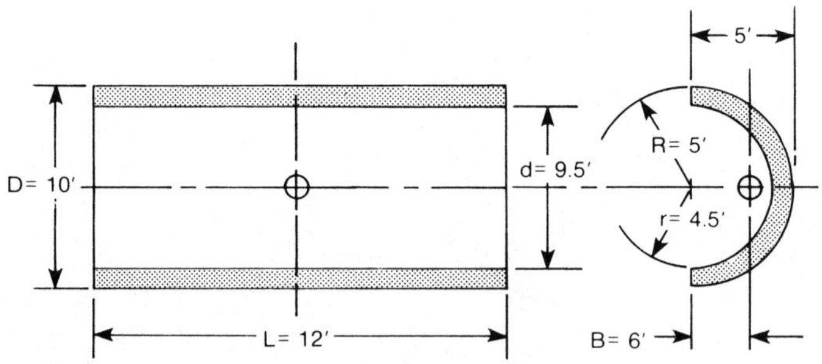

Fig. 13-9: Figuring the volume of a half hollow cylinder.

The time taken to calculate the approximate weight of any object, whether steel plates, columns, girders, boiler components, tanks, etc., is time well spent and may prevent a serious accident through failure of lifting gear. The tables of weights of various materials given in the appendices should enable any rigger to compute the approximate weight of a given load. When in doubt, do not hesitate to seek advice. Do not take a chance where the safety of workers is involved.

DETERMINING CENTER OF GRAVITY

It is always important in rigging practice to rig the load so that it is stable. A stable load is one in which the center of gravity of the load is directly below the main hook and below the lowest point of attachment of the slings (Fig. 13-10A).

The center of gravity of an object is that point at which the object will balance. The entire weight may be considered as concentrated at this point. A suspended object will always move so that the center of gravity is below the point of support (Figs. 13-10B and C). In order to make a level or stable lift, the crane or hook block must be directly above this point. Thus, a load which is slung above and through the center of gravity will be stable and will not tend to topple or slide out of the slings.

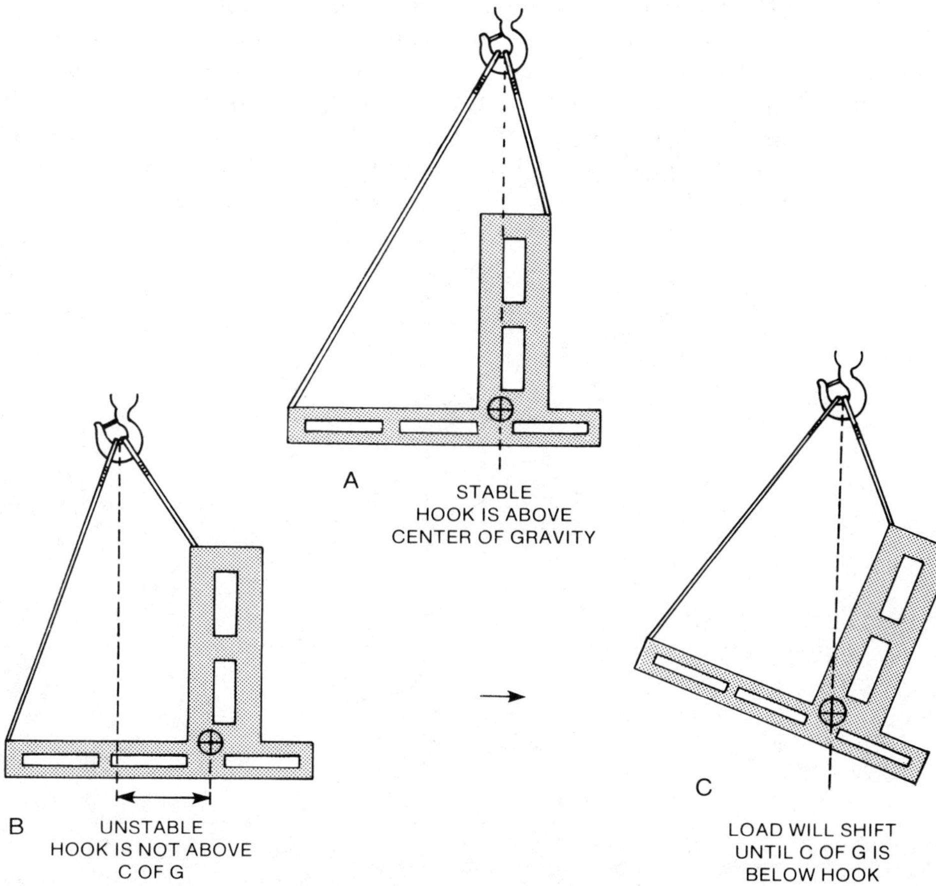

Fig. 13-10: Center of gravity always hangs below the hook.

On objects having uniform shapes and compositions there is no problem in determining where the center of gravity lies since it is at the center of the shape or object. The center of gravity of a uniform round rod is at the center of its diameter halfway along its length. The center of gravity of a cube or sphere is at its exact center. For example, a steel tank such as the one in Fig. 13-11 presents no problem. The tank is symmetrical in any plane passing through its centerline. The center of gravity lies in the center of both views.

On odd shaped bodies where the center of gravity cannot be easily determined, the rigger may guess where it lies, try to lift with the hook over that point, and then correct it by trial and error methods, moving the hook, load, and sling suspension a little at a time until a satisfactory result is obtained. The object will usually tilt until the center of gravity is directly beneath the load hook (Fig. 13-10C), so there is an indication of the direction in which to shift the slings. Remember, too, that when the center of gravity is closer to one point of sling attachment than to the other, the sling legs will have to be of unequal length. This means that their angles and loads will also be unequal. The rigger may also calculate the center of gravity for an odd shaped object. Determining the center of gravity mathematically will be discussed later.

If a load tilts when it is lifted and it is not corrected, then one leg of the sling will see a large load increase and the load on the other leg will decrease. If any load tilts more than 5° after it is lifted clear of the ground, it should be landed and rigged differently.

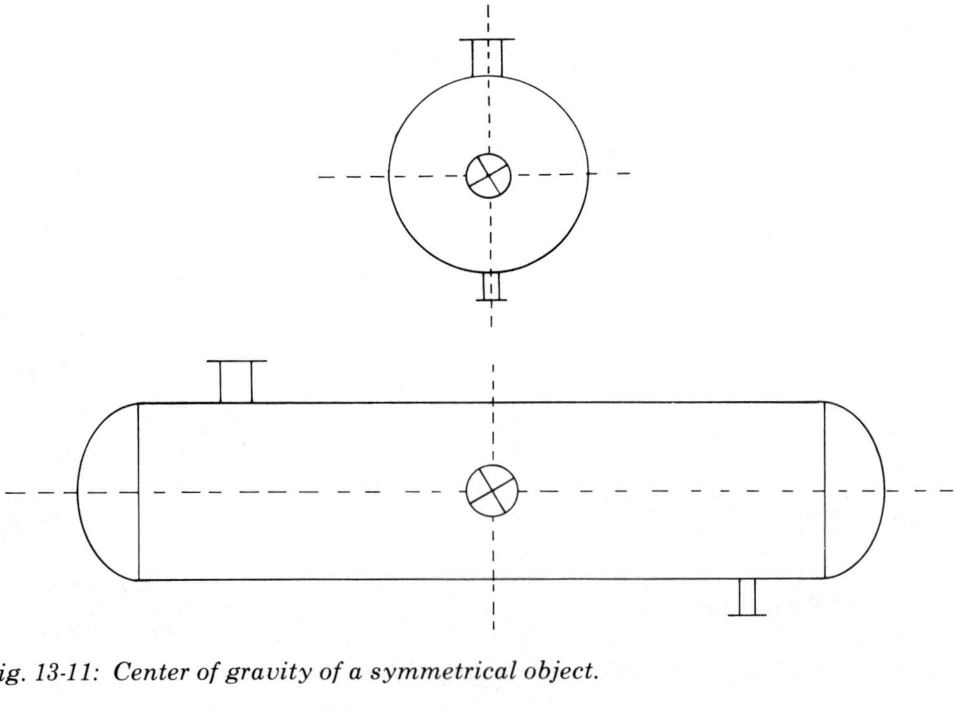

Fig. 13-11: Center of gravity of a symmetrical object.

It is equally important to ensure that the points of support of a load (i.e., where the slings are attached to the load) lie above the center of gravity and not below it (Fig. 13-12) because the center of gravity will always tend to move to the lowest point possible below the point of support. This precaution is especially applicable whenever lifting from pallets, skids, or the base of any object since they all have a tendency to topple. The greatest stability will be achieved when the sling angles are much larger than the angle formed between the plane of support and the line through the center of gravity (Fig. 13-13), and when the attachments are above the center of gravity.

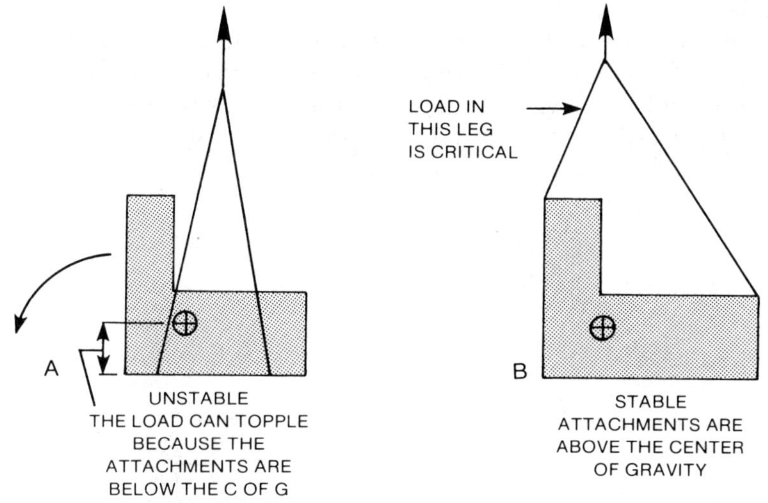

Fig. 13-12: Placing the pick-points above the center of gravity.

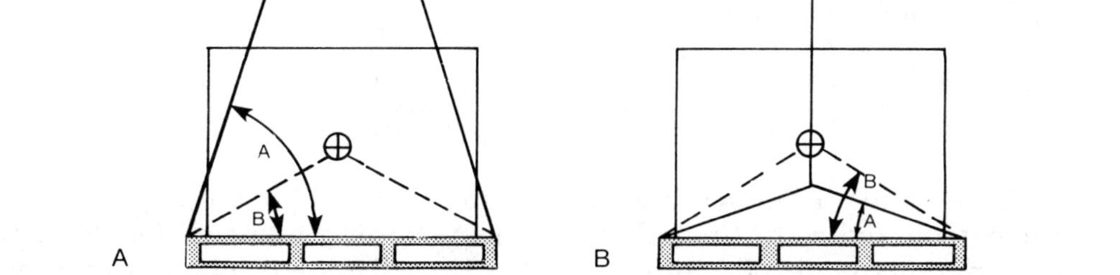

Fig. 13-13: Increasing the sling angle increases the stability.

Calculating Center of Gravity of Asymmetrical Objects

For bodies that are not symmetrical, the center of gravity may be found experimentally or may be calculated by using formulas. Volume and center of gravity equations are contained in Appendix K for the most common asymmetrical shapes encountered.

Many objects are symmetrical in at least one plane. In a 6′ length of any shape, such as channel, I-beam, L-shape, or even railroad rail, the center of gravity will be at the midpoint of its length (3′ from either end). These shapes may or may not be symmetrical in other planes. This requires the use of formulas.

To show how the center of gravity might be calculated, a shape such as a half hollow cylinder is considered in Fig. 13-14. In the left hand, or side view, of this half hollow cylinder, visual inspection indicates that the center of gravity is halfway from either end and at the midpoint between the top and bottom since the shape is symmetrical in two planes.

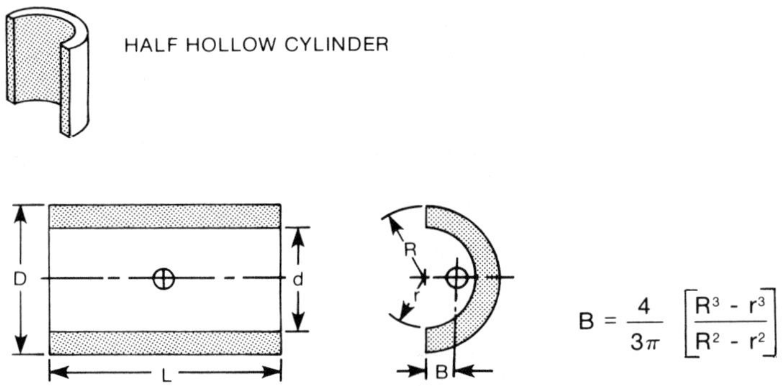

Fig. 13-14: Find the center of gravity of an asymmetrical object.

Referring to the lettered designations, the center of gravity would be at one-half "L" and one-half "D." In the other plane, or end view, the piece is not symmetrical, so it is necessary to use the formula for finding the center of gravity of a half cylinder. The distance that the center of gravity lies from the flat side is indicated by "B."

The formula for "B" is given as:

$$B = \frac{4}{3\pi}\left[\frac{R^3 - r^3}{R^2 - r^2}\right]$$

with the small "r" indicating the inside radius and the large "R" indicating the outside radius of the half cylinder. Assigning numbers for the lettered designations of the half cylinder (Fig. 13-15) will allow calculation of the value of "B," such as:

$$B = \frac{4}{3\pi}\left[\frac{R^3 - r^3}{R^2 - r^2}\right] = \frac{4}{9.42}\left[\frac{125 - 91.125}{25 - 20.25}\right]$$

$$= \frac{4}{9.42}\left[\frac{33.875}{4.75}\right]$$

$$= .424\ [7.132]$$

$$= 3.02$$

A visual inspection of the half hollow cylinder, based on a sense of proportion, might have indicated that the center of gravity would be approximately in the center of the end view, or that "B" would equal 2.5′. Such an estimate would have been reasonably accurate. The actual center of gravity is 3.02′ from the flat side of the half cylinder.

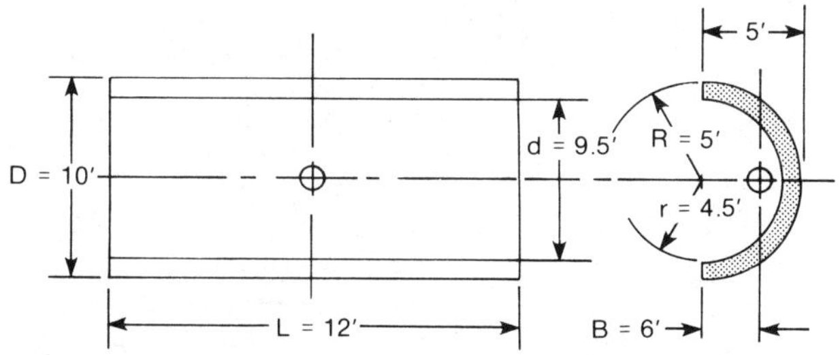

Fig. 13-15: Figuring the center of gravity of a half hollow cylinder.

Calculating Center of Gravity of Complex Shapes

The appearance of the ductwork made of 1″ steel plate in Fig. 13-16 might indicate that it would be difficult to establish the center of gravity location and weight. Usually it is possible to separate a complex object into parts that make it easier to calculate both the weight and center of gravity location (Fig. 13-17).

The center of gravity location of the object in this illustration was calculated by separating the object into three simple shapes as indicated by "A," "B," and "C." Each of these parts is symmetrical, so their center of gravity locations are at their exact centers. Use of these three locations in formulas makes it possible to determine the center of gravity location of the entire object.

Imaginary lines, axes "Z," "Y," and "X," placed on the sketch will help to establish the three planes necessary to calculate the center of gravity for each plane or surface. Visual inspection of the end view indicates that the center of gravity will be on the "Z" axis since the object is symmetrical in this plane.

In the side view, "X" and "Y" axes have been assigned to the object. These axes could have been placed anywhere on the sketch since their only purpose is to serve as reference lines for establishing the stress each part imposes on the entire object.

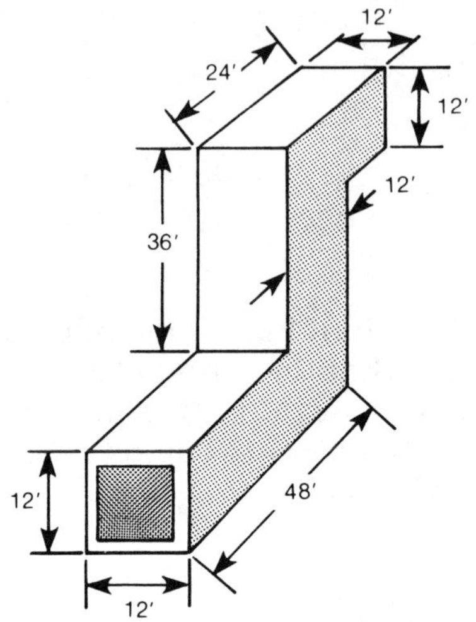

Fig. 13-16: Figuring the center of gravity of a section of ductwork.

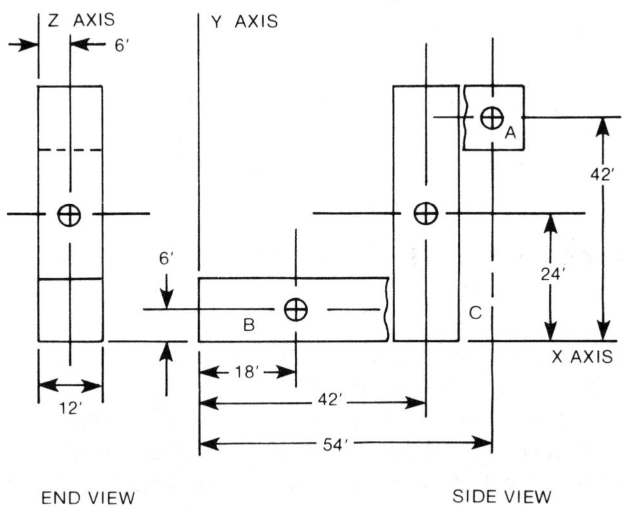

Fig. 13-17: Divide the asymmetrical object into symmetrical parts, and measure their centers from a base line.

One more item of information is required before the center of gravity can be calculated. The weight of each part and the distance that their centers of gravity are from the "X" and "Y" axes must be determined.

The weight of each section may be found by determining the area and multiplying the area by the weight of steel 1″ thick (40 pounds). For example, mentally unfold section A into a large rectangle. The area of section A is 48′ × 12′ = 576 sq ft. The weight is 48 × 12 × 40 = 23,040 lbs.

The center of gravity of each part is at the center because each part is symmetrical. The distance from each axis can be determined by inspecting the initial sketch of the complex object which has dimensions indicated. An easy to use chart can be made summarizing this information as shown in Table 13-1. The equation following Table 13-1 will locate from each axis the center of gravity of the entire object.

Table 13-1: Summary of Dimensions

Object	Width in feet	Height in feet	Depth in feet	Area in feet^2	Weight in lbs	CG Distance from Y Axis	CG Distance from X Axis
A	12	12	12	576	23,040	54	42
B	36	12	12	1152	46,080	18	6
C	12	48	12	1440	57,600	42	24

CG Axis =

$$\frac{(\text{Weight A} \times \text{CG Distance A}) + (\text{Weight B} \times \text{CG Distance B}) + (\text{Weight C} \times \text{CG Distance C})}{\text{Weight A} + \text{Weight B} + \text{Weight C}}$$

This charted information can now be applied to the given formula.

CG Y Axis:

$$= \frac{(23{,}040 \text{ lb} \times 54') + (46{,}080 \text{ lb} \times 18') + (57{,}600 \text{ lb} \times 42')}{23{,}040 \text{ lb} + 46{,}080 \text{ lb} + 57{,}600 \text{ lb}}$$

$$= \frac{1{,}244{,}160 + 829{,}440 + 2{,}419{,}200}{126{,}720}$$

$$= \frac{4{,}492{,}800}{126{,}720} = 35.45$$

CG X Axis:

$$= \frac{(23{,}040 \text{ lb} \times 42') + (46{,}080 \text{ lb} \times 6') + (57{,}600 \text{ lb} \times 24')}{23{,}040 \text{ lb} + 46{,}080 \text{ lb} + 57{,}600 \text{ lb}}$$

$$= \frac{967{,}680 + 276{,}480 + 1{,}382{,}400}{126{,}720}$$

$$= \frac{2{,}626{,}560}{126{,}720} = 20.73$$

The center of gravity is, then, 35.45′ from the Y axis and 20.73′ from the X axis (Fig. 13-18).

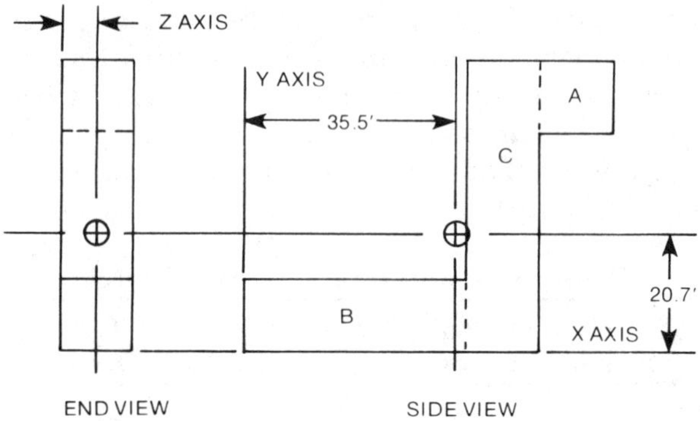

Fig. 13-18: Location of the center of gravity.

Calculating Center of Gravity of Other Complex Shapes

Many complex shapes are not symmetrical in even one plane. Like the previous examples discussed, the center of gravity location of the more complex shapes can be calculated in a similar manner using formulas.

In Fig. 13-19, another section of ductwork made of 1″ steel plate is shown. Three axes or planes must be considered so that the final center of gravity can be determined. Three axes, indicated on the illustration as "Z," "Y," and "X," are assigned in the end and side view (Fig. 13-20). The final center of gravity location is also indicated (Fig. 13-21) as found by the following procedures.

Separating the complex object into parts makes it easier to use the center of gravity of each part to find the final center of gravity of the entire object. In

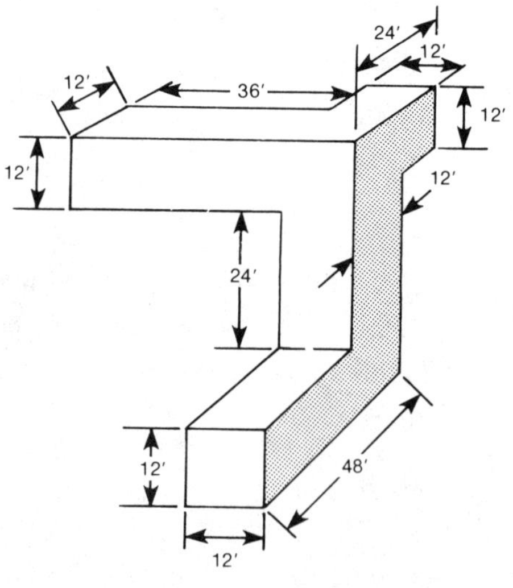

Fig. 13-19: Finding the center of gravity of a complex shape.

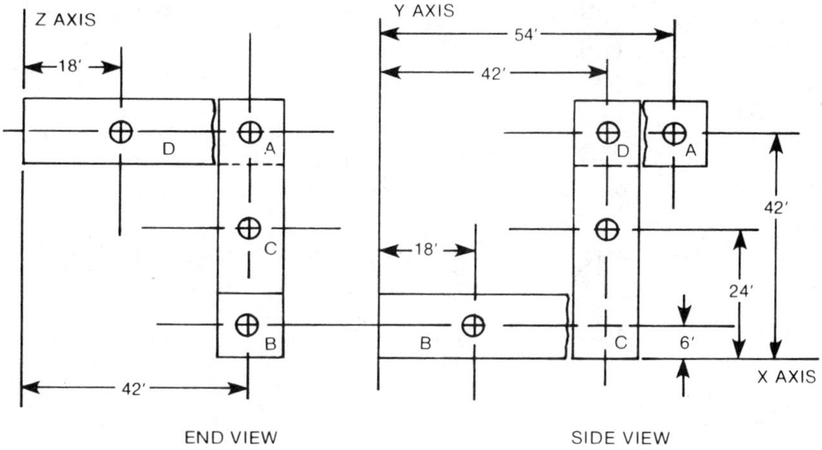

Fig. 13-20: Divide into symmetrical parts and measure centers from base lines.

Fig. 13-20, dimensions have been added to show the center of gravity locations of each part in relation to the three axes, "Z," "Y," and "X." Note that each part, "A," "B," "C," and "D," is symmetrical and that its center of gravity location will be at its geometrical center. Dimensions have been added to this sketch. They are determined by referring to the dimensions of the initial sketch of this object.

It is also easy to find the area of each separate part so that the weight can be determined for use in the formulas. Mentally unfold each section and determine the weight by the area.

Table 13-2, which summarizes the information on this illustration, will make it easier to use the information for setting up the equations for calculations. This information will indicate the center of gravity location of the entire object. It is now possible to correctly rig the load since the load moment can be calculated in each of the three planes.

Transferring the charted information to the following formula will allow solving for the location of the CG of the entire object.

Table 13-2: Summary of Dimensions

Object	Width in feet	Height in feet	Depth in feet	Area in feet^2	Weight in lbs	CG Distance from X Axis	CG Distance from Y Axis	CG Distance from Z Axis
A	12	12	12	576	23,040	42	54	42
B	36	12	12	1152	46,080	6	18	42
C	12	48	12	1440	57,600	24	42	42
D	12	36	12	1152	46,080	42	42	18

The center of gravity, then, is 26.4′ from the X axis, 37.2′ from the Y axis, and 35.6′ from the Z axis (Fig. 13-21).

$$\text{CG Axis} = \frac{(\text{Weight A} \times \text{CG Distance A}) + (\text{Weight B} \times \text{CG Distance B}) + (\text{Weight C} \times \text{CG Distance C}) + (\text{Weight D} \times \text{CG Distance D})}{(\text{Weight A}) + (\text{Weight B}) + (\text{Weight C}) + (\text{Weight D})}$$

CG X Axis:

$$= \frac{(23{,}040 \text{ lb} \times 42') + (46{,}080 \text{ lb} \times 6') + (57{,}600 \text{ lb} \times 24') + (46{,}080 \text{ lb} \times 42')}{23{,}040 \text{ lb} + 46{,}080 \text{ lb} + 57{,}600 \text{ lb} + 46{,}080 \text{ lb}}$$

$$= \frac{967{,}680 + 276{,}480 + 1{,}382{,}400 + 1{,}935{,}360}{172{,}800}$$

$$= \frac{4{,}561{,}920}{172{,}800} = 26.4$$

CG Y Axis:

$$= \frac{(23{,}040 \text{ lb} \times 54') + (46{,}080 \text{ lb} \times 18') + (57{,}600 \text{ lb} \times 42') + (46{,}080 \text{ lb} \times 42')}{23{,}040 \text{ lb} + 46{,}080 \text{ lb} + 57{,}600 \text{ lb} + 46{,}080 \text{ lb}}$$

$$= \frac{1{,}244{,}160 + 829{,}440 + 2{,}419{,}200 + 1{,}935{,}360}{172{,}800}$$

$$= \frac{6{,}428{,}160}{172{,}800} = 37.2$$

CG Z Axis:

$$= \frac{(23{,}040 \text{ lb} \times 42') + (46{,}080 \text{ lb} \times 42') + (57{,}600 \text{ lb} \times 42') + (46{,}080 \text{ lb} \times 18')}{23{,}040 \text{ lb} + 46{,}080 \text{ lb} + 57{,}600 \text{ lb} + 46{,}080 \text{ lb}}$$

$$= \frac{967{,}680 + 1{,}935{,}360 + 2{,}419{,}200 + 829{,}440}{172{,}800}$$

$$= \frac{6{,}151{,}680}{172{,}800} = 35.6$$

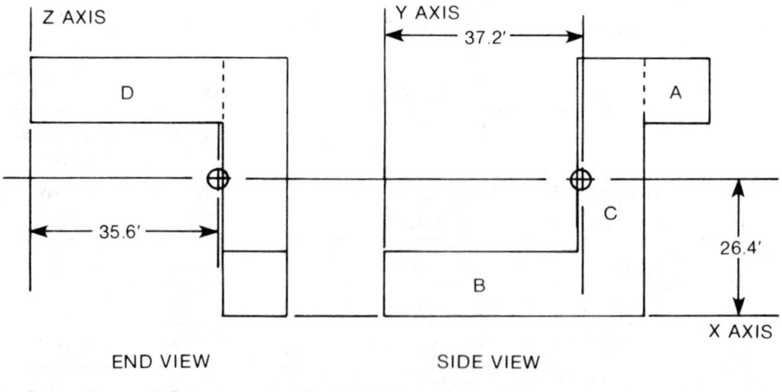

Fig. 13-21: Location of the center of gravity.

14

Basic Methods of Equalizing and Distributing Loads

INTRODUCTION

There are very many different ways to distribute weight or to equalize the forces in a rigging system. A subject so varied cannot be covered exhaustively in this chapter. Instead, a few basic principles are given that will apply to most rigging procedures involving equalizing and distributing weight. Three general methods—equalizer beams, multiple block reeving systems, and interconnected jacks—and a sample rigging procedure are explained in detail.

EQUALIZING LOADS WITH AN EQUALIZER BEAM

Figure 14-1 illustrates both a spreader beam and an equalizer beam. The spreader beam widens the angle of the sling and increases the stability of the load, eliminating the hazard of load tipping, sliding, or bending. Equalizer beams are used to equalize the load in sling legs and to keep equal loads on dual hoist lines when making tandem lifts (Fig. 14-2).

For example, Fig. 14-3 shows a nuclear steam generator with only one lifting lug attached to the top flange. Assume the generator weighs 100 tons and will be lifted with two cranes each having a capacity of 75 tons. Neither crane is capable of making the lift alone; so, the weight must be divided between the two cranes. One way to do this is with an equalizer beam. In order for the cranes to bear an equal amount of weight, the moment arms (the horizontal distance between the pin and the crane attachment) must be of equal length. If the beam weighs 2 tons and the load blocks, slings, shackles, and so forth weigh an additional 4 tons, the gross weight of the lift is 106 tons. Divided equally, each crane will lift 53 tons, well within their lifting capacity (Fig. 14-4).

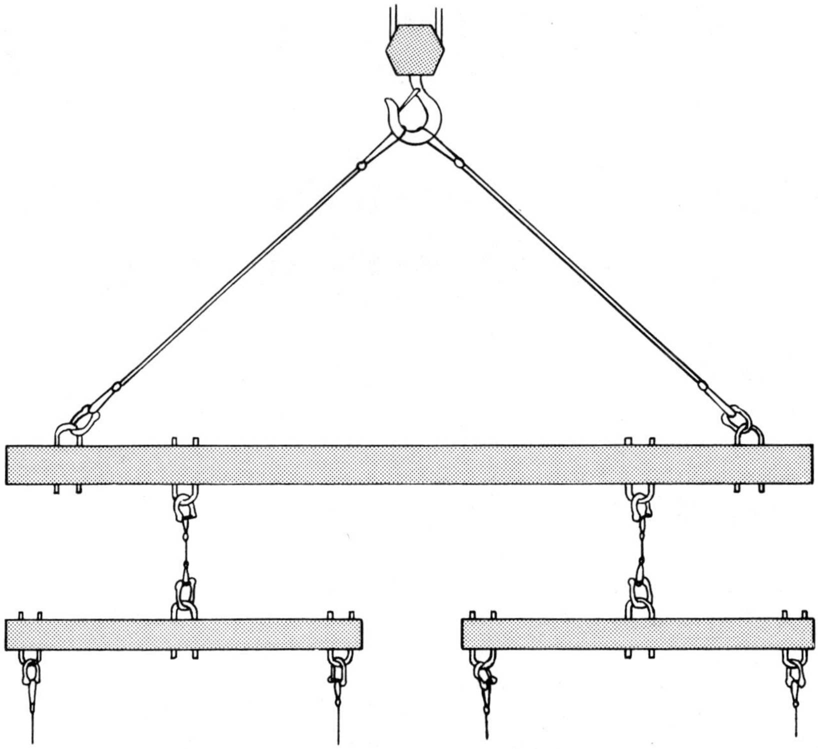

Fig. 14-1: Combination of spreader beam and equalizer beams.

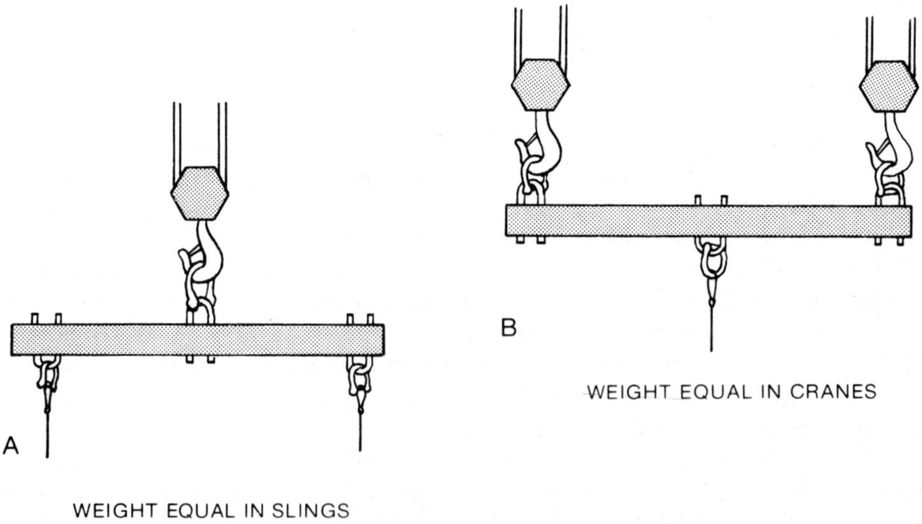

Fig. 14-2: An equalizer beam may distribute weight between slings or between cranes.

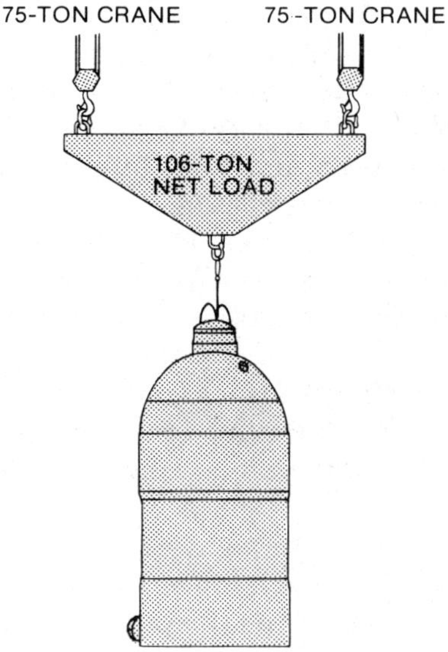

Fig. 14-3: Distributing the weight of a nuclear steam generator between two cranes of equal capacity.

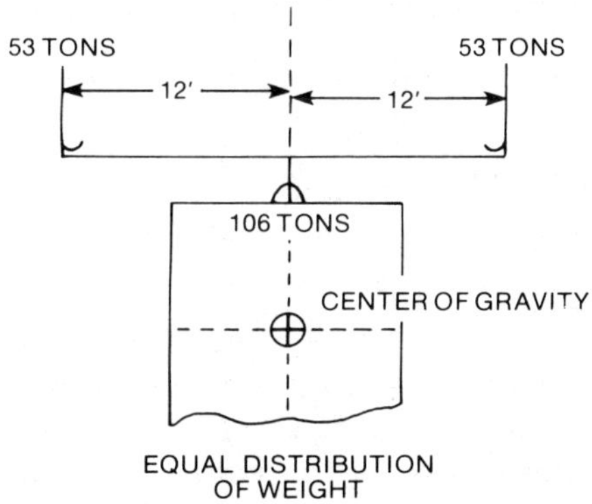

Fig. 14-4: Equal distribution of weight; equal length of moment arms.

DISTRIBUTING WEIGHT UNEQUALLY WITH AN EQUALIZER BEAM

Often two large cranes of the same capacity are not available when both are needed to raise a heavy component. Instead, some combination of a large crane and a smaller crane must be used. When this is the case, the load must be distributed unevenly so that the larger crane carries most of the load.

For example, the same 100-ton steam generator in Fig. 14-3 must be lifted with a 75-ton capacity crane and a 50-ton capacity crane. Although the smaller crane has the capacity to lift half the net weight of the generator, the weight of the equalizer beam and the other rigging accessories must be taken into consideration. The actual lifting capacity of a 50-ton crane (after subtracting 1 ton for half the weight of the equalizer beam, and 1 ton for the weight of the block, and 1 ton for the weight of the slings, shackles, hoisting lines, etc.) might be about 47 tons. The same would be true for the 75-ton capacity crane. After subtracting 3 tons from its gross capacity, the net weight that the crane can safely lift is 72 tons.

It is obvious that the weight of the generator cannot be distributed evenly between the two cranes. The smaller crane would be overloaded; thus, the weight of the lift must be distributed unevenly. This will require some very simple calculations.

First determine the total net capacity of both cranes: 47 + 72 = 119 tons. Then divide the net capacity of each crane by the total net capacity of both and multiply by 100. This will determine the percent of the load each crane should carry.

For the 50-ton crane, $\frac{47}{119} \times 100 = 39.5\%$. Round off to 40%.

For the 75-ton crane, $\frac{72}{119} \times 100 = 60.5\%$. Round off to 60%.

By using these figures, we can determine the percentage of the load to be lifted by each crane:

For the 50-ton crane, 106 tons × 0.40 = 42.4 tons
For the 75-ton crane, 106 tons × 0.60 = 63.6 tons

In order to distribute the weight as divided in these steps, the spreader beam in Fig. 14-5 has been modified. By changing the position of either the main lifting lug or one of the attachment lugs, one moment arm is made shorter than the other. The shorter arm will carry the greater weight. The question which must be answered is how close to the center should the attachment lug be moved to ensure that it will support at least 60% of the load. The distance, d, can be found by working a simple problem of proportion:

$$\frac{60 \text{ tons}}{40 \text{ tons}} = \frac{12 \text{ feet}}{d}$$

$$60\,d = 480$$

$$d = 8$$

The proportionate weight, 60:40 or 3 to 2, is equal to the proportionate distance 12:8 or 3 to 2. But the proportions are inverse because the larger weight is supported by the smaller distance (length of moment arm) and the smaller weight is supported by the longer arm. This inverse proportion will always hold true. The equalizer beam could be of a different length as long as the ratio is the same.

The new rigging arrangement is pictured in Fig. 14-6. Now the 100-ton generator can be safely lifted with the 75- and 50-ton cranes.

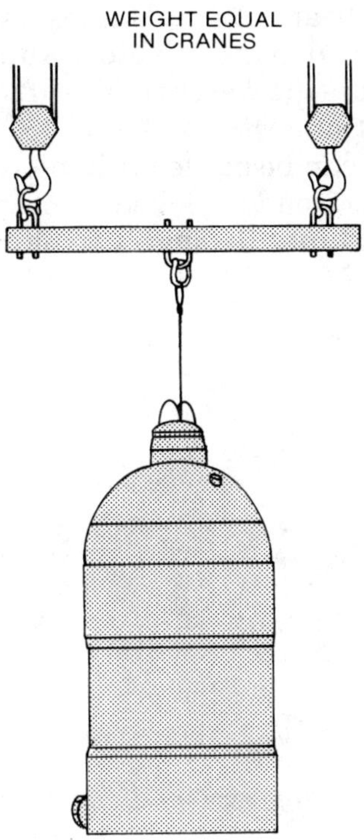

Fig. 14-5: Distributing the weight between two cranes of unequal capacity.

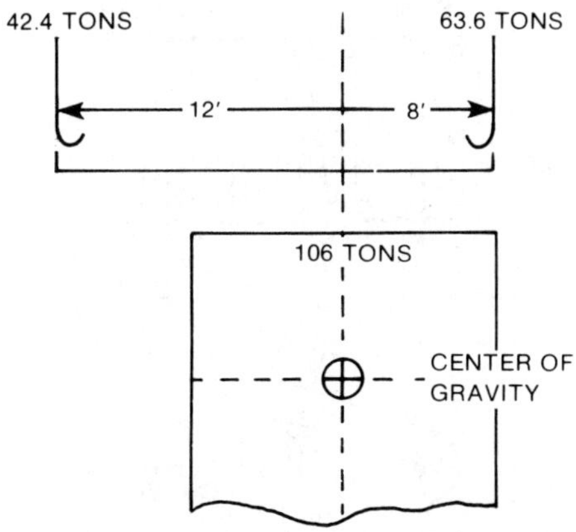

Fig. 14-6: Unequal distribution of weight; unequal lengths of moment arms.

USING MULTIPLE BLOCK REEVING SYSTEMS

Another method of equalizing a load is with a multiple block reeving system. In the example in Fig. 14-7, a 30-ton load is being raised with two head blocks, a fall block, and a spreader beam. A single line passes through the three blocks. The total weight is supported in the fall block and divided equally between the two head blocks. This places the weight of the lift on the two columns where the support structure is the strongest.

A similar arrangement can be made utilizing the reeving system illustrated in Fig. 14-8. In this case the load is picked from two points. A single rope runs between two hoisting machines through two pair of sheaves and an equalizer sheave.

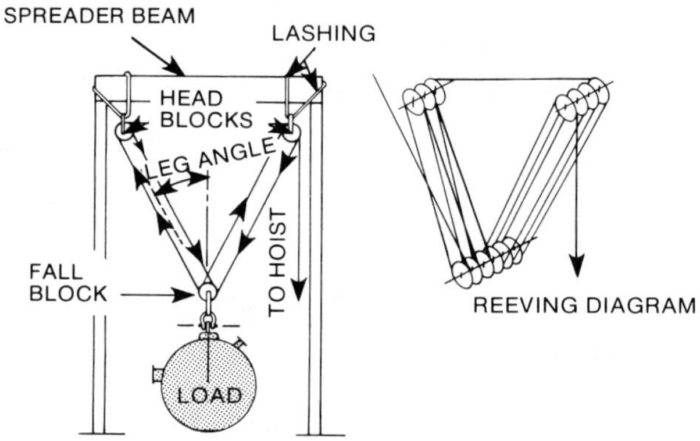

Fig. 14-7: A multiple block reeving system distributes load equally between two columns or poles.

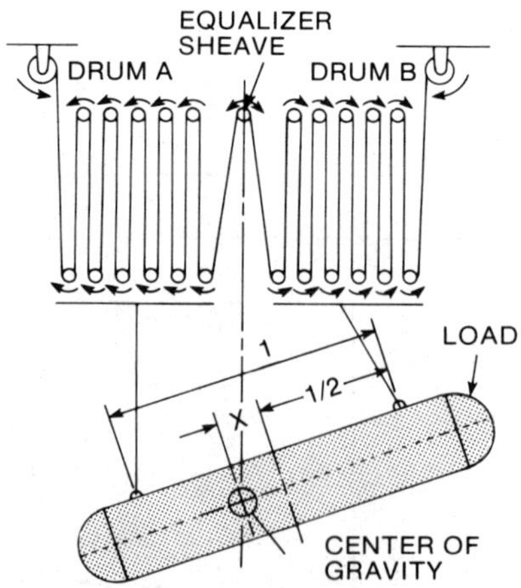

Fig. 14-8: The equalizer sheave allows the sheaves to adjust to the shift of the object so that each sheave carries an equal load.

The center of gravity will always shift to a point directly below the equalizer. As it does so, the length of rope in one sheave will lengthen while the other is reduced. When the rope is freely suspended in this way, each part of the rope will equally share the load.

Another method of equalizing, using the multiple block reeving system, is illustrated in Fig. 14-9. The principle in this example is the same as in Fig. 14-5 where an equalizer beam was used to allow the sheaves to adjust to the weight of the load. In this example, though, an equalizer beam adjusts to the weight difference to equalize the load in the ropes and permit the load to tilt.

A method of equalizing heavy loads is shown in Fig. 14-10. Although the sheaves A and B act independently of each other, sheaves C and D share the same rope which passes from one drum, through both systems of sheaves, and to the other drum. As long as the center of gravity falls directly below the center of the four sheaves, the weight will be carried equally by all four ropes. If the center of gravity lies off-center, only the sheaves C and D will equalize.

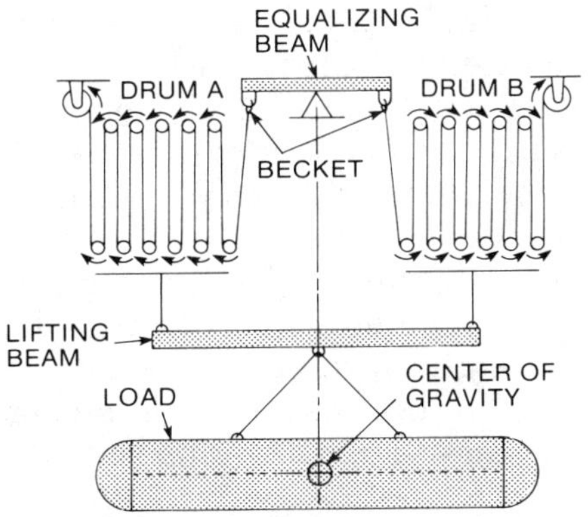

Fig. 14-9: An equalizer beam allows the weight to distribute between the sheaves.

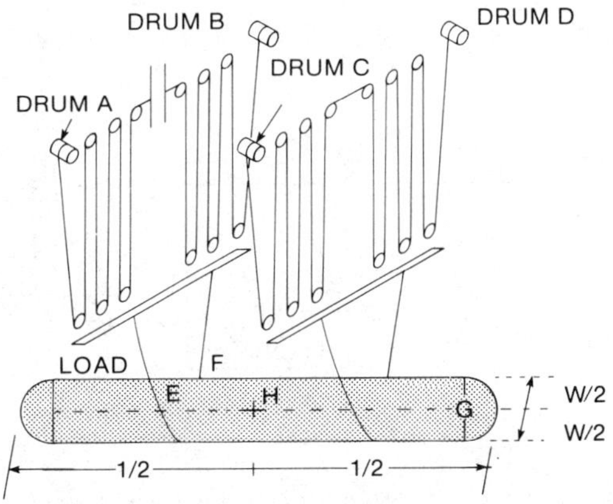

Fig. 14-10: Equalizing heavy loads with a multiple block reeving system.

DISTRIBUTING WEIGHT TO TEMPORARY RIGGING STEEL

In addition to building and boiler steel supports, temporary steelwork must usually be erected to support the rigging. This requires a distribution of weight to steel beams and columns according to their load bearing capacity. Normally, the building steel is much weaker than boiler steel, and when building steel is used to support rigging steel, care must be taken to prevent overloading any member of the support steel.

A good example of this is the rigging procedures necessary to install a heavy steam drum. Very often there will not be sufficient clearance between boiler support steel and the final position of the drum to allow for the blocks and rigging. Temporary rigging steel is installed above the boiler supports and is supported by portions of the building steel.

When distributing the weight on a temporary beam between columns, the position of the support point is calculated in the same way that a lifting lug is positioned on an equalizer beam (Fig. 14-11).

The beam is divided into moment arms (one on either side of the point of support), and the length of the moment arm is inversely proportionate to the load it will carry. The short arm will carry the larger portion of the load; the long arm will carry the smaller portion.

For example, assume that a tank or drum weighing 45 tons must be lifted at the center point between four columns, as illustrated in Fig. 14-12. In addition to the load weight, the weight of the rigging and rigging steelwork must be added to the load weight. For the sake of simplicity, assume that these accessories add 5 tons to the load weight, bringing the net weight to 50 tons.

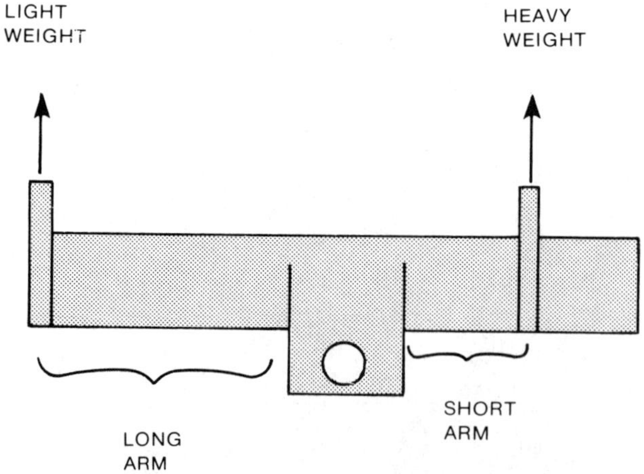

Fig. 14-11: Inverse proportion. Long arm carries light weight; short arm carries heavy weight.

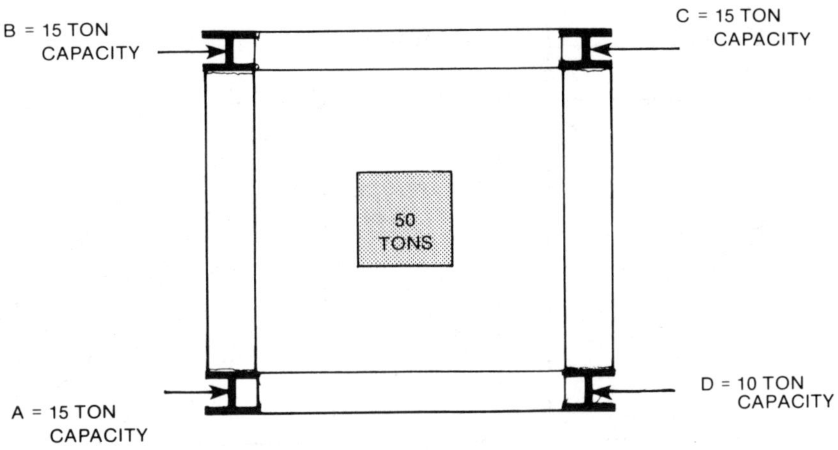

Fig. 14-12: Permanent columns that must support temporary rigging steel.

In the example, column A has a load bearing capacity of 15 tons; column B, 15 tons; column C, 15 tons; and column D, 10 tons. The obvious solution might be to lay a temporary beam from the center of AB to the center of CD. However, the moment on each column would be 12.5 tons (50 tons ÷ 4 = 12.5 tons). This situation would overload column D.

The problem of distributing the weight so that column D is not overloaded could be solved in several ways. Here is one example of finding a solution.

The diagram in Fig. 14-13 shows the four columns. The outside dimension between each post is 20′. The center point G of the four columns is 14.14′ from each column. The starting point for solving this problem begins at the weakest column, column D. It is known that the column can support only 10 of the 50 tons and the point of support for the load is 14.14′ away from column D. When the 50-ton load hangs from that point, only 10 tons can be imposed on the 14.14′ of temporary beam between G and D. The temporary beam must be supported at a point beyond point G. Remember that the length of the moment arms of a beam are inversely proportionate to the divided weight of the load. The short arm supports the heavier load, the long arm supports the lighter load. If the 14.14′ moment arm supports 10 tons, the remainder of the beam must support 40 tons. How long must the beam be? Set up a ratio like this:

$$\frac{40 \text{ tons}}{10 \text{ tons}} = \frac{14.14}{x \text{ ft}}$$

$$40x = 141.4$$

$$x = 3.5$$

By cross multiplying and dividing, it is found that an additional 3.5′ is necessary to support the 40 tons. Thus, the length of the first temporary rigging beam is 17.64′ (Fig. 14-14).

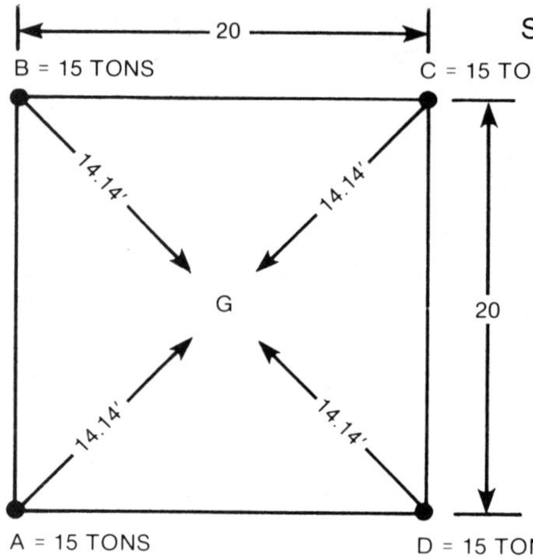

Fig. 14-13: Dimension of columns and the center of gravity.

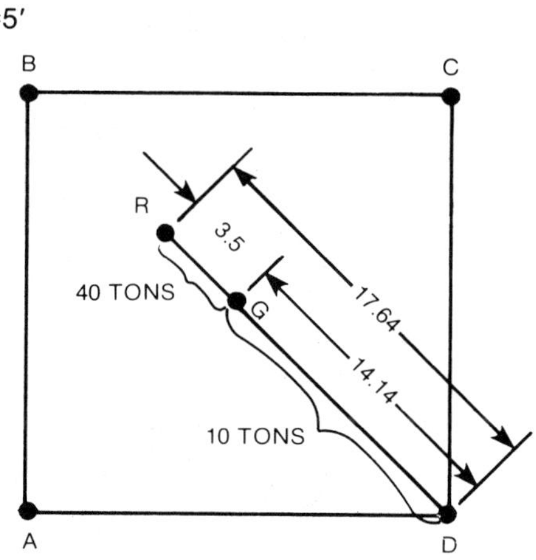

Fig. 14-14: Length of moment arms DG and GR.

At this stage, point R, the end of the temporary beam is supported by thin air. An additional beam is needed. This second temporary rigging beam will run from column C through point R to some point on the beam AB which will be labeled K. The length of this second beam can be determined by drawing a line from column C through point R to beam AB. Measure the line and convert the measurement according to the scale of the drawing (in this case 1″ = 5′). By doing this it becomes obvious that moment arm RC = 15′ and moment arm KR = 8.125′ and the total length of the temporary beam KC = 23.125′ (Fig. 14-15).

Formulating a weight/length ratio for either arm, determine the portion of the 40 tons at point R each arm bears. For example:

$$\frac{23.125'}{15'} = \frac{40 \text{ tons}}{x \text{ tons}}$$

By using the length of the long arm (15′) in the ratio, solve for the load carried by the short arm (remember the proportion is inverse). Therefore, arm KR carries 25.9 tons.

The load imposed on the opposite arm can be determined by subtracting 25.9 tons from 40 tons. Therefore, arm RC carries 14.1 tons (Fig. 14-16).

At this point, it is known that column D will carry 10 tons and column C will support 14.1 tons. The loads imposed on columns A and B must still be calculated. Beam AB supports 25.9 tons at point K. Measure the moment arms AK and KB and convert according to the scale of 1″ to 5′. The actual dimensions are indicated in Fig. 14-17.

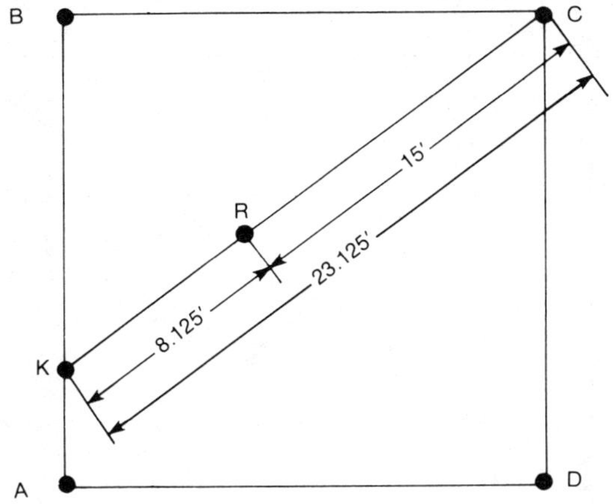

Fig. 14-15: Length of moment arms KR and RC.

Again set up a ratio. Solve for either

$$\frac{20}{8.75} = \frac{25.9}{x} \quad \text{or} \quad \frac{20}{11.25} = \frac{25.9}{x}$$

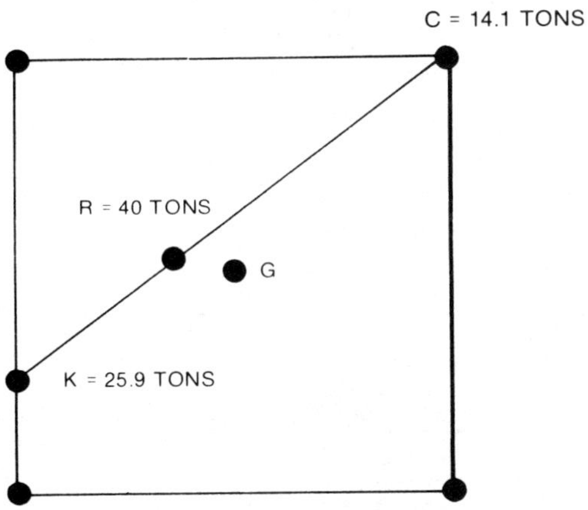

Fig. 14-16: Weight imposed at points K, R, and C.

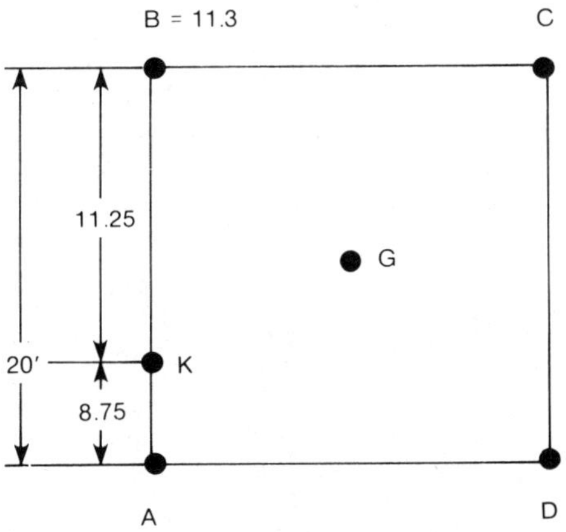

Fig. 14-17: Length of moment arms AK and KB.

By solving the ratios it is found that the load imposed on column A is 14.6 tons and the load carried by column B is 11.3 (Fig. 14-18).

The final diagram of this particular distribution of weight is given in Fig. 14-19. The combined weight supported by each column is 50 tons. The load may be hoisted from center point G without any of the supporting columns.

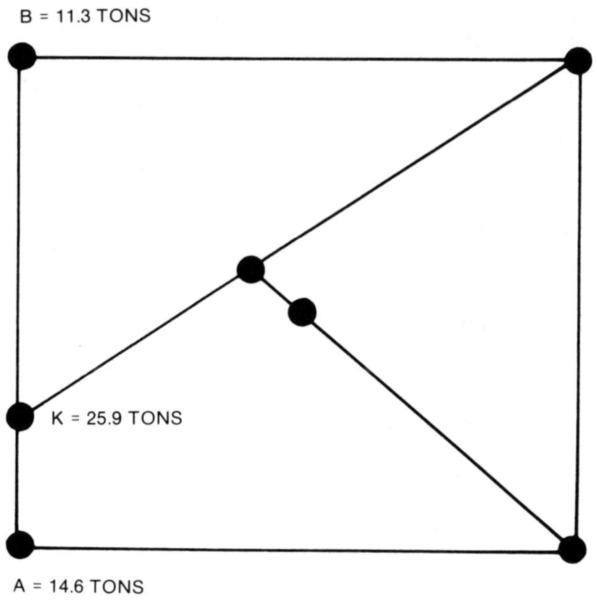

Fig. 14-18: Weight imposed on points A, K, and B.

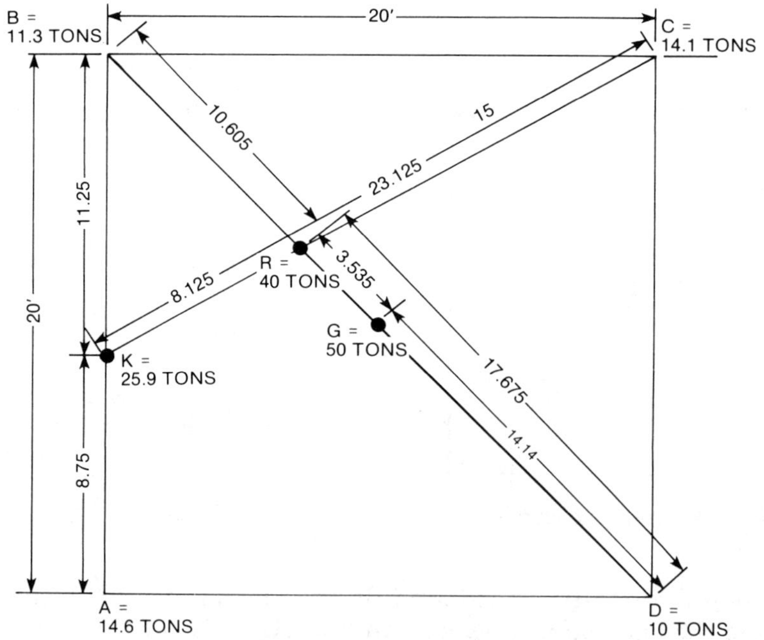

Fig. 14-19: Final diagram showing the length of each moment arm and the weight carried by each column.

EQUALIZING LOADS WITH JACKS

Jacks are seldom used in lifting and rigging. In the absence of rigs with sufficient lifting capacity, a number of jacks may be used to lift a load until a truck bed or some sort of transporter may be positioned underneath. A jack is more often used as a brace or as a pushing force instead of a lifting force. Yet, occasionally a specially designed rigging operation that calls for jacks is encountered. A mention of them here is warranted.

The equalization of loads with hydraulic jacks is simply accomplished. By utilizing a system of interconnected jacks, all of equal bore diameter, each jack will exert an equal force on the load. This is true because:

$$\text{Force} = \text{Pressure} \times \text{Area}$$

Take, for example, Fig. 14-20. The jacks pictured here are connected and each operates from the same hydraulic pump. As pressure is applied to the pump, that same pressure is transmitted throughout the interconnected jacks. Because each jack has the same piston area or bore diameter, an equal amount of pressure is produced by each jack.

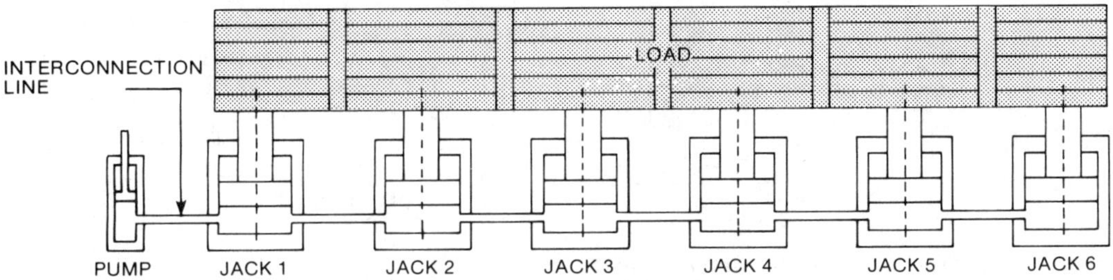

Fig. 14-20: Equalized system of interconnected jacks.

Whenever lifting with a number of interconnected jacks, it is important to know the center of gravity of an object. Not only is pressure produced by the pump brought to bear against the load, but the load is a source of pressure against the jacks. As long as the center of gravity is in the center of the jacks, the lift will be stable; but when the center of gravity shifts or is miscalculated, tipping can occur.

To simplify the forces, think of two children of equal weight sitting on opposite ends of a seesaw. Because each child is exerting an equal pressure on the board at an equal distance from the balancing bar, the board is level or the load equalized. Increase the weight of one child or move one child closer to the balancing bar, and the equilibrium of forces will be disturbed. The heavier or longer side will exert the greater force, forcing one side down and the other up.

Jacks act in much the same way. When a disproportionately greater pressure is exerted on one jack, the pressure forces hydraulic oil to flow to a jack experiencing less pressure. The heavily pressed jack depresses, and the less pressed jack extends. If this action affects a large enough area of the total area, the load will tip over (Fig. 14-21).

The center of gravity of an object can be determined by the methods already discussed. The center of gravity is that point where all forces or moments are in equilibrium or cancel each other out. There is also a point where the combined forces of the jacks are in equilibrium. These two points must coincide if the lift is to be balanced. The point of equilibrium among a cluster of jacks is called the center of action.

For example, assume that the center of gravity for the object in Fig. 14-22 is located as indicated by the shaded X. Place five jacks so that their combined forces will be centered at the same point as the center of gravity—5′ from the left side and 8′ from the front side.

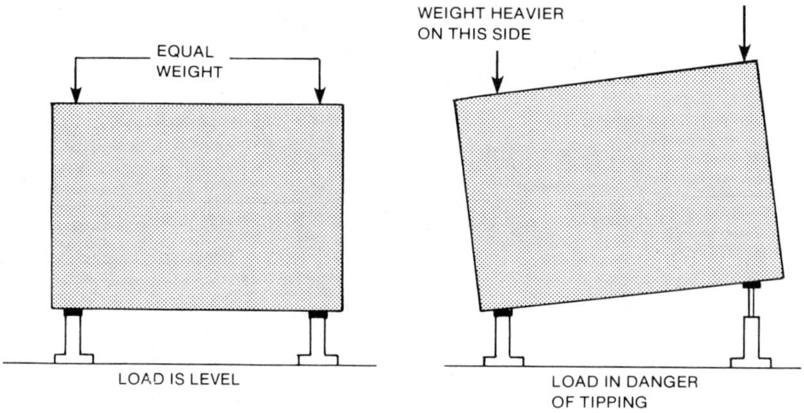

Fig. 14-21: The out-of-center object tips.

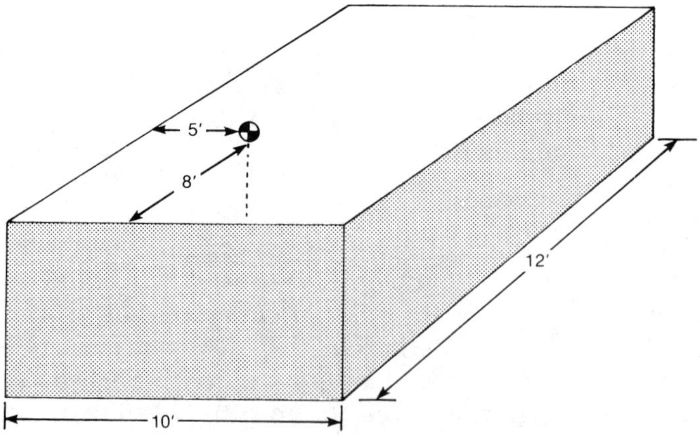

Fig. 14-22: The center of gravity of this object does not lie at its center.

The jacks must be positioned so that the sum of their distances from the left side divided by the number of jacks equals 5. Also they must be positioned so that the sum of their distances from the front side divided by the number of jacks equals 8.

This exercise is drawn out in Fig. 14-23. The left side is given a base line X. The front side is given a base line Y. The distance of each jack is measured from both base lines and their measurements divided by five. The distance from the center of action to base line X is:

$$\frac{10 + 10 + 5}{5} = \frac{25}{5} = 5'$$

This distance from the center of action to base line Y is:

$$\frac{12 + 12 + 10 + 6}{5} = \frac{40}{5} = 8'$$

One other point of load distribution can be discussed in regard to jacks. Seldom will the soil have the load bearing capacity equal to the load being handled by a jack; therefore, the load must be distributed over a wider area of soil. This is often done with timber crib piles (Fig. 14-24). The pressure exerted by a jack on its base support extends downward and outward at a 45° angle in all directions. It does no good to widen the base beyond the area included in the 45° angle. The cribbing should be made of alternating lengths of timber planed

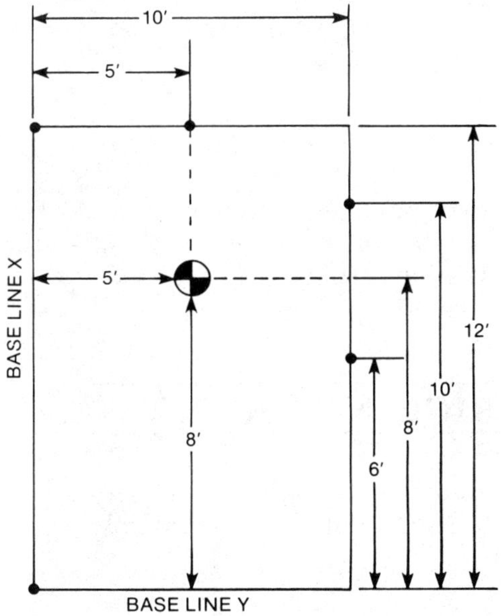

Fig. 14-23: Determining the center of action of a cluster of five jacks.

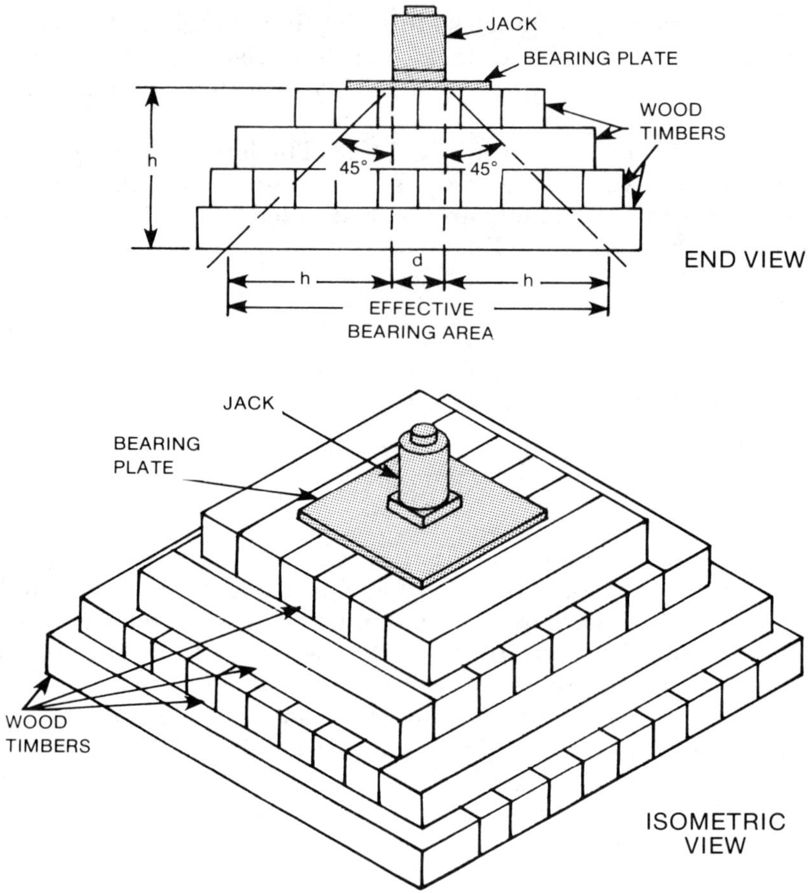

Fig. 14-24: Distributing weight with a crib pile.

on two sides to ensure a level, equal support. A steel bearing plate should be placed between the jack and the top layer of timbers to prevent the timber directly beneath the jack from being crushed.

A SAMPLE RIGGING DRAWING

In Fig. 14-25, an actual reproduction of a rigging blueprint is given. The drawing illustrates the transfer of a reactor head from a guyed derrick to a crawler crane. The drawing is divided into three diagrams. In the left third of the drawing, the position and type of lifting lug that must be attached to the reactor head is given. The center diagram identifies the various rigging accessories and gives their load bearing capacity. Also given are the length and radius of the boom. In the right third of the drawing, the transfer point between derrick and crane is laid out. The following paragraphs will explain each diagram in detail.

In the situation pictured in Fig. 14-25, the existing reactor head must be removed from the reactor building. To make the lift possible, 12 lifting lugs must be welded to the head. Figure 14-26 shows the position of the lugs: six lugs lie on a radius 4′ 4-3/4″ from the center point of the head. Six more lugs are on a radius 9′ 0″ from the center. The lugs are equally spaced on lines converging at the center and forming 60° angles. The lugs are evenly spaced in this manner so that the weight of the head will be evenly distributed to the legs of the sling. Each lug is also identified as MC T 15.

The T 15 lugs are shown in detail in Fig. 14-26. Section A-A shows the distance of the lugs from the centerline of the head, identifies the size of take-up jaw and jaw turnbuckle, and gives the capacity of the shackle. Details B and C give the instructions for welding the T 15 lugs to the head. As indicated by the welding symbol, each lug is to be fillet welded on both sides. The welds have a 3/8″ throat and will be 3″ long spaced 7-1/2″ apart, center to center.

The center diagram pictures the load suspended by the crane (Fig. 14-27). A Manitowoc crawler is used for the lift. The boom is extended to 160′ and has a radius of 28′. In this position, the crane has a capacity to lift 180,200 pounds. The combined weight of the load and rigging is 136,000 pounds or 68 tons.

The rigging consists of a 100-ton transfer lug, a 130-ton lifting cone, six 25-ton snatch blocks connected to the cone with six 25-ton shackles, and six 1′ diameter steel ropes, each rope 30′ in length. The load transfer in the transfer lug is also pictured. As the load is transferred from derrick to crane, the lug swivels until the crane totally supports the weight.

Instructions are also given to keep the bottom of the cyclones as close to the ground or surface as possible during the transfer. If any part of the crane rigging fails, the load can be set down quickly.

The last diagram is an overhead view of the position of derrick and crane (Fig. 14-28). In the upper right corner of the picture, the center of the existing reactor is located. In the center of the drawing is the square outline of the guyed derrick. The derrick must boom out 84.5′ to make the pick and then boom in to 50′ before turning with the load. The necessity of first lifting with the derrick and then transferring to a crawler is evident in the large pipe 82′ high running between the two rigs. A great deal of planning has gone into this lift and cooperation is essential.

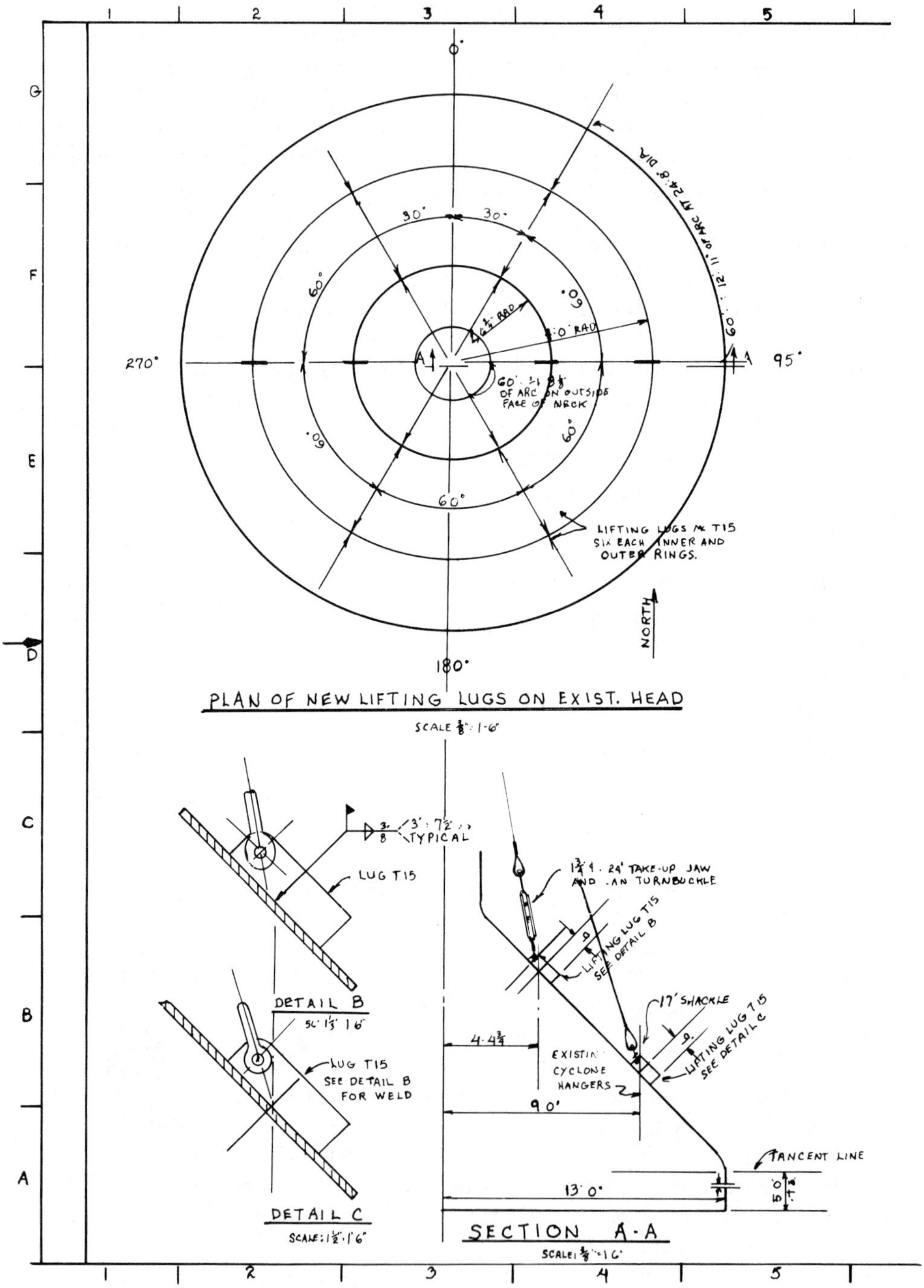

Fig. 14-25: A sample of a rigging drawing.

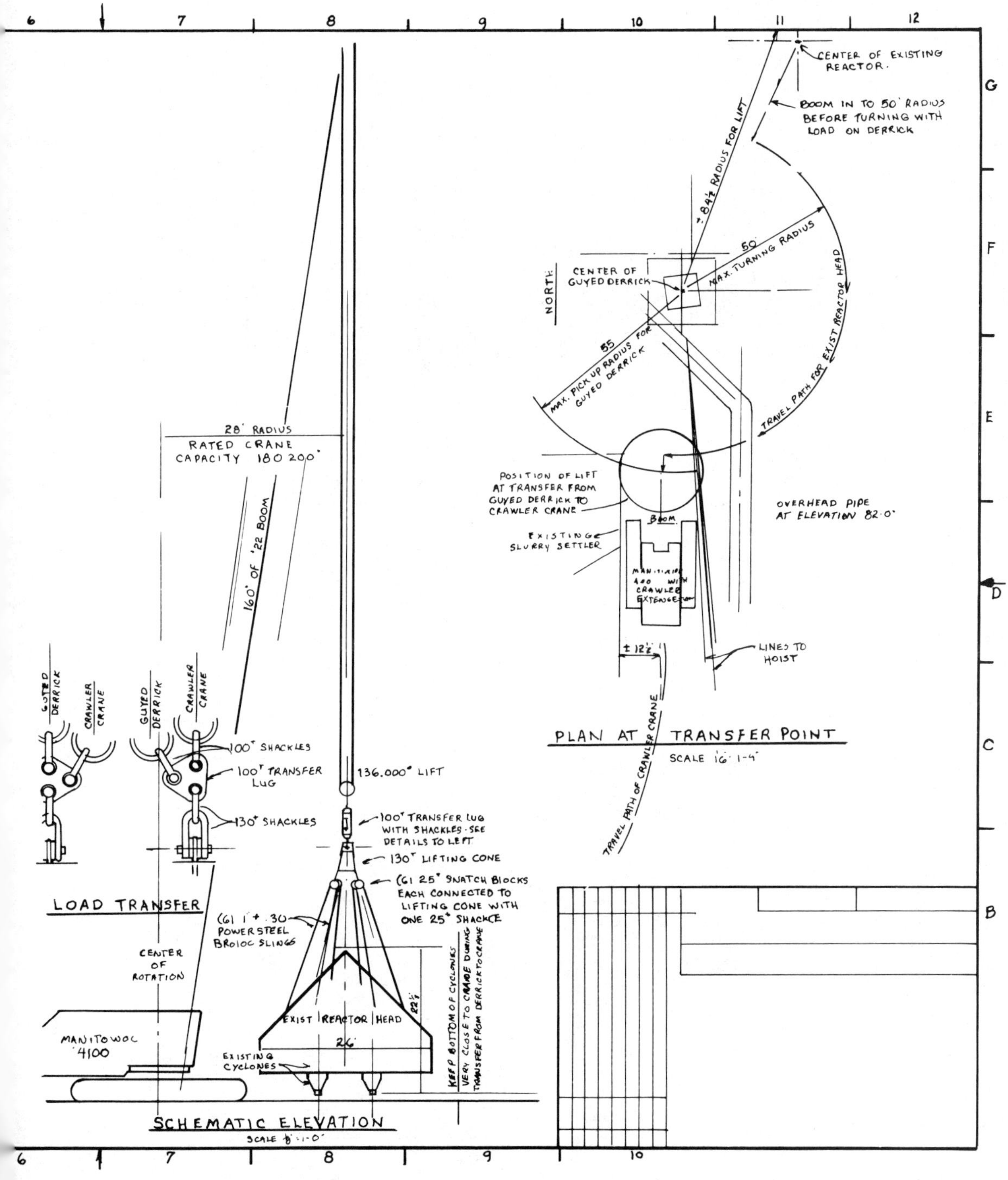

Fig. 14-25: (Continued)

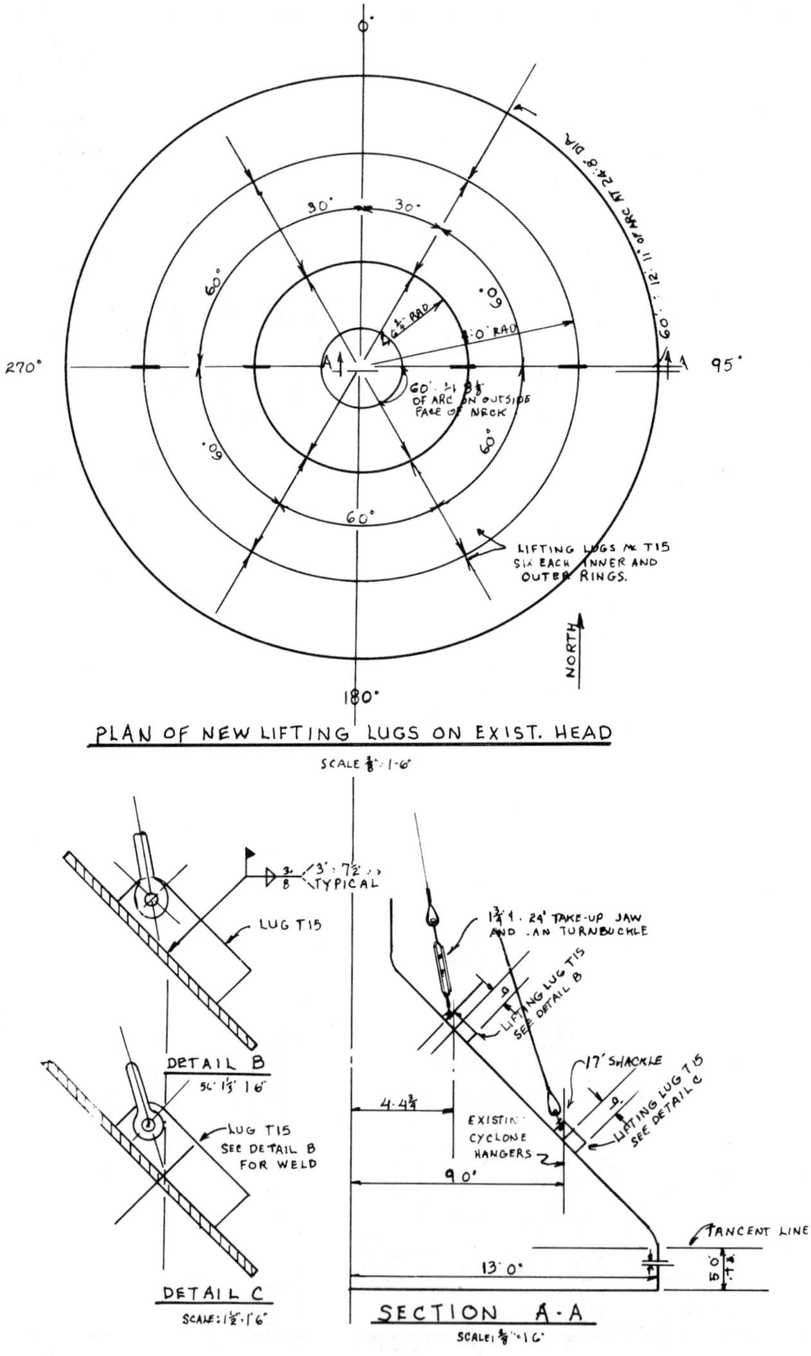

Fig. 14-26: Sample drawing of detailed placement of lifting lugs.

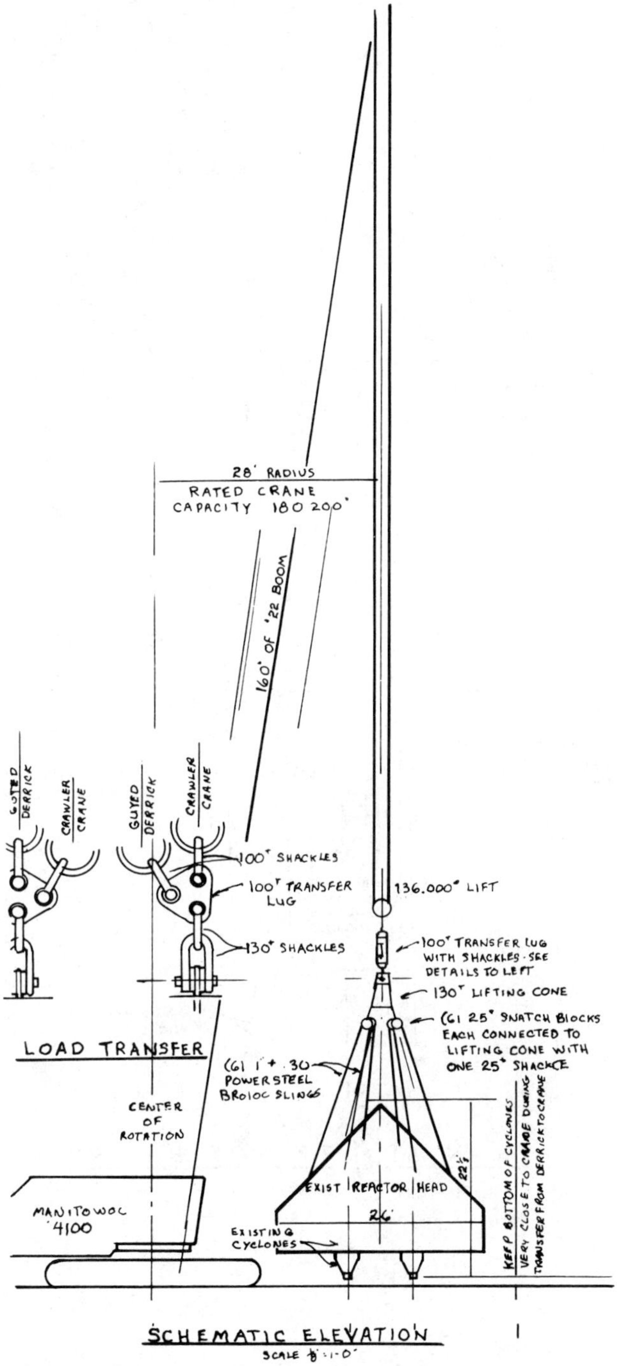

Fig. 14-27: Sample drawing of details of crane and rigging accessories.

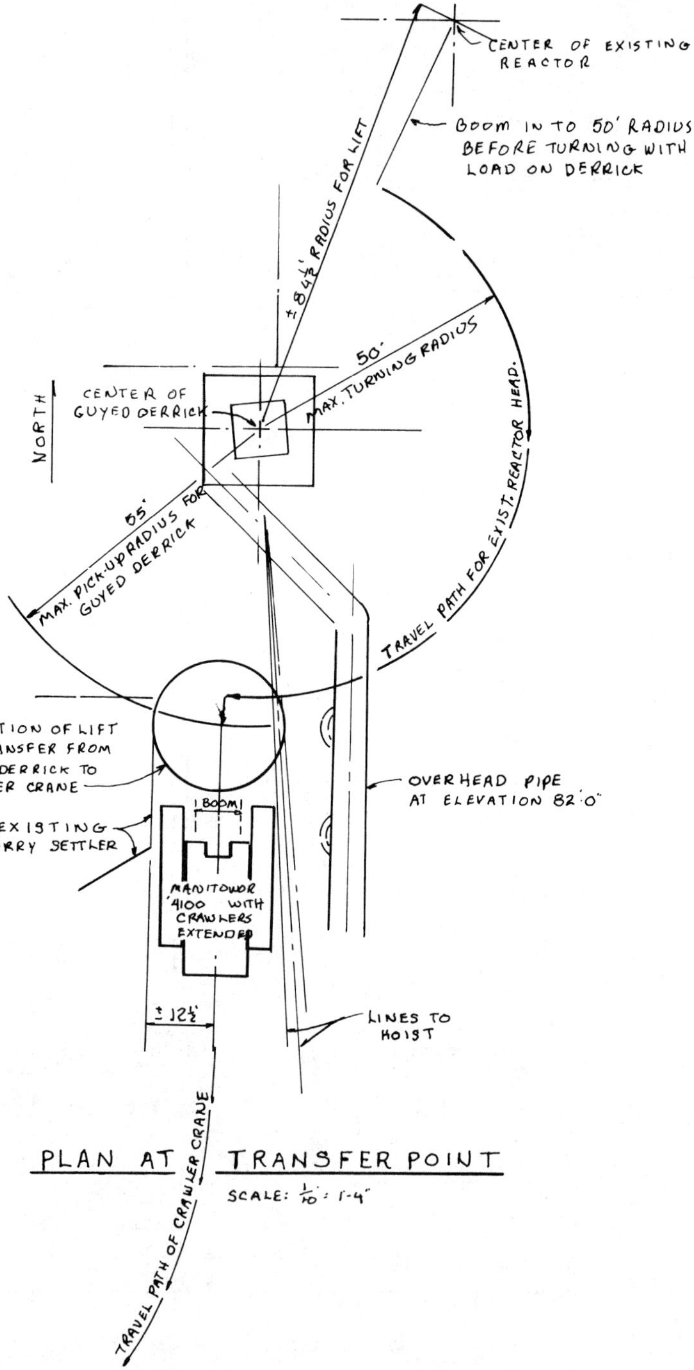

Fig. 14-28: Sample drawing of placement of crane for transfer.

15

Strength and Stability During Operation

INTRODUCTION

The distribution of weight among a load's rigging is not the only factor which affects a lift's success. The proper distribution of weight as it affects the crane is also vital. The rated capacity of the crane must be taken into account when planning a lift, no matter how light or small the load may be. It is important that personnel are familiar with the factors the operator must consider any time a lift is made. Additional information is included in Chapter 9.

MAXIMUM LOAD CAPACITY

Before the machine is put into operation, the operator must know the capacity under all conditions and configurations. Load chart capacities are based on almost ideal conditions seldom achieved under actual working conditions. As such it is extremely important to not only know how to determine the capacity from the chart but also to recognize the factors which can reduce the capacity below what the chart says.

The rated capacities of mobile cranes are based on both strength and stability. It is extremely important to know the difference because in one case, one of the structural components of the crane will break and in the other case, the crane will tip over.

Too many operators still operate by the "seat of the pants" in that they use the stability of the crane to determine the maximum capacity. When the machine "goes light" (i.e., begins to tip) they assume that to be the most it can lift. This method worked fine on old truck cranes and on some crawler machines, but it is terribly hazardous on modern carrier mounted units. They are de-

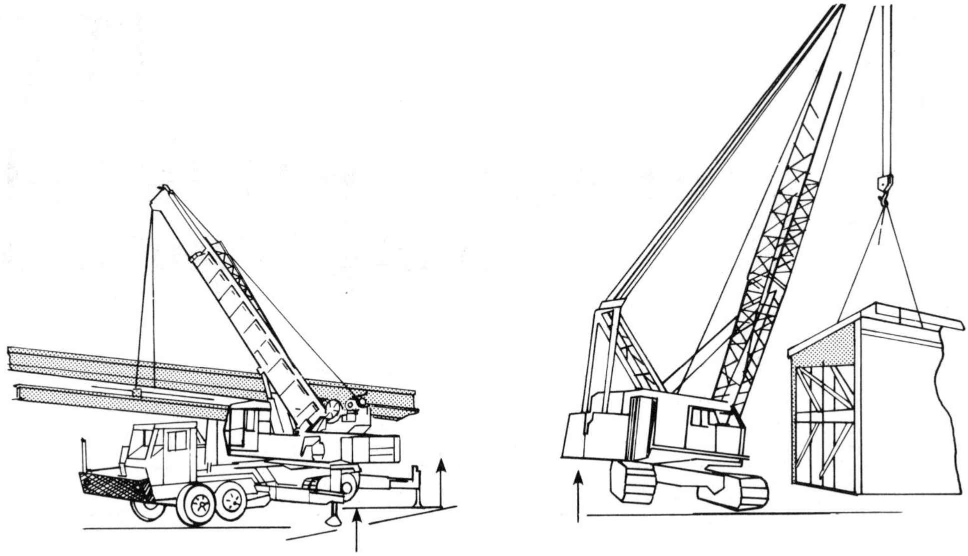

Fig. 15-1: Do not use stability to determine capacity.

signed and manufactured to provide an absolutely solid operating base and when they "go light" they are already grossly overloaded. The machine is either in the process of tipping over or failing its outriggers, hook, rollers, boom, or pendants (Fig. 15-1).

Figure 15-2 shows pendant, boom, outrigger, overturning, and rated loads for a 200-ton crane equipped with a 180′ main boom lifting over the rear. The graph shows how much these loads vary and why it is so difficult to determine exactly what will break and when. The outrigger loads are highest at short radii. The boom loads drop as the radius is increased then they rise again as the boom approaches horizontal. The pendant loads increase as the radius is increased and are highest when the boom is horizontal. The loads causing overturning drop as the radius is increased. In summary, this graph indicates that at short radii the ratings are based on boom or outrigger strength, while at long radii stability and pendant tension are the governing factors.

The manufacturer accounts for these variable factors by providing a load chart that limits the maximum allowable loads to values below the critical loads. If the operator exceeds these limits he/she could be in trouble. Hydraulic cranes, unlike friction machines, do not all have load ratings based on stability. In most cases their ratings depend on strength and hydraulic pressure limits. Therefore, the hydraulic crane operator who waits for signs of tipping to warn of an overloaded condition will often bend the boom, exceed the pressure limits, or cause severe damage to the machine before any signs of tipping occur (Fig. 15-3).

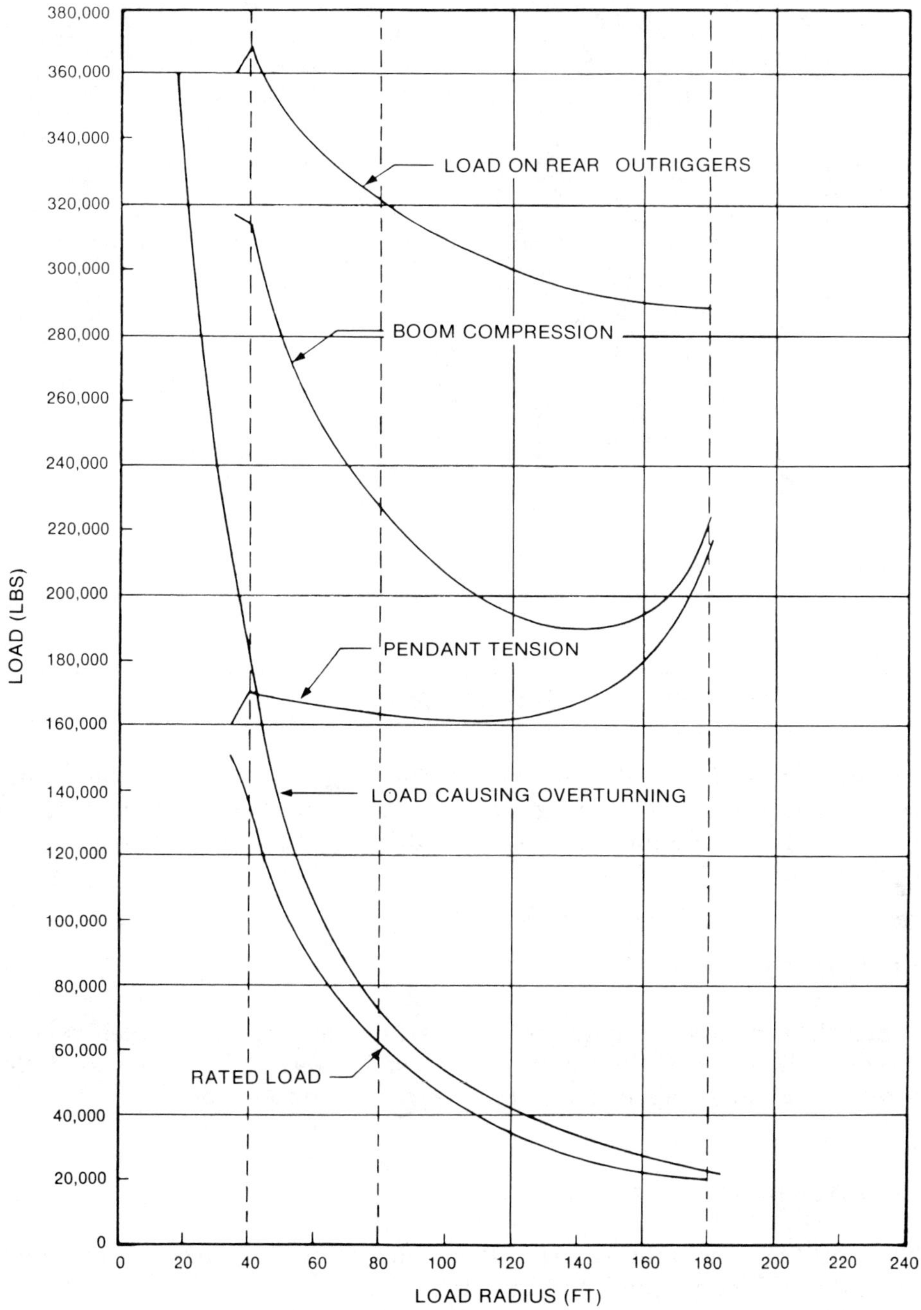

Fig. 15-2: Component loadings.

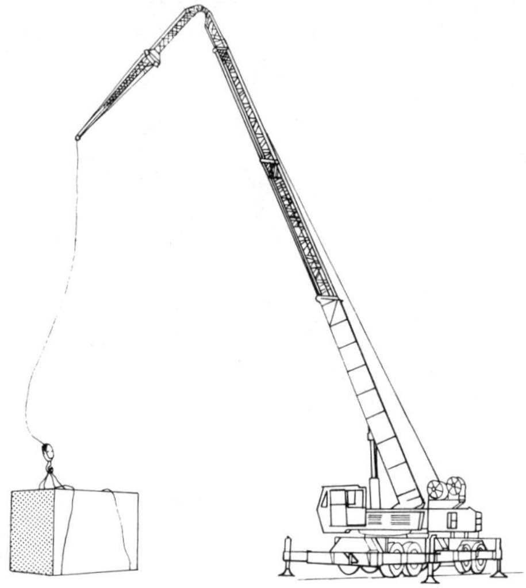

Fig. 15-3: Overloading may cause tipping or boom failure.

FACTORS AFFECTING CAPACITY

The operators of all hydraulic cranes must be guided solely by the rating plate recommendations concerning load, boom length, boom radius, lifting quadrants, and other factors listed on the rating plate, such as outrigger position and terrain.

Outriggers

When the outriggers are not used, the strength of the crane itself very seldom governs the capacity, but new factors such as axle loadings, tire deflection, strength of chassis, and ground pressure gain importance.

Quadrants

The ability of a mobile crane to lift is also affected by the quadrant in which it is working. Lifting quadrants are defined as "over the front, over the side, or over the rear." The load which can be safely lifted will vary considerably from quadrant to quadrant. It is the operator's responsibility to see to it that rating plate ratings are not exceeded, regardless of the quadrant in which he/she is operating (Fig. 15-4).

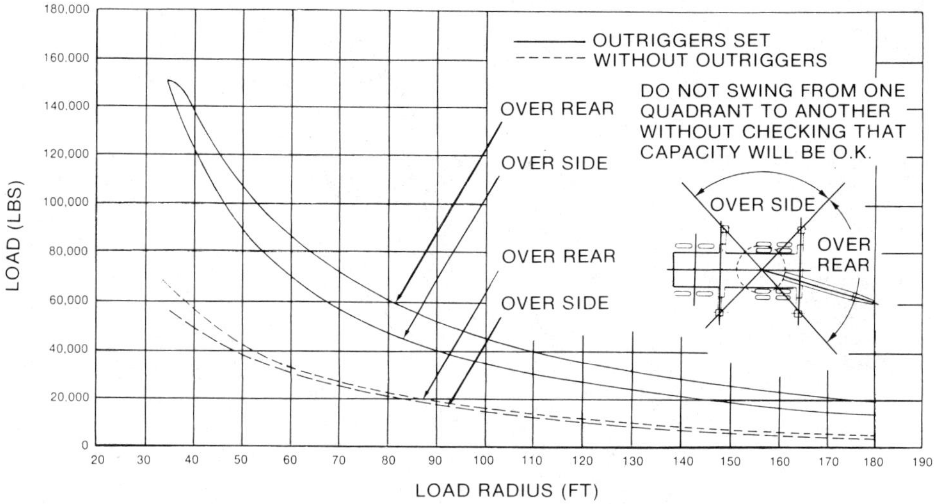

Fig. 15-4: The capacity changes from quadrant to quadrant.

It is also especially important when swinging through quadrants not to carry loads from an area of higher rating into an area of lower rating without first raising the boom to bring the load in within a safe operating radius. Never, unless specifically allowed by the manufacturer, lift over the front (Fig. 15-4).

Boom Length and Angle

The load carrying capacity of a crane is also significantly affected by the length of boom, its angle, and the angle the pendants make with the boom. As the pendant angle decreases, the boom stresses increase. It is, therefore, recommended that the pendant angle be kept at maximum by operating the crane with its gantry and/or live mast in the highest position (Fig. 15-5).

Counterweights

The crane's lifting capacity is also dependent upon the counterweights. Operating with too little will permit the crane to tip forward and operating with too much will cause the crane to tip backward more readily. Increasing the counterweight to lift loads in excess of chart capacities will reduce the possibility of tipping forward, but the lift will increase the structural loads to the point of collapse. The counterweights must be correct for the machine, the boom length, and the type of operation (Fig. 15-6).

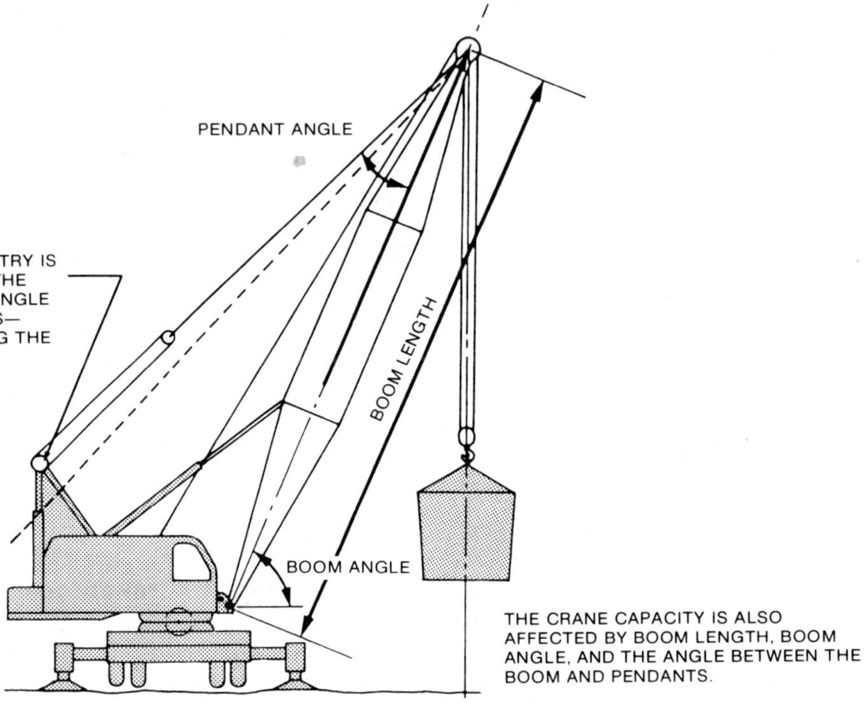

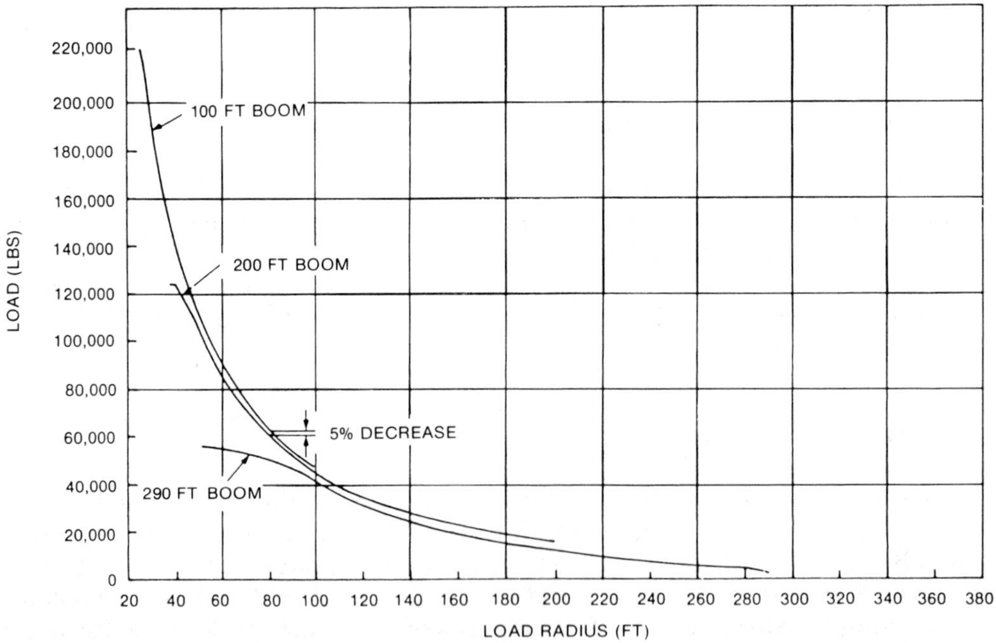

Fig. 15-5: Effect of boom length on capacity.

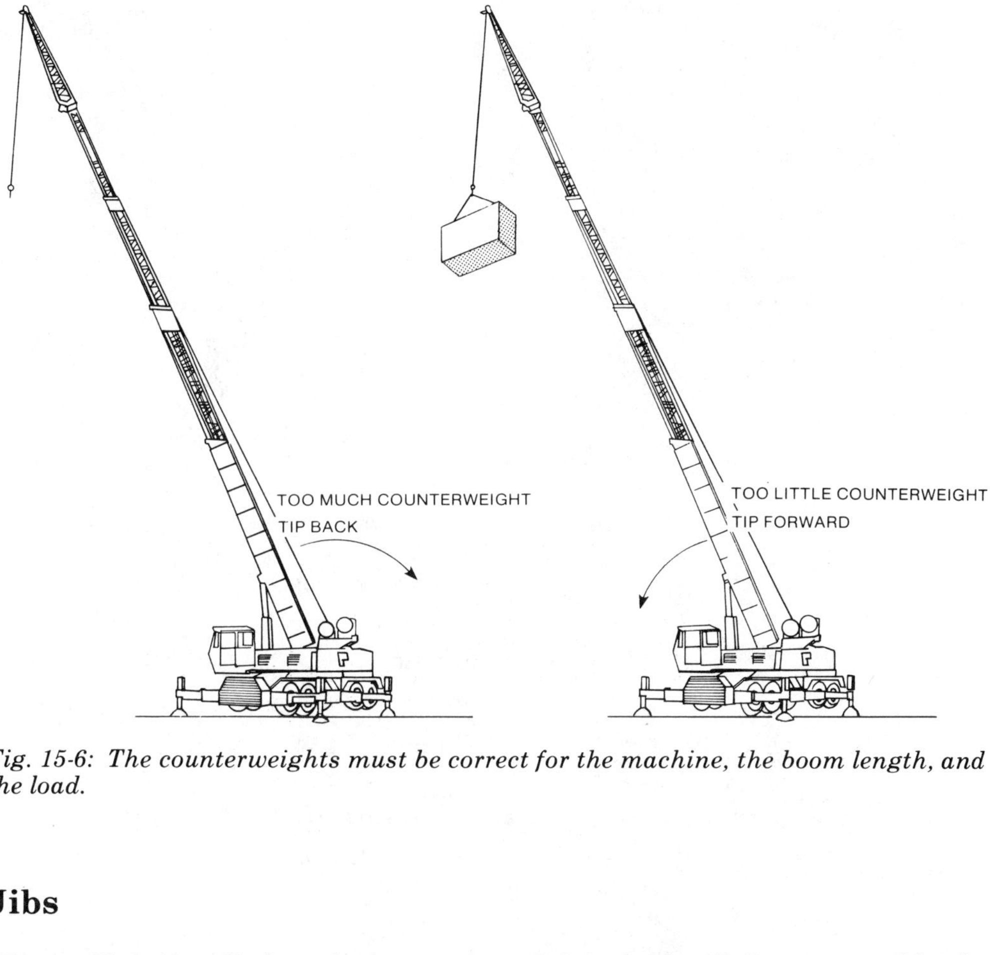

Fig. 15-6: The counterweights must be correct for the machine, the boom length, and the load.

Jibs

When a jib is installed on the boom, the tensions in the jib forestay and backstay ropes depend on the angles they make with the jib mast. For equal loading on both, the jib mast should bisect the angle between the jib and the boom. Unfortunately, only the length of the backstay lines is normally adjustable and therefore the tensions vary considerably with the jib offset. These variations can be kept within safe limits by making the length of the backstays equal or longer than the forestays (Fig. 15-7A).

The capacity of a crane equipped with a jib is affected by and can only be determined by calculating the effect of (Fig. 15-7B):

1. The length of the main boom.
2. The main boom angle.
3. The jib length.
4. The jib angle with respect to the boom (jib offset).
5. The weight of the rigging and hooks on both the jib and main boom.

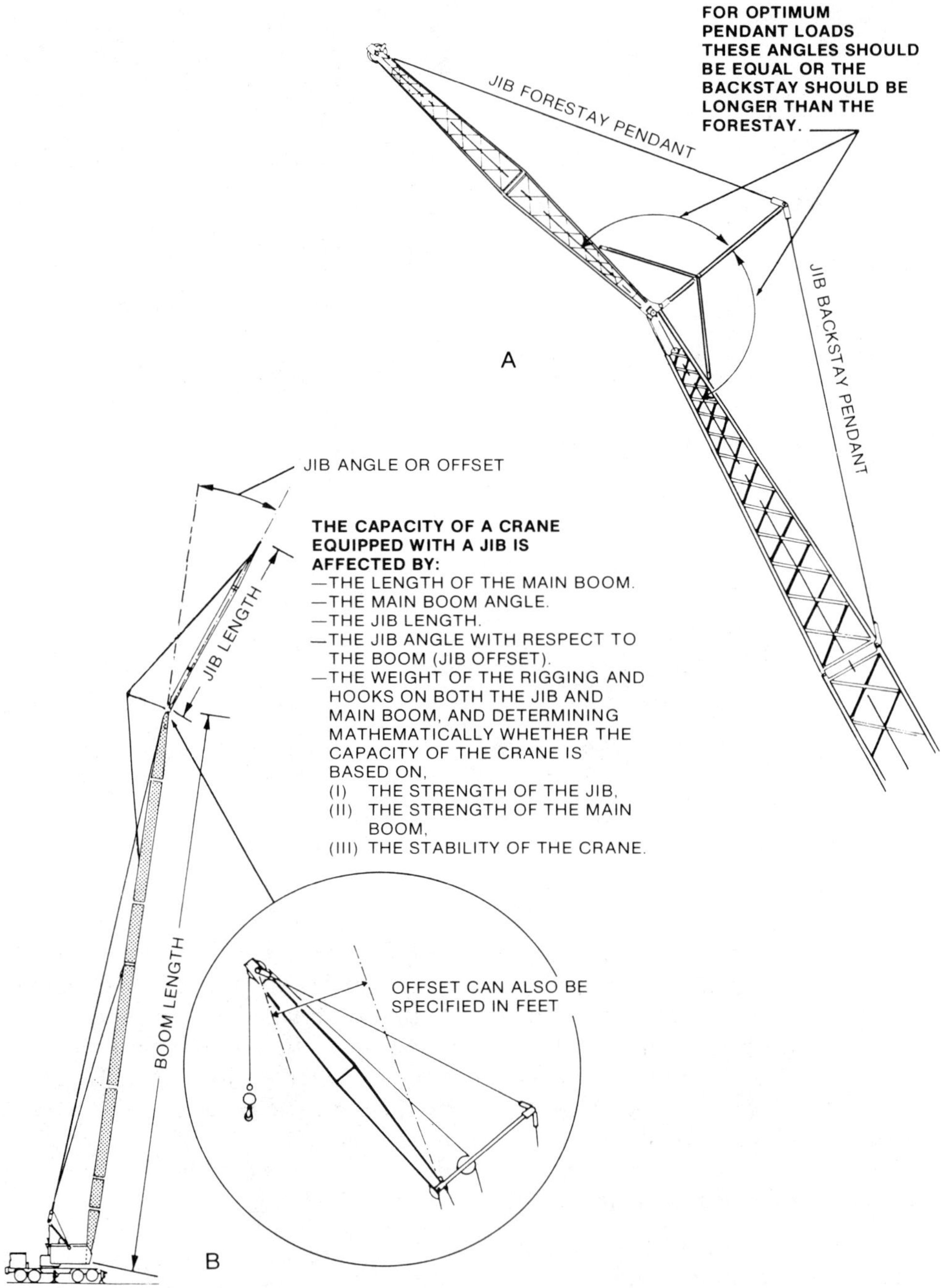

Fig. 15-7: (A) Stability of the pendant. (B) Factors which affect capacity of crane with jib.

Alterations

The cranes' rated loads are also detrimentally affected whenever:

1. Any machine other than a "Sky Horse" or derrick type of rig is tied down or guyed back in an effort to increase stability and capacity (Fig. 15-8).
2. The hydraulic relief valve settings on hydraulic machines are altered in attempts to increase the lifting capacity.

Side Loads

The chart loads apply only to freely-suspended loads on plumb hoist lines. If the hoist line is not plumb at all times when handling loads, then additional side loads will hazard the stability and introduce stresses into the boom for which it has not been designed. In circumstances such as this, structural failures can result without any warning. This precaution is especially important on hydraulic cranes because the boom has very little lateral restraint at the boom point and is, therefore, not capable of supporting any significant side loads. These side loads also produce high forces on the turret and the boom elements at the boom foot connection (Fig. 15-9).

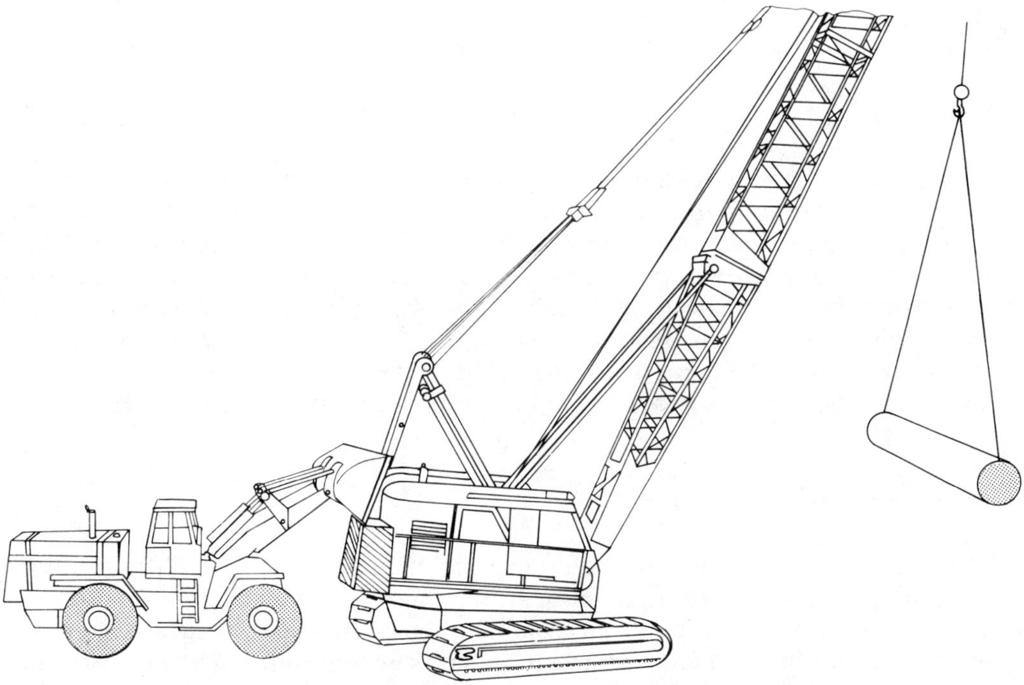

Fig. 15-8: Do not tie down any machine in an effort to increase stability.

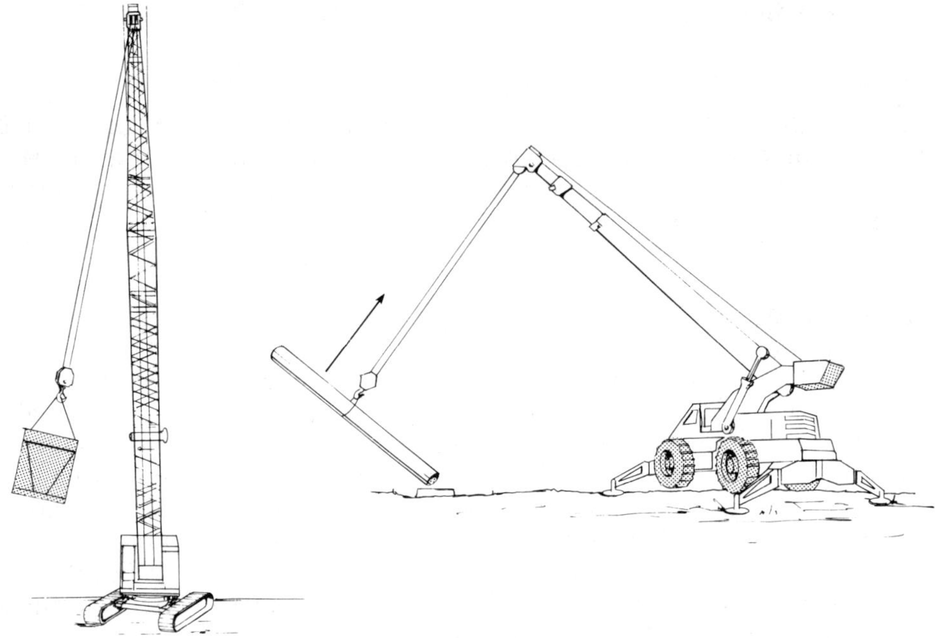

Fig. 15-9: The load must be directly below the boom tip. The hoist line must be plumb at all times.

Swinging

Rapid swinging of a suspended load causes the load to drift away from the machine, thus increasing the radius (Fig. 15-10). This can cause the crane to tip forward or can collapse the boom. The same effect can be generated by rapid swinging of long booms, with or without loads. The inertial effect of the boom can overturn the carrier or collapse the boom. At high boom angles and low load, overturning backwards could occur (Fig. 15-11). These effects are greatly multiplied when the swing is started or stopped suddenly since the boom is side loaded. The load must always be kept directly below the boom point (Fig. 15-12).

Mechanical Conditions

The safe working loads also apply only to cranes in good condition, having undamaged and unkinked structural members. The strength of the boom is very sensitive to local irregularities. The main chords and lattice members must be straight, and the connections between sections must not introduce eccentricity (Fig. 15-13). If a crane becomes damaged in service, such as might

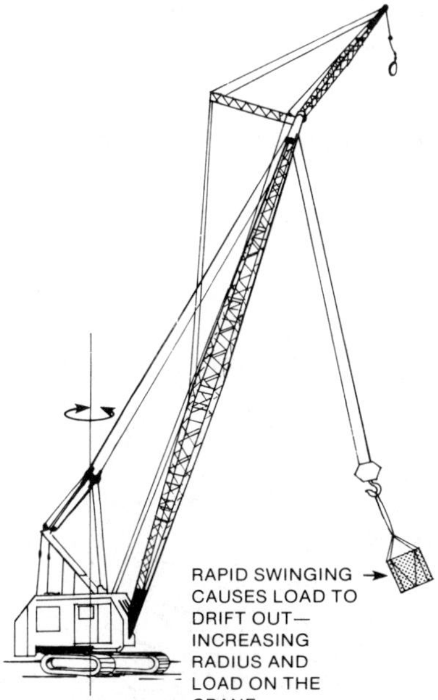

Fig. 15-10: Effects of rapid swinging.

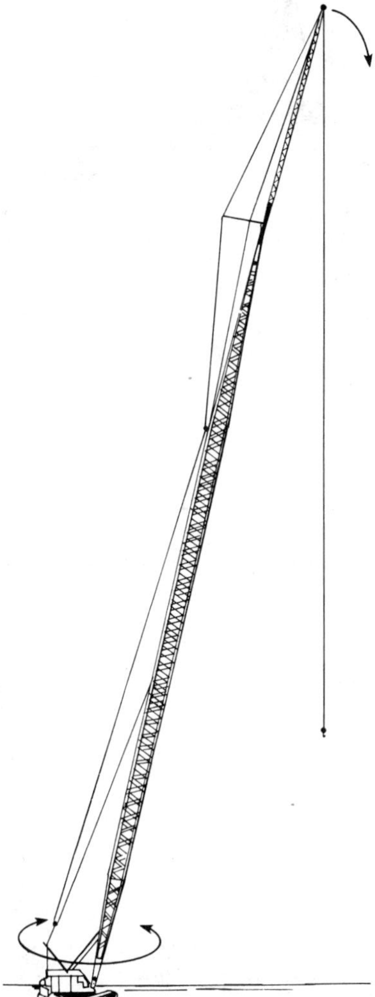

Fig. 15-11: Rapid swinging of long booms with or without load can upset the crane or collapse the boom.

occur if the load strikes the boom, report the damage at once. The crane should then be taken out of service until necessary repairs have been completed.

Dynamic Forces

The maximum safe working loads of cranes are determined from static loads. No account is taken for dynamic motions of the load or crane or of the effect

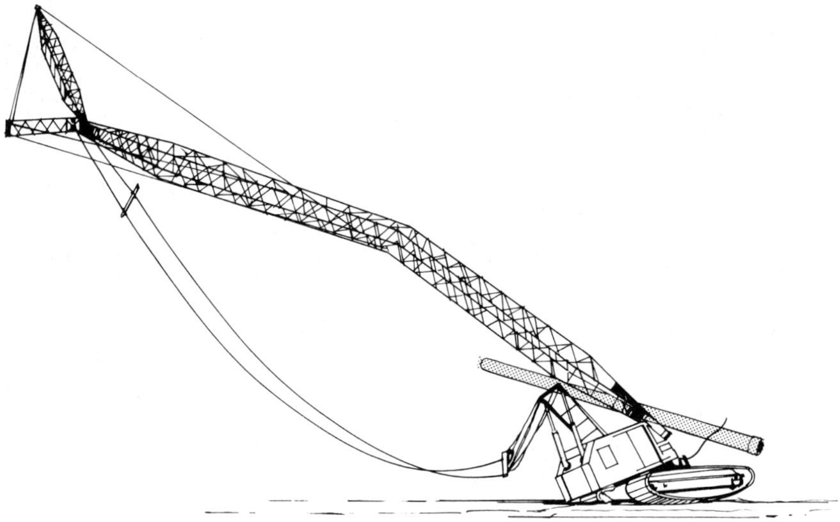

Fig. 15-12: The boom can topple backwards if the swing is started or stopped suddenly.

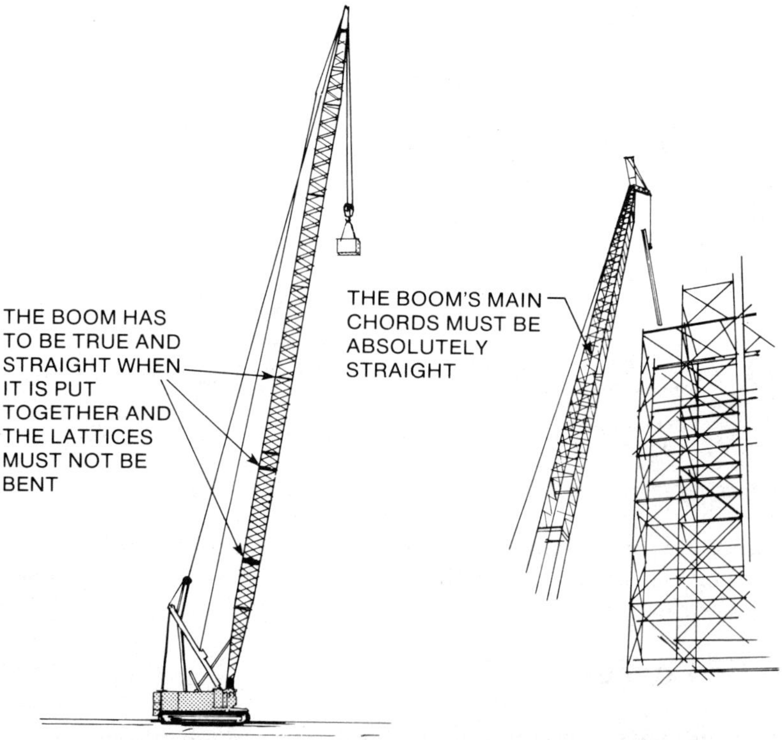

Fig. 15-13: The boom must be structurally sound.

of the wind on the boom or the load. In order to ensure that the safe working load is not exceeded during operation, allowances must be made for wind loading and dynamic forces set up by the normal operational movements of the crane and load. It is essential to avoid sudden snatching, swinging, and stopping of suspended loads since rapid acceleration and deceleration will greatly increase the boom stresses. This is particularly important on cranes having long slender booms where wind gusts or sudden movements could lead to instability or structural failure due to side loads.

Rough handling of the load will affect the capacity of the crane. Sudden stopping of the load by using the hoist brakes harshly could produce forces well in excess of the weight being handled causing failure of the hoist rope, boom pendants, the boom, or some other less significant load supporting component (Fig. 15-14). Excessive pulling on a load which is caught or frozen to the ground could also produce similar results if it suddenly breaks loose. At high boom angles this could make the boom whip backwards and collapse over the cab.

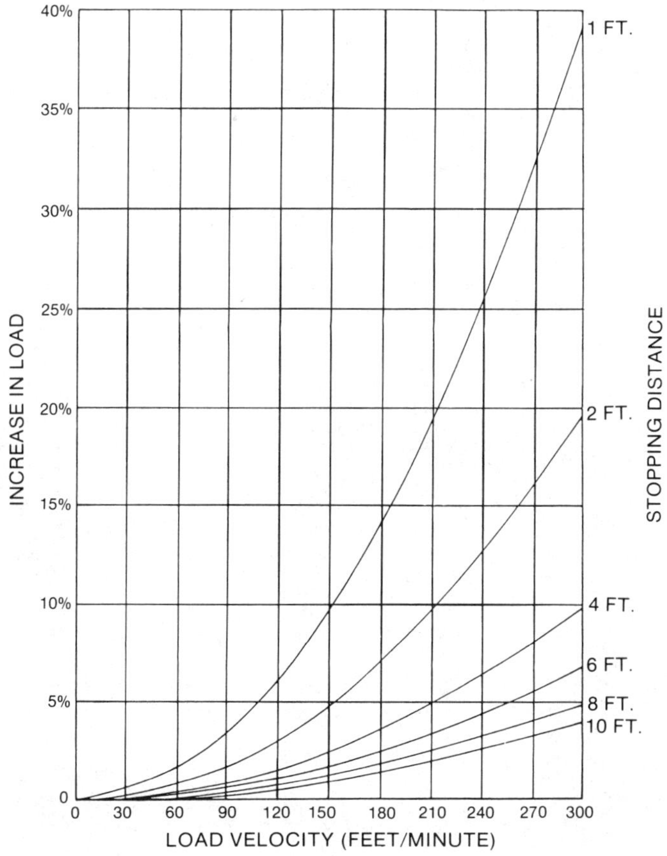

Fig. 15-14: Increase in load due to rapid braking.

Ground Conditions

The load chart supplied with every crane is based on the machine being perfectly level. Any deviation from absolute level reduces the crane's lifting ability. The degree of the reduction depends upon the degree of level of the machine. It is also important to note that when a crane is set up on a grade, swinging from the high side to the low side increases the operating radius. It also increases the load on the turntable, the loading on the outriggers, and the supporting frame structure. A boom at maximum elevation (minimum radius) on the low side cannot be swung over to the high side without risking collapse of the boom over the cab. Working on a slope also produces a severe side load on the boom. Tipping can also occur if the crane is swung to the side when no boom is attached or by suddenly starting or stopping the swing when the boom is elevated at a high angle and the load is near the boom tip (Fig. 15-15).

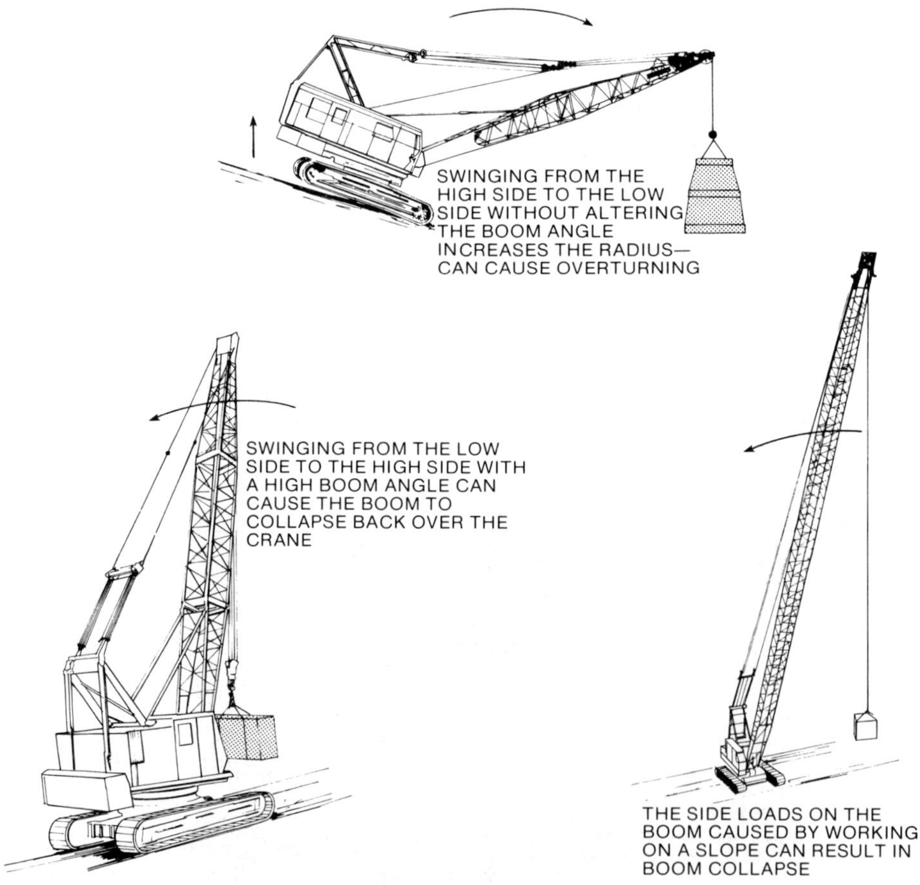

Fig. 15-15: Operating on a slope.

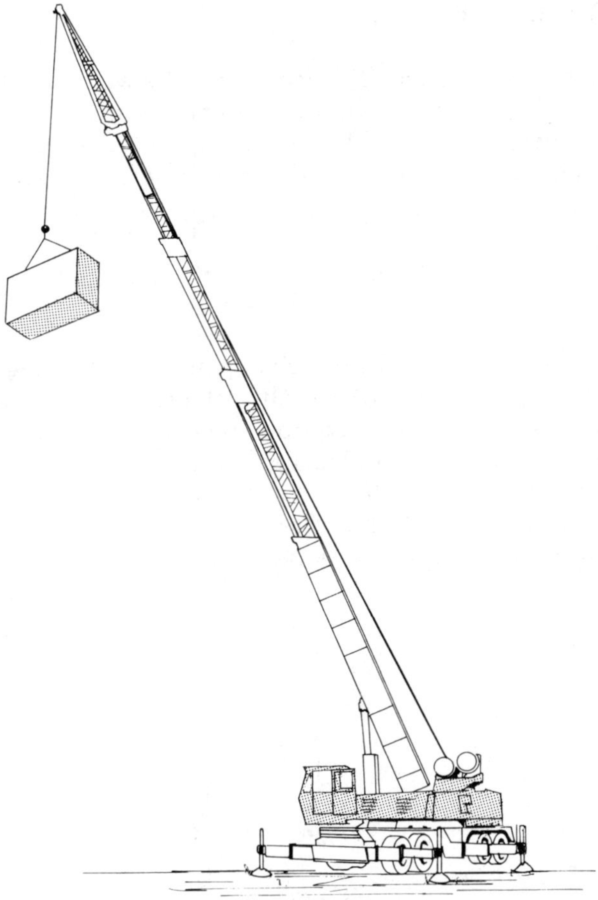

Fig. 15-16: Unless the rig is level, the outrigger beams are out all the way, and the wheels are clear of the ground, you do not have full capacity.

Mobile cranes must never be used on soft or uneven surfaces or on slopes in circumstances in which their stability is likely to be affected unless adequate precautions are taken. Firm footing, uniformly level, should be provided in accordance with the recommendations given in the section dealing with machine set up. In addition, they must never be operated close to trenches as the ground pressure and vibration could cause the walls to collapse.

The safe working loads of machines equipped with outriggers are generally stated two ways. They are quoted as either "on tires" (or crawlers) or "on outriggers." The "on outriggers" rating is based on all outrigger beams being fully extended and all tires or crawlers within the outrigger boundary being completely off the ground. Unless these two conditions are met the "on outrigger" rating cannot be used and the safe working load drops to the "on tires" rating. There is no in-between (Fig. 15-16).

Rigging Accessories

The rated loads do not generally account for the weight of hook blocks, hooks, slings, equalizer beams, material handling equipment, concrete buckets, jibs, and other elements of lifting tackle. Their combined weight must be subtracted from the load chart capacity when determining the maximum allowable load to be lifted (Fig. 15-17).

Load Weight

The rated capacities of all cranes are dependent upon the weight of the load and the radius of operation and as such both factors must be accurately known at all times. Unless the crane is equipped with a load measuring device, there should be a competent individual assigned on a full time basis to be responsible for determining the weight of loads to be lifted or lowered. The operator should not make a lift unless he/she has first determined the weight of the load or is told the weight by the person responsible for the operation.

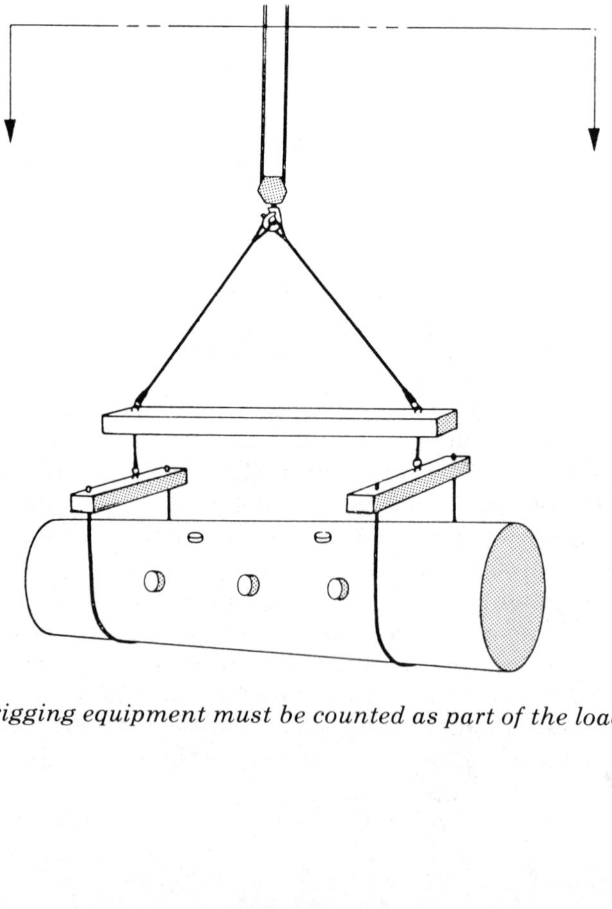

Fig. 15-17: All rigging equipment must be counted as part of the load.

Boom Radius

The radius is the horizontal distance between the center of rotation of the machine and the hoisting hook. All load charts are based on this measurement and any errors can substantially affect the stress level in the boom and the stability of the crane. When utilizing the full capacity of the crane, the radius must be accurately determined. Reaching beyond the vertical plane tends to draw the boom forward and must not be practiced. Consideration must also be given to the fact that the radius increases whenever a heavy load is lifted off the ground. The boom suspension ropes stretch and the boom deflects forward. This increase can be substantial on long boom machines (Fig. 15-18).

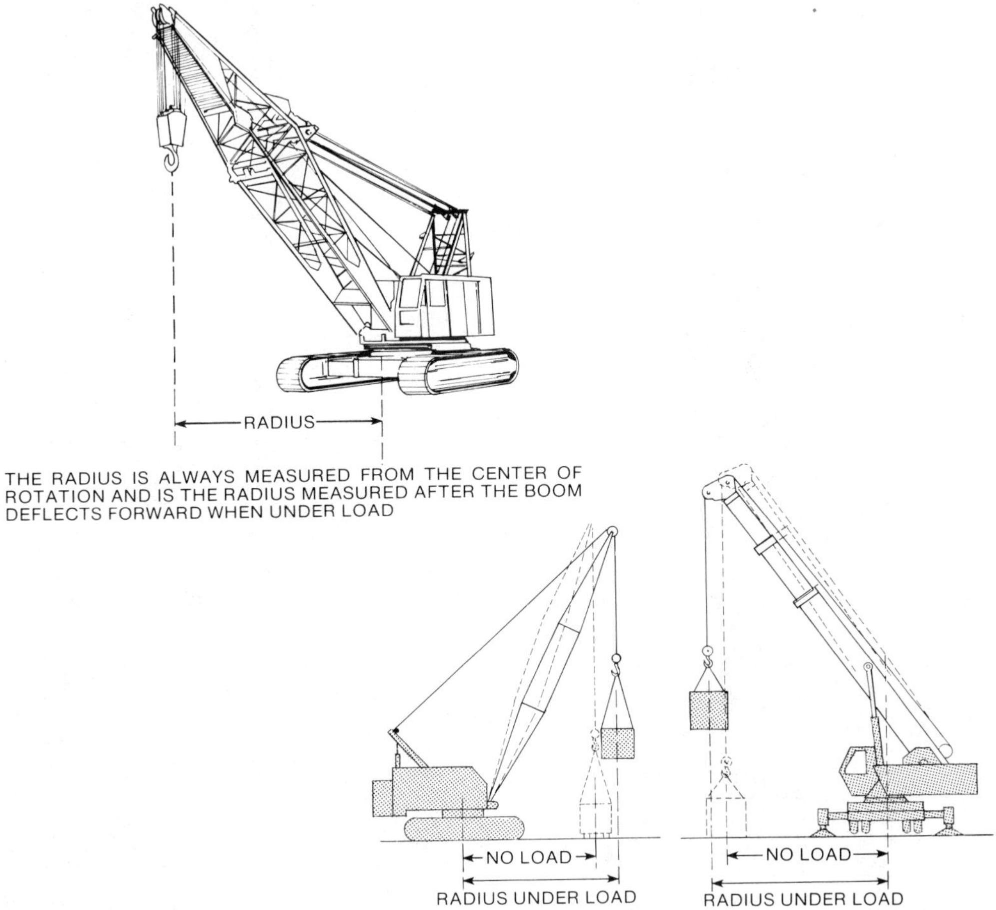

Fig. 15-18: The radius will increase when a load is lifted.

Torsional Loading

The rated load does not account for torsional loading of the boom which occurs whenever the main hoist line is reeved to one side of the boom tip. The boom tip must always be symmetrically rigged (Fig. 15-19).

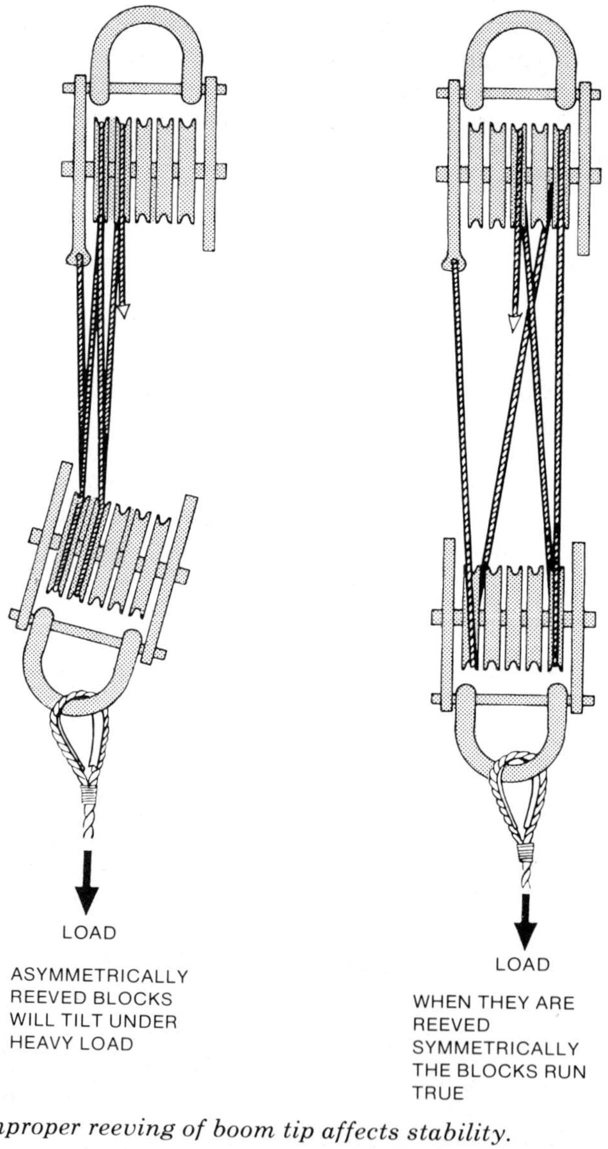

Fig. 15-19: Improper reeving of boom tip affects stability.

16

General Operation Procedures and Precautions

INTRODUCTION

In any lift utilizing a crane or any other lifting rig, it is important to know the capacity of the machine, the maneuvers it must make to accomplish the lift, and the dangers inherent in its operation. Precautions must be taken to reduce the probability of an accident. Careful observance of the following recommendations will prevent the majority of common accidents and, as such, it is strongly recommended that everyone associated with crane operations be aware of these fundamental considerations.

PRIOR TO OPERATION

There are several functions the operator must perform to ensure that the crane is operable. The main considerations are the mechanical condition of the machine and the possible limitations of the environment.

Mechanical Condition

No lifting rig should be operated without first being inspected for mechanical difficulties. Never use or allow the use of any crane not in perfect mechanical condition. No one who has reasonable cause to believe that any crane or part of the machinery or structure is unsafe because of its condition or suitability should use or operate it until he/she has reported the defect to his/her supervisor and safe conditions have been assured and all hazards rectified. Whenever there is any doubt as to safety, no operator should be requested to operate a crane until safety has been assured.

Weather Conditions

Never use or allow the use of any crane when the weather conditions are such that hazards to personnel, property, or the public are created. The most adverse weather condition is high winds. The size and shape of the loads being lifted must be carefully examined to determine if a safety hazard exists when they are to be lifted during conditions of high wind speeds.

It is advisable to avoid handling loads presenting large wind catching surfaces which could result in loss of control of the load and crane even though the weight of the load is within the normal capacity of the crane. Consider the safety of the workers handling the loads. Workers have been killed when knocked off buildings by loads that were blown out of control by the wind. Wind loading can be critical depending on such factors as the boom length, boom angle, bulkiness of the load, and direction and velocity of the wind. In the absence of the crane manufacturer's instructions regarding maximum permissible wind speeds for operation, use 25 to 30 mph as the range to consider removal of the crane from service.

Two other weather conditions to consider are visibility and temperature. If the visibility of the operator is impaired by dust, darkness, snow, fog, or rain, strict supervision of the crane operation must be exercised, and if necessary, the crane should be withdrawn from service. In low temperatures (generally below freezing), extreme caution must be exercised to ensure that no part of the crane's structure is shock loaded or impacted, as brittle fracture of the steel can result.

STARTING THE ENGINE

When satisfied that the rig is operable, the engine may be cranked. Before starting the engine make sure that all personnel are well clear of the machine. Normally, the operator will allow the engine to warm up for a few minutes to give the oil in the pumps a chance to circulate. The machine should not be operated under load until the hydraulic oil has warmed to the minimum operating temperature specified in the manufacturer's manual. The operator will operate the winches, boom hoist cylinders, boom extension cylinders, and swing mechanism to circulate the oil. Failure to warm the hydraulic system prior to operation may result in pump failure or erratic operation due to cold viscous oil. It is extremely important to engage boom cylinders slowly in cold weather, since severe shock loading could occur if cold oil in the cylinders is not displaced gently.

After starting the engine and before putting the crane into operation, check that it has not moved and that it remains level. Ensure also that all guards, controls, clutches, brakes, gears, and the like are properly set and that putting the equipment into operation will not endanger property, the public, or personnel.

PLACING THE MACHINE IN POSITION

Exercise extreme caution when moving the crane into position. With the exception of those involved in the operation of the crane, no one should be permitted to get on, leave, or ride on the equipment when it is in motion or in operation. Be very careful when the machine must be backed into position. If the operator does not have a clear view of the area behind the crane, use a signalperson. The operator should sound the horn before moving the machine and intermittently during travel, especially when approaching personnel. Do not allow anyone to divert the attention of the operator from the operation of the crane.

Know the exact location of utility lines, pipelines, sewers, and other underground obstructions, and avoid them with room to spare. Mark locations clearly to prevent having to make estimations (Fig. 16-1).

Ensure that, whenever possible, the machine is operated in its most stable position and in the area of highest capacity. In general this means that most of the lifting should be done in the rear quadrant. Never, unless otherwise specified by the manufacturer, lift or swing over the front of the machine.

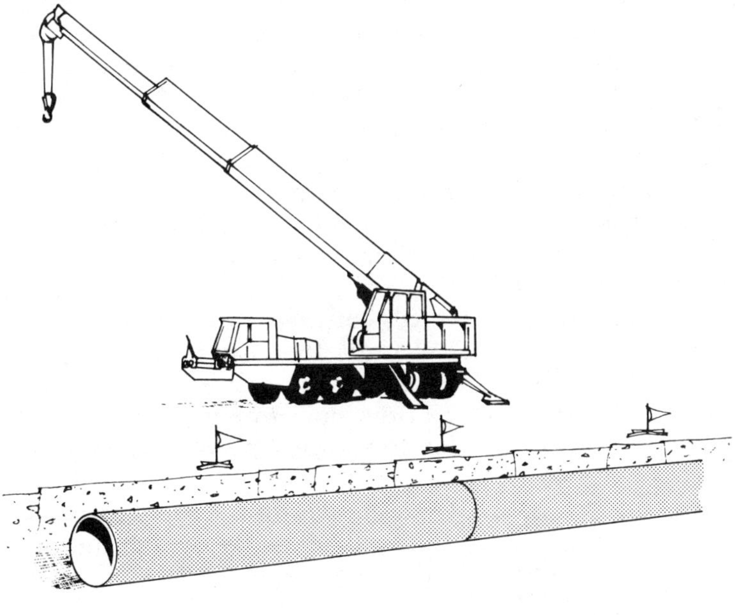

Fig. 16-1: Mark the location of all underground services that could be crushed by the weight of the crane.

Position the crane as close to the load as possible and in such a way as to minimize the swing. Be sure there is adequate clearance for tail swing of the revolving frame, especially when people or vehicles may enter the area. The crane should be positioned so that no part of the superstructure comes within 2′ of any obstruction in which a workperson could be trapped and crushed. If this is not possible, then entry to the obstructed area must be prevented by barriers or fences.

Regardless of the size or weight of the load to be lifted, use the crane's outriggers. Extend the beams fully and get the wheels off the ground. The revolving portion of the superstructure of the crane must be dead level before making the lift.

PRIOR TO RIGGING THE LOAD

When organizing to make a lift, ensure that no one is within the radius of rotation of any part of the crane or load unless he/she is authorized by the person in charge of the work to be in that area. Insofar as is practical, ensure that no one is ever directly beneath the load (Fig. 16-2).

Before performing any type of crane work, be sure that the machine is secured against travel. If necessary, block the crane to prevent movement. Remove blocking before attempting to travel.

Before rigging the load, the operator should make a dry run, especially in areas which are really tight. Go through all the motions without a load, anticipating what actions should be made to make a safe and smooth operation.

One of the most important precautions is to determine the weight of all loads before rigging them or attempting to lift them, making ample allowances for unknown factors and determining the available capacity of the equipment being used. In cases where the assessment of load weight is difficult, safe load indicators or weighing devices should be fitted.

In figuring the total weight of the lift, be sure to include the block, hook, and any slings or other rigging devices between the boom peak and the load. When making near capacity lifts, calculate the entire load carefully and check it against the rated lifting capacity of the crane. It takes less time to check clearances and loads ahead of time than to try to lift and not make it, then having to boom up to reposition the crane and load for another try. It is much safer, too.

Before making any lift that exceeds 75% of the rated capacity of the crane, it is recommended, in addition to the usual precautions, that the load radius be measured to avoid any possibility of error (Fig. 16-3).

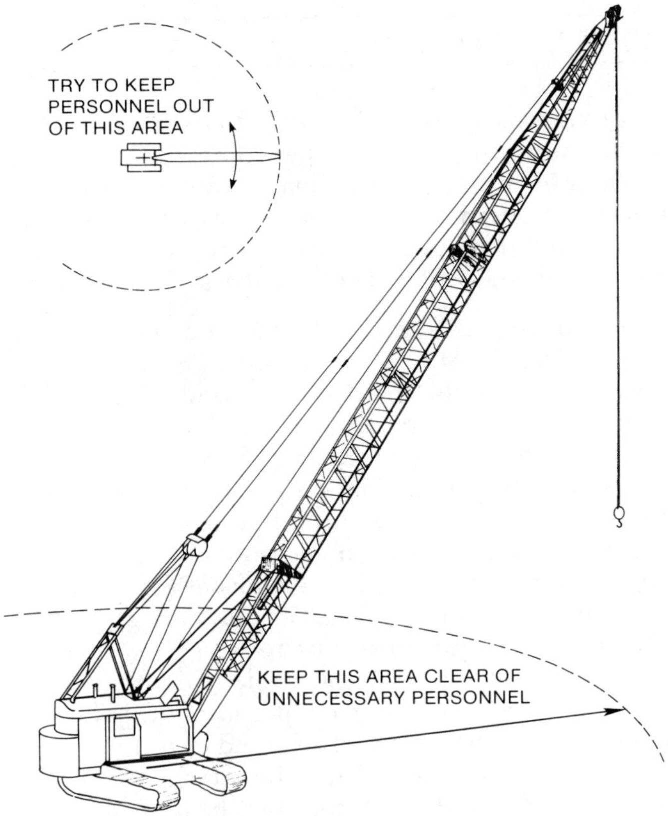

Fig. 16-2: Avoid lifting over personnel.

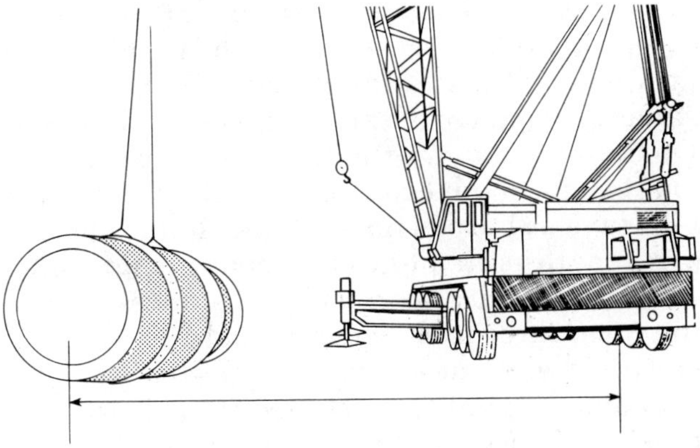

Fig. 16-3: Measure the load radius before making capacity lifts.

RIGGING THE LOAD

The importance of rigging the load properly is as great as determining the weight of the load. The quality of these operations depends, to a large extent, on the crane operator even though he/she may not actually rig the load. The crane operator is in a position to provide guidance and control for the rigging crews and, as such, should ensure that the following procedures are observed by not only himself/herself but all others concerned with the operation.

1. Know the safe working load of the tackle and rigging equipment being used and never exceed this limit.
2. Examine all hardware, equipment, tackle, and slings before using them and *destroy* defective components. Discarded pieces of equipment could be used by someone not aware of the hazards or defects.
3. Never use kinked or damaged slings. In order to provide maximum operating efficiency and safety, all slings and fittings should be given thorough periodic inspections as well as daily inspections for signs of wear and abrasion, broken wires, worn or cracked fittings, loose seizings and splices, kinking, crushing, flattening, and corrosion. Special care should be taken in inspecting the areas around thimbles and fittings.
4. All slings should be identified with an identification number and their maximum capacity on a flat ferrule or ring, permanently attached to the sling. Mark the capacity of the sling for a vertical load or at an angle of 45°, but be certain that all employees are aware of how the rating system works (Fig. 16-4).
5. Sharp bends, pinching, and crushing must be avoided. Corner pads that prevent the sling from being sharply bent or cut can consist of large diameter split pipe sections, corner saddles, padding, or blocking (Fig. 16-5). A good rule to follow is to make sure that the length of the arc of contact of the rope is at least equal to one rope lay (about seven times the rope diameter). When the bend is of this length, each of the strands has been on the inner and outer sides of the rope bend, and the slippage of the strands relative to each other minimizes the stress. On the other hand, if the bend is very short, the strands on the outer side of the rope bend will have to stretch, and this stretch will usually be permanent and will leave a sharp bend or "kink" in the sling.
6. Never allow wire rope to lie on the ground for any length of time or on damp or wet surfaces, rusty steel, or near corrosive substances.
7. Avoid dragging rope slings from beneath loads.
8. Keep all rope away from flame cutting and electric welding operations.
9. Lifting beams should be plainly marked with their weight and designed working loads and should only be used for the purpose for which they were designed.
10. The hoist rope must never be wrapped around the load. The load should be attached to the hook by slings or other rigging devices that are adequate for the load being lifted.

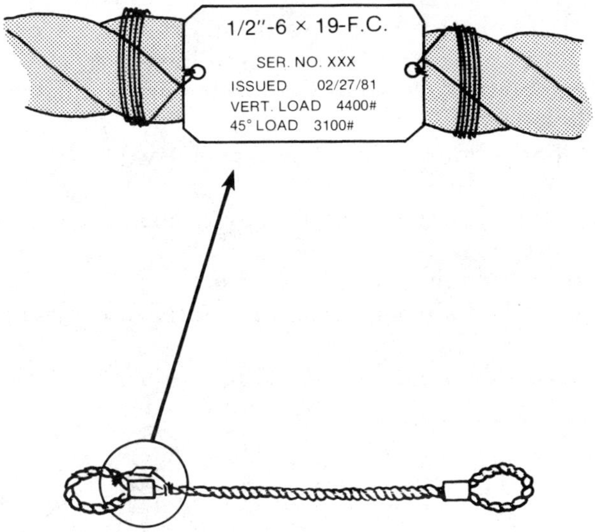

Fig. 16-4: All slings should carry identification tags.

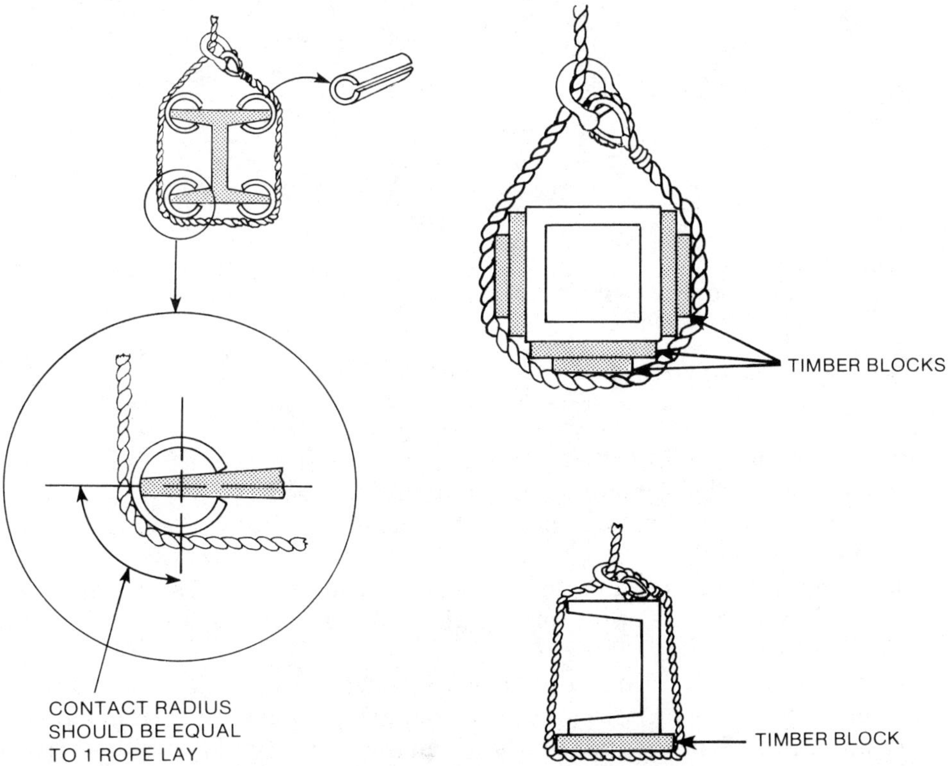

Fig. 16-5: Ensure that slings are protected from sharp corners on heavy items.

11. Never make temporary repairs to a sling. Procedures for proper repairs to a damaged sling should be established and followed.
12. Never lift loads with one leg of a multi-leg sling until the unused legs are made secure (Fig. 16-6).
13. Never point load a hook unless it is especially designed and rated for such use.
14. When using two or more slings on a load, ensure that all slings are made from the same materials.
15. Avoid contact with solvents and chemicals.
16. Knotted and kinked slings are permanently damaged and should not be used.
17. Never use discarded hoist rope as sling material.

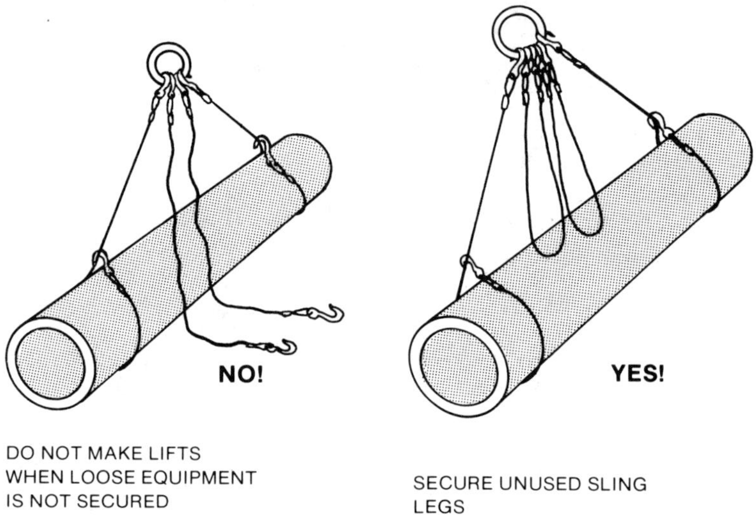

Fig. 16-6: Loose rigging equipment is a hazard.

18. Avoid using single leg wire rope slings with hand spliced eyes, because the load can spin, causing the rope to unlay, and allowing the splice to pull out. Slings having Flemish spliced eyes should be used.
19. Never wrap a wire rope completely around a hook; the sharp radius will damage the sling.
20. Avoid bending the eye section of wire rope slings around corners as the bend will weaken the splice or swaging. There must be no bending near any attached fitting (Fig. 16-7).
21. Ensure that the sling angle is always greater than 45°. One of the easiest ways to make sure that the angle is adequate once a load is rigged is by checking that the horizontal distance between the points of attachment of the load is less than the length of the shortest sling leg. If this is the case, then the angle is greater than 60° (Fig. 16-8).

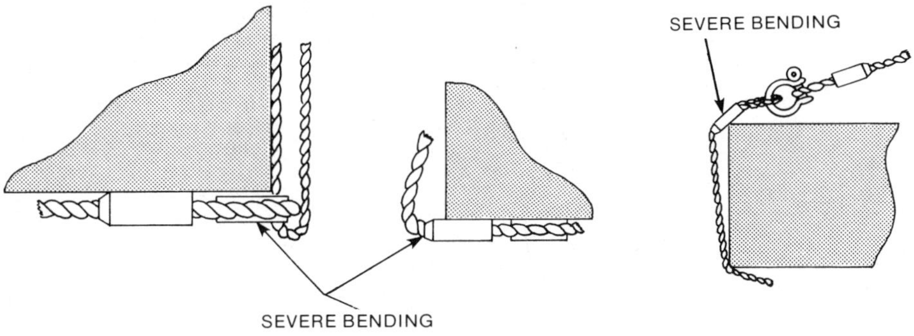

Fig. 16-7: Do not permit bending near any splice or attached fitting.

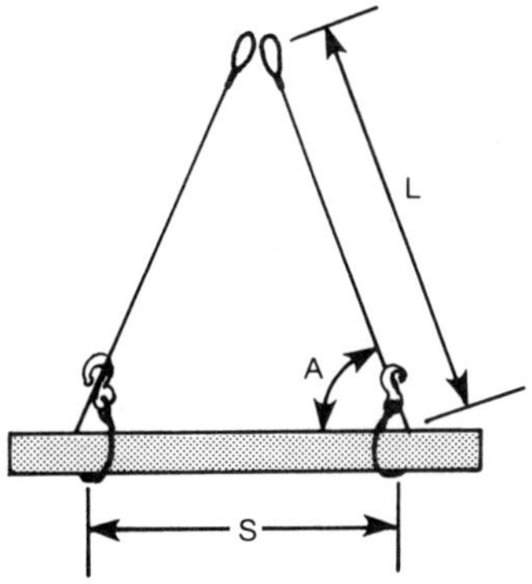

Fig. 16-8: If L is greater than S, then the sling angle (A) is large enough to be safe.

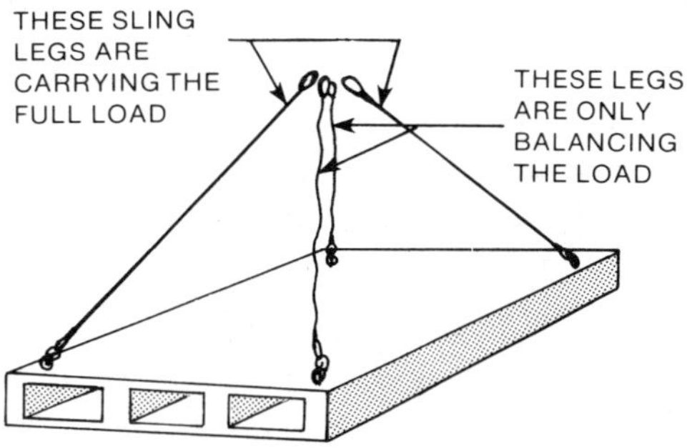

Fig. 16-9: On a rigid object being hoisted with a four-leg bridle hitch (or three-leg hitch), it is possible for the total weight to be taken by two legs while the other legs only serve to balance the load.

22. Do not assume that a multi-legged bridle sling will safely lift a load equal to the safe load on one leg multiplied by the number of legs, because there is no way of knowing that each leg is carrying its share of the load. With slings having more than two legs and a rigid load, it is possible for two of the legs to take practically the full load while the others only balance it (Fig. 16-9).

23. When lifting rigid objects with slings having three or four legs, any two of the slings must be capable of supporting the total load. In other words, after considering the angle between the legs, the slings should be considered as having only two legs when estimating the size required. Where the object is flexible and able to bend to adjust itself to the sling legs, each leg can be assumed to take its own share of the load.

24. When using multi-legged slings to lift loads in which one end is much heavier than the other, the tension in the most heavily loaded leg is much more important than the total weight and the sling must be selected to suit the most heavily loaded leg rather than the total weight (Fig. 16-10).

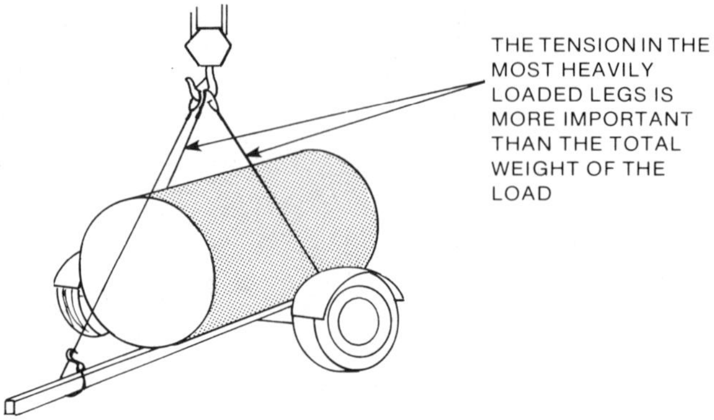

Fig. 16-10: Unequal distribution of weight in sling legs.

25. When using choker hitches, do not force the eye down towards the load once tension is applied, as rope damage is the invariable result (Fig. 16-11).
26. Whenever two or more rope eyes must be placed over a hook, install a shackle on the hook with the shackle pin resting in the hook and attach the rope eyes to the shackle. This will prevent the spread of the sling legs from opening up the hook and also prevent the eyes from damaging each other when under load (Fig. 16-12).

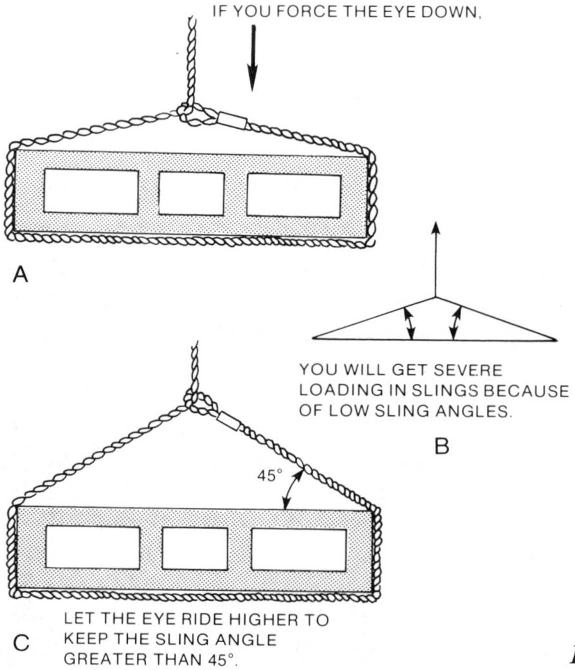

Fig. 16-11: Eyes and sling angles.

Fig. 16-12: Use a shackle whenever two or more ropes are to be placed over a hook.

HANDLING THE LOAD

Thus far, the crane has been inspected, positioned, and rigged. Before beginning the actual lift, the operator must make sure that the following conditions exist:

1. The crane is rigged with sufficient parts of line to make the lift.
2. The load has been properly rigged and well secured to prevent the movement or dislodgement of any part. All loose objects must be removed from the load and machine (Fig. 16-13).
3. Taglines are used to control the load.
4. The load is free before lifting (i.e., it is not bolted, tied, or fastened to the ground or another object) and that all sling legs are taking the load.
5. The boom peak is directly over the load, the load line is directly over the center of gravity of the load, and the hoist line is vertical (Fig. 16-14).

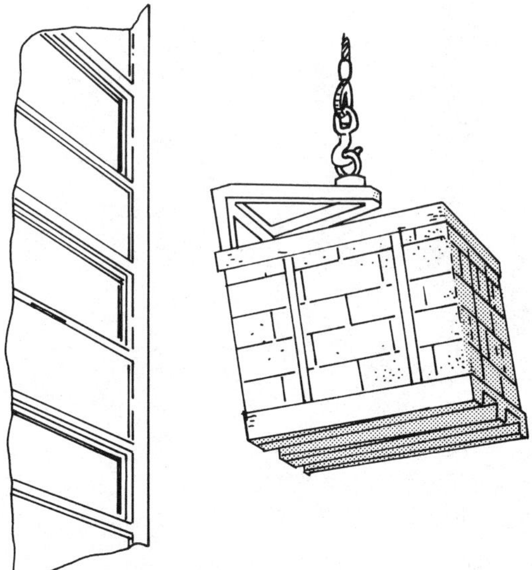

Fig. 16-13: Load and secure all materials so as to prevent any movement or possibility of dislodgement.

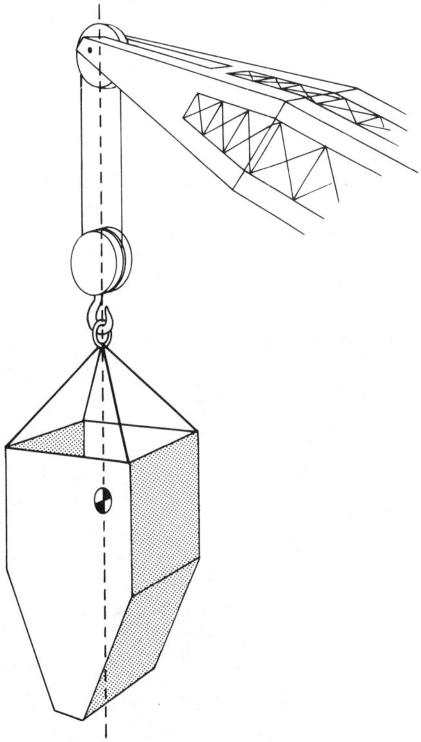

Fig. 16-14: Keep center of gravity directly under hook and boom tip.

6. Multiple part lines are not twisted around each other (Fig. 16-15).
7. If there has been a slack rope condition, check that the rope is properly seated on the drum and in the sheaves. Check the rope and rope spooling on the drum if it has become slack or overwound. A loose coil of rope can become cross-coiled on the drum or can even come off the drum altogether. Both cases can lead to premature and/or sudden failure of the rope with disastrous consequences.
8. All nonoperating personnel have been warned or told to leave the immediate area before making the lift.

The next important aspect in handling the load is taking up the slack. The hoist line should be raised gradually until the slack is eliminated. Doing this will avoid impact loading caused by sudden jerking when lifting or lowering. Be sure that everyone keeps their hands away from pinch points as the slack is being taken up.

After the slack has been snugged up, the operator will raise the load a few inches and hold it there while testing the winch brakes. Make sure that everyone stands well back from the load in case a hook slips or a sling breaks and suddenly flies free. If it is obvious that the load does not ride properly, lower it and have it readjusted.

Fig. 16-15: Keep the hoist lines untangled.

Once the supervisor in charge of the lift is confident that everything is properly rigged, the lift may be made. During the lift, follow these basic safety precautions:

1. Never allow anyone to ride on a load that is being lifted.
2. Never allow anyone to work under a suspended load unless the load has been adequately supported from the floor and all conditions have been approved by the supervisor in charge of the operation.
3. Do not hoist two or more separately rigged loads in one lift, even though the combined load is within the crane's rated capacity.
4. Never use a crane boom to drag a load sideways. The boom was not designed nor intended for such use, and it may collapse under side strains. Make vertical lifts only.
5. Do not lift loads with excess slings hanging loose. This is a very dangerous condition. Never allow any excess cable or sling to be suspended beneath the crane.
6. Always keep the load as near to the carrier and as close to the ground as possible (Fig. 16-16).
7. Keep speed slow in lifting and lowering loads.
8. Swing carefully and slowly. Be sure to avoid boom or jib "whipping."
9. Do not let the load strike boom or outriggers (Fig. 16-17).
10. Allow maximum clearance between hook block and head sheaves.
11. Keep near capacity loads as close to the ground as possible.
12. Avoid hitting nearby structures with the boom. If you do, inspect boom prior to next lift.

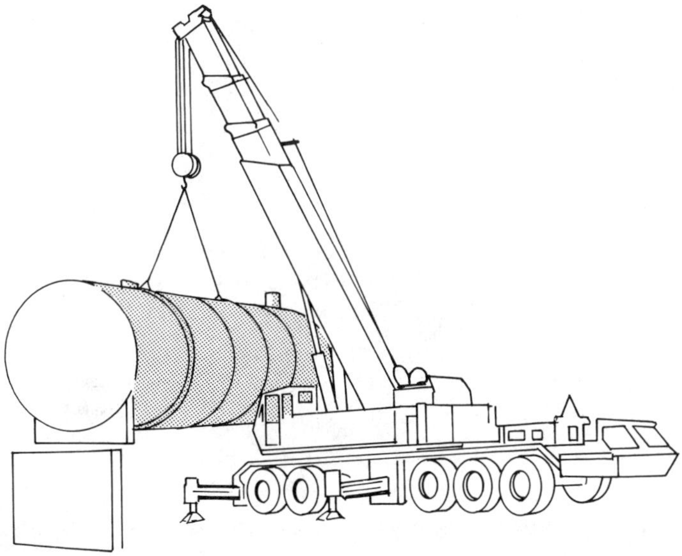

Fig. 16-16: Keep the load as close to the ground and to the carrier as possible.

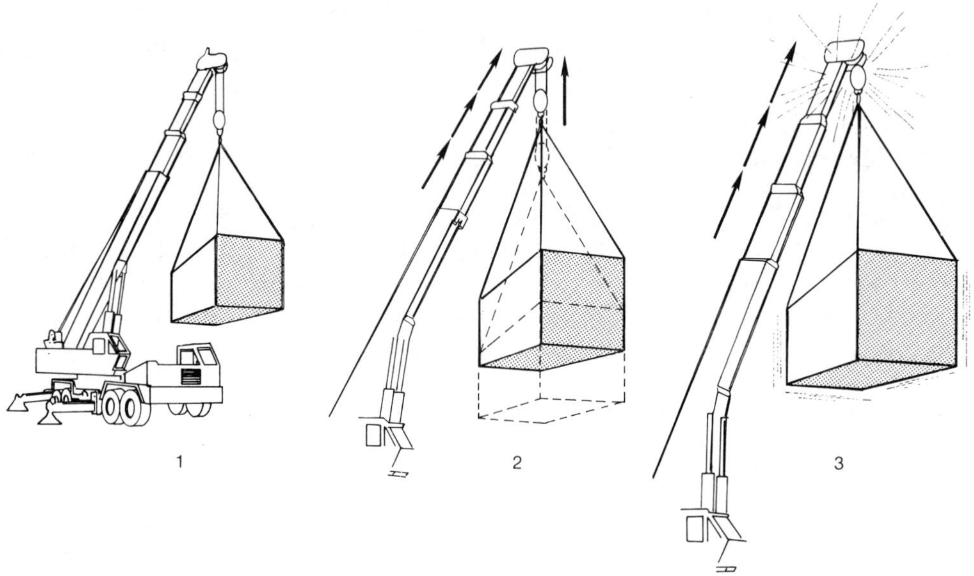

Fig. 16-17: If the boom is extended without paying out more rope, the load will hit the boom tip and the load line will fail.

13. When landing loads, they must be set on adequate blocking to prevent damage to the slings. They must also be safely landed and properly blocked before being unhooked and unslung (Fig. 16-18).

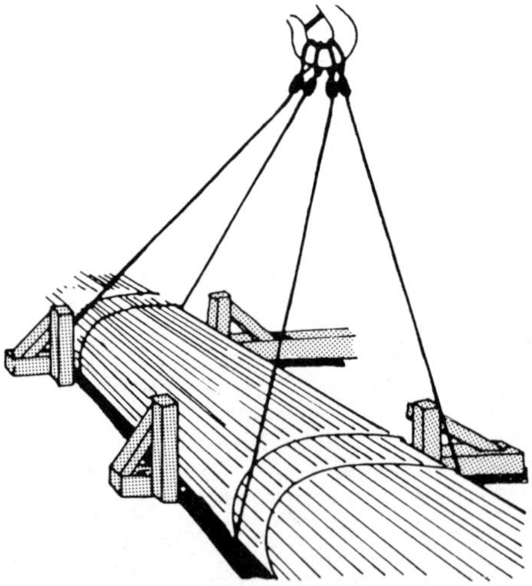

Fig. 16-18: All loads must be safely landed and properly blocked before being unhooked.

MORE THAN ONE CRANE OPERATING IN THE SAME AREA

Whenever two or more cranes are operating in the same area, there is always a possibility of collision or of the hoisting ropes or loads fouling each other or the structure of the other crane even when the booms are of different lengths. To minimize the probability of this occurrence, the cranes should always be sited in such a way that the operators have a clear view of the other cranes operating in the collision danger area. The operators should also have direct communication with each other so that one operator may alert the other to impending danger. It is also recommended that their overall lifting program be set out and controlled by one worker who is in contact with all operators and riggers and assigns priority of operation for the cranes (Fig. 16-19).

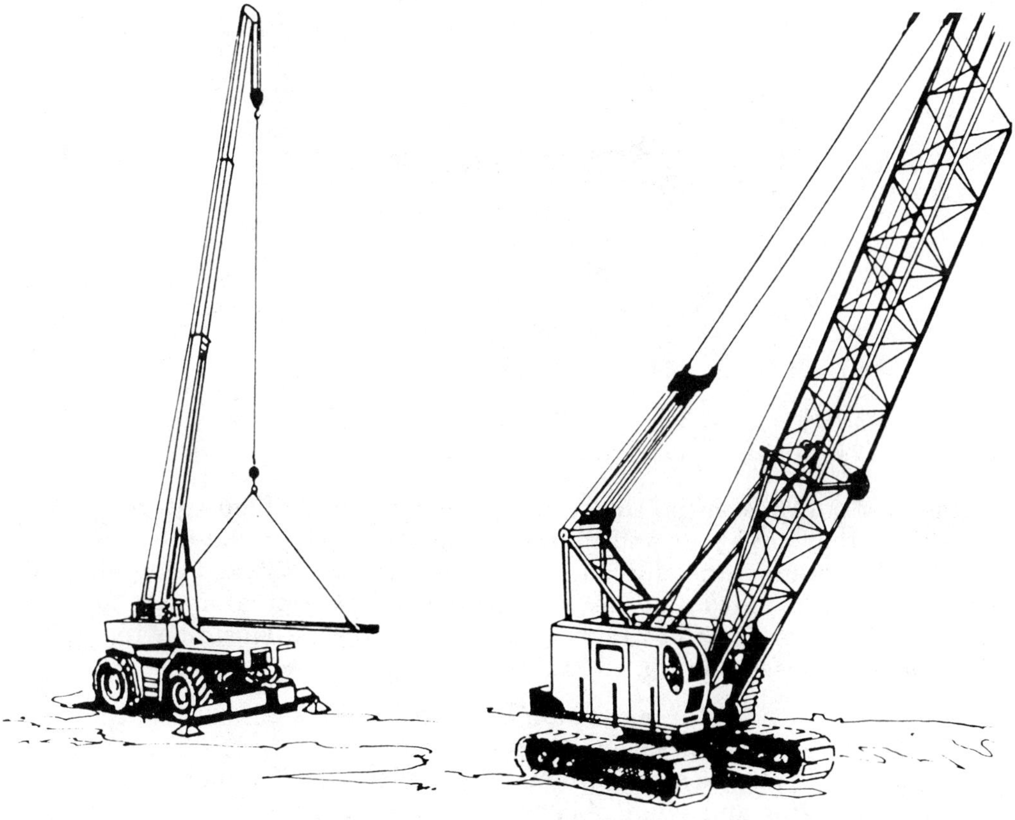

Fig. 16-19: Whenever two or more cranes are working in the same area, the operators should be in radio contact with each other.

LIFTING WITH MORE THAN ONE CRANE

Tandem lifts are extremely complex operations that need a great deal of planning before actually doing them. It is possible to lose not only the load, but also all the cranes that are involved (Fig. 16-20).

One qualified person should be made responsible for the operation. He/she should analyze the operation and instruct all personnel involved in the proper positioning, rigging of the load, and the movements to be made.

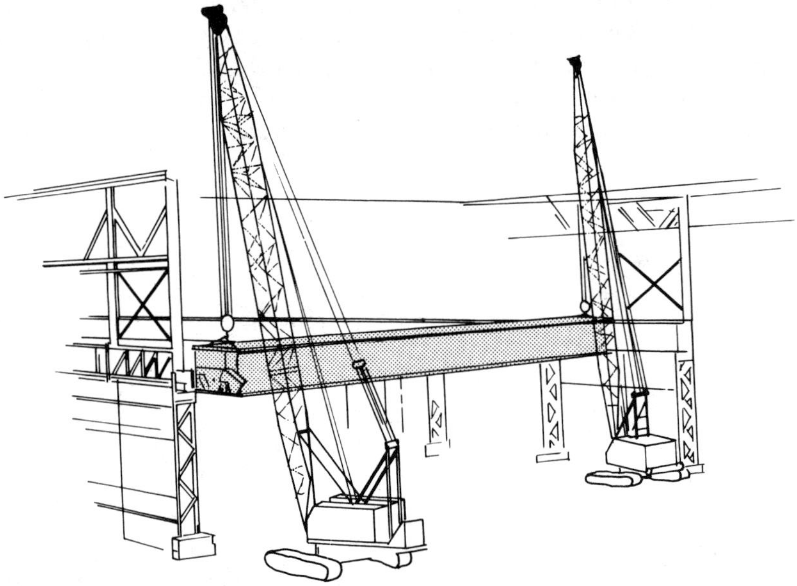

Fig. 16-20: Tandem lifts must be carefully planned.

It is absolutely essential that the signalperson and all operators know exactly what they have to do and what movements have to be made before the lift is started. The machines have to be coordinated and worked as a unit. It is strongly recommended that the signalperson and the operators be able to talk to each other over portable radios as the lift progresses. The following precautions must be observed (Fig. 16-21):

1. Use only the best rigging equipment and cranes.
2. Make all movements as slowly as possible.
3. Keep engine speeds at maximum governed speeds to avoid stalling.
4. Match the line and swing speeds of all cranes.
5. Do not apply clutches or brakes suddenly.
6. Use only one signalperson and ensure that he/she is in radio contact with the operators.
7. Know how much of the load is to be carried by each crane and be absolutely sure that the slings are arranged to divide the load as planned.

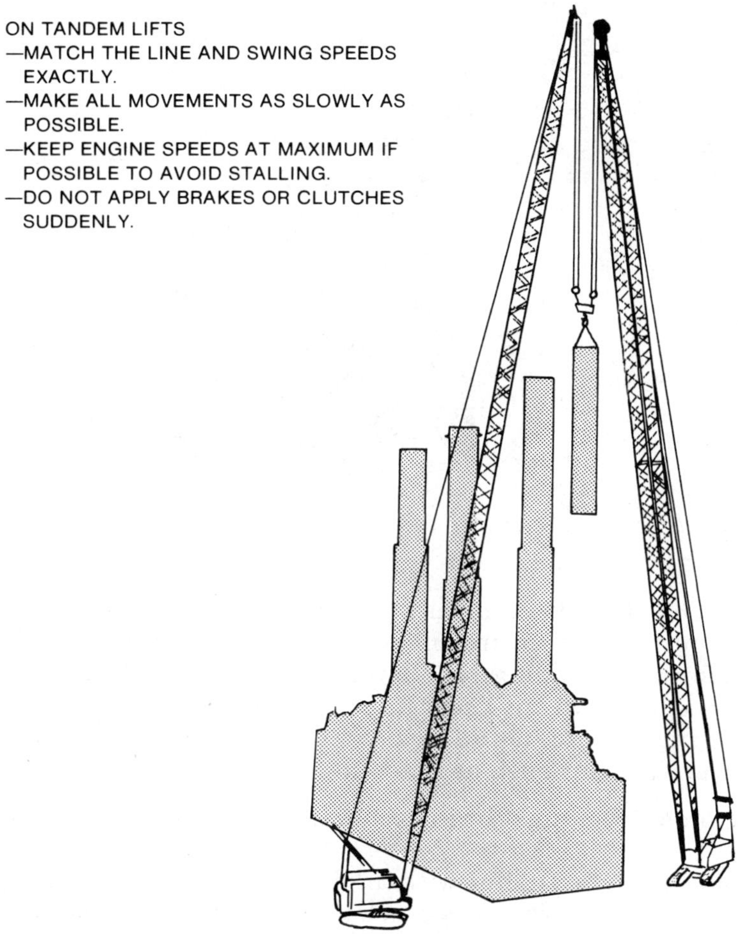

Fig. 16-21: Guidelines for tandem lifts.

8. Bear in mind that machine capacities are based on freely suspended and balanced loads, with the boom tip directly over the center of gravity of the load. The use of two or more machines introduces dangerous elements which include side loading of the boom, overloading, operator error, differing ground conditions, and many other hazards not normally encountered.

PICK AND CARRY OPERATIONS (OR TRAVELING WITH ERECTED BOOM)

The load ratings of every mobile crane are based on the machine's carrier being stationary and level. There are no ratings published for pick and carry operations.

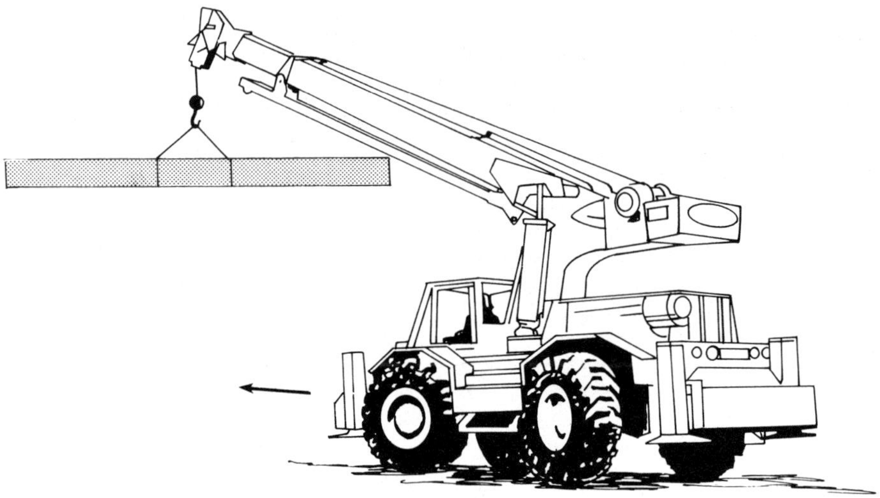

Fig. 16-22: When traveling with a load the boom should always be carried in line with the direction of motion.

Travel with suspended loads involves so many variables such as ground conditions, boom length, side, forward and rearward swinging of the load, and momentum in starting and stopping, etc., that it is impossible to devise a single standard rating procedure with any assurance of safety. If such an operation is attempted the user must evaluate the local conditions, determine safe practices, and exercise precautions such as the following:

1. The traveling procedures must be in accordance with the manufacturer's instructions.
2. The boom should be carried in line with the direction of motion (Fig. 16-22).
3. Both the swing brake and the swing lock (house lock) should be set. The swing lock is necessary because the brake is not positive and may allow the upper to swing. If swing during travel is necessary, engage the swing clutch before releasing the lock and the brakes.
4. Never travel with capacity or near capacity loads. As a rule of thumb, in the absence of manufacturer's recommendations, the maximum load that can be carried over the rear of a carrier mounted crane should never exceed the rating given for over the side on rubber (Fig. 16-23).
5. Extreme caution should be exercised when traveling with over 50% capacity loads. Ground irregularities will subject the carrier, boom, and suspension to additional shock loads which must be compensated for by reducing the hook load (Fig. 16-24).
6. Where permitted by the design, the crane operator must remain in the crane cab to control the load and a second worker should be used to drive the vehicle. The signalperson should coordinate the operation, walk ahead of the load, and warn of hazards.

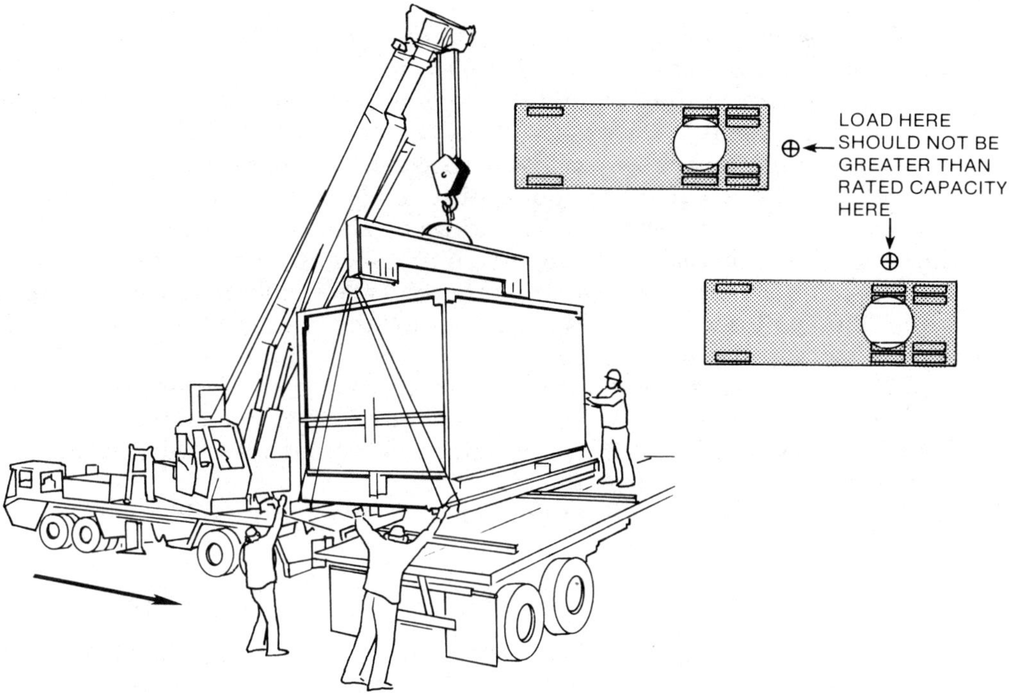

Fig. 16-23: Never travel with capacity loads. The load over the rear should never exceed the rating for over the side while on rubber.

Fig. 16-24: Extreme caution should be exercised when traveling with over 50% capacity loads. Ground irregularities will subject the carrier, boom, and suspension to additional shock loads which must be compensated for by reducing the hook load.

7. Travel speed must be as low as possible (Fig. 16-25).
8. Sudden starts and stops should be avoided.
9. Tag or restraint lines should be provided to snub and control swinging of the load (Fig. 16-25).
10. The load should be kept as close to the ground as possible so that the length of hoisting cable between the boom point and load is as great as possible. This will assist in reducing the impact loading on the boom due to the elasticity of the cable which increases in proportion to its length (Fig. 16-26).
11. Never move the crane away from the load when carrying over 50% capacity loads. This will tend to increase the radius and may tip the machine. Move slowly toward the load (Fig. 16-27).
12. When using a crawler crane in soft ground, it may be necessary to move the load behind in order to raise the leading end of the crawlers. Snub the load to the machine to prevent it from swinging away, and travel with extreme caution as it is possible to upset the crane (Fig. 16-28).
13. Never travel a rubber-tired unit with a load over the side.
14. Use the shortest boom possible.

Fig. 16-25: When traveling with a load, keep the speed low, avoid sudden starts and stops, and use a tagline to control the load.

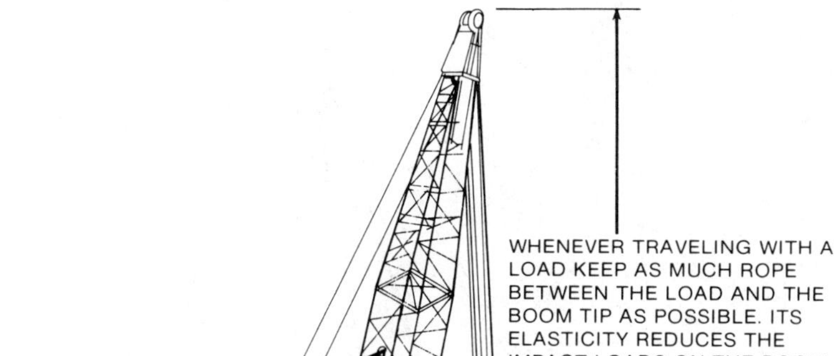

Fig. 16-26: Reducing impact loads when traveling with a load.

Fig. 16-27: Always move toward the load carrying over 50% of the crane's rated on rubber capacity.

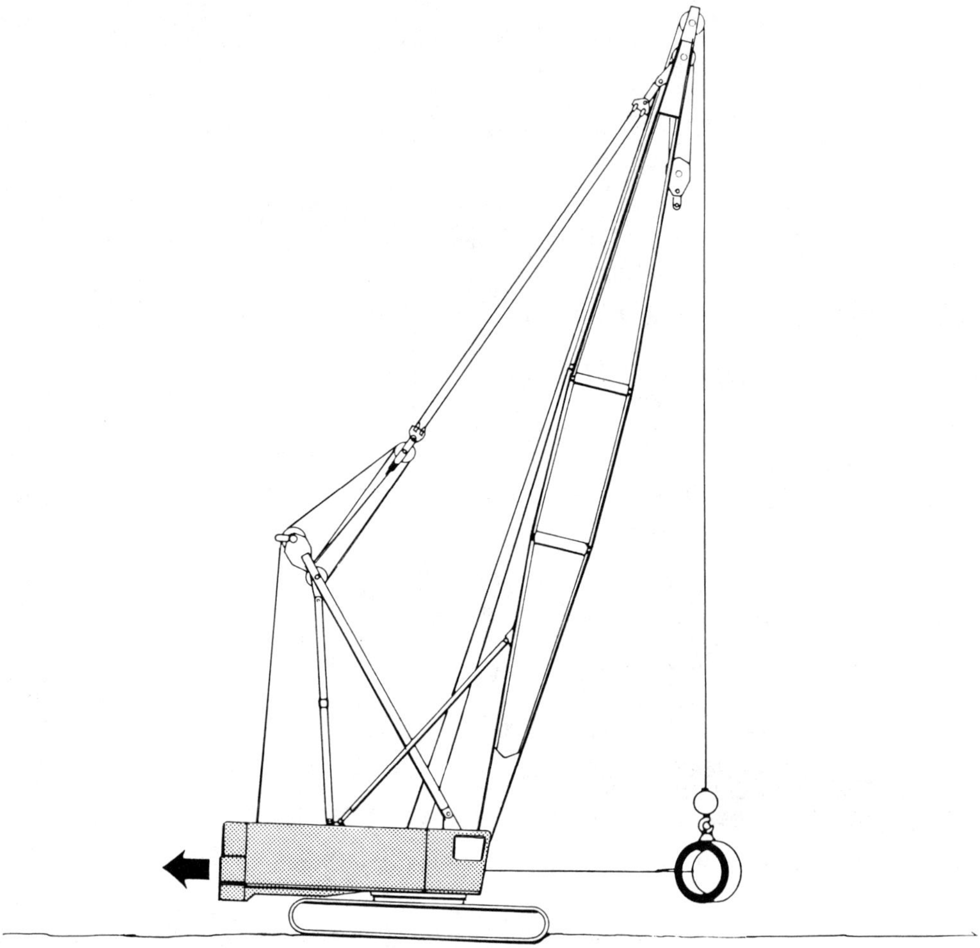

Fig. 16-28: When traveling on soft ground, snug the load and carry it behind the crane.

15. Do not carry the boom at its highest angle. Many accidents occur when the boom topples backward over the machine (Fig. 16-29).
16. When the conditions appear to be hazardous due to such factors as steepness of a slope, ground conditions, etc., the crane should be snubbed to prevent it from running away if the clutch or brakes fail. Block the treads on crawler machines when moving uphill. Be sure they are blocked to prevent downhill movement before shifting steering clutches.
17. Do not move the machine until everyone is clear of the path of travel.
18. If traveling with extended outrigger beams, check to ensure that they do not strike any obstructions.
19. Never travel over uneven, unstable, or uncompacted ground if the machine stability and strength is likely to be endangered.

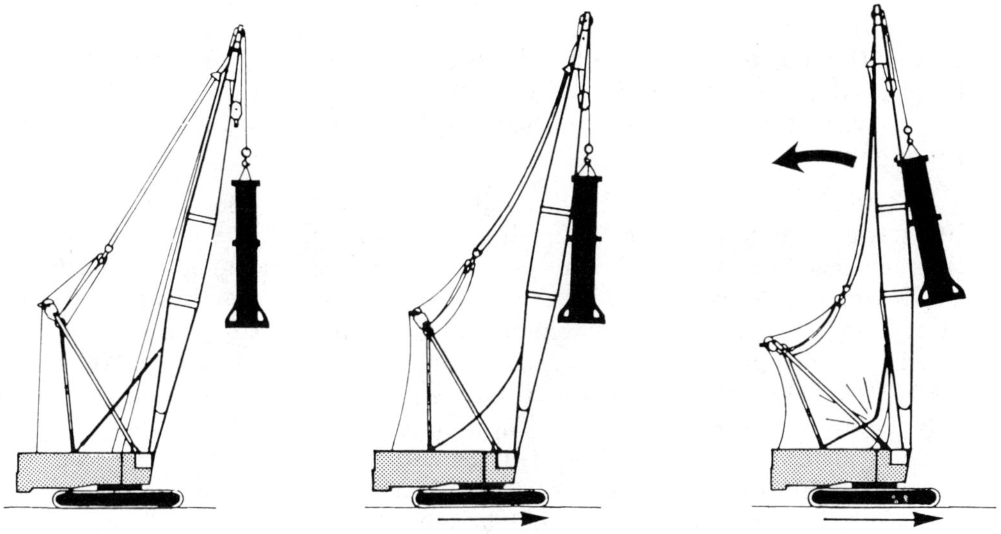

Fig. 16-29: If the boom is high and the carrier moves forward suddenly, it can literally walk out from under the boom.

ELECTRICAL HAZARDS

A discussion on cooperation between construction worker and operator would not be complete without a warning concerning power lines.

Electrocutions are among the most frequently repeated of all construction accidents, and most of them are caused by cranes contacting overhead power lines. *Power line contact is the largest single cause of fatalities associated with cranes.*

When it comes to cranes and power lines, safety is everybody's business. The ideal solution is to have the electricity turned off; but when that is not feasible, keep the machine as far from the lines as possible. When close proximity is necessary, various codes establish an absolute limit of approach. The American National Standard Institute (ANSI) B30.15-1973 states it this way:

Except where the electrical distribution and transmission lines have been deenergized and visibly grounded at point of work, or where insulating barriers not a part of or an attachment to the crane have been erected to prevent physical contact with the lines, cranes shall be operated proximate to, under, over, by, or near power lines only in accordance with the following:

1. For lines rated 50kV or below, minimum clearance between the lines and any part of the crane or load shall be 10′.

2. For lines rated over 50kV, minimum clearance between the lines and any part of the crane or load shall be 10′ plus 0.4″ for each 1kV over 50kV, or use twice the length of the line insulator but never less than 10′.
3. In transit with no load and boom lowered the clearance shall be a minimum of 4′.

In addition to maintaining a healthy distance between crane and power lines, the crane must have a signalperson whose sole duty and responsibility is to observe from the best vantage point and to warn the operator whenever the boom, the load, or any of the load lines are approaching the "limit of approach." This is necessary whenever the crane is within a boom's length of the limit. This means that if a crane has a 200′ (60.96 m) boom and it is located 180′ (54.88 m) from a power line, then a signalperson is necessary; but if it is 210′ (64 m) or more from the line, then none is needed (Fig. 16-30).

There is no substitute for these requirements. All commercial cage type boom guards, insulating links, and proximity warning devices have serious limitations and the use of them can lead to a feeling of false security. The use of them does not alter the previous requirements.

There are other precautions that can lead to a higher degree of safety when working around wires:

1. The operator must pay close attention to the signalperson's signals and warnings.
2. Slow down the operating cycle of the machine by reducing hoisting, booming, swinging, and travel speeds.
3. Issue strict warnings to all ground personnel to stand clear of the machine at all times. If the load must be guided into place by hand, a special check should be made with the signalperson before the rigger touches the load. Use taglines whenever possible. The best material for taglines in an area such as this is dry polypropylene rope. Both manila and nylon retain moisture and can conduct electricity.
4. Post warning signs prominently on the crane that will effectively keep job personnel on the alert for accidental contact.
5. Do not stockpile, load, or unload any material near power lines.
6. Do not locate access roads or ramps near power lines.
7. If it is known that craning will have to be done in an area, plan for it, and have the lines insulated or relocated before the crane arrives.
8. Treat all wires as being hot until you have reliable information to the contrary.
9. Be especially careful when traveling the crane as uneven ground can cause the boom to weave or bob into the lines.
10. Ensure that whenever cranes must repeatedly travel beneath power lines a route is plainly marked and "rider poles" are erected on each side of the crossing approach to ensure that the crane structure is lowered to a safe position.

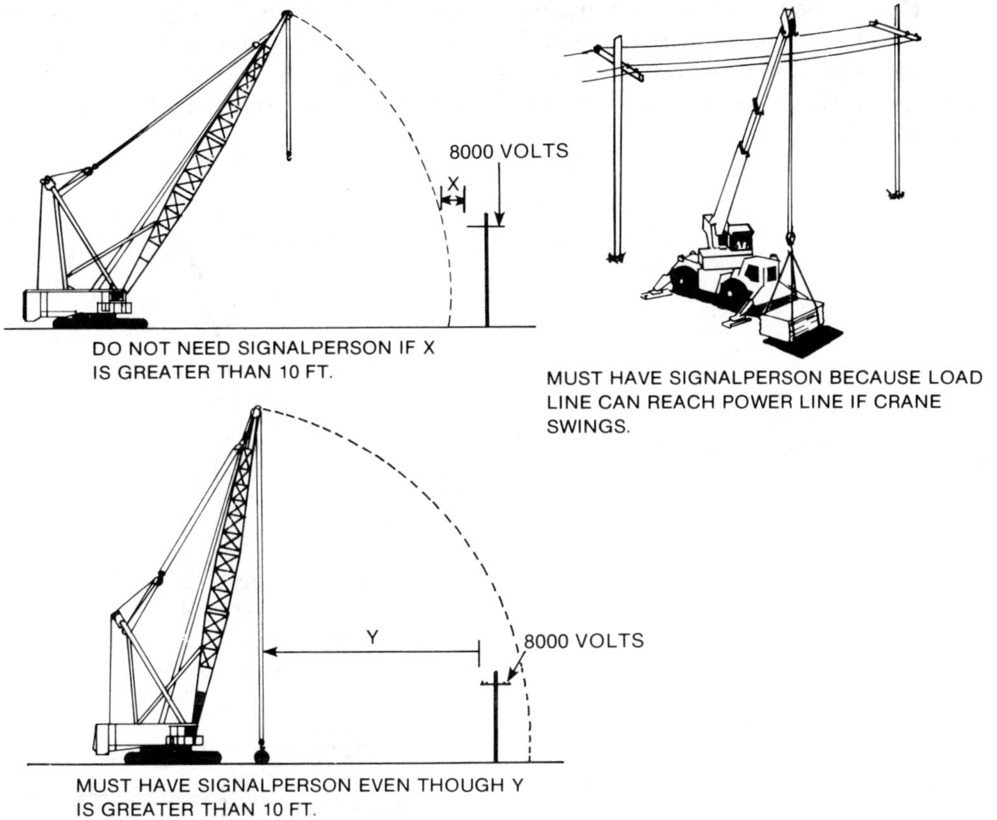

Fig. 16-30: Use a signalperson when boom radius falls within absolute limit of approach.

The routes should be located as close to the power line support tower or pole as possible in order to take advantage of the greater ground clearance.

11. Ensure that if a crane will be working in a relatively fixed location which is near electrical hazards that it is electrically grounded with a ground rod, permanent grounding bus, or ground mat. Ground rods should be placed as close to the machine as possible. The ground lead should first be connected to the rod and then to the machine. All personnel must be instructed to stay clear of both the lead and rod at all times. *Note:* Effective grounds are difficult to achieve, and the use of them does not alter the requirements regarding the absolute limit of approach or the signalperson. When removing the ground, the lead should first be disconnected from the machine. If a ground mat is used, the equipment must be placed on the mat and bonded to it. The ground mat area should be enclosed with a fence to prevent personnel from stepping on and off the mat during operation of the machine.

12. Exercise caution when working near overhead lines having long spans as they tend to swing laterally due to the wind and accidental contact could occur.
13. Use extreme caution when working in proximity to radio transmission towers as the crane's boom can act like an antenna and become electrically charged (Fig. 16-31).

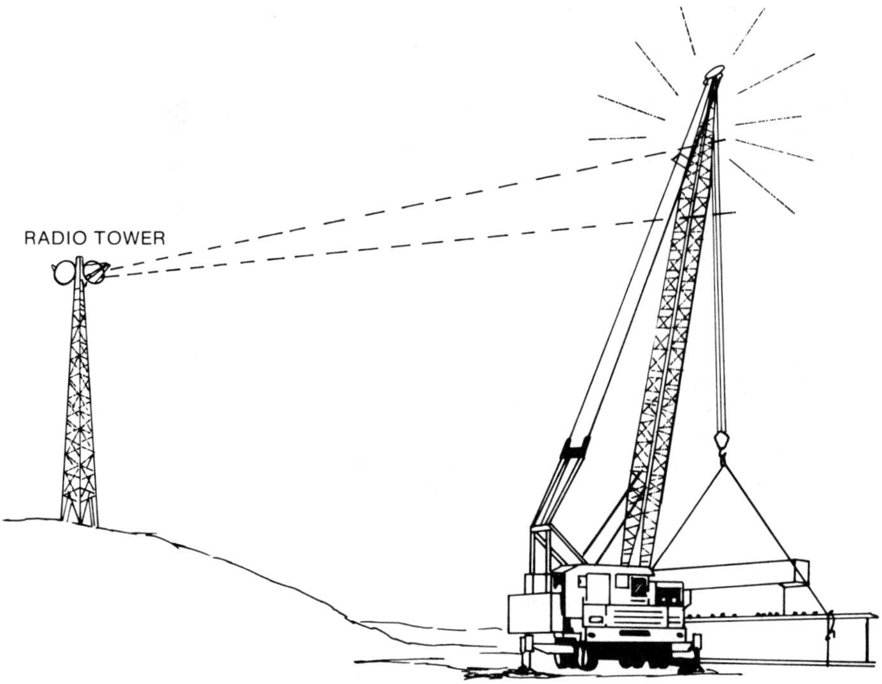

Fig. 16-31: The crane's boom can act like an antenna and pick up an electrical charge. Install a proper ground rod to prevent shock.

APPENDIX

A

Rigging Terms

Abrasion. Frictional surface wear on the wires of a wire rope.

A-Frame. An A-shaped framework from which a load can be suspended. A-frames are usually made from wood, metal tubing, or structural shapes.

Aggregate Breaking Strength. The breaking strength derived by totalling the individual breaking strengths of the elements of the strand or rope. This strength does not give recognition to the reduction in strength resulting from the angularity of the elements in the rope nor other factors which may affect efficiency.

Anchorage. Natural, temporary, or permanent installation on the ground for anchoring the end of a guyline.

Angle of Loading. Inclination of a leg or branch of a sling that may be measured from the horizontal or vertical plane. When angle of loading is less than 5° from the vertical, the load may be considered a vertical load.

Articulated Jib. A tower crane jib that, in general, has a pivot somewhere in its middle area, also called pivoted luffing jib.

Axis of Rotation. The vertical axis around which the crane jibs rotate.

Back Splice. A method of braiding the end of a rope to keep the strands from unraveling. It is started with a crown knot; then, several additional tucks are made proceeding back from the end of the rope.

Back Stay. Guy used to support a boom or mast; or that section of a main cable, as on a suspension bridge, cableway, etc., leading from the tower to the anchorage.

Bail. A U-shaped member of a bucket, or a U-shaped portion of a socket or other fitting used on wire rope.

Bailing Line. In well drilling, it is the wire rope that operates the bailer that removes water and drill cuttings.

Ballast. Weight added to a crane base to create additional stability; it does not rotate when the crane swings.

Base. The mounting flanges or feet for attachment of a hoist to its supporting structure or foundation.

Base Mounting. The structure forming the lowest element of a crane or derrick; it transmits loads to the ground or other supporting surface. For mobile cranes this is synonymous with carrier or crawler mounting. For tower cranes, the term includes a travel base, knee frame base, or fixed base (footing).

Base Section. The lowermost section of a telescopic boom; it does not telescope but contains the boom foot pin mountings and the boom-hoist-cylinder upper end mountings.

Basket of Socket. The conical portion of a socket into which a broomed-rope-end is inserted and then secured either with zinc or resin.

Becket. A small eye or strap on a block for fastening rope.

Becket Loop. A loop of small rope or strand fastened to the end of a large wire rope. Its function is to facilitate wire rope installation.

Bend. A knot by which one rope is made fast to another rope, a hook, ring, or eye.

Bending Stress. Stress that is imposed on the wires of a wire rope by a bending or curving action.

Bight. A bend or U-shaped curve in a rope.

Binder. A lever- or ratchet-actuated device employed to apply tension to ropes or to pull two objects together for a short distance.

Birdcaging. The deformation of a section of wire rope imparting to such section a birdcage-like appearance.

Block. An apparatus consisting of a frame holding a sheave over which a rope can operate.

Block and Tackle. Two single- or multiple-pulley blocks complete with load hooks and with rope reeved between to give a mechanical advantage in lifting or pulling. Same as block and fall.

Boatswain's Chair. Simply rigged sling on which one person can be suspended.

Bogie. An assembly of two or more axles arranged to permit both vertical wheel displacement and an equalization of loading on the wheels.

Boom. A spar fastened to a fixed structure at an angle.

Boom Angle. The angle between the horizontal and the longitudinal centerline of the boom, boom base, or base section.

Boom Base. The lowermost section of a sectional latticed boom having the attachment or boom foot pins mounted at its lower end; also called boom butt or butt section.

Boom Foot Mast. A component of some mobile-crane boom suspensions. It consists of a frame hinged at or near the boom foot that serves to increase the height of the inboard end of the fixed-boom suspension ropes, thereby increasing the angle those ropes make with the boom while being itself controlled by the boom hoist ropes. Its purpose is to decrease the axial compressive force on the boom; also called hi-height gantry.

Boom Harness. The block and sheave arrangement on the boom point to which the topping lift cable is reeved for changing the boom angle.

Boom Head. The portion of a boom that houses the upper load sheaves.

Boom Hoist. The rope drum(s), drive(s), and reeving controlling the derricking motion of the boom.

Boom Hoist Line. Wire rope that operates the boom hoist of derricks, cranes, draglines, shovels, etc.

Boom Line. A wire rope for supporting or operating the boom on derricks, cranes, draglines, shovels, etc.

Boom Point. The outward end of the top section of the boom.

Boom Stop. A device intended to limit the maximum angle to which the boom should be derricked.

Boom Suspension. A system of ropes and fittings, either fixed or variable in length, that supports the boom and controls the boom angle.

Boom Tip Section. The uppermost section of a sectional latticed boom, which usually includes the weldment mounting the upper load sheaves as an integral part; also called **Boom Point,** head section, or tapered tip.

Brake. A device, other than a motor, used for retarding or stopping motion by friction or power means.

Drag. A friction brake which provides a continuous retarding force having no external control.

Emergency. A brake for bridge and trolley held released during normal operation and arranged to apply a retarding force when initiated by the operator during an emergency or to automatically apply a retarding force upon loss of power.

Holding. A friction brake for a hoist that is automatically applied and prevents motion when power is off.

Mechanical Load. An automatic type of friction brake used for controlling loads in a lowering direction. This unidirectional device requires torque from the motor to lower a load but does not impose any additional load on the motor when hoisting a load.

Parking. A brake for bridge and trolley which may be automatically or manually applied in an attempt to prevent horizontal motion by restraining wheel rotation.

Service. A brake for bridge or trolley used by operator, during normal operation, to apply a retarding force.

Breaking Strength. It is the ultimate load at which a tensile failure occurs in the sample of wire rope being tested. (*Note*: The term breaking strength is synonymous with actual strength.)

Minimum Acceptance Strength. That strength which is 2-1/2% lower than the catalog or nominal strength. This tolerance is used to offset testing variables which exist when the test is made to determine the breaking strength of a specific sample of wire rope. Its use originated with the basic government wire rope specification.

Nominal Strength. This is the published (catalog) strength calculated by a standard procedure and accepted by the wire rope industry. The wire rope manufacturer designs wire rope to this strength, and the user should consider this to be the minimum strength when making design calculations.

Bridge. That part of a crane consisting of two or more girders, trucks, end ties, footwalks, and drive mechanism which carries the trolley or trolleys.

Bridge Cable. The all-metallic wire ropes or strands used as the catenary and suspenders on a suspension bridge.

Bridge Socket. A wire rope fitting of forged or cast steel that is designed with baskets—having adjustable bolts—for securing rope ends. There are two styles: (1) the *closed type* has a U-bolt with or without a bearing block in the U of the bolt, and (2) the *open type* has two eye bolts and a pin.

Bridge Travel. The crane movement in a direction parallel to the crane runway.

Bridle Cable. The all-metallic wire ropes or strands used as the catenary and suspenders on a suspension bridge.

Bridle Line. A rope fastened to the load being hoisted at several points and used for controlling the motion of the load.

Bridle Sling. A multipart wire rope sling.

Bright Rope. Wire rope fabricated from wires that are not metallic coated.

Bronze Rope. Wire rope fabricated from bronze wire.

Bull Pole. A pole, generally of steel pipe, mounted to project laterally from the base of a derrick mast. It is used to swing the derrick manually.

Bull Wheel. A term applied to a large-diameter wire rope sheave.

Button Conveyor Rope. Wire ropes to which buttons or discs are attached at regular intervals to move material as in a trough.

Cab. An operator's compartment attached to a crane or carrier.

Cable. A term loosely applied to wire ropes, wire strand, and electrical conductors.

Cable Crowd Rope. A wire rope used to force the bucket of a power shovel into material handled.

Cable-Laid. A kind of 9-stranded left lay rope, with the strands made up in 3 right-handed ropes of 3 strands each.

Cable-Laid Wire Rope. A type of wire rope consisting of several wire ropes laid into a single wire rope. Example, 6x6x7 tiller rope.

Cable Tool Drilling Line. The wire rope used to operate the cutting tools in the "cable tool" drilling method (i.e., rope drilling).

Cableway. Aerial conveying system for transporting single loads along a suspended track cable.

Cantilever. A structural member which supports the trolley of a wall crane.

Cantilever Frame. A structural member which supports the trolley of a wall crane.

Carbody. The part of a crawler crane base mounting which carries the upperstructure and to which the crawler side frames are attached.

Cargo Net. A square rope net for slinging materials.

Carrier. A wheeled chassis that is the base mounting for mobile truck and rough terrain cranes.

Carrier, Cab-Operated. A carrier controlled by an operator in a cab attached to the carrier.

Carrier (trolley). A unit that travels on the bottom flange of a monorail track or a bridge girder to transport a load.

Cathead. A material-hoisting device normally designed as a horizontal cantilever having, at its unsupported end, a single load sheave extending over the structure on which it is supported.

Chafe. To wear the surface of a rope or spar by rubbing it.

Chain Gauge. A positive measuring device for the wear or elongation of chain links.

Choker. A sling or attaching device that has a loop in one or both ends. When the other end or both ends of the rope passes through the eye of the loop, it tightens itself or chokes the load.

Choker Rope. A short wire rope sling that forms a slip noose around an object that is to be moved or lifted.

Circumference. Measured perimeter of a circle that circumscribes either the wires of a strand or the strands of a wire rope.

Clamps, Strand. A fitting for forming a loop at the end of a length of strand, consisting of two grooved plates and bolts.

Classification. Group or family designation based on wire rope constructions with strengths and weights jointly listed under the broad designation.

Clearance. Distance from any part of the crane to a point of the nearest obstruction.

Cleat. A small wooden plate, usually nailed to a member.

Cleavis. See **Shackle.**

Climbing Ladder. A steel member with crossbars (used in pairs) suspended from a climbing frame and used as jacking support points when some tower cranes climb.

Climbing System. A system which can be attached to the mast by various means and is used for extending the height of the crane.

Clip. Fitting for clamping two parts of wire rope to each other.

Closed Socket. A wire rope end fitting consisting of basket and bail made integral.

Closing Line. Wire rope that performs two functions: (1) closes a clamshell or orange peel bucket, and (2) operates as a hoisting rope.

Clutch. A means for engagement or disengagement of power.

Coil. To lay rope down in a series of circular turns.

Combination Ladder. A portable ladder capable of being used either as a stepladder or as a single or extension ladder. It may also be capable of being used as a trestle ladder or a stairwell ladder. Its components may be used as single ladders.

Come Along. Device for making a temporary grip on a wire rope.

Conical Drum. Grooved hoisting drum with a varying diameter.

Continuous Bead. Reeving of wire rope over sheaves and drums so that it bends in one direction, as opposed to reverse bend.

Contractor. The company, firm, individual, or other business enterprise which contracts with an operator to perform work.

Cord. Term applied to small size wire ropes or strands.

Core. The central member of a wire rope about which the strands are laid. It can be made of fiber, a wire strand, or an independent wire rope.

Corrosion. Chemical decomposition of the wires in a rope through the action of moisture, acids, alkalines, or other destructive agents.

Corrugated. Term used to describe the grooves of a sheave or drum after these have been worn down to a point where they show an impression of a wire rope.

Counter Jib. A horizontal member of a tower crane on which the counterweights and usually the hoisting machinery are mounted; also called counterweight jib.

Counterweight. Weight used to provide stability for lifting working loads.

Coupling (track strand). Device for joining the ends of two lengths of track strand.
Cover Wires. Outer layer of wires.
Cracker. Manila rope spliced or otherwise attached to the end of a wire drilling line.
Crane. A machine for lifting and lowering a load and moving it horizontally, with the hoisting mechanism an integral part of the machine. Cranes, whether fixed or mobile, are driven manually, by power, or by a combination of both.

Automatic Crane. A crane which, when activated, operates through a preset cycle or cycles.

Bridge Crane. A fabricated structural crane that operates on elevated straight tracks that bridge over the area of lifting. The hoist mechanism is generally located on a movable trolley.

Cab-Operated Crane. A crane controlled by an operator in a cab attached to the bridge or trolley.

Cantilever Gantry Crane. A gantry or semi-gantry crane in which the bridge girders or trusses extend transversely beyond the crane runway on one or both sides.

Crawler Crane. A crane consisting of a rotating superstructure with power plant, operating machinery, and boom, mounted on a base and equipped with crawler treads for travel. Its function is to hoist, lower, and swing loads at various radii.

Double-Girder Crane. A crane having two bridge girders mounted between and supported from the end trucks.

Floor-Operated Crane. A crane which is controlled by a means suspended from the crane by an operator on the floor or an independent platform.

Gantry Crane. A crane similar to an overhead crane except that the bridge for carrying the trolley or trolleys is rigidly supported on two or more legs running on fixed rails or other runway.

Hammerhead Tower Crane. A lifting machine consisting of a mast with an upper rotating member to which is attached a jib boom extending horizontally from the rotating member, with a counterweighted jib boom extending from the rotating member in the opposite direction, neither of which jib booms are arranged or rigged for luffing. The main jib boom carries a trolley which travels the length of the jib and which makes up the upper load block and from which the lower load block is suspended. The counterweight jib boom contains the counterweight for the main jib and working load and sometimes provides the mounting for the hoisting and trolley travel motors and drums.

Hot Molten Material Handling Crane. An overhead crane used for transporting or pouring molten material.

Hydraulic Crane. Full revolving crane, usually of the mobile type, that uses hydraulic power to operate the majority of the basic functions. They may be crawler mounted cranes, wheel mounted cranes of both truck and self-propelled wheel type, and any variations thereof which retain the same fundamental characteristics.

Interlocking Crane. A crane with an interlock mechanism on one or both ends, enabling it to be mechanically locked to another crane, fixed transfer section, or spur track for the purpose of transferring a carrier from one to another.

Jib Crane. A crane which features a horizontal or inclined arm (jib), attached to a rotatable vertical post. The post is anchored to both the floor and the ceiling. The load rope runs from a winch mounted on this post and over a sheave at the end of the jib.

Lattice Boom Crane. General term for all cranes whose booms are of lattice-design structural framework construction.

Manually-Operated Crane. A crane whose hoist mechanism is driven by pulling an endless chain or whose travel mechanism is driven in the same manner or by manually moving the load or hook.

Mobile Crane. Self-propelled crane consisting of a rotating superstructure with a power plant, operating machinery, and boom on a truck type mounting equipped with a power plant for travel on crawler tracks or wheels. When equipped with wheels it is called a *truck crane.*

Overhead Crane. A crane with a multiple girder movable bridge carrying a movable or fixed hoisting mechanism and traveling on an overhead fixed runway structure.

Pillar Crane. A fixed crane consisting of a vertical member held in position at the base to resist overturning moment, with constant-radius revolving boom supported at the outer end by a tension member.

Polar Crane. A bridge or gantry type crane which travels on a circular runway.

Portal Crane. A gantry crane without trolley motion, which has a boom attached to a revolving crane mounted on a gantry with the boom capable of being raised or lowered at its head (outer end). Portal cranes may be fixed or mobile.

Power-Operated Crane. A crane whose mechanism is driven by electric, air, hydraulic, or internal combustion means.

Pulpit-Operated Crane. A crane operated from a fixed operator station not attached to the crane.

Remote-Operated Crane. A crane controlled by an operator not in a pulpit or in the cab attached to the crane, by any method other than a means suspended from a crane.

Semi-Gantry Crane. A gantry crane with one end of the bridge rigidly supported on one or more legs that run on a fixed rail or runway, the other end of the bridge being supported by a truck running on an elevated rail or runway. Also called a *half-gantry crane.*

Single-Girder Crane. A crane having one bridge girder mounted between and supported from the end trucks.

Standby Crane. A crane which is not in regular service but which is used occasionally or intermittently as required.

Tower Crane. A portal crane, with or without an opening between the legs of its supporting structure, adapted to hoist and swing loads over high obstructions and mounted upon a fixed or mobile tower-like gantry. The revolving crane may be supported on the tower by a revolving mast or by a turntable.

Wall Crane. A crane having a cantilever frame with or without trolley and supported from a side wall or line of columns of a building. It is a traveling type and operates on a runway attached to the side wall or columns.

Wheel Mounted Crane (wagon crane). A crane consisting of a rotating superstructure with power plant, operating machinery and boom, mounted on a base or platform equipped with axles and rubber-tired wheels for travel. The base may be propelled by an engine in the superstructure or with a separate engine controlled from the superstructure. Its function is to hoist, lower, and swing loads at various radii.

Crawler Frames. Part of the base mounting of a crawler crane attached to the carbody and supporting the crawler treads, the track rollers, and the drive and idler sprockets. Crawler frames transmit crane weight and operational loadings to the ground; also called *side frames.*

Creep. The unique movement of a wire rope with respect to a drum surface or sheave surface resulting from the asymmetrical load between one side of the sheave (drum) and the other. It is not dissimilar from the action of a caterpillar moving over a flat surface. It should be distinguished from slip which is yet another type of relative movement between rope and the sheave or drum surface.

Cribbing. A strong framework or structure, usually made of timber, employed to support a load being moved or raised.

Critical Diameter. For any given wire rope, it is the diameter of the smallest bend that permits both wires and strands to adjust themselves by relative movement while retaining their normal cross section position.

Crossover Points. Points of rope contact where one layer of rope on a rope drum crosses over the previous layer.

Cylindrical Drum. A hoisting drum of uniform diameter. See **Drum.**

Dead End. The point of fastening of one rope end in a running rope system, the other (live) end being fastened at the rope drum.

Dead-Line. In drilling, it is the end of the rotary drilling line fastened to the anchor or dead-line clamp.

Deadman. Log or beam buried in the ground to serve as an anchor.

Dead Part. The part of a rope that supports a load which is not under direct tension from the load because it is passed through a lug and doubled back against the rope part or is wrapped around a post or other similar member.

Dead Weight. The total weight of all the suspended rigging employed in a lifting operation.

Deflection. (1) The sag of a rope in a span, usually measured at mid-span as the depth from the chord joining the tops of the two supports. (2) Any deviation from a straight line.

Derrick. An apparatus for lifting or lowering loads consisting of a mast or equivalent member held at the head by guys or braces, with or without a boom, for use with hoists and ropes.

A-Frame Derrick. A derrick in which the boom is hinged from a cross member between the bottom ends of two upright members spread apart at the lower ends and joined at the top; the boom point is secured to the junction of the side members, and the side members are braced or guyed from this junction point.

Basket Derrick. A derrick without a boom, similar to a gin pole, with its base supported by ropes attached to corner posts or other parts of the structure. The base is at a lower elevation than its supports. The top of the pole is secured with multiple reeved guys to position the top of the pole to the desired location by varying the length of the upper guy lines. The load is lifted and lowered by ropes through a sheave or block secured to the top of the pole.

Breast Derrick. A derrick without a boom. The mast consists of two side members spread farther apart at the base than at the top and tied together at top and bottom by rigid members. The mast is prevented from tipping forward by guys connected to its top. The load is lifted and lowered by ropes through a sheave or block secured to the top crosspiece.

Chicago Boom Derrick. A boom which is attached to a structure, an outside upright member of the structure serving as the mast, and the boom being stepped in a fixed socket clamped to the upright. The derrick is complete with load, boom, and boom point swing line falls.

Gin Pole Derrick. A derrick without a boom. Its guys are so arranged from its top to permit leaning the mast in any direction. The load is lifted and lowered by ropes reeved through sheaves or blocks at the top of the mast and the load block (lower).

Guy Derrick. A fixed derrick consisting of a mast capable of being rotated, supported in a vertical position by guys, and a boom whose bottom end is hinged or pivoted to move in a vertical plane with a reeved rope between the head of the mast and the boom harness for lifting and lowering the boom and a reeved rope from the boom point for lifting and lowering the load.

Shearleg Derrick. A derrick without a boom and similar to a *Breast Derrick.* The mast, wide at the bottom and narrow at the top, is hinged at the bottom and has its top secured by a multiple reeved guy to permit handling loads at various radii by means of load tackle suspended from the mast top.

Stiffleg Derrick. A derrick similar to a guy derrick except that the mast is supported or held in place by two or more stiff members, called stifflegs, which are capable of resisting either tensile or compressive forces. Sills are generally provided to connect the lower ends of the stifflegs to the foot of the mast.

Derrick Bullwheel. A horizontal ring or wheel, fastened to the mast base of a derrick, for the purpose of swinging the derrick by means of ropes leading from this wheel to a powered drum.

Derricking. Changing boom angle by varying the length of the boom suspension ropes; also called *luffing, booming in (out),* or *topping.*

Design Factor. In a wire rope, it is the ratio of the listed breaking strength to the total working load.

Diameter. A line segment which passes through the center of a circle and whose end points lie on the circle. As related to wire rope, it would be the diameter of a circle which circumscribes the wire rope.

Differential Hoist. A chain hoist consisting of two sheaves of different diameter at the top and a mounting hook and one sheave at the load end.

Dog. A pawl used in conjunction with a ratchet built into one flange of a rope drum to lock the drum from rotation in the spooling-out direction.

Dogged Off. The condition of a rope drum when its dog is engaged.

Dog-Leg. Permanent short bend or kink in a wire rope caused by improper use or handling.

Dolly. A number of different types of devices usually consisting of a base to which a number of rollers, wheels, or casters are attached and upon which loads are placed for movement along runways or floors.

Double Block. A block with two sheaves.

Dragline. (1) Wire rope used for pulling excavating or drag buckets, and (2) name applied to a specific type of excavator.

Drift. The distance along the ground from a line through the load being lifted to the base of a gin pole or shears.

Drifting. Pulling a suspended load laterally to change its horizontal position.

Drum. A cylindrical flanged barrel, either of uniform or tapering diameter, on which rope is wound either for operation or storage. Its surface may be smooth or grooved.

Drum Capacity, Rope. The length of a specific diameter of rope which can be wound on a drum.

Drum Hoist. A hoisting mechanism incorporating one or more rope drums; also called *hoist, winch,* or *hoisting engine.*

Duty Cycle Work. Steady work at a fairly constant short cycle time with fairly constant loading levels for one or more daily shifts.

Dynamic Loading. Loads introduced into the machine or its components by forces in motion.

Ease Off. To slacken.

Efficiency. Ratio of a wire rope's measured breaking strength and the aggregate strength of all individual wires tested separately—usually expressed as a percentage.

Elastic Limit. Stress limit above which permanent deformation will take place within the material.

Elliptic Spool. An endless-rope drive drum with a face in the shape of an elliptic arc.

Endless Rope. Rope whose two ends are spliced together to form a single continuous loop.

End Seizing. A seizing at the end of a rope.

End Tie. A structural member which connects the ends of the bridge girders to maintain squareness of the bridge.

Equalizer. A device which compensates for unequal length or stretch of a rope.

Equalizing Sheave. The sheave at the center of a rope system over which no rope movement occurs other than equalizing movement. It is frequently overlooked during crane inspections, with disastrous consequences. It can be the source of severe fatigue deterioration.

Equalizing Slings. Multiple-leg slings composed of wire rope and fittings that are designed to help distribute the load equally. See **Sling.**

Equalizing Thimbles. Special type of load-distributing fitting used as a component of certain wire rope slings.

Exposed. Applies to hazardous objects not adequately guarded or isolated (capable of being contacted inadvertently).

Extension Cylinder. Hydraulic ram used to extend a section of a telescopic boom; the most common but not the only means for power-extending boom sections.

Extension Ladder. A non-self-supporting portable ladder, adjustable in length. It consists of two or more sections traveling in guides or brackets or the equivalent and so arranged as to permit length adjustment.

Eye. A loop formed at the end of a rope by securing the dead end to the live end of the base of the loop.

Eyebolt. A bolt terminating in an eye, used to secure a line.

Eye Splice. An eye formed in the end of a rope by splicing the rope.

Factor of Safety. A term originally used in the wire rope industry to state the ratio of nominal strength to the total working load. The term is no longer used since it implies the permanent existence of this ratio when, in actuality, the rope strength begins to reduce the moment it is placed in service.

Fairlead Sheave. In a hoisting system, the sheave employed to keep the fleet angle of the rope within the required limits with respect to the winding drum or another sheave.

Fall Line. The line from a tackle system on which the pull will be exerted.

Fall Rope. The rope used in a fall or block and tackle.

Fatigue. In a wire rope, the term is usually applied to the process of progressive fracture, from bending, of the individual wires. These fractures may and usually do occur at bending stresses well below the ultimate strength of the material; it is not an abnormality although it may be accelerated due to conditions in the rope such as rust or lack of lubrication.

Ferrule. A metallic button, usually cylindrical in shape, normally fastened to a wire rope by swaging but sometimes by spelter socketing.

Fiber Centers. Cords or rope made of vegetable or synthetic fiber that are used as the center of a strand.

Fiber Cores. Cords or rope made of vegetable or synthetic fiber that are used as the center of a wire rope.

Fiber Rope. Rope made of vegetable fibers twisted together. Sometimes called *cordage.*

Fiddle Block. A block consisting of sheaves held in place by the same cheek plates; sheaves are not necessarily on the same shaft.

Fitting. Any functional accessory attached to a wire rope.

Fixed Ladder. A ladder permanently attached to a structure, building, or equipment. Ladders referred to in this standard shall be construed to be fixed ladders.

Flag. Marker placed on a rope so as to locate the load position.

Flange Point. The point of contact between the rope and drum flange where the rope changes layers on a rope drum.

Flat Rope. Wire rope that is made of a series of parallel, alternating right-lay and left-lay ropes, sewn together with relatively soft wires.

Fleet Angle. The angle between rope being wound on a drum and a line from the centerline of the drum to the nearest sheave.

Fleeting Sheave. Sheave mounted on a shaft parallel to the rope-drum shaft and arranged so that it can slide laterally as the rope spools, permitting close sheave placement without excessive fleet angle.

Floating Harness. A frame, forming part of the boom suspension, supporting sheaves for the live suspension ropes and attached to the fixed suspension ropes (pendants), also called bridle, spreader, upper spreader, live spreader, or spreader bar.

Foot Bearing. The lower support of a derrick on which the mast rotates.

Footblock. A steel weldment or assembly serving as the base mounting for a guy derrick, gin pole, or Chicago boom derrick.

Footing. A broad foundation of concrete, stone, or other building material which is dug into the ground for supporting loads and structures.

Footwalk. A walkway with handrail on, or attached to, the bridge or trolley for access purposes.

Fouled. Jammed or tangled, not clear to operate.

Frap. To pass tight wrappings of rope.

Free Fall. Lowering the hook (load) or boom by gravity; the lowering speed is controlled only by a retarding device such as a brake.

Front End Attachments. Optional load-supporting members for use on mobile cranes, e.g., taper tip boom, hammerhead boom, guy derrick attachment, and tower attachment.

Fulcrum. A fixed point on a boom lever or beam around which the boom lever or beam rotates.

Galvanized. Hot-dipped (occasionally electrochemical) zinc coating for corrosion resistance.

Galvanized Rope. Wire rope made up of galvanized wire.

Gantry (A-Frame). A structural frame, extending above a base structure, to which boom support ropes are reeved.

Gantry Leg. The structural member which supports a bridge girder or end tie from the sill.

Gear. A general term for ropes, blocks, tackle, fittings, etc.

Gooseneck Boom. A boom with an upper section projecting at an angle to the longitudinal centerline of the lower section.

Grab Bar. An individual handhold placed adjacent to or as an extension above a ladder for the purpose of providing safe access/egress for a ladder.

Grade. Wire rope or strand classification by strength and/or type of material—i.e., Improved Plow Steel, Type 302 Stainless, Phosphor Bronze, etc. It does not imply a strength of the basic wire used to meet the rope's nominal strength.

Grades, Rope. Classification of wire rope by the wire's metallic composition and the rope's breaking strength.

Grommet. An endless 6-strand wire rope with a strand core made of one continuous length of strand.

Grooved Drum. Drum with a grooved surface that accommodates the rope and guides it for proper winding and unwinding.

Grooves. Depressions—helical or parallel—in the periphery of a sheave or drum that are shaped to position and support the rope.

Gudgeon Pin. A pin connecting the mast cap to the mast, allowing rotation of the mast.

Guyline. A rope from a structure to an anchorage for steadying the structure. Also called a *guyrope.*

Handle Eye. An opening in the handle shaped to accept a hook, shackle, or other lifting device.

Haul. To pull on a line, particularly the fall line of tackle.

Haulage Rope. Wire rope used for pulling movable devices such as cars that roll on a track.

Hauling Line. The fall line of tackle.

Hawser. Wire rope, usually galvanized, used for towing or mooring marine vessels.

Headache Ball. An iron ball placed above the hook on a lead line which furnishes weight to allow the line to be lowered when no load is attached.

Helicopter. A rotorcraft which derives its horizontal and vertical flight ability primarily from its engine driven rotors; in this book it will be the basic vehicle for lifting, hoisting, pulling, towing, and moving cargo.

Hog Rods. The rigid members of a crane which are employed to support the boom at a fixed angle of inclination.

Hoist. A power driven drum or drums capable of lifting and lowering loads.

Drum. A hoist with hoisting drum(s), and with or without a swinger.

Direct Geared. A hoist with drum(s) geared directly to its power source.

Friction Drum. A hoist with drum(s) controlled by friction clutches and brakes and provided with drum ratchets and pawls.

Hoist Chain. The load bearing chain in a hoist.

Hoist Motion. That motion which lifts or lowers a load.

Holding Line. Wire rope on a clamshell or orange peel bucket that acts as a restraint on the bucket while the closing line is released to dump its load.

Hook. A curved metal piece used on the end of a rope or chain, or on a block. There are several types used in rigging including the *barrel hook, choker hook, clevis hook, duplex (sister) hook, eye hook, foundry hook,* and *shank hook.*

Hydraulic. A method of controlling or powering drive or braking by means of displacement of a liquid.

Idler. Sheave or roller used to guide or support a rope.

Improved Plow Steel Rope. A specific grade of wire rope.

Independent Wire Rope Core. Wire rope used as the core of a larger rope.

Interlocking Mechanism. A mechanical device to lock together the adjacent ends of two cranes, or a crane to a fixed transfer section of spur track, to permit the transfer of carriers from one crane or track to the other.

Jib. A structural nonluffing member, extending horizontally from the mast about the axis of rotation, and containing the trolley and necessary sheaves for lifting and moving loads by means of ropes.

Jib Boom. A horizontal cantilever track for supporting the carrier.

Jib, Counterweight. A structural nonluffing member extending from the mast in the opposite direction from the load carrying jib boom and containing the ballast or counterweight necessary to counterbalance the working load(s). The counterweight jib frequently contains the hoisting machinery.

Jib Line. The load line from the jib of a crane or derrick which is equipped with both a main boom with the main load lines and a jib boom with a single auxiliary load line.

Jib Mast. A short strut or frame mounted on the boom head to provide a means for attachment of the jib support ropes; also called *jib strut, rooster, horse,* or *A-frame.*

Jib Point. The outward end of the load bearing jib.

Jumping. Raising a derrick from one operational mounting height to another.

Jury Rig. Any kind of improvised or expedient rigging.

Kink. A sharp bend in wire rope, usually causing the wires to separate.

Knot. The intertwining of the end of a rope within a portion of the rope; in general, nearly any type of knot, splice, hitch, or bend made with rope.

Ladder. A metal, or wooden, combination of two side braces holding rungs for climbing.

Lagging. (1) External wood covering on a reel to protect the wire rope or strand, or (2) the grooved shell of a drum.

Landing. Any area such as the ground, roof, or platform which provides access/egress for a fixed ladder.

Lashing. A rope wrapping two pieces to fasten them together.

Latch, Hook. A device used to bridge the throat opening of a hook.

Lay. (1) The manner in which the wires in a strand or the strands in a rope are helically laid. (2) The length, parallel to the longitudinal axis, in which a wire makes one complete turn about the axis of the strand or a strand about the axis of a rope. In this connection lay is also referred to as *lay length* or *pitch.*

Lay, Types. There are various lay types.

Right Lay. Product in which the elements are laid in a right-hand helix.

Left Lay. Product in which the elements are laid in a left-hand helix.

Cross Lay. Product in which one or more laying or closing operations are performed in opposite directions. A multiple operation product is named according to the direction of the outside layer.

Regular Lay. Wire rope in which the wires in the strands and the strands in the rope are laid in opposite directions. The crowns of the wires appear to line up with the axis of the rope.

Lang Lay. Wire rope in which the wires in the strands and the strands in the rope are laid in the same direction. The crowns of the wires make an angle with the axis of the rope.

Alternate Lay. Lay of a wire rope in which the strands are alternately regular and lang lay.

Alberts Lay. An old, now rarely used, term for lang lay.

Reverse Lay. Another term for alternate lay.

Spring Lay. This is not actually a unique lay and more properly refers to a wire rope construction. See **Mooring Lines.**

Herringbone. An unusual construction consisting of four lang lay strands, each pair of which is separated by a regular lay strand. See **Lay.**

Layer. A series of wraps of wire rope around a rope drum barrel, extending full from flange to flange.

Leading Block. A single block used to change the direction of pull on a line without changing the mechanical advantage.

Lead Line. That part of a rope tackle leading from the first, or fast, sheave to the drum.

Level Luffing. An automatic arrangement whereby the crane or derrick hook does not significantly change elevation as the boom derricks.

Life Belt. A web type belt, worn by a worker, that has a line attached from it to a stationary object.

Life Net. A rope type net which is located directly below an elevated work area to catch any worker or equipment which might fall.

Lift Angle. The angle between an imaginary line vertical to the load to be lifted and an imaginary line between the load and the hoist or lifting device.

Lifting Devices. Devices which are not reeved onto the hoist ropes such as hook on buckets, magnets, grabs, and other supplemental devices used for ease of handling certain types of loads. The weight of these devices is to be considered part of the rated load.

Lift Section. A mechanism which will lift a section of track out of alignment with a stationary track(s).

Limiting Device. A mechanical device which is operated by some part of a power-driven machine or equipment to control motions of the machine or equipment.

Limit Switch. A switch which is operated by some part of a power-driven machine or equipment to alter the electric circuit associated with the machine or equipment.

Line. A thread, string, cord, or rope, especially a comparatively slender and strong cord.

Line Pull. The pulling force attainable in a rope leading off a rope drum or lagging at a particular pitch diameter (number of layers).

Line Speed. The speed attainable in a rope leading off a rope drum or lagging at a particular pitch diameter (number of layers).

Link. One of the individual units in a piece of chain.

Live Part. The part of a rope or line that supports a load when it is under direct tension from the load.

Load. The total superimposed weight on the load block or hook.

Load Bar. A load-carrying member between carriers.

Load Block. The assembly of hook or shackle, swivel, bearing, sheaves, pins, and frame suspended by the hoisting rope. This shall include any appurtenances reeved in the hoisting ropes.

Load Block, Lower. The assembly of hook or shackle, swivel, sheaves, pins, and frame suspended by the hoisting ropes.

Load Block, Upper. The assembly of hook or shackle, swivel, sheaves, pins, and frame suspended from the boom point.

Load Capacity, Maximum. The maximum load capacity is the capacity for which the tractor is designed.

Load, Working. The external load, in pounds (kg) or short tons (metric tons), applied to a hoist, including the weight of auxiliary load attaching equipment.

Longitudinal. Running in a direction from end to end.

Loop. A fold or doubling of a thread, cord, or rope through which another thread, cord, or rope can be passed.

Lower Spreader. A frame, forming part of the boom suspension, supporting sheaves for the live suspension ropes and attached to the gantry or upperstructure.

Lug. A projection from an object designed to be used as an attachment for lifting.

Main Hoist. The primary hoist mechanism provided for lifting and lowering the rated load.

Man Trolley. A trolley having an operator's cab attached thereto.

Marline. A prelubricated fiber material.

Marline Spike. Tapered steel pin used as a tool for splicing wire rope.

Mast. The upright member of a derrick.

Mast Cap. The fitting at the top of a derrick mast to provide attachment points for the guy ropes. Also called a *spider*.

Material Elevator. A mechanism consisting of a tower with vertical members which guide a platform that is lifted and lowered by means of a hoist.

Messenger Strand. Galvanized strand used to support telephone and electrical cables.

Midsection. On a telescopic boom, the intermediate powered telescoping section(s) mounted between the base and fly sections.

Milking. Sometimes called ironing, it is the progressive movement of strands along the axis of the rope that results from its movement through a restricted passage such as a tight sheave.

Modulus of Elasticity. Mathematical quantity expressing the ratio, within the elastic limit, between a definite range of unit stress on a wire rope and the corresponding unit elongation.

Moment. The product of a force or loading and the normal distance to a particular or arbitrary axis, producing in some instances rotational effects (or loading in a member resisting rotation) and in other instances bending effects.

Monkey Tail. Any device employed to keep a load block from twisting or to prevent a turnbuckle from twisting while being tightened or after it has been tightened.

Monorail. A single run of overhead track upon which carriers travel.

Mooring Lines. Galvanized wire rope, usually 6x12, 6x24, of spring lay construction, for holding ships to dock.

Mousing. A wrapping across a hook to prevent it from unhooking.

Nonrotating Wire Rope. 18x7 wire rope consisting of a 6x7 left lay lang lay inner rope covered by twelve 7-wire strands right lay regular lay.

Open Socket. A wire rope fitting that consists of a "basket" and two "ears" with a pin.

Operating Sectors. Portions of a horizontal circle about the axis of rotation of a mobile crane providing the limits of zones where over-the-side, over-the-rear, and over-the-front ratings are applicable.

Outriggers. Extensible arms attached to a crane base mounting, which include means for relieving the wheels (crawlers) of crane weight; used to increase stability.

Overhaul. To separate the blocks in tackle by a greater distance.

Overload. Any load greater than the rated load.

Overturning Moment. The moment of the load plus the boom weight about the tipping fulcrum. Wind and dynamic effects can be included when appropriate.

Parachute Harness. A web type of harness which fits snugly around the body and is used to lower a person through confined areas.

Parts of Line. A number of running ropes supporting a load or force; also called *parts* or *falls.*

Pawl (dog). A device for positively holding machinery against undesired rotation.

Paying Out. Adding slack to a line or relieving load on a line by letting (spooling) out rope.

Peening. Permanent distortion resulting from cold plastic flow of the outer wires. Usually caused by pounding against a sheave or machine member or by heavy operating pressure between rope and sheave, rope and drum, or rope and adjacent wrap of rope.

Pendant. A fixed-length rope forming part of the boom-suspension system; also called *boom guy line, hog line, boom stay, standing line,* or *stay rope.*

Pigtail. A short section of wire rope of the same diameter as the lead line which has one end attached by wire rope clips to the lead line and the other end fastened to some firm anchorage. It is used to secure the lead line temporarily in an emergency situation or when maintenance or repair work must be performed on the hoisting or rigging device.

Pitch Diameter. The diameter of a sheave or rope drum measured at the centerline of the rope; tread diameter plus rope diameter.

Plow Steel. A steel used in making wire rope that is tougher and stronger than mild grades of steel.

Pole. Any structural member supported vertically or at an angle from the vertical which is employed by itself or in combination with other poles and girders to hoist loads.

Power Controlled Lowering. A system or device in the power train, other than the load hoist brake, which can control the lowering rate of speed of the load hoist mechanism.

Power Hoist. A hoist with high gear reduction that is powered by an electric motor or gasoline engine.

Preformed Strand. Strand in which the wires are permanently shaped, before fabrication in the strands, to the helical form they assume in the strand.

Preformed Wire Rope. Wire rope in which the strands are permanently shaped, before fabrication into the rope, to the helical form they assume in the wire rope.

Pressed Fittings. Fittings which are attached to wire rope by cold forming the fitting onto the rope by means of a rotary swager or a press.

Prestressing. Stressing a wire rope or strand before use under such a tension and for such a time that the constructional stretch is largely removed.

Prestretching. Subjecting a wire rope or strand to tension prior to its intended application, for an extent and over a period of time sufficient to remove most of the constructional stretch.

Preventer. In rigging practice, a means, usually comprising but not limited to a wire rope, for preventing an unwanted movement or occurrence or acting as a saving device when an anchorage or attachment fails.

Radio-Controlled. A crane or carrier operated from a radio transmitter located at a point not mechanically attached to the device being controlled.

Radius, Load (operating). Nominally, the horizontal distance from the axis of rotation to the center of gravity of a lifted load. In mobile crane practice, this is more specifically defined as the horizontal distance from the projection to the ground of the axis of rotation before loading to the center of a loaded but vertical hoist line.

Rail Sweep. A mechanical device attached to the crane but located in front of the crane leading wheels to throw off any obstructions.

Ratchet. A toothed member, securely attached to the drum, for engagement with the pawl.

Rated Capacity. The load which a wire rope or wire rope sling, when new, may handle under given operating conditions and at an assumed design factor.

Rated Load. The maximum load for which a crane or individual hoist is designed and built by a manufacturer. The rated load is shown on the equipment nameplate(s).

Reach. Distance from the axis of rotation of a crane or derrick; sometimes used synonymously with radius.

Reel. A flanged spool on which wire rope or strand is wound for storage or shipment.

Reeve. To thread the end of a rope through blocks in making up tackle.

Reeving. A rope system in which the rope travels around drums and sheaves.

Reverse Bend. Reeving a wire rope over sheaves and drums so that it bends in opposing directions.

Rollers. Long pieces of hardwood or pipe placed under heavy objects to facilitate rolling along a flat surface. Also, small-diameter cylinders or wide-faced sheaves for supporting ropes.

Rooster. Vernacular term for one or more struts at the top of a boom or mast, such as a jib strut, a tower-crane top tower, or the struts at the top of the mast of a mobile-crane tower attachment.

Rope. A large, stout cord made of strands of fiber or wire twisted or braided together (often called a *line*).

Rope Drum. That part of a drum hoist which consists of a rotating cylinder with side flanges on which hoisting rope is spooled in or out (wrapped).

Running End. The free or working end of rope.

Running Line. A rope that moves over sheaves or drums.

Running Sheave. A sheave which rotates as the load block is raised or lowered.

Runway. An assembly of rails, beams, girders, brackets, and framework on which the crane travels.

Saddle Jib. The horizontal live-load supporting member of a hammerhead type tower crane having the load falls supported from a trolley that traverses the jib; also called load jib.

Safety Belt. A high-strength belt or harness attached to a fall-limiting device which is worn by individuals working at elevated locations, and is specifically designed to secure, suspend, or retrieve the user.

Safe Working Load. Potentially misleading, this term is in disfavor. Essentially, it means the portion of the nominal rope strength that can be applied to move or sustain a load. It is misleading because it is valid only when the rope is new and equipment is in good condition.

Scaffolding. An elevated platform for supporting materials, tools, and/or workers.

Seale. The name for a type of strand construction that is characterized by: (1) cover wires of a single size; (2) the same number of one size of wires in the adjacent layer; and (3) each layer having the same length and direction of lay. The most common construction of this type has one center wire, nine inner wires, and nine cover wires.

Seize. To bind with wire or light cord.

Serve. To cover the surface of a wire rope or strand with a fiber cord or wire wrapping.

Shackle. A U- or anchor-shaped fitting with pin.

Sheave. A wheel or pulley with a circumferential groove designed for a particular size of wire rope; used to change direction of a running rope.

Shoring. A temporary framework of timber, planking, or other structural materials erected to support floors or walls to prevent them from caving in.

Shroud-Laid. Four stranded right-hand rope.

Side Boom Tractor. A track-type or wheel-type tractor having a boom mounted on the side of the tractor, used for lifting, lowering, or transporting a load suspended on the load hook. The boom or hook can be lifted or lowered in a vertical direction only.

Side Loading. A loading applied at any angle to the vertical plane of the boom.

Side Pull. The portion of the hoist pull acting horizontally when the hoist lines are not operated vertically.

Sills. Horizontal structural members which connect the lower ends of two or more legs of a gantry crane on one runway.

Single-Block. A block with a single sheave.

Skid. A heavy timber used under machinery or other equipment that is being moved on rollers.

Slew, or Slewing. In the operation of a crane or derrick, to rotate the boom horizontally.

Slings. Wire rope or ropes made into forms, with or without fittings, for handling loads and designed to permit the attachment of an operating rope.

Smooth-Faced Drum. Drum with a plain face, not grooved.

Snatch Block. A single sheave block hinged so that it can be opened to insert a loop of rope over the sheave.

Snub. To check or slow the progress of.

Snubber. A rope wrapped around a stationary object to hold or slowly lower a heavy load.

Socket. Generic name for a type of wire rope fitting.

Span. The horizontal distance center-to-center of runway rails.

Spiral Groove. A continuous helical groove that follows a path on and around a drum face, similar to a screw thread.

Splice. To join two ropes together by tucking strands of each into the other.

Splicing. Making a loop or eye in the end of a rope by tucking the ends of the strands back into the main body of the rope. The term is also applied to the formation of loops or eyes in a rope by means of mechanical attachments pressed onto the rope. The term also relates to the joining of two rope ends so as to form a long or short splice in two pieces of rope.

Spooling, Rope. Winding of rope on a cylindrical drum in evenly-spaced, uniform layers to provide maximum capacity of rope on the drum.

Spreader. A horizontal bar placed in a sling to prevent crushing of the load being hoisted.

Stabilizers. Devices for increasing stability of a crane; they are attached to the crane base mounting but are incapable of relieving the wheels (crawlers) of crane weight.

Standing Block. In a tackle system, the block which is fixed and does not move.

Standing Line. A fixed-length line that supports loads without being spooled on or off a drum; a line of which both ends are dead; also called *standing guy line, stay rope,* or *pendant.*

Standing Part. The balance of a rope excluding the running end.

Static Base. Tower-crane support (base mounting) where the crane mast is set on or into a foundation.

Steel Clad Rope. Rope with individual strands spirally wrapped with flat steel wire.

Stepladder. A self-supporting portable ladder, nonadjustable in length, with flat steps and a hinged back.

Step-Surface. The clear portion of steps and rungs on which a person may step while ascending or descending a ladder.

Stiffleg. Rigid member supporting the mast at the head.

Stirrup. The U-bolt or eyebolt attachment on a bridge socket.

Stop. A device to limit travel of a trolley or crane bridge. This device normally is attached to a fixed structure and normally does not have energy absorbing ability.

Strand. A symmetrically arranged and helically-wound assembly of wires.

Strength Margin. The ratio of structural failure load (or stress) to actual or permitted load (or stress).

Stress. The force or resistance within any solid body to alteration of form; in the case of a solid wire it would be the load on the rope divided by the area of the wire.

Stretch. The elongation of a wire rope under load.

Swaged Fittings. Fittings into which wire rope can be inserted and then permanently attached by cold pressing (swaging) the shank that encloses the rope.

Swing. Rotating of a boom structure for movement of loads in a horizontal direction about the axis of rotation.

Swingers.

Attached. A reversible drum unit arranged to rotate or swing a derrick mast and boom or some other structure which supports a load lifting or lowering boom and which is attached to and receives its power from a hoist.

Independent. A swinger direct geared to its own power.

Rope. A swinger provided with one or two reversible drums for winding the rope used for rotating or swinging a bull wheel of a boom supporting structure.

Switch. A device for making, breaking, or for changing the connections in an electric circuit.

Tackle. An assembly of ropes and sheaves arranged for lifting, lowering, and pulling.

Tag Line. A light rope fastened to the load being hoisted and used for controlling the motion of the load.

Taking Up. The process of removing slack from a line or drawing (spooling) in on a line; loading a line by drawing in on it.

Tapering and Welding. Reducing the diameter of a wire rope at its end, and then welding the wires so as to facilitate reeving.

Telescoping. A process whereby the height of a traveling or freestanding tower crane is increased by adding sections at the top of the outer tower and then raising the inner tower. There are cranes that are telescoped by adding to the inner tower from below.

Thimble. An iron reinforcement, grooved on the outside for a rope eye.

Tinned Wire. Wire coated with tin.

Tipping Fulcrum. The horizontal line about which a crane or derrick will rotate should it overturn; the point(s) on which the entire weight of a crane or derrick will be imposed during tipping.

Tipping Load. The load for a particular operating radius that brings the crane or derrick to the point of incipient overturning.

Topping Lift. The block-and-tackle or wire rope arrangement located at the tip of a crane or derrick boom which is used to change the vertical angle of the boom.

Track. The structural member which supports the carrier or crane wheels.

Tramway. An aerial conveying system for transporting multiple loads.

Travel. Movement of a mobile or wheel-mounted crane about a jobsite under its own power.

Travel Base. The base mounting for a wheel-mounted (traveling) tower crane.

Travel Mechanism. The machinery involved in providing travel.

Tread Diameter. The diameter of a sheave or grooved rope drum measured at the base of the groove; the diameter of a smooth barrel on a rope drum.

Triple Block. A block with three sheaves.

Trolley. The unit which travels on the bridge rails and supports the load block.

Trolley Travel. The trolley movement.

Truck. The unit consisting of a frame, wheels, bearings, and axles which support the bridge girders, end ties of an overhead crane, or the sill of a gantry crane.

Turn. Same as a loop, but usually describes the placing of a rope around a specific object such as a post, rail, or ring with the running end continuing in a direction opposite to the standing part.

Turnbuckle. Device attached to wire rope for making limited adjustments in length. It consists of a barrel and right- and left-hand threaded bolts.

Turntable. A track device with a movable inner frame containing a straight section of track which can be rotated with a loaded carrier on it to align the section of track with other tracks for the transfer of carriers from one track to another.

Two Blocking. Excessive taking in of tackle rope, which causes the blocks to make contact with each other.

Upper Block. A fixed block located on a trolley which through a system of sheaves, bearings, pins, and frame, supports the load block and its load.

Upperstructure. On a mobile crane, the entire rotating structure less the front end attachment; also called *upper, superstructure,* or *revolving superstructure.*

Vangs (vang lines). Side lines reeved to a derrick boom and used to swing the boom.

Warrington. The name for a type of strand construction that is characterized by having one of its wire layers (usually the outer) made up of an arrangement of alternately large and small wires.

Weathervane. To swing with the wind when out of service so as to expose a minimal area to the wind.

Wedge Socket. Wire rope fittings wherein the rope end is secured by a wedge.

Whip Line. A secondary or auxiliary hoist line.

Whipping. A binding of light cord on the end of a rope to prevent it from untwisting.

Winch. A power source for hoisting or moving, usually consisting of a cable drum with a gear-reduction unit for hand operation or with a power drive.

Winch Head (gypsy head). A rotatable cylindrical shaped drum with curved end flanges, used for load handling by means of fiber rope coiled about its barrel with hand tension applied to the non-load end.

Wire Rope. A designed assembly of symmetrically arranged and helically wound strands. See **Core** and **Strand.**

Wire (round). A single, continuous length of metal, with a circular cross section, that is cold-drawn from rod.

Wire (shaped). A single, continuous length of metal with a Z- or H-shaped cross section that is either cold-drawn or cold-rolled from rod.

Wrap. One circumferential turn of wire rope around a rope-drum barrel.

Yoke. A frame on which a pair of load-carrying wheels are mounted.

APPENDIX

B

Safe Working Loads of Wire Rope

The following tables of safe working loads are provided for the most commonly used wire rope classifications. The tables are for reference purposes only. Check your rope manufacturer's ratings before determining the safe working loads as they may differ from the tables. If the manufacturer quotes his rope's capacity in terms of breaking strength, you must divide that figure by the factor of safety to get the safe working load (SWL).

$$\text{Max. SWL} = \frac{\text{Catalogue Breaking Strength}}{\text{Factor of Safety}}$$

Because of the difficulty in remembering the safe working loads of the most common wire ropes, the following rule of thumb should be used to compute the SWL in tons.

SWL = Rope Diameter × Rope Diameter × 8

Examples:

(a) 1/2 inch diameter rope.
SWL = 1/2 × 1/2 × 8 = 2 tons.
(b) 5/8 inch diameter rope.
SWL = 5/8 × 5/8 × 8 = 3.125 tons.
(c) 1 inch diameter rope.
SWL = 1 × 1 × 8 = 8 tons.

Table 1

6 × 7 Classification Round Strand Group: The ropes included in this group are composed of large wires, have limited flexibility, are resistant to abrasive wear, and are not suited to frequent bending.

Rope Diameter (inches)	Maximum Safe Working Load (tons) Safety Factor = 5			
	Grade 100/110 Plow		Grade 110/120 Improved Plow	
	Fiber Core	Steel Core	Fiber Core	Steel Core
1/4	.44	.47	.50	.54
5/16	.86	.92	.94	1.00
3/8	1.08	1.16	1.14	1.23
7/16	1.60	1.72	1.70	1.83
1/2	2.08	2.24	2.28	2.45
9/16	2.56	2.75	2.84	3.05
5/8	3.20	3.44	3.50	3.76
3/4	4.60	4.95	5.00	5.38
7/8	6.40	6.88	7.00	7.53
1	8.20	8.82	8.80	9.46
1 1/8	10.20	10.97	11.20	12.00
1 1/4	12.60	13.55	13.80	14.84
1 3/8	15.80	17.00	17.40	18.70
1 1/2	18.80	20.20	20.20	21.72

Table 2

6 × 19 Classification Round Strand Group: The ropes included in this group have good resistance to abrasion and better flexibility than the 6 × 7 group. This classification covers ropes with 6 strands having 8 to 26 wires per strand.

Rope Diameter (inches)	Maximum Safe Working Load (tons) Safety Factor = 5			
	Grade 100/110 Plow		Grade 110/120 Improved Plow	
	Fiber Core	Steel Core	Fiber Core	Steel Core
3/16	.26	.28	.30	.32
1/4	.48	.52	.54	.58
5/16	.76	.82	.82	.88
3/8	1.08	1.16	1.20	1.28
7/16	1.40	1.50	1.60	1.72
1/2	2.00	2.14	2.20	2.36
9/16	2.34	2.50	2.66	2.86
5/8	3.00	3.22	3.30	3.54
3/4	4.30	4.62	4.76	5.10
7/8	5.66	6.08	6.40	6.89
1	7.60	8.17	8.34	8.96
1 1/8	9.70	10.42	10.60	11.38
1 1/4	12.00	12.90	13.10	14.10
1 3/8	14.70	15.80	16.20	17.40
1 1/2	17.70	19.00	19.20	20.64
1 5/8	20.60	22.14	22.60	24.28
1 3/4	23.80	25.58	26.00	27.94
1 7/8	27.60	29.66	30.40	32.68
2	30.80	33.10	33.80	36.32
2 1/4	38.60	41.48	42.00	45.14
2 1/2	47.00	50.52	52.00	55.90
2 3/4	56.00	60.20	61.00	65.56

Table 3

6 × 25 Classification Flattened Strand Wire Rope: This classification covers ropes with 6 strands closed around a fiber or independent wire rope core and includes ropes having 8 to 12 outer wires per strand, 10 to 12 inner wires per strand with either a solid or built-up triangular center. Following values are for fiber core ropes; add 7-1/2% for wire rope core.

Rope Diameter (inches)	Maximum Safe Working Load (tons) Safety Factor = 5		
	Grade 110/120 Improved Plow	Grade 115/125 Special Improved Plow Type I	Grade 120/130 Special Improved Plow Type II
1/2	2.40	—	—
9/16	3.00	—	—
5/8	3.70	—	—
3/4	5.34	5.54	5.74
7/8	7.20	7.46	7.74
1	9.40	9.74	10.10
1 1/8	11.90	12.34	12.80
1 1/4	14.74	15.28	15.84
1 3/8	18.20	18.80	19.40
1 1/2	21.60	22.40	23.20
1 5/8	25.40	26.30	27.20
1 3/4	29.30	30.30	31.40
1 7/8	33.90	35.10	36.40
2	38.00	39.40	40.80

Table 4

6 × 37 Classification Round Strand Group: The ropes included in this group have excellent flexibility, reasonable resistance to crushing, and are well adapted to high speed and multiple reeving applications. This classification covers ropes with 6 strands having 27 to 49 wires per strand.

Rope Diameter (inches)	Maximum Safe Working Load (tons) Safety Factor = 5			
	Grade 100/110 Plow		Grade 110/120 Improved Plow	
	Fiber Core	Steel Core	Fiber Core	Steel Core
1/4	.44	.46	.50	.52
5/16	.76	.80	.80	.86
3/8	1.00	1.06	1.10	1.18
7/16	1.38	1.48	1.50	1.60
1/2	1.84	1.96	2.00	2.14
9/16	2.28	2.44	2.50	2.68
5/8	2.90	3.10	3.20	3.44
3/4	4.04	4.34	4.44	4.76
7/8	5.50	5.90	6.04	6.48
1	7.20	7.74	7.90	8.48
1 1/8	8.80	9.46	9.80	10.52
1 1/4	11.00	11.82	12.20	13.10
1 3/8	13.70	14.72	14.90	16.00
1 1/2	16.40	17.62	18.00	19.74
1 5/8	19.30	20.74	21.10	22.68
1 3/4	22.00	23.64	24.20	26.00
1 7/8	25.80	27.72	28.40	30.52
2	28.40	30.52	31.00	33.32
2 1/4	36.40	39.12	40.20	43.20
2 1/2	45.00	48.36	49.00	52.66
2 3/4	53.80	57.82	58.60	63.00

Table 5

8 × 19 Classification Round Strand Group: This classification covers ropes with 8 strands each having 8 to 26 wires per strand.

Rope Diameter (inches)	Maximum Safe Working Load (tons) Safety Factor = 5			
	Grade 100/110 Plow		Grade 110/120 Improved Plow	
	Fiber Core	Steel Core	Fiber Core	Steel Core
1/4	.40	.43	.44	.47
5/16	.62	.67	.68	.73
3/8	.92	.99	1.02	1.10
7/16	1.26	1.35	1.40	1.51
1/2	1.64	1.76	1.80	1.94
9/16	2.06	2.21	2.30	2.47
5/8	2.50	2.69	2.82	3.03
3/4	3.56	3.83	4.00	4.30
7/8	4.80	5.16	5.40	5.81
1	6.24	6.71	7.00	7.53
1 1/8	7.88	8.47	8.86	9.52
1 1/4	9.68	10.40	10.90	11.72
1 3/8	11.66	12.53	13.10	14.08
1 1/2	13.80	14.84	15.60	16.77

Table 6

18 × 7 Nonrotating Construction Round Strand Wire Rope: This construction consists of an outer rope of 12 strands laid Right Lay over an inner rope of 6 strands laid Left Lay around fiber core. All strands have 7 wires.

Rope Diameter (inches)	Maximum Safe Working Load (tons) Safety Factor = 5	
	Grade 100/110 Plow	Grade 110/120 Improved Plow
5/16	.69	.77
3/8	1.02	1.12
7/16	1.38	1.52
1/2	1.78	1.95
9/16	2.26	2.48
5/8	2.80	3.06
3/4	3.97	4.35
7/8	5.38	5.89
1	6.99	7.66
1 1/8	8.80	9.60
1 1/4	10.80	11.79
1 3/8	13.01	14.19
1 1/2	15.41	16.80

APPENDIX

C

Breaking Strengths of Wire Rope

The breaking strength is the ultimate load registered on a wire rope sample during a tension test.

The nominal strengths given (Tables 1 through 19), have been calculated by a standardized, industry-accepted procedure, and manufacturers design wire rope to these strengths. When making design calculations, it should be noted that the given figures are the static strengths. All discussion of strength is predicated on the assumption of there being a gradually applied load less than 1″/minute. Designers should base their calculations on these strengths.

A minimum acceptance strength, 2-1/2% lower than the published nominal breaking strengths, was established as the industry tolerance. It serves to offset testing variables that occur during the actual physical test of a wire rope sample. This tolerance is used in the basic wire rope governmental specifications.

Wire rope testing, whether it is performed for the purpose of determining grade or for adherence to specifications, requires the sample to be tested to meet certain standards. For example: the sample's length must not be less than 3′ (0.91 m) between sockets for ropes with diameters of from 1/8″ (3.2 mm) through 3″ (77 mm); on ropes with larger (over 3″) diameters, the clear length must be at least 20 times the rope diameter. The test is considered valid only if failure occurs 2″ (51 mm) or more from either of the sockets, or from the holding mechanism.

Table 1: 6 × 7 Classification/Bright (Uncoated), Fiber Core

Nominal Diameter		Approximate Mass		Nominal Strength*	
				Improved Plow Steel	
inches	mm	lb/ft	kg/m	tons	metric tons
1/4	6.5	0.09	0.14	2.64	2.4
5/16	8	0.15	0.22	4.10	3.72
3/8	9.5	0.21	0.31	5.86	5.32
7/16	11.5	0.29	0.43	7.93	7.2
1/2	13	0.38	0.57	10.3	9.35
9/16	14.5	0.48	0.71	13.0	11.8
5/8	16	0.59	0.88	15.9	14.4
3/4	19	0.84	1.25	22.7	20.6
7/8	22	1.15	1.71	30.7	27.9
1	26	1.50	2.23	39.7	36.0
1 1/8	29	1.90	2.83	49.8	45.2
1 1/4	32	2.34	3.48	61.0	55.3
1 3/8	35	2.82	4.23	73.1	66.3
1 1/2	38	3.38	5.03	86.2	78.2

*To convert to kilonewtons (kN), multiply tons (nominal breaking strength) by 8.896; 1 lb = 4.448 newtons (N).

Table 2: 6 × 7 Classification/Bright (Uncoated), IWRC

Nominal Diameter		Approximate Mass		Nominal Strength*	
				Improved Plow Steel	
inches	mm	lb/ft	kg/m	tons	metric tons
1/4	6.5	0.10	0.15	2.84	2.58
5/16	8	0.16	0.24	4.41	4.0
3/8	9.5	0.23	0.34	6.30	5.72
7/16	11.5	0.32	0.48	8.52	7.73
1/2	13	0.42	0.63	11.1	10.1
9/16	14.5	0.53	0.79	14.0	12.7
5/8	16	0.65	0.97	17.1	15.5
3/4	19	0.92	1.37	24.4	22.1
7/8	22	1.27	1.89	33.0	29.9
1	26	1.65	2.46	42.7	38.7
1 1/8	29	2.09	3.11	53.5	48.5
1 1/4	32	2.57	3.82	65.6	59.5
1 3/8	35	3.12	4.64	78.6	71.3
1 1/2	38	3.72	5.54	92.7	84.1

*To convert to kilonewtons (kN), multiply tons (nominal breaking strength) by 8.896; 1 lb = 4.448 newtons (N).

Table 3: 6 × 19 Classification/Bright (Uncoated), Fiber Core

Nominal Diameter		Approximate Mass		Nominal Strength*	
				Improved Plow Steel	
inches	mm	lb/ft	kg/m	tons	metric tons
1/4	6.5	0.11	0.16	2.74	2.49
5/16	8	0.16	0.24	4.26	3.86
3/8	9.5	0.24	0.35	6.10	5.53
7/16	11.5	0.32	0.48	8.27	7.50
1/2	13	0.42	0.63	10.7	9.71
9/16	14.5	0.53	0.79	13.5	12.2
5/8	16	0.66	0.98	16.7	15.1
3/4	19	0.95	1.41	23.8	21.6
7/8	22	1.29	1.92	32.2	29.2
1	26	1.68	2.5	41.8	37.9
1 1/8	29	2.13	3.17	52.6	47.7
1 1/4	32	2.63	3.91	64.6	58.6
1 3/8	35	3.18	4.73	77.7	70.5
1 1/2	38	3.78	5.63	92.0	83.5
1 5/8	42	4.44	6.61	107	97.1
1 3/4	45	5.15	7.66	124	112
1 7/8	48	5.91	8.8	141	128
2	51	6.72	10.0	160	145
2 1/8	54	7.59	11.3	179	162
2 1/4	57	8.51	12.7	200	181
2 3/8	61	9.48	14.1	222	201
2 1/2	64	10.5	15.6	244	221
2 5/8	67	11.6	17.3	268	243
2 3/4	70	12.7	18.9	292	265

*To convert to kilonewtons (kN), multiply tons (nominal breaking strength) by 8.896; 1 lb = 4.448 newtons (N).

Table 4: 6 × 19 Classification/Bright (Uncoated), IWRC

Nominal Diameter		Approximate Mass		Nominal Strength*			
				Improved Plow Steel		Extra Imp. Plow Steel	
inches	mm	lb/ft	kg/m	tons	metric tons	tons	metric tons
¼	6.5	0.12	0.17	2.94	2.67	3.40	3.08
5/16	8	0.18	0.27	4.58	4.16	5.27	4.78
⅜	9.5	0.26	0.39	6.56	5.95	7.55	6.85
7/16	11.5	0.35	0.52	8.89	8.07	10.2	9.25
½	13	0.46	0.68	11.5	10.4	13.3	12.1
9/16	14.5	0.59	0.88	14.5	13.2	16.8	15.2
⅝	16	0.72	1.07	17.7	16.2	20.6	18.7
¾	19	1.04	1.55	25.6	23.2	29.4	26.7
⅞	22	1.42	2.11	34.6	31.4	39.8	36.1
1	26	1.85	2.75	44.9	40.7	51.7	46.9
1⅛	29	2.34	3.48	56.5	51.3	65.0	59.0
1¼	32	2.89	4.30	69.4	63.0	79.9	72.5
1⅜	35	3.5	5.21	83.5	75.7	96.0	87.1
1½	38	4.16	6.19	98.9	89.7	114	103
1⅝	42	4.88	7.26	115	104	132	120
1¾	45	5.67	8.44	133	121	153	139
1⅞	48	6.5	9.67	152	138	174	158
2	51	7.39	11.0	172	156	198	180
2⅛	54	8.35	12.4	192	174	221	200
2¼	57	9.36	13.9	215	195	247	224
2⅜	61	10.4	15.5	239	217	274	249
2½	64	11.6	17.3	262	238	302	274
2⅝	67	12.8	19.0	288	261	331	300
2¾	70	14.0	20.8	314	285	361	327

*To convert to kilonewtons (kN), multiply tons (nominal breaking strength) by 8.896; 1 lb = 4.448 newtons (N).

Table 5: 6 × 37 Classification/Bright (Uncoated), Fiber Core

Nominal Diameter		Approximate Mass		Nominal Strength* Improved Plow Steel	
inches	mm	lb/ft	kg/m	tons	metric tons
1/4	6.5	0.11	0.16	2.74	2.49
5/16	8	0.16	0.24	4.26	3.86
3/8	9.5	0.24	0.35	6.10	5.53
7/16	11.5	0.32	0.48	8.27	7.50
1/2	13	0.42	0.63	10.7	9.71
9/16	14.5	0.53	0.79	13.5	12.2
5/8	16	0.66	0.98	16.7	15.1
3/4	19	0.95	1.41	23.8	21.6
7/8	22	1.29	1.92	32.2	29.2
1	26	1.68	2.50	41.8	37.9
1 1/8	29	2.13	3.17	52.6	47.7
1 1/4	32	2.63	3.91	64.6	58.6
1 3/8	35	3.18	4.73	77.7	70.5
1 1/2	38	3.78	5.63	92.0	83.5
1 5/8	42	4.44	6.61	107	97.1
1 3/4	45	5.15	7.66	124	112
1 7/8	48	5.91	8.8	141	128
2	51	6.72	10.0	160	145
2 1/8	54	7.59	11.3	179	162
2 1/4	57	8.51	12.7	200	181
2 3/8	61	9.48	14.1	222	201
2 1/2	64	10.5	15.6	244	221
2 5/8	67	11.6	17.3	268	243
2 3/4	70	12.7	18.9	292	265
2 7/8	74	13.9	20.7	317	287
3	77	15.1	22.5	344	312
3 1/8	80	16.4	24.4	371	336
3 1/4	83	17.7	26.3	399	362

*To convert to kilonewtons (kN), multiply tons (nominal breaking strength) by 8.896; 1 lb. = 4.448 newtons (N).

Table 6: 6 × 37 Classification/Bright (Uncoated), IWRC

Nominal Diameter		Approximate Mass		Nominal Strength*			
				Improved Plow Steel		Extra Imp. Plow Steel	
inches	mm	lb/ft	kg/m	tons	metric tons	tons	metric tons
1/4	6.5	0.12	0.17	2.94	2.67	3.4	3.08
5/16	8	0.18	0.27	4.58	4.16	5.27	4.78
3/8	9.5	0.26	0.39	6.56	5.95	7.55	6.85
7/16	11.5	0.35	0.52	8.89	8.07	10.2	9.25
1/2	13	0.46	0.68	11.5	10.4	13.3	12.1
9/16	14.5	0.59	0.88	14.5	13.2	16.8	15.2
5/8	16	0.72	1.07	17.9	16.2	20.6	18.7
3/4	19	1.04	1.55	25.6	23.2	29.4	26.7
7/8	22	1.42	2.11	34.6	31.4	39.5	35.9
1	26	1.85	2.75	44.9	40.7	51.7	46.9
1 1/8	29	2.34	3.48	56.5	51.3	65.0	59.0
1 1/4	32	2.89	4.30	69.4	63.0	79.9	72.5
1 3/8	35	3.50	5.21	83.5	75.7	96.0	87.1
1 1/2	38	4.16	6.19	98.9	89.7	114	103
1 5/8	42	4.88	7.26	115	104	132	120
1 3/4	45	5.67	8.44	133	121	153	139
1 7/8	48	6.5	9.67	152	138	174	158
2	51	7.39	11.0	172	156	198	180
2 1/8	54	8.35	12.4	192	174	221	200
2 1/4	57	9.36	13.9	215	195	247	224
2 3/8	61	10.4	15.5	239	217	274	249
2 1/2	64	11.6	17.3	262	238	302	274
2 5/8	67	12.8	19.0	288	261	331	300
2 3/4	70	14.0	20.8	314	285	361	327
2 7/8	74	15.3	22.8	341	309	392	356
3	77	16.6	24.7	370	336	425	386
3 1/8	80	18.0	26.8	399	362	458	415
3 1/4	83	19.5	29.0	429	389	492	446
3 3/8	86	21.0	31.3	459	416	529	480
3 1/2	90	22.7	33.8	491	445	564	512
3 5/8	96	24.3	36.2	523	458	602	528
3 3/4	103	26.0	38.7	557	505	641	581

*To convert to kilonewtons (kN), multiply tons (nominal breaking strength) by 8.896; 1 lb = 4.448 newtons (N).

Table 7: 6 × 61 Classification/Bright (Uncoated), Fiber Core

Nominal Diameter		Approximate Mass		Nominal Strength* Improved Plow Steel	
inches	mm	lb/ft	kg/m	tons	metric tons
1	26	1.68	2.5	39.8	36.1
1⅛	29	2.13	3.17	50.1	45.4
1¼	32	2.63	3.91	61.5	55.8
1⅜	35	3.18	4.73	74.1	67.2
1½	38	3.78	5.63	87.9	79.7
1⅝	42	4.44	6.61	103	93.4
1¾	45	5.15	7.66	119	108
1⅞	48	5.91	8.80	136	123
2	51	6.77	10.1	154	140
2⅛	54	7.59	11.3	173	157
2¼	57	8.51	12.7	193	175
2⅜	61	9.48	14.1	214	194
2½	64	10.5	15.6	236	214
2⅝	67	11.6	17.3	260	236
2¾	70	12.7	18.9	284	258
2⅞	74	13.9	20.7	309	280
3	77	15.1	22.5	335	304
3¼	83	17.7	26.3	390	354
3⅜	86	19.1	28.4	419	380
3½	90	20.6	30.7	449	407
3¾	96	23.6	35.1	511	464
4	103	26.9	40.0	577	523
4¼	109	30.3	45.1	646	586
4½	115	34.0	50.6	719	652
4¾	122	37.9	56.4	794	720
5	128	42.0	62.5	872	791

*To convert to kilonewtons (kN), multiply tons (nominal breaking strength) by 8.896; 1 lb = 4.448 newtons (N).

Table 8: 6 × 61 Classification/Bright (Uncoated), IWRC

Nominal Diameter		Approximate Mass		Nominal Strength*			
				Improved Plow Steel		Extra Imp. Plow Steel	
inches	mm	lb/ft	kg/m	tons	metric tons	tons	metric tons
1	26	1.85	2.75	42.8	38.8	49.1	44.5
1⅛	29	2.34	3.48	53.9	48.9	61.9	56.2
1¼	32	2.89	4.30	66.1	60.0	76.1	69.0
1⅜	35	3.50	6.59	79.7	72.3	91.7	83.2
1½	38	4.16	6.19	94.5	85.7	109	98.9
1⅝	42	4.88	7.62	111	101	127	115
1¾	45	5.67	8.44	128	116	146	132
1⅞	48	6.50	9.67	146	132	168	152
2	51	7.39	11.0	165	150	190	172
2⅛	54	8.35	12.4	186	169	214	194
2¼	57	9.36	13.9	207	188	239	217
2⅜	61	10.40	15.5	230	209	264	240
2½	64	11.6	17.3	254	230	292	265
2⅝	67	12.8	18.3	279	253	321	291
2¾	70	14.	20.8	305	277	350	318
2⅞	74	15.3	22.8	333	302	382	347
3	77	16.6	24.7	360	327	414	376
3¼	83	19.5	29.0	419	380	483	438
3⅜	86	21.0	31.3	451	409	518	470
3½	90	22.7	33.8	483	438	555	503
3¾	96	26.0	38.7	549	498	632	573
4	103	29.6	44.1	620	562	713	647
4¼	109	33.3	49.6	694	630	799	725
4½	115	37.4	55.7	772	700	888	806
4¾	122	41.7	62.1	853	774	981	890
5	128	46.2	68.8	937	850	1078	978

*To convert to kilonewtons (kN), multiply tons (nominal breaking strength) by 8.896; 1 lb = 4.448 newtons (N).

Table 9: 6 × 91 Classification/Bright (Uncoated), Fiber Core

Nominal Diameter		Approximate Mass		Nominal Strength* Improved Plow Steel	
inches	mm	lb/ft	kg/m	tons	metric tons
2	51	6.77	10.1	146	132
2⅛	54	7.59	11.3	164	149
2¼	57	8.51	12.7	183	166
2⅜	61	9.48	14.1	203	184
2½	64	10.5	15.6	225	204
2⅝	67	11.6	17.3	247	224
2¾	70	12.7	18.9	270	245
3	77	15.1	22.5	318	288
3¼	83	17.7	26.3	371	337
3½	90	20.6	30.7	426	386

*To convert to kilonewtons (kN), multiply tons (nominal breaking strength) by 8.896; 1 lb = 4.448 newtons (N).

Table 10: 6 × 91 Classification/Bright (Uncoated), IWRC

Nominal Diameter		Approximate Mass		Nominal Strength*			
				Improved Plow Steel		Extra Imp. Plow Steel	
inches	mm	lb/ft	kg/m	tons	metric tons	tons	metric tons
2	51	7.39	11.0	157	142	181	164
2⅛	54	8.35	12.4	176	160	203	184
2¼	57	9.36	14.0	197	179	227	206
2⅜	61	10.4	15.5	218	198	251	228
2½	64	11.6	17.3	242	220	277	251
2⅝	67	12.8	19.0	265	240	305	277
2¾	70	14.0	20.8	290	263	333	302
3	77	16.6	24.7	342	310	393	357
3¼	83	19.5	29.0	399	362	458	415
3½	90	22.7	33.8	458	415	527	478
3¾	96	26.0	38.7	522	474	600	544
4	103	29.6	44.1	589	534	677	614
4¼	109	33.3	49.6	660	599	759	689
4½	115	37.4	55.7	734	666	844	766
4¾	122	41.7	62.1	810	787	932	846
5	128	46.2	68.7	891	808	1024	929
5¼	134	49.8	74.1	974	884	1120	1016
5½	140	54.5	81.1	1060	962	1219	1106
5¾	146	59.6	88.7	1148	1041	1320	1198
6	153	65.0	96.7	1240	1125	1426	1294

*To convert to kilonewtons (kN), multiply tons (nominal breaking strength) by 8.896; 1 lb = 4.448 newtons (N).

Table 11: 6 × 30, 6 × 30G, 6 × 25B, & 6 × 27H Flattened Strand/Fiber Core

Nominal Diameter		Approximate Mass		Nominal Strength* Improved Plow Steel	
inches	mm	lb/ft	kg/m	tons	metric tons
½	13	0.45	0.67	11.8	10.8
9⁄16	14.5	0.57	0.85	14.9	13.5
⅝	16	0.70	1.04	18.3	16.6
¾	19	1.01	1.50	26.2	23.8
⅞	22	1.39	2.07	35.4	32.1
1	26	1.80	2.68	46.0	41.7
1⅛	29	2.28	3.39	57.9	52.5
1¼	32	2.81	4.18	71.0	64.4
1⅜	35	3.40	5.06	85.5	77.6
1½	38	4.05	6.03	101	91.6
1⅝	42	4.75	7.07	118	107
1¾	45	5.51	8.20	136	123
1⅞	48	6.33	9.42	155	141
2	51	7.20	10.70	176	160

*To convert to kilonewtons (kN), multiply tons (nominal breaking strength) by 8.896; 1 lb = 4.448 newtons (N).

Table 12: 6 × 30, 6 × 30G, 6 × 25B, 6 × 27H Flattened Strand/IWRC

Nominal Diameter		Approximate Mass		Nominal Strength*			
				Improved Plow Steel		Extra Imp. Plow Steel	
inches	mm	lb/ft	kg/m	tons	metric tons	tons	metric tons
½	13	0.47	0.70	12.6	11.4	14	12.7
9/16	14.5	0.60	0.89	16.0	14.5	17.6	16
5/8	16	0.73	1.09	19.6	17.8	21.7	19.7
¾	19	1.06	1.58	28.1	25.5	31	28.1
7/8	22	1.46	2.17	38.0	34.5	41.9	38
1	26	1.89	2.83	49.4	44.8	54.4	49.4
1 1/8	29	2.39	3.56	62.2	56.4	68.5	62.1
1¼	32	2.95	4.39	76.3	69.2	84	76.2
1 3/8	35	3.57	5.31	91.9	83.4	101	91.6
1½	38	4.25	6.32	108	98	119	108
1 5/8	42	4.98	7.41	127	115	140	127
1¾	45	5.78	8.60	146	132	161	146
1 7/8	48	6.65	9.90	167	152	184	167
2	51	7.56	11.3	189	171	207	188

*To convert to kilonewtons (kN), multiply tons (nominal breaking strength) by 8.896; 1 lb = 4.448 newtons (N).

Table 13: 8 × 19 Classification/Bright (Uncoated), Fiber Core

Nominal Diameter		Approximate Mass		Nominal Strength* Improved Plow Steel	
inches	mm	lb/ft	kg/m	tons	metric tons
¼	6.5	0.10	0.15	2.35	2.13
5⁄16	8	0.15	0.22	3.65	3.31
⅜	9.5	0.22	0.33	5.24	4.75
7⁄16	11.5	0.30	0.45	7.09	6.43
½	13	0.39	0.58	9.23	8.37
9⁄16	14.5	0.50	0.74	11.6	10.5
⅝	16	0.61	0.91	14.3	13.0
¾	19	0.88	1.31	20.5	18.6
⅞	22	1.20	1.79	27.7	25.1
1	26	1.57	2.34	36.0	32.7
1⅛	29	1.99	2.96	45.3	41.1
1¼	32	2.45	3.65	55.7	50.5
1⅜	35	2.97	4.42	67.1	60.7
1½	38	3.53	5.25	79.4	72.0

*To convert to kilonewtons (kN), multiply tons (nominal breaking strength) by 8.896; 1 lb = 4.448 newtons (N).

Table 14: 8 × 19 Classification/Bright (Uncoated), IWRC

Nominal Diameter		Approximate Mass		Nominal Strength*			
				Improved Plow Steel		Extra Imp. Plow Steel	
inches	mm	lb/ft	kg/m	tons	metric tons	tons	metric tons
½	13	0.47	0.70	10.1	9.16	11.6	10.5
9/16	14.5	0.60	0.89	12.8	11.6	14.7	13.3
⅝	16	0.73	1.09	15.7	14.2	18.1	16.4
¾	19	1.06	1.58	22.5	20.4	25.9	23.5
⅞	22	1.44	2.14	30.5	27.7	35.0	31.8
1	26	1.88	2.80	39.6	35.9	45.5	41.3
1⅛	29	2.39	3.56	49.8	45.2	57.3	51.7

*To convert to kilonewtons (kN), multiply tons (nominal breaking strength) by 8.896; 1 lb = 4.448 newtons (N).

Table 15: 6 × 7 Classification/Galvanized, Fiber Core*

Nominal Diameter		Approximate Mass		Nominal Strength** Improved Plow Steel	
inches	mm	lb/ft	kg/m	tons	metric tons
1/4	6.5	0.09	0.14	2.38	2.16
5/16	8	0.15	0.22	3.69	3.35
3/8	9.5	0.21	0.31	5.27	4.78
7/16	11.5	0.29	0.43	7.14	6.48
1/2	13	0.38	0.57	9.27	8.41
9/16	14.5	0.48	0.71	11.7	10.6
5/8	16	0.59	0.88	14.3	13.0
3/4	19	0.84	1.25	20.4	18.5
7/8	22	1.15	1.71	27.6	25.0
1	26	1.50	2.23	35.7	32.4
1 1/8	29	1.90	2.83	44.8	40.6
1 1/4	32	2.34	3.48	54.9	49.8
1 3/8	35	2.84	4.23	65.8	59.7
1 1/2	38	3.38	5.03	77.6	70.4

*For ropes with an IWRC, add 7-1/2% to their respective nominal strengths and 10% to their approximate mass (weights). Fiber cores consist either of polypropylene or natural fiber.

**To convert to kilonewtons (kN), multiply tons (nominal breaking strength) by 8.896; 1 lb = 4.448 newtons (N).

Table 16: 6 × 12 Construction/Galvanized, Fiber Core*

Nominal Diameter		Approximate Mass		Nominal Strength** Improved Plow Steel	
inches	mm	lb/ft	kg/m	tons	metric tons
5/16	8	0.10	0.14	2.34	2.12
3/8	9.5	0.15	0.22	3.36	3.05
7/16	11.5	0.20	0.30	4.55	4.13
1/2	13	0.26	0.39	5.91	4.71
9/16	14.5	0.33	0.49	7.45	6.76
5/8	16	0.41	0.61	9.16	8.31
3/4	19	0.59	0.88	13.1	11.9
13/16	21	0.69	1.03	15.3	13.9
7/8	22	0.80	1.19	17.7	16.1
1	26	1.05	1.56	23.0	20.9

*Fiber cores consist either of polypropylene or natural fiber.

**To convert to kilonewtons (kN), multiply tons (nominal breaking strength) by 8.896; 1 lb = 4.448 newtons (N).

Table 17: 18 × 7 Construction/Bright (Uncoated) or Drawn-Galvanized Wire, Fiber Core

Nominal Diameter		Approximate Mass		Nominal Strength*			
				Improved Plow Steel		Extra Imp. Plow Steel	
inches	mm	lb/ft	kg/m	tons	metric tons	tons	metric tons
½	13	0.43	0.64	9.85	8.94	10.8	9.8
9⁄16	14.5	0.55	0.82	12.4	11.2	13.6	12.3
⅝	16	0.68	1.01	15.3	18.9	16.8	15.2
¾	19	0.97	1.44	21.8	19.8	24.0	21.8
⅞	22	1.32	1.96	29.5	26.8	32.5	29.5
1	26	1.73	2.57	38.3	34.7	42.2	38.3
1⅛	29	2.19	3.26	48.2	43.7	53.1	48.2
1¼	32	2.70	4.02	59.2	53.7	65.1	59.1
1⅜	35	3.27	4.87	71.3	64.7	78.4	71.1
1½	38	3.89	5.79	84.4	76.6	92.8	84.2

*To convert to kilonewtons (kN), multiply tons (nominal breaking strength) by 8.896; 1 lb = 4.448 newtons (N).

APPENDIX

D

Fiber and Wire Rope Sling Capacities

Table 1: Manila Rope Slings (Spliced Eyes in Both Ends)						
Rope Diameter (Inches)	Maximum Safe Working Loads—Pounds (Safety Factor = 5)					
	Single Vertical Hitch 	Single Choker Hitch 	Single Basket Hitch (Vertical Legs) 	2-Leg Bridle Hitch & Single Basket Hitch with Legs Inclined  		
				60°	45°	30°
3/16	100	75	200	170	140	100
1/4	120	90	240	210	170	120
5/16	200	150	400	350	280	200
3/8	270	200	540	470	380	270
1/2	530	400	1,060	920	750	530
9/16	680	500	1,360	1,180	960	680
5/8	880	660	1,760	1,520	1,240	880
3/4	1,080	800	2,160	1,870	1,530	1,080
13/16	1,300	980	2,600	2,250	1,840	1,300
7/8	1,540	1,150	3,080	2,670	2,180	1,540
1	1,800	1,350	3,600	3,100	2,550	1,800
1 1/16	2,100	1,580	4,200	3,600	2,970	2,100
1 1/8	2,400	1,800	4,800	4,150	3,400	2,400
1 1/4	2,700	2,000	5,400	4,670	3,800	2,700
1 5/16	3,000	2,250	6,000	5,200	4,200	3,000
1 1/2	3,700	2,800	7,400	6,400	5,250	3,700
1 5/8	4,500	3,400	9,000	7,800	6,350	4,500
1 3/4	5,300	4,000	10,600	9,200	7,500	5,300
2	6,200	4,650	12,400	10,700	8,800	6,200
2 1/8	7,200	5,400	14,400	12,500	10,200	7,200
2 1/4	8,200	6,150	16,400	14,200	11,600	8,200
2 1/2	9,300	7,000	18,600	16,100	13,200	9,300
2 5/8	10,400	7,800	20,800	18,000	14,700	10,400

Note: If used with Choker Hitch multiply above values by 3/4, and for Double Basket Hitch multiply by 2. For safe working loads of Endless or Grommet Slings, multiply above values by 2.

Table 2: Polypropylene Rope Slings (Spliced Eyes in Both Ends)

Rope Diameter (Inches)	Maximum Safe Working Loads—Pounds (Safety Factor = 5)					
	Single Vertical Hitch 	Single Choker Hitch 	Single Basket Hitch (Vertical Legs) 	2-Leg Bridle Hitch & Single Basket Hitch with Legs Inclined  		
				60°	45°	30°
3/16	150	110	300	260	210	150
1/4	250	190	500	430	350	250
5/16	400	300	800	700	560	400
3/8	500	375	1,000	860	700	500
1/2	830	620	1,660	1,400	1,200	830
9/16	960	720	1,920	1,700	1,350	960
5/8	1,300	975	2,600	2,250	1,800	1,300
3/4	1,700	1,275	3,400	2,900	2,400	1,700
13/16	1,900	1,425	3,800	3,300	2,700	1,900
7/8	2,200	1,650	4,400	3,800	3,100	2,200
1	2,900	2,175	5,800	5,000	4,100	2,900
1 1/16	3,000	2,250	6,000	5,200	4,200	3,000
1 1/8	3,750	2,800	7,500	6,500	5,300	3,750
1 1/4	4,200	3,150	8,400	7,300	5,900	4,200
1 5/16	4,400	3,300	8,800	7,600	6,200	4,400
1 1/2	6,000	4,500	12,000	10,400	8,500	6,000
1 5/8	7,300	5,500	14,600	12,600	10,300	7,300
1 3/4	8,700	6,500	17,400	15,100	12,300	8,700
2	10,400	7,800	20,800	18,000	14,700	10,400
2 1/8	11,500	8,600	23,000	19,900	16,300	11,500
2 1/4	13,200	9,900	26,400	22,900	18,700	13,200
2 1/2	15,100	11,300	30,200	26,200	21,400	15,100
2 5/8	17,000	12,750	34,000	29,400	24,000	17,000

If used with Choker Hitch multiply above values by 3/4.

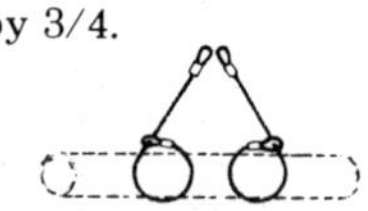

For Double Basket Hitch multiply above values by 2.

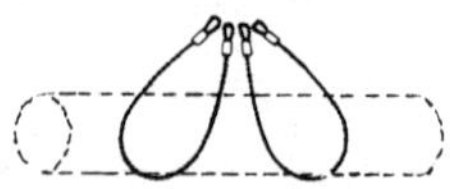

Note: For safe working loads of Endless or Grommet Slings, multiply above values by 2.

Table 3: Nylon Rope Slings (Spliced Eyes in Both Ends)

Rope Diameter (Inches)	Maximum Safe Working Loads—Pounds (Safety Factor = 5)					
	Single Vertical Hitch 	Single Choker Hitch 	Single Basket Hitch (Vertical Legs) 	2-Leg Bridle Hitch & Single Basket Hitch with Legs Inclined  		
				60°	45°	30°
3/16	200	150	400	350	280	200
1/4	300	225	600	520	420	300
5/16	500	375	1,000	870	700	500
3/8	700	525	1,400	1,200	1,000	700
1/2	1,250	940	2,500	2,200	1,770	1,250
9/16	1,500	1,125	3,000	2,600	2,100	1,500
5/8	2,000	1,500	4,000	3,500	2,800	2,000
3/4	2,800	2,100	5,600	4,850	4,000	2,800
13/16	3,200	2,400	6,600	5,500	4,500	3,200
7/8	3,800	2,850	7,600	6,600	5,400	3,800
1	4,800	3,600	9,600	8,300	6,800	4,800
1 1/16	5,500	4,125	11,000	9,500	7,800	5,500
1 1/8	6,300	4,725	12,600	10,900	8,900	6,300
1 1/4	7,200	5,400	14,400	12,500	10,200	7,200
1 5/16	8,200	6,150	16,400	14,200	11,600	8,200
1 1/2	10,200	7,650	20,400	17,700	14,400	10,200
1 5/8	12,400	9,300	24,800	21,500	17,500	12,400
1 3/4	15,000	11,250	30,000	26,000	21,200	15,000
2	17,900	13,400	35,800	31,000	25,300	17,900
2 1/8	20,200	15,150	40,400	35,000	28,600	20,200
2 1/4	23,800	17,850	47,600	41,200	33,700	23,800
2 1/2	26,600	20,000	53,200	46,100	37,600	26,600
2 5/8	30,700	23,000	61,400	53,200	43,400	30,700

If used with Choker Hitch multiply above values by 3/4.

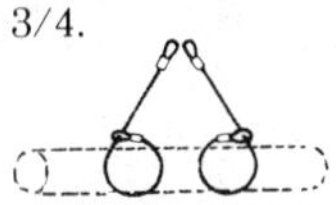

For Double Basket Hitch multiply above values by 2.

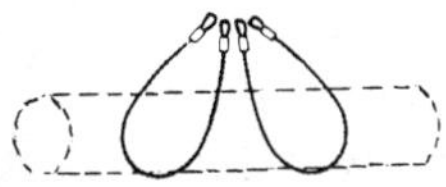

Note: For safe working loads of Endless or Grommet Slings, multiply above values by 2.

Table 4: Metal (Wire or Chain) Mesh Slings

Sling Width (Inches)	Maximum Safe Working Loads—Pounds (Safety Factor = 5)					
	Single Vertical Hitch 	Single Choker Hitch 	Single Basket Hitch (Vertical Legs) 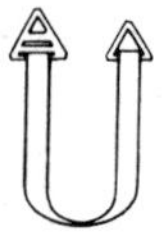	2-Leg Bridle Hitch & Single Basket Hitch Nonvertical Legs 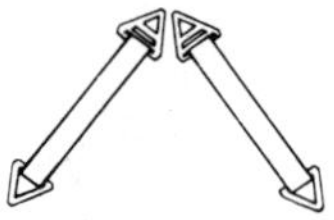 		
				60°	45°	30°
Heavy Duty Classification (10 gauge mesh)						
2	1,500	1,100	3,000	2,600	2,100	1,500
3	2,700	2,000	5,400	4,700	3,800	2,700
4	4,000	3,000	8,000	6,900	5,600	4,000
6	6,000	4,500	12,000	10,400	8,500	6,000
8	8,000	6,000	16,000	13,800	11,300	8,000
10	10,000	7,500	20,000	17,300	14,100	10,000
12	12,000	9,000	24,000	20,800	17,000	12,000
Medium Duty Classification (12 gauge mesh)						
2	1,350	1,000	2,700	2,300	1,900	1,350
3	2,000	1,500	4,000	3,500	2,800	2,000
4	2,700	2,000	5,400	4,700	3,800	2,700
6	4,500	3,400	9,000	7,800	6,400	4,500
8	6,000	4,500	12,000	10,400	8,500	6,000
10	7,500	5,600	15,000	13,000	10,600	7,500
12	9,000	6,750	18,000	15,600	12,700	9,000
Light Duty Classification (14 gauge mesh)						
2	900	700	1,800	1,600	1,300	900
3	1,400	1,000	2,800	2,400	2,000	1,400
4	2,000	1,500	4,000	3,500	2,800	2,000
6	3,000	2,250	6,000	5,200	4,200	3,000
8	4,000	3,000	8,000	6,900	5,700	4,000
10	5,000	3,750	10,000	8,600	7,100	5,000
12	6,000	4,500	12,000	10,400	8,500	6,000
				If used with Choker Hitch multiply above values by 3/4. 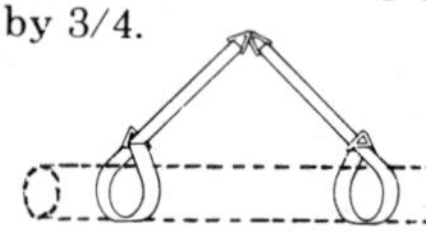 For Double Basket Hitch multiply above values by 2. 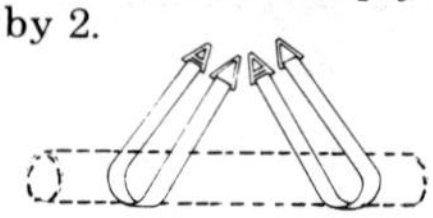		

Table 5: Wire Rope Slings
(6 × 19 Classification Group, Improved Plow Steel, Fiber Core)

Maximum Safe Working Loads—Pounds (Safety Factor = 5)

Rope Diameter (Inches)	Single Vertical Hitch 	Single Choker Hitch 	Single Basket Hitch (Vertical Legs) 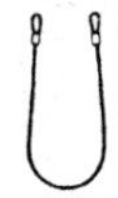	2-Leg Bridle Hitch & Single Basket Hitch with Legs Inclined   60°	45°	30°
3/16	600	450	1,200	1,050	850	600
1/4	1,100	825	2,200	1,900	1,550	1,100
5/16	1,650	1,250	3,300	2,850	2,350	1,650
3/8	2,400	1,800	4,800	4,150	3,400	2,400
7/16	3,200	2,400	6,400	5,550	4,500	3,200
1/2	4,400	3,300	8,800	7,600	6,200	4,400
9/16	5,300	4,000	10,600	9,200	7,500	5,300
5/8	6,600	4,950	13,200	11,400	9,350	6,600
3/4	9,500	7,100	19,000	16,500	13,400	9,500
7/8	12,800	9,600	25,600	22,200	18,100	12,800
1	16,700	12,500	33,400	28,900	23,600	16,700
1 1/8	21,200	15,900	42,400	36,700	30,000	21,200
1 1/4	26,200	19,700	52,400	45,400	37,000	26,200
1 3/8	32,400	24,300	64,800	56,100	45,800	32,400
1 1/2	38,400	28,800	76,800	66,500	54,300	38,400
1 5/8	45,200	33,900	90,400	78,300	63,900	45,200
1 3/4	52,000	39,000	104,000	90,000	73,500	52,000
1 7/8	60,800	45,600	121,600	105,300	86,000	60,800
2	67,600	50,700	135,200	117,100	95,600	67,600
2 1/4	84,000	63,000	168,000	145,500	118,800	84,000
2 1/2	104,000	78,000	208,000	180,100	147,000	104,000
2 3/4	122,000	91,500	244,000	211,300	172,500	122,000

If used with Choker Hitch multiply above values by 3/4.

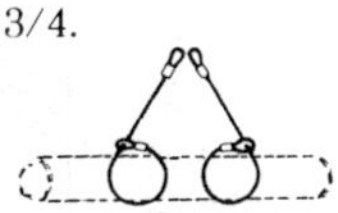

For Double Basket Hitch multiply above values by 2.

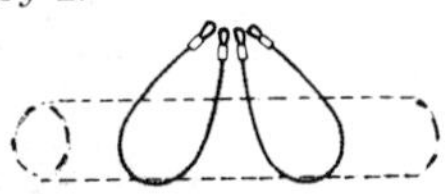

Note: Table values are for slings with eyes and thimbles in both ends, Flemish Spliced Eyes and mechanical sleeves.

Hand tucked spliced eyes—reduce loads according to *Table 1-10: Efficiencies of Tucked or Hand Spliced Eyes.*

Eyes formed by cable clips—reduce loads by 20%.

Table 6: Wire Rope Slings
(6 × 37 Classification Group, Improved Plow Steel, Fiber Core)

Rope Diameter (Inches)	Maximum Safe Working Loads—Pounds (Safety Factor = 5)					
	Single Vertical Hitch	Single Choker Hitch	Single Basket Hitch (Vertical Legs)	2-Leg Bridle Hitch & Single Basket Hitch with Legs Inclined		
				60°	45°	30°
1/4	1,000	750	2,000	1,750	1,400	1,000
5/16	1,600	1,200	3,200	2,750	2,250	1,600
3/8	2,200	1,650	4,400	3,800	3,100	2,200
7/16	3,000	2,250	6,000	5,200	4,250	3,000
1/2	4,000	3,000	8,000	6,900	5,650	4,000
9/16	5,000	3,750	10,000	8,650	7,100	5,000
5/8	6,400	4,800	12,800	11,100	9,050	6,400
3/4	8,900	6,700	17,800	15,400	12,600	8,900
7/8	12,100	9,100	24,200	21,000	17,100	12,100
1	15,800	11,900	31,600	27,400	22,300	15,800
1 1/8	19,600	14,700	39,200	33,900	27,700	19,600
1 1/4	24,400	18,300	48,800	42,300	34,500	24,400
1 3/8	29,800	22,400	59,600	51,600	42,100	29,800
1 1/2	36,000	27,000	72,000	62,400	50,900	36,000
1 5/8	42,200	31,700	84,400	73,100	59,700	42,200
1 3/4	48,400	36,300	96,800	83,800	68,400	48,400
1 7/8	56,800	42,600	113,600	98,400	80,300	56,800
2	62,000	46,500	124,000	107,400	87,700	62,000
2 1/4	80,400	60,300	160,800	139,300	113,700	80,400
2 1/2	98,000	73,500	196,000	169,700	138,600	98,000
2 3/4	117,200	87,900	234,400	203,000	165,700	117,200

If used with Choker Hitch multiply above values by 3/4.

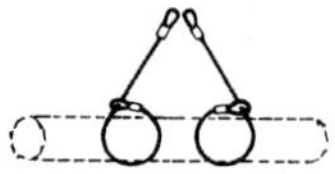

For Double Basket Hitch multiply above values by 2.

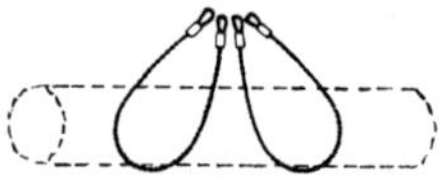

Note: Table values are for slings with eyes and thimbles in both ends, Flemish Spliced Eyes and mechanical sleeves.

Hand tucked spliced eyes—reduce loads according to *Table 1-10: Efficiencies of Tucked or Hand Spliced Eyes.*

Eyes formed by cable clips—reduce loads by 20%.

Table 7: Braided Wire Rope Slings
8-Part Braided Rope
(6 × 19 Classification Group, Improved Plow Steel)

Component Rope Diameter (Inches)	Maximum Safe Working Loads—Pounds (Safety Factor = 5)					
	Single Vertical Hitch 	Single Choker Hitch 	Single Basket Hitch (Vertical Legs) 	2-Leg Bridle Hitch & Single Basket Hitch with Legs Inclined  		
				60°	45°	30°
3/16	3,400	2,550	6,800	5,900	4,800	3,400
1/4	6,200	4,650	12,400	10,700	8,750	6,200
5/16	9,600	7,200	19,200	16,600	13,600	9,600
3/8	13,600	10,200	27,200	23,600	19,200	13,600
7/16	18,600	14,000	37,200	32,200	26,300	18,600
1/2	24,000	18,000	48,000	41,600	33,900	24,000
9/16	30,000	22,500	60,000	52,000	42,400	30,000
5/8	38,000	28,500	76,000	65,800	53,700	38,000
3/4	54,000	40,500	108,000	93,500	76,400	54,000
7/8	72,000	54,000	144,000	124,700	101,800	72,000
1	94,000	70,500	188,000	162,800	132,900	94,000

If used with Choker Hitch multiply above values by 3/4.

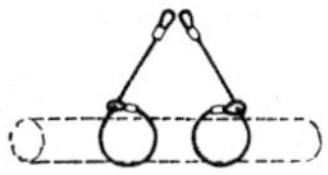

For Double Basket Hitch multiply above values by 2.

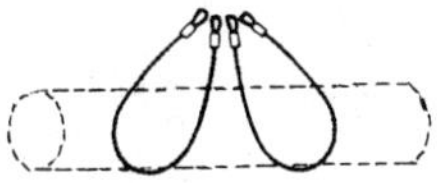

Note: Table values are for braided slings with eyes and thimbles in both ends. Eyes formed by hand tucking ropes and securing with mechanical sleeves.

Table 8: Wire Rope Slings
(6 × 19 Classification Group, Improved Plow Steel, IWRC)

Maximum Safe Working Loads—Pounds (Safety Factor = 5)

Rope Diameter (Inches)	Single Vertical Hitch 	Single Choker Hitch 	Single Basket Hitch (Vertical Legs) 	2-Leg Bridle Hitch & Single Basket Hitch with Legs Inclined  		
				60°	45°	30°
3/16	650	480	1,300	1,100	900	650
1/4	1,150	860	2,300	2,000	1,600	1,150
5/16	1,750	1,300	3,500	3,000	2,500	1,750
3/8	2,550	1,900	5,100	4,400	3,600	2,550
7/16	3,450	2,600	6,900	6,000	4,900	3,450
1/2	4,700	3,500	9,400	8,150	6,650	4,700
9/16	5,700	4,200	11,400	9,900	8,050	5,700
5/8	7,100	5,300	14,200	12,300	10,000	7,100
3/4	10,200	7,650	20,400	17,700	14,400	10,200
7/8	13,750	10,300	27,500	23,800	19,400	13,750
1	17,950	13,450	35,900	31,100	25,400	17,950
1 1/8	22,750	17,000	45,500	39,400	32,200	22,750
1 1/4	28,200	21,200	56,400	48,800	39,900	28,200
1 3/8	34,800	26,100	69,600	60,300	49,200	34,800
1 1/2	41,300	31,000	82,600	71,500	58,400	41,300
1 5/8	48,600	36,400	97,200	84,200	68,700	48,600
1 3/4	55,900	41,900	111,800	96,800	79,000	55,900
1 7/8	65,400	49,000	130,800	113,300	92,500	65,400
2	72,600	54,500	145,200	125,700	102,700	72,600
2 1/4	90,300	67,600	180,600	156,400	127,700	90,300
2 1/2	111,800	83,700	223,600	193,600	158,100	111,800
2 3/4	131,100	98,200	262,200	227,000	185,400	131,100

If used with Choker Hitch multiply above values by 3/4.

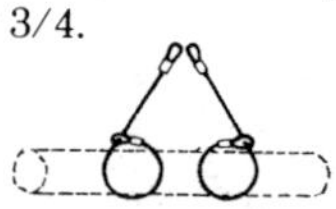

For Double Basket Hitch multiply above values by 2.

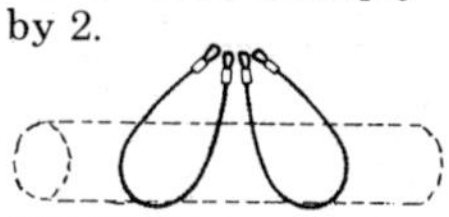

Note: Table values are for slings with eyes and thimbles in both ends, Flemish Spliced Eyes and mechanical sleeves.

Hand tucked spliced eyes—reduce loads according to *Table 1-10: Efficiencies of Tucked or Hand Spliced Eyes.*

Eyes formed by cable clips—reduce loads by 20%.

Table 9: Braided Wire Rope Slings
6-Part Braided Rope
(6 × 19 Classification Group, Improved Plow Steel)

Component Rope Diameter (Inches)	Maximum Safe Working Loads—Pounds (Safety Factor = 5)					
	Single Vertical Hitch 	Single Choker Hitch 	Single Basket Hitch (Vertical Legs) 	2-Leg Bridle Hitch & Single Basket Hitch with Legs Inclined  		
				60°	45°	30°
3/16	2,600	1,950	5,200	4,500	3,700	2,600
1/4	4,600	3,450	9,200	7,950	6,500	4,600
5/16	7,200	5,400	14,400	12,500	10,200	7,200
3/8	10,200	7,650	20,400	17,700	14,400	10,200
7/16	13,800	10,400	27,600	23,900	19,500	13,800
1/2	18,000	13,500	36,000	31,200	25,500	18,000
9/16	22,000	16,500	44,000	38,100	31,100	22,000
5/8	28,000	21,000	56,000	48,500	39,600	28,000
3/4	40,000	30,000	80,000	69,300	56,600	40,000
7/8	54,000	40,500	108,000	93,500	76,400	54,000
1	70,000	52,500	140,000	121,200	99,000	70,000

If used with Choker Hitch multiply above values by 3/4.

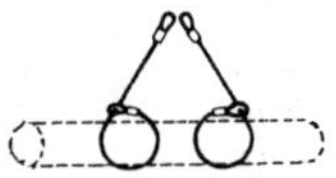

For Double Basket Hitch multiply above values by 2.

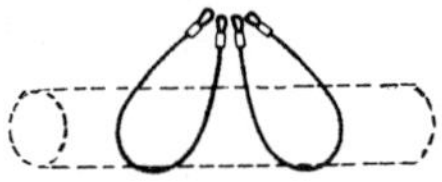

Note: Table values are for braided slings with eyes and thimbles in both ends. Eyes formed by hand tucking ropes and securing with mechanical sleeves.

Table 10: Wire Rope Slings
(6 × 37 Classification Group, Improved Plow Steel, IWRC)

Rope Diameter (Inches)	Maximum Safe Working Loads—Pounds (Safety Factor = 5)					
	Single Vertical Hitch 	Single Choker Hitch 	Single Basket Hitch (Vertical Legs) 	2-Leg Bridle Hitch & Single Basket Hitch with Legs Inclined  		
				60°	45°	30°
1/4	1,050	800	2,100	1,800	1,500	1,050
5/16	1,700	1,300	3,400	2,950	2,400	1,700
3/8	2,350	1,750	4,700	4,100	3,300	2,350
7/16	3,200	2,400	6,400	5,550	4,500	3,700
1/2	4,300	3,200	8,600	7,450	6,100	4,300
9/16	5,350	4,000	10,700	9,250	7,550	5,350
5/8	6,900	5,200	13,800	11,950	9,750	6,900
3/4	9,500	7,100	19,000	16,450	13,400	9,500
7/8	13,000	9,750	26,000	22,500	18,400	13,000
1	17,000	12,750	34,000	29,450	24,000	17,000
1 1/8	21,000	15,750	42,000	36,400	29,700	21,000
1 1/4	26,200	19,650	52,400	45,400	37,000	26,200
1 3/8	32,000	24,000	64,000	55,400	45,200	32,000
1 1/2	39,500	29,600	79,000	68,400	55,900	39,500
1 5/8	45,400	34,000	90,800	78,600	64,200	45,400
1 3/4	52,000	39,000	104,000	90,000	73,500	52,000
1 7/8	61,000	45,750	122,000	105,700	86,300	61,000
2	66,600	49,950	133,200	115,400	94,200	66,600
2 1/4	86,400	64,800	172,800	149,600	122,200	86,400
2 1/2	105,300	79,000	210,600	182,400	148,900	105,300
2 3/4	126,000	94,500	252,000	218,200	178,200	126,000

If used with Choker Hitch multiply above values by 3/4.

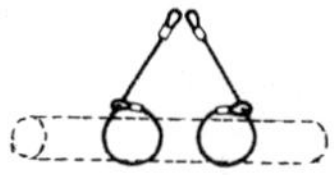

For Double Basket Hitch multiply above values by 2.

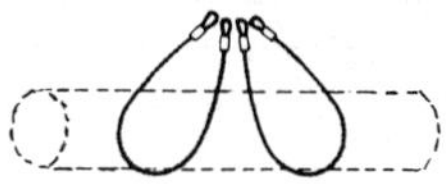

Note: Table values are for slings with eyes and thimbles in both ends, Flemish Spliced Eyes and mechanical sleeves.

Hand tucked spliced eyes—reduce loads according to *Table 1-10: Efficiencies of Tucked or Hand Spliced Eyes.*

Eyes formed by cable clips—reduce loads by 20%.

Table 11: Cable Laid Wire Rope Slings (Mechanical Splices)

Cable Body Diameter (Inches)	Maximum Safe Working Loads—Pounds (Safety Factor = 5)					
	Single Vertical Hitch 	Single Choker Hitch 	Single Basket Hitch (Vertical Legs) 	2-Leg Bridle Hitch & Single Basket Hitch with Legs Inclined  		
				60°	45°	30°
1/4	1,000	750	2,000	1,750	1,400	1,000
3/8	2,200	1,650	4,400	3,800	3,100	2,200
1/2	3,600	2,700	7,200	6,250	5,100	3,600
5/8	5,600	4,200	11,200	9,700	7,900	5,600
3/4	7,600	5,700	15,200	13,200	10,700	7,600
7/8	10,000	7,500	20,000	17,300	14,100	10,000
1	12,800	9,600	25,600	22,200	18,100	12,800
1 1/8	15,400	11,600	30,800	26,700	21,800	15,400
1 1/4	18,400	13,800	36,800	31,900	26,000	18,400
1 3/8	22,000	16,500	44,000	38,100	31,100	22,000
1 1/2	26,000	19,500	52,000	45,000	36,800	26,000

If used with Choker Hitch multiply above values by 3/4.

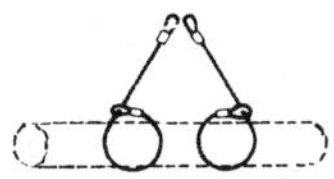

For Double Basket Hitch multiply above values by 2.

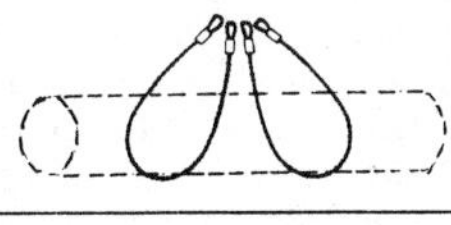

Note: Cable laid wire rope slings with mechanical splices are slings made from cable and having eyes fabricated by pressing or swaging one or more metal sleeves over the rope junction at the eyes. The cable consists of 6 individual wire ropes wrapped around a 7th wire rope which forms the core. For cable body diameters of 1/4 to 5/8″ the construction is 7 × 7 × 7 (i.e., 7 wire ropes of 7 × 7 construction) and for diameters above 5/8″ it is 7 × 6 × 19 IWRC (i.e., 7 wire ropes of 6 × 19 IWRC construction).
The table values are for slings with eyes and thimbles in both ends.

Table 12: Cable Laid Grommet Slings (Hand Tucked, Improved Plow Steel)

Cable Body Diameter (Inches)	Maximum Safe Working Loads—Pounds (Safety Factor = 5)					
	Single Vertical Hitch 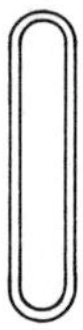	Single Choker Hitch 	Single Basket Hitch (Vertical Legs) 	Single Basket Hitch with Legs Inclined 		
				60°	45°	30°
3/8	2,600	1,950	5,200	4,500	3,700	2,600
9/16	5,600	4,200	11,200	9,700	7,900	5,600
5/8	7,800	5,850	15,600	13,500	11,000	7,800
3/4	10,200	7,650	20,400	17,700	14,400	10,200
15/16	15,800	11,900	31,600	27,400	22,300	15,800
1 1/8	22,000	16,500	44,000	38,100	31,100	22,000
1 5/16	30,000	22,500	60,000	52,000	42,400	30,000
1 1/2	38,000	28,500	76,000	65,800	53,700	38,000
1 11/16	48,000	36,000	96,000	83,100	67,900	48,000
1 7/8	60,000	45,000	120,000	103,900	84,800	60,000
2 1/4	84,000	63,000	168,000	145,500	118,800	84,000
2 5/8	112,000	84,000	224,000	194,000	158,400	112,000
				For Double Basket Hitch multiply above values by 2. 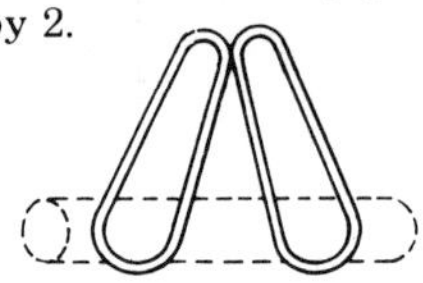		

Note: A cable laid, hand tucked grommet is an endless wire rope sling fabricated from a single wire rope that is wrapped six times. After wrapping, the two free ends of the rope are tucked inside the wraps to form the core of the grommet. It is very similar to the strand laid grommet except that a complete rope rather than a strand is used to form the grommet. For cable body diameters of 3/8 and 9/16″ the construction is 7 × 6 × 7 (the first figure denotes 6 rope wraps plus 1 rope core, and the second two digits indicate that the rope used is 6 × 7). For diameters of 5/8″ and up the construction is 7 × 6 × 19 (i.e., 6 rope wraps plus 1 rope core, all of 6 × 19 rope).

Table 13: Strand Laid Endless Slings (Mechanical Splice, Improved Plow Steel)

Rope Body Diameter (Inches)	Maximum Safe Working Loads—Pounds (Safety Factor = 5)					
	Single Vertical Hitch 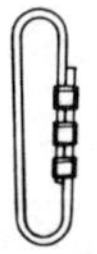	Single Choker Hitch 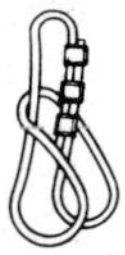	Single Basket Hitch (Vertical Legs) 	Single Basket Hitch with Legs Inclined 		
				60°	45°	30°
1/4	1,850	1,400	3,700	3,200	2,600	1,850
3/8	4,000	3,000	8,000	6,900	5,650	4,000
1/2	7,200	5,400	14,400	12,500	10,200	7,200
5/8	11,200	8,400	22,400	19,400	15,800	11,200
3/4	16,000	12,000	32,000	27,700	22,600	16,000
7/8	22,000	16,500	44,000	38,100	31,100	22,000
1	28,000	21,000	56,000	48,500	39,600	28,000
1 1/8	36,000	27,000	72,000	62,400	50,900	36,000
1 1/4	42,000	31,500	84,000	72,700	59,400	42,000
1 3/8	50,000	37,500	100,000	86,600	70,700	50,000
1 1/2	58,000	43,500	116,000	100,500	82,000	58,000
				For Double Basket Hitch multiply above values by 2. 		

Note: A strand laid endless sling with mechanical splice is a sling fabricated from a single wire rope that is bent into a loop and whose ends are attached to each other by means of one or more mechanical fittings. For rope diameters up to and including 1-1/8″ the rope used is 6 × 19 IWRC and for diameters greater than 1-1/8″ it is 6 × 37 IWRC.

Table 14: Cable Laid Endless Slings (Mechanical Splice)

Cable Body Diameter (Inches)	Maximum Safe Working Loads—Pounds (Safety Factor = 5)					
	Single Vertical Hitch 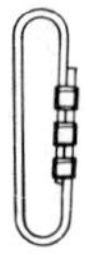	Single Choker Hitch 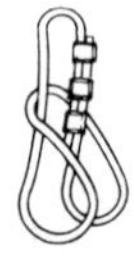	Single Basket Hitch (Vertical Legs) 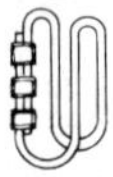	Single Basket Hitch with Legs Inclined 		
				60°	45°	30°
1/4	1,650	1,250	3,300	2,850	2,350	1,650
3/8	3,600	2,700	7,200	6,250	5,100	3,600
1/2	6,000	4,500	12,000	10,400	8,500	6,000
5/8	9,000	6,750	18,000	15,600	12,700	9,000
3/4	12,400	9,300	24,800	21,500	17,500	12,400
7/8	16,600	12,500	33,200	28,800	23,500	16,600
1	20,000	15,000	40,000	34,600	28,300	20,000
1 1/8	26,000	19,500	52,000	45,000	36,800	26,000
1 1/4	32,000	24,000	64,000	55,400	45,200	32,000
1 3/8	36,000	27,000	72,000	62,400	50,900	36,000
1 1/2	44,000	33,000	88,000	76,200	62,200	44,000

For Double Basket Hitch multiply above values by 2.

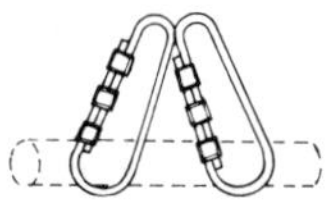

Note: Cable laid endless slings with mechanical splices are endless slings fabricated from cable that is bent into a loop and whose ends are attached to each other by means of one or more mechanical fittings. The cable is composed of 6 individual wire ropes wrapped around a 7th wire rope which forms the core. For cable body diameters of 1/4 to 5/8″ the construction is 7 × 7 × 7 (i.e., 7 wire ropes of 7 × 7 construction) and for diameters above 5/8″ it is 7 × 6 × 19 IWRC (i.e., 7 wire ropes of 6 × 19 IWRC construction).

Table 15: Strand Laid Grommet Slings (Hand Tucked, Improved Plow Steel)

Rope Body Diameter (Inches)	Maximum Safe Working Loads—Pounds (Safety Factor = 5)					
	Single Vertical Hitch 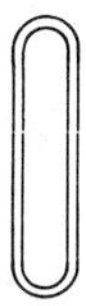	Single Choker Hitch 	Single Basket Hitch (Vertical Legs) 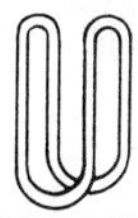	Single Basket Hitch with Legs Inclined 		
				60°	45°	30°
1/4	1,700	1,275	3,400	2,950	2,400	1,700
5/16	2,600	1,950	5,200	4,500	3,700	2,600
3/8	3,800	2,850	7,600	6,600	5,400	3,800
7/16	5,200	3,900	10,400	9,000	7,350	5,200
1/2	7,000	5,250	14,000	12,100	9,900	7,000
9/16	8,400	6,300	16,800	14,500	11,900	8,400
5/8	10,400	7,800	20,800	18,000	14,700	10,400
3/4	14,800	11,100	29,600	25,600	20,900	14,800
7/8	20,000	15,000	40,000	34,600	28,300	20,000
1	26,000	19,500	52,000	45,000	36,800	26,000
1 1/8	29,800	22,400	59,600	51,600	42,100	29,800
1 1/4	35,000	26,300	70,000	60,600	49,500	35,000
1 3/8	42,600	32,000	85,200	73,800	60,200	42,600
1 1/2	51,600	38,700	103,200	89,400	73,000	51,600
1 5/8	56,600	42,500	113,200	98,000	80,000	56,600
1 3/4	65,000	48,800	130,000	112,600	91,900	65,000
1 7/8	76,000	57,000	152,000	131,600	107,500	76,000
2	83,000	62,300	166,000	143,800	117,400	83,000
2 1/4	101,000	75,800	202,000	174,900	142,800	101,000
				For Double Basket Hitch multiply above values by 2. 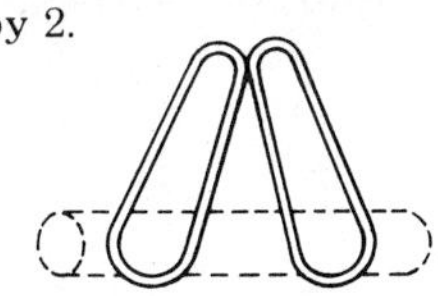		

Note: A strand laid grommet is an endless wire rope sling fabricated from a single strand that is wrapped six times. After wrapping the two free strand ends are tucked inside the wraps to form the core of the grommet. Grommet diameters up to and including 1″ are fabricated from strands of 6 × 19 rope and grommet diameters of 1-1/8″ and greater are fabricated from strands of 6 × 37 rope.

APPENDIX

E

Synthetic Webbing and Chain Sling Capacities

Table 1: Nylon Web Slings (8000 lb/in. Material)

Web Width (Inches)	Maximum Safe Working Loads—Pounds (Safety Factor = 5) (Eye & Eye, Twisted Eye, Triangle Fittings, Choker Fittings)					
	Single Vertical Hitch 	Single Choker Hitch 	Single Basket Hitch (Vertical Legs) 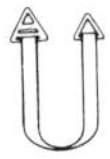	2-Leg Bridle Hitch & Single Basket Hitch with Legs Inclined 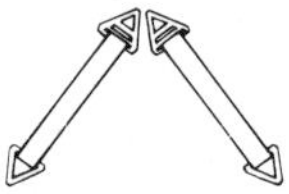 		
				60°	45°	30°
1	1,600	1,200	3,200	2,770	2,260	1,600
2	3,200	2,400	6,400	5,550	4,520	3,200
3	4,800	3,600	9,600	8,300	6,800	4,800
4	6,400	4,800	12,800	11,100	9,050	6,400
5	8,000	6,000	16,000	13,850	11,300	8,000
6	9,600	7,200	19,200	16,600	13,600	9,600
7	11,200	8,400	22,400	19,400	15,800	11,200
8	12,800	9,600	25,600	22,200	18,100	12,800
9	14,400	10,800	28,800	25,000	20,400	14,400
10	16,000	12,000	32,000	27,700	22,600	16,000
11	17,600	13,200	35,200	30,500	24,900	17,600
12	19,200	14,400	38,400	33,300	27,200	19,200

If used with Choker Hitch multiply above values by 3/4.

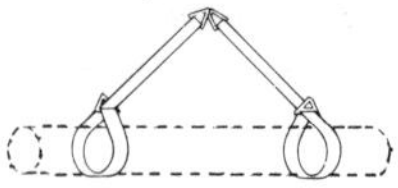

For Double Basket Hitch multiply above values by 2.

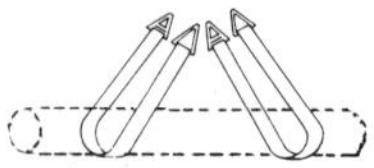

Note: For safe working loads of Endless or Grommet Slings, multiply above values by 2.

Table 2: Nylon Web Slings (6000 lb/in. Material)

Web Width (Inches)	Maximum Safe Working Loads—Pounds (Safety Factor = 5) (Eye & Eye, Twisted Eye, Triangle Fittings, Choker Fittings)					
	Single Vertical Hitch 	Single Choker Hitch 	Single Basket Hitch (Vertical Legs) 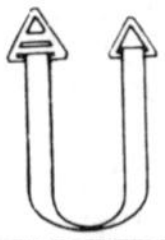	2-Leg Bridle Hitch & Single Basket Hitch with Legs Inclined 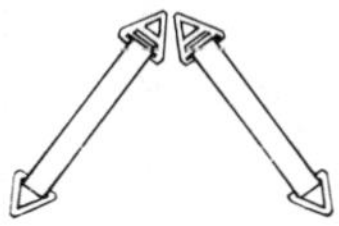 		
				60°	45°	30°
1	1,200	900	2,400	2,080	1,700	1,200
2	2,400	1,800	4,800	4,160	3,400	2,400
3	3,600	2,700	7,200	6,240	5,100	3,600
4	4,800	3,600	9,600	8,300	6,800	4,800
5	6,000	4,500	12,000	10,400	8,500	6,000
6	7,200	5,400	14,400	12,500	10,200	7,200
7	8,400	6,300	16,800	14,550	11,900	8,400
8	9,600	7,200	19,200	16,600	13,600	9,600
9	10,800	8,100	21,600	18,700	15,300	10,800
10	12,000	9,000	24,000	20,800	17,000	12,000
11	13,200	9,900	26,400	22,900	18,650	13,200
12	14,400	10,800	28,800	25,000	20,400	14,400

If used with Choker Hitch multiply above values by 3/4.

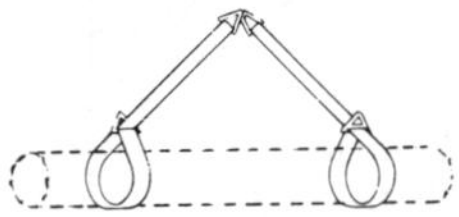

For Double Basket Hitch multiply above values by 2.

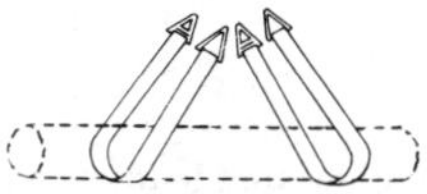

Note: For safe working loads of Endless or Grommet Slings, multiply above values by 2.

Table 3: Dacron Web Slings (5000 lb/in. Material)

Web Width (Inches)	Maximum Safe Working Loads—Pounds (Safety Factor = 5) (Eye & Eye, Twisted Eye, Triangle Fittings, Choker Fittings)					
	Single Vertical Hitch 	Single Choker Hitch 	Single Basket Hitch (Vertical Legs) 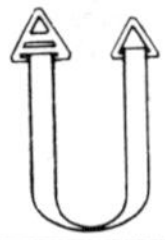	2-Leg Bridle Hitch & Single Basket Hitch with Legs Inclined 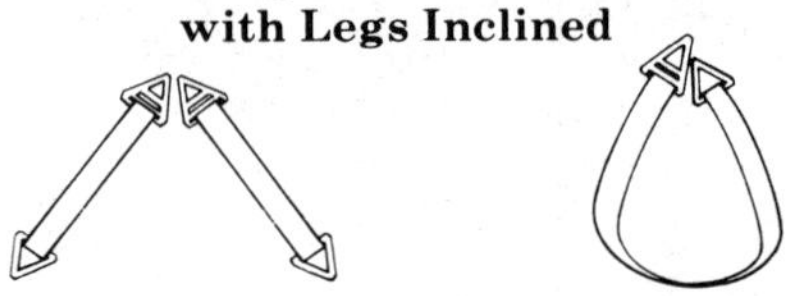		
				60°	45°	30°
1	1,000	750	2,000	1,730	1,400	1,000
2	2,000	1,500	4,000	3,460	2,830	2,000
3	3,000	2,250	6,000	5,200	4,250	3,000
4	4,000	3,000	8,000	6,950	5,650	4,000
5	5,000	3,750	10,000	8,660	7,070	5,000
6	6,000	4,500	12,000	10,400	8,500	6,000
7	7,000	5,250	14,000	12,100	9,900	7,000
8	8,000	6,000	16,000	13,850	11,300	8,000
9	9,000	6,750	18,000	15,600	12,700	9,000
10	10,000	7,500	20,000	17,350	14,100	10,000
11	11,000	8,250	22,000	19,100	15,500	11,000
12	12,000	9,000	24,000	20,800	17,000	12,000
				If used with Choker Hitch multiply above values by 3/4. 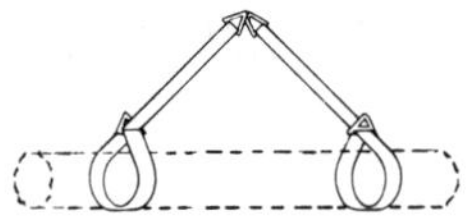 For Double Basket Hitch multiply above values by 2. 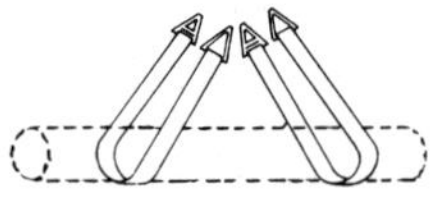		

Note: For safe working loads of Endless or Grommet Slings, multiply above values by 2.

Table 4: Chain Slings (Alloy Steel)

Chain Size (Inches)	Maximum Safe Working Loads—Pounds					
	Single Vertical Hitch 	Single Choker Hitch 	Single Basket Hitch (Vertical Legs) 	2-Leg Bridle Hitch & Single Basket Hitch with Legs Inclined  		
				60°	45°	30°
1/4	3,250	2,440	6,500	5,600	4,600	3,250
3/8	6,600	4,950	13,200	11,400	9,300	6,600
1/2	11,250	8,400	22,500	19,500	15,900	11,250
5/8	16,500	12,400	33,000	28,600	23,300	16,500
3/4	23,000	17,200	46,000	39,800	32,500	23,000
7/8	28,750	21,500	57,500	49,800	40,600	28,750
1	38,750	29,000	77,500	67,100	54,800	38,750
1 1/8	44,500	33,400	89,000	77,000	63,000	44,500
1 1/4	57,500	43,000	115,000	99,500	81,000	57,500
1 3/8	67,000	50,000	134,000	116,000	94,500	67,000
1 1/2	80,000	60,000	160,000	138,000	113,000	80,000
1 3/4	100,000	75,000	200,000	173,000	141,000	100,000

If used with Choker Hitch multiply above values by 3/4.

For Double Basket Hitch multiply above values by 2.

APPENDIX

F

Rigging Hardware

Table 1: Eye Hooks, Shank Hooks, Swivel Hooks, (Forged Alloy Steel) (Safety Factor = 5)

Eye Hook

Swivel Hook

Shank Hook

Throat Opening (Inches)	Maximum Safe Working Load (Pounds)
5/8	600
11/16	800
1	1,500
1 1/16	2,000
1 1/8	2,500
1 1/4	4,000
1 3/8	4,500
1 13/32	5,000
1 1/2	5,500
1 17/32	6,000
1 11/16	6,800
1 25/32	8,000
1 7/8	8,400
1 15/16	10,000
2 1/16	10,400
2 1/8	11,000
2 1/4	12,500
2 5/16	13,000
2 1/2	16,000
2 9/16	18,000
3	19,200
3 1/16	20,000
3 3/8	24,000
3 7/16	26,000
4	33,400

Table 2: Typical Sorting Hook (Forged Alloy Steel)

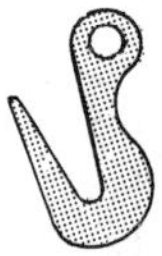

I.D. of Eye	1 1/4″
Opening at Top of Hook	2 13/16″
Safe Working Load 2½″ From Tip	2 Tons
Safe Working Load at Bottom of Hook	7 1/2 Tons

Table 3: Chain Grab Hooks (Clevis Type and Eye Type) (Forged Alloy Steel)

Clevis Type
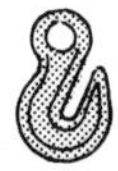
Eye Type

Throat Opening (Inches)	For Size of Chain (Inches)	Maximum Safe Working Load (Pounds)
11/32	1/4	2,750
7/16	5/16	4,300
1/2	3/8	5,250
9/16	7/16	7,000
21/32	1/2	9,000
25/32	5/8	13,500
15/16	3/4	19,250
1 1/16	7/8	26,000
1 3/16	1	34,000

Table 4: Chain Slip Hooks
(Clevis Type and Eye Type)
(Forged Alloy Steel)
(Safety Factor = 4)

Throat Opening (Inches)	For Size of Chain (Inches)	Maximum Safe Working Load (Pounds)
15/16	1/4	2,750
1 1/16	5/16	4,300
1 5/16	3/8	5,250
1 9/16	7/16	7,000
1 11/16	1/2	9,000
2	5/8	13,500
2 1/8	3/4	19,250
2 3/4	7/8	26,000
3	1	34,000

Table 5: Sliding Choker Hooks
Forged Alloy Steel
(Safety Factor = 5)

Throat Opening (Inches)	For Rope Size (Inches)	Maximum Safe Working Load (Pounds)
1/2	1/4 - 5/16	1,500
5/8	3/8	2,600
7/8	1/2	3,400
1 1/8	5/8	5,100
1 1/8	3/4	8,000
1 7/16	7/8 - 1	15,000
1 3/4	1 1/8 - 1 1/4	23,000
2 3/16	1 3/8 - 1 1/2	30,000

Table 6: Rings
(Weldless Construction)
(Forged Alloy Steel)

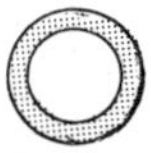

Stock Diameter (Inches)	Inside Diameter (Inches)	Maximum Safe Working Load (Pounds)
7/8	4	7,200
7/8	5 1/2	5,600
1	4	10,800
1 1/8	6	10,400
1 1/4	5	17,000
1 3/8	6	19,000

Table 7: Pear Shaped Links
(Sling Links)
(Weldless Construction)
(Forged Alloy Steel)

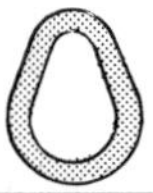

Stock Diameter (Inches)	Inside Length (Inches)	Maximum Safe Working Load (Pounds)
3/8	2 1/4	1,800
1/2	3	2,900
5/8	3 3/4	4,200
3/4	4 1/2	6,000
7/8	5 1/4	8,300
1	6	10,800
1 1/4	7 3/4	16,750
1 3/8	8 1/4	20,500

Table 8: Master Links
(Weldless Construction)
(Forged Alloy Steel)

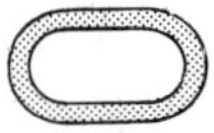

Stock Diameter (Inches)	Inside Width (Inches)	Maximum Safe Working Load (Pounds)
1/2	2 1/2	3,250
5/8	3	4,400
3/4	2 3/4	7,000
1	3 1/2	16,500
1 1/4	4 3/8	25,000
1 1/2	5 1/4	35,500
1 3/4	6	44,500
2	7	57,500

Table 9: End Links
(Weldless Construction)
(Forged Alloy Steel)

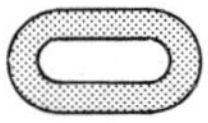

Stock Diameter (Inches)	Inside Width (Inches)	Maximum Safe Working Load (Pounds)
5/16	1/2	2,500
3/8	9/16	3,800
1/2	3/4	6,500
5/8	1	9,300
3/4	1 1/8	14,000
7/8	2	12,000
1	2 1/4	15,200
1 1/4	2 1/2	26,400
1 3/8	2 3/4	30,000

Table 10: Swivels (All Types) (Weldless Construction) (Forged Alloy Steel)

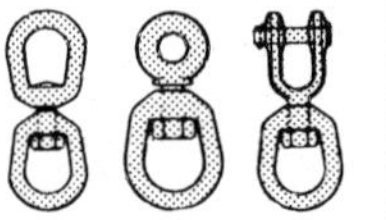

Stock Diameter (Inches)	Max. Safe Working Load (Pounds)
1/4	850
5/16	1,250
3/8	2,250
1/2	3,600
5/8	5,200
3/4	7,200
7/8	10,000
1	12,500
1 1/8	15,200
1 1/4	18,000
1 1/2	45,200

Table 11: Double Clevis Links (Weldless Construction) (Forged Alloy Steel)

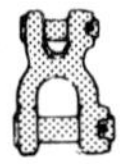

Small Pin Diameter (Inches)	Large Pin Diameter (Inches)	Maximum Safe Working Load (Pounds)
5/16	1/2	3,250
7/16	5/8	6,600
9/16	11/16	8,750
5/8	3/4	11,250

APPENDIX

G

Block and Tackle Rigging

Table 1: Simple Block and Tackle Rigging for Plow Steel Wire Rope (Factor of Safety 6)

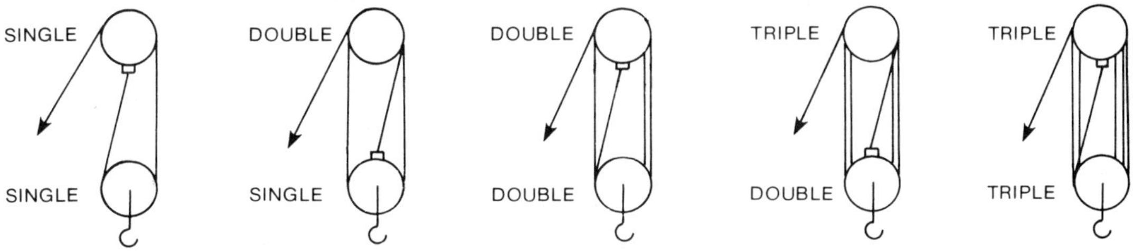

Load to be Lifted (tons)		Total Number of Sheaves in Blocks				
		2 (2-single blocks)	3 (1-single 1-double)	4 (2-double blocks)	5 (1-double 1-triple)	6 (2-triple blocks)
1	Smallest permissible rope diameter (inch)	3/8	3/8	3/8	3/8	3/8
	Lead line pull (lbs)	1,000	720	560	460	400
2	Rope	1/2	3/8	3/8	3/8	3/8
	Pull	2,100	1,400	1,100	920	800
4	Rope	5/8	1/2	1/2	3/8	3/8
	Pull	4,200	2,900	2,200	1,800	1,600
6	Rope	3/4	5/8	5/8	1/2	1/2
	Pull	6,200	4,300	3,400	2,800	2,400
8	Rope	7/8	3/4	5/8	5/8	5/8
	Pull	8,300	5,800	4,500	3,700	3,200
10	Rope	1	7/8	3/4	5/8	5/8
	Pull	10,400	7,200	5,600	4,600	4,000
15	Rope	1-1/8	1	7/8	3/4	3/4
	Pull	15,600	10,800	8,400	6,900	6,000
20	Rope	1-1/2	1-1/8	1	7/8	7/8
	Pull	20,800	14,400	11,200	9,200	8,000

Table 2: Recommended Sizes of Tackle Blocks

Wire Rope		Manila Rope	
Rope Diameter (inches)	**Outside-Diameter of Sheave (inches)**	**Rope Diameter (inches)**	**Length of Shell (inches)**
3/8	6—8	1/2	4
1/2	8—10	5/8	6
5/8	10—12	3/4	6—7
3/4	12—16	7/8	7—8
7/8	14—18	1	8—10
1	14—20	1-1/8	8—10
		1-1/4	10—12
		1-1/2	12—14
		1-3/4	14—16

Note: Largest diameter of sheave for a given size of rope is preferred, when available, except that for 6 × 37 wire rope the smaller diameter of sheave is suitable.

Table 3: Multiple Sheave Blocks—Efficiency of Wire Rope Tackle (Lead Line and Efficiency Factors for Derricks, Booms, etc.)

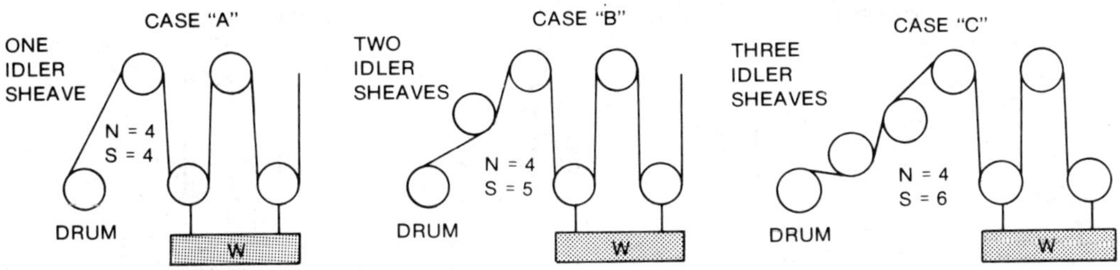

N	Plain Bearing Sheaves						Roller Bearing Sheaves					
	K = 1.09						K = 1.04					
	Efficiency			Lead Line Factor			Efficiency			Lead Line Factor		
	Case A	Case B	Case C	Case A	Case B	Case C	Case A	Case B	Case C	Case A	Case B	Case C
2	.880	.807	.740	.568	.620	.675	.943	.907	.872	.530	.551	.574
3	.844	.774	.710	.395	.431	.469	.925	.889	.855	.360	.375	.390
4	.810	.743	.682	.309	.336	.367	.908	.873	.839	.275	.286	.298
5	.778	.714	.655	.257	.280	.305	.890	.856	.823	.225	.234	.243
6	.748	.686	.629	.223	.243	.265	.874	.840	.808	.191	.198	.206
7	.719	.660	.605	.199	.216	.236	.857	.824	.793	.167	.173	.180
8	.692	.635	.582	.181	.197	.215	.842	.809	.778	.148	.154	.161
9	.666	.611	.561	.167	.182	.198	.826	.794	.764	.135	.140	.145
10	.642	.589	.540	.156	.170	.185	.811	.780	.750	.123	.128	.133
11	.619	.568	.521	.147	.160	.175	.796	.766	.736	.114	.119	.124
12	.597	.547	.502	.140	.152	.166	.782	.752	.723	.106	.111	.115
13	.576	.528	.485	.133	.145	.159	.768	.739	.710	.100	.104	.108
14	.556	.510	.468	.128	.140	.153	.755	.725	.698	.095	.099	.102
15	.537	.493	.452	.124	.135	.147	.741	.713	.685	.090	.094	.097

Lead Line Stress = Lead Line Factor × Load

$$\text{Efficiency} = \frac{(K^N - 1)}{K^S N(K - 1)} \qquad \text{Lead Line Factor} = \frac{1}{N \times \text{Efficiency}}$$

Notes: The above cases apply also where the rope is dead ended at the lower or traveling block.
W = Load
S = Number of sheaves (not counting equalizer)
N = Number of rope parts supporting load

Table 4: Multiple Sheave Blocks—Efficiency of Wire Rope Tackle (Lead Line and Efficiency Factors for Overhead Cranes Only)

CASE "D"

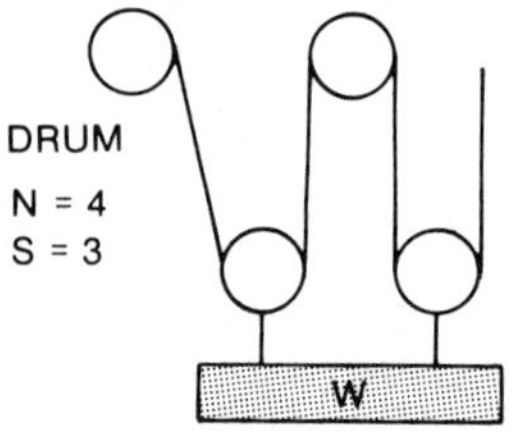

SINGLE DRUM

CASE "E"

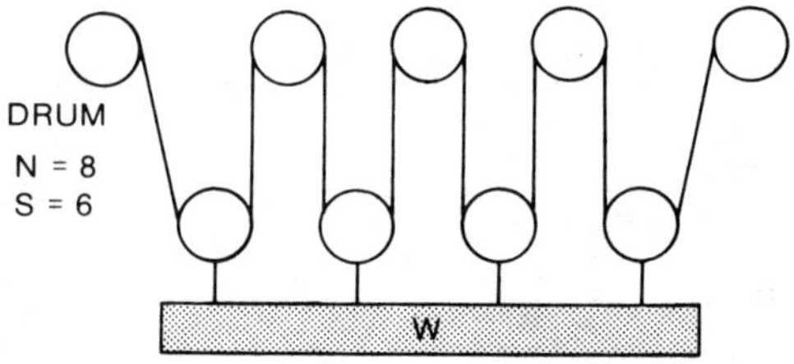

DOUBLE DRUM WITH EQUALIZER

N	Plain Bearing Sheaves				Roller Bearing Sheaves			
	K = 1.09				K = 1.04			
	Efficiency		Lead Line Factor		Efficiency		Lead Line Factor	
	Case D	Case E	Case D	Case E	Case D	Case E	Case D	Case E
2	.959	1.000	.522	.500	.981	1.000	.510	.500
3	.920		.362		.962		.346	
4	.883	.959	.283	.261	.944	.981	.265	.255
5	.848		.236		.926		.216	
6	.815	.920	.204	.181	.909	.962	.183	.173
7	.784		.182		.892		.160	
8	.754	.883	.166	.141	.875	.944	.143	.132
9	.726		.153		.859		.130	
10	.700	.848	.143	.118	.844	.926	.119	.108
11	.674		.135		.828		.110	
12	.650	.815	.128	.102	.813	.909	.101	.091
13	.628		.122		.799		.096	
14	.606	.784	.118	.091	.785	.892	.091	.080
15	.586		.114		.771		.086	

Lead Line Stress = Lead Line Factor × Load

CASE "D" $$\text{Efficiency} = \frac{(K^N - 1)}{K^S N (K - 1)}$$

$$\text{Lead Line Factor} = \frac{1}{N \times \text{Efficiency}}$$

CASE "E" $$\text{Efficiency} = \frac{2(K^{\frac{N}{2}} - 1)}{K^{\frac{S}{2}} N (K - 1)}$$

$$\text{Lead Line Factor} = \frac{1}{N \times \text{Efficiency}}$$

Notes: The above cases apply also where the rope is dead ended or the equalizer is located at the lower or traveling block.
W = Load
S = Number of sheaves (not counting equalizer)
N = Number of rope parts supporting load.

APPENDIX

H

Safe Loads for Timber

Table 1: Planks
(Loads Given in Pounds Concentrated at Center of Span Douglas Fir No. 1 Common or Better)

Size in Inches	Span in Feet 4	6	8	10	12	14	16	18
2 × 6	300	195						
2 × 8	400	260	190					
2 × 10	500	325	235	180				
2 × 12	600	390	285	220	165			
3 × 10	1245	820	600	465	370	305		
3 × 12	1495	980	720	555	445	370	300	
3 × 14	1740	1145	840	650	520	430	350	290

Table 2: Beams

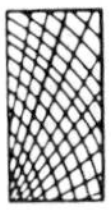

VERTICAL

HORIZONTAL

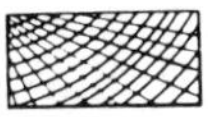

(Loads Given in Pounds Concentrated at Center of Span Douglas Fir Structural Grades)

Size in Inches	Span in Feet 4	6	8	10	12	14	16	18	20	24
4 × 4	990	650	480	380	310	260	220			
4 × 6 Hor.	1530	1010	750	590	480	400	340			
6 × 6	3440	2290	1700	1340	1110	930	800	690	610	
6 × 8 Hor.	4700	3120	2320	1830	1510	1270	1090	950	830	
6 × 8 Vert.	6430	4260	3180	2520	2080	1760	1520	1330	1170	940
8 × 8	6690	6660	4330	3440	2840	2400	2070	1810	1600	1280
8 × 12 Vert.	10250	10200	10160	8150	6740	5730	4970	4370	3890	3160
8 × 14 Vert.	12050	12000	11940	11260	9320	7940	6890	6070	5410	4410
8 × 16 Vert.	13820	13760	13690	13630	12310	10500	9130	8040	7180	5870
10 × 10	10740	10690	10640	7020	5800	4930	4270	3740	3320	2680
10 × 12 Vert.	12960	12900	12840	10320	8540	7260	6300	5540	4930	4000
10 × 14 Vert.	15220	15150	15080	14280	11810	10050	8730	7690	6860	5590

APPENDIX

I

Weights of Materials (Based on Volume)

Material	Approximate Weight (lbs. per cubic foot)
METALS	
Aluminum	165
Brass	535
Bronze	500
Copper	560
Iron	480
Lead	710
Steel	490
Tin	460
MASONRY	
Ashlar masonry	140-160
Brick masonry, soft	110
Brick masonry, common (about 3 tons per thousand)	125
Brick masonry, pressed	140
Clay tile masonry, average	60
Rubble masonry	130-155
Concrete, cinder, haydite	100-110
Concrete, slag	130
Concrete, stone	144
Concrete, stone, reinforced (4050 lbs. per cu. yd.)	150
ICE AND SNOW	
Ice	56
Snow, dry, fresh fallen	8
Snow, dry, packed	12-25
Snow, wet	27-40
MISCELLANEOUS	
Asphalt	80
Tar	75
Glass	160
Paper	60
TIMBER, AIR-DRY	
Cedar	22
Fir, Douglas, seasoned	34
Fir, Douglas, unseasoned	40
Fir, Douglas, wet	50
Fir, Douglas, glue laminated	34
Hemlock	30
Pine	30
Poplar	30
Spruce	28
LIQUIDS	
Alcohol, pure	49
Gasoline	42
Oils	58
Water	62
EARTH	
Earth, wet	100
Earth, dry (about 2050 lbs. per cu. yd.)	75
Sand and gravel, wet	120
Sand and gravel, dry	105
River sand (about 3240 lbs. per cu. yd.)	120
VARIOUS BUILDING MATERIALS	
Cement, portland, loose	94
Cement, portland, set	183
Lime, gypsum, loose	53-64
Mortar, cement-lime, set	103
Crushed rock (about 2565 lbs. per cu. yd.)	90-110

APPENDIX

J

Weights of Materials (Based on Surface Area)

Table 1

Material	Approximate Weight (lbs. per square foot)	Material	Approximate Weight (lbs. per square foot)
CEILINGS (per inch of thickness)		**FLOORING (per inch of thickness)**	
Plasterboard	5	Hardwood	5
Acoustic and fire resistive tile	2	Sheathing	2.5
Plaster, gypsum-sand	8	Plywood, fir	3
Plaster, light aggregate	4	Wood block, treated	4
Plaster, cement sand	12	Concrete, finish or fill	12
ROOFING		Mastic base	12
Three-ply felt and gravel	5.5	Mortar base	10
Five-ply felt and gravel	6.5	Terrazzo	12.5
Three-ply felt, no gravel	3	Tile, vinyl 1/8″	1.5
Five-ply felt, no gravel	4	Tile, linoleum 3/16″	1
Shingles, wood	2	Tile, cork, per 1/16″	0.5
Shingles, asbestos	3	Tile, rubber or asphalt 3/16″	2
Shingles, asphalt	2.5	Tile, ceramic or quarry 3/4″	11
Shingles, 1/4″ slate	10	Carpeting	2
Shingles, tile	14	**DECKS AND SLABS**	
PARTITIONS		Steel roof deck 1-1/2″—14 ga.	5
Steel partitions	4	—16 ga.	4
Solid 2″ gypsum-sand plaster	20	—18 ga.	3
Solid 2″ gypsum-light agg. plaster	12	—20 ga.	2.5
Metal studs, metal lath, 3/4″ plaster both sides	18	—22 ga.	2
Metal or wood studs, plaster board and 1/2″ plaster both sides	18	Steel cellular deck 1-1/2″—12/12 ga.	11
Plaster 1 / 2″	4	—14/14 ga.	8
Hollow clay tile 2″	13	—16/16 ga.	6.5
3″	16	—18/18 ga.	5
4″	18	—20/20 ga.	3.5
5″	20	Steel cellular deck 3″—12/12 ga.	12.5
6″	25	—14/14 ga.	9.5
Hollow slag concrete block 4″	24	—16/16 ga.	7.5
6″	35	—18/18 ga.	6
		—20/20 ga.	4.5
		Concrete, reinforced, per inch	12.5

Material	Approximate Weight (lbs. per square foot)	Material	Approximate Weight (lbs. per square foot)
Hollow gypsum block 3″	10	Concrete, gypsum, per inch	5
4″	13	Concrete, lightweight, per inch	5-10
5″	15.5	**MISCELLANEOUS**	
6″	16.5	Windows, glass, frame	8
Solid gypsum block 2″	9.5	Skylight, glass, frame	12
3″	13	Corrugated asbestos 1/4″	3.5
MASONRY WALLS (per 4″ of thickness)		Glass, plate 1/4″	3.5
		Glass, common	1.5
Brick	40	Plastic sheet 1/4″	1.5
Glass brick	20	Corrugated steel sheet, galv.	
Hollow concrete block	30	—12 ga.	5.5
Hollow slag concrete block	24	—14 ga.	4
Hollow cinder concrete block	20	—16 ga.	3
Hollow haydite block	22	—18 ga.	2.5
Stone, average	55	—20 ga.	2
Bearing hollow clay tile	23	—22 ga.	1.5
		Wood Joists—16″ ctrs. 2 × 12	3.5
		2 × 10	3
		2 × 8	2.5
		Steel plate (per inch of thickness)	40

Table 2

Approximate Weight per Foot of Length of Round Steel Bars and Rods

Diameter (inches)	Weight (lbs.) Per Ft. of Length	Diameter (inches)	Weight (lbs.) Per Ft. of Length
3/16	.094	1-3/8	5.05
1/4	.167	1-1/2	6.01
5/16	.261	1-5/8	7.05
3/8	.376	1 3/4	8.18
7/16	.511	1-7/8	9.39
1/2	.668	2	10.68
9/16	.845	2-1/8	12.06
5/8	1.04	2-1/4	13.52
3/4	1.50	2-3/8	15.06
7/8	2.04	2-1/2	16.69
1	2.67	2-5/8	18.40
1-1/8	3.38	2-3/4	20.20
1-3/16	3.77	2-7/8	22.07
1-1/4	4.17	3	24.03

APPENDIX

K

Center of Gravity and Volume of Odd Shaped Objects

The following formulas are given to determine the center of gravity and volume of odd shaped objects when no specifications are available. Solve V for volume, B for center of gravity.

1. Cylinder

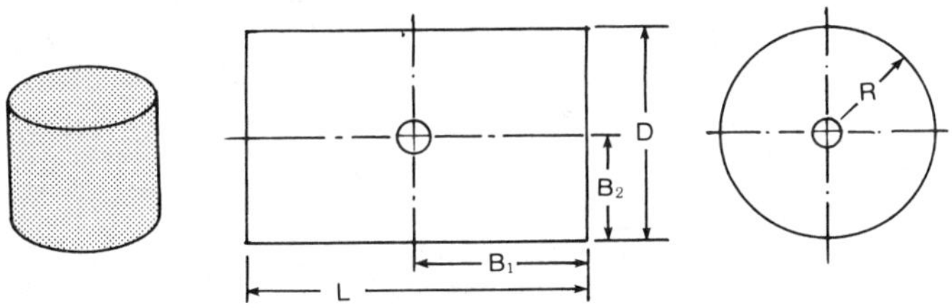

$$V = \frac{\pi}{4} D^2L = 0.7854D^2L$$

$$B_1 = L/2$$

$$B_2 = R$$

2. Half cylinder

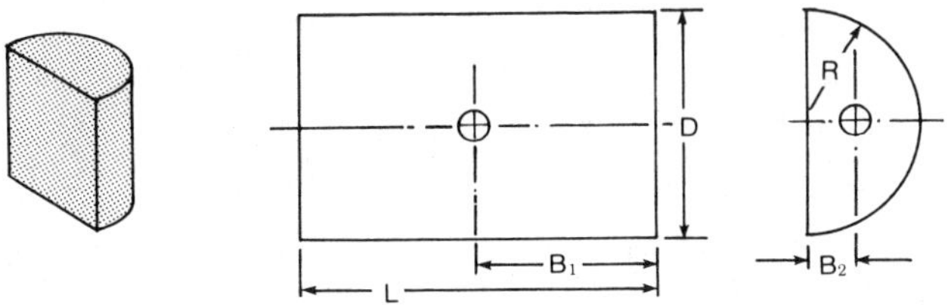

$$V = \frac{\pi}{8} D^2L = 0.3927D^2L$$

$$B_1 = L/2$$

$$B_2 = \frac{4R}{3\pi} = 0.4244R$$

3. Quadrant of cylinder

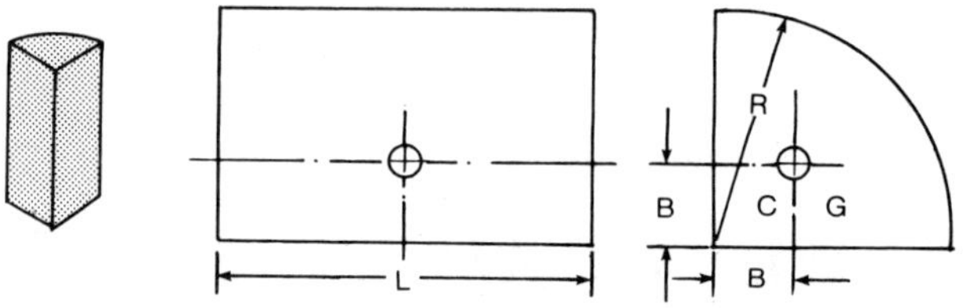

$$V = \frac{\pi}{4} R^2L = 0.7854R^2L \qquad B = \frac{4R}{3\pi} = 0.4244R$$

4. Fillet or spandrel

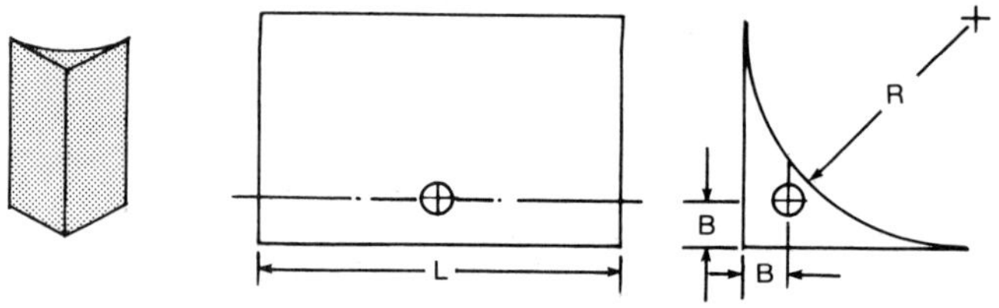

$$V = \left(1 - \frac{\pi}{4}\right) R^2L = 0.2146R^2L$$

$$B = \frac{10 - 3\pi}{12 - 3\pi} R = 0.2234R$$

5. Shell of spherical sector

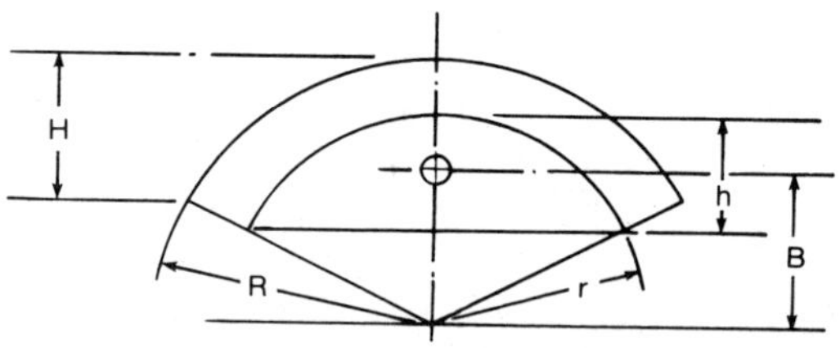

$$V = \frac{2\pi}{3} (R^2H - r^2h)$$

$$B = 0.375 \left\{\frac{[R^2H(2R - H)] - [r^2h\,(2r - h)]}{R^2H - r^2h}\right\}$$

6. Shell of spherical segment

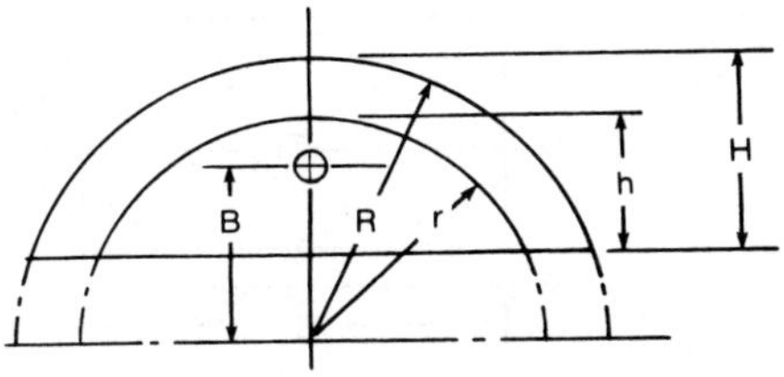

$$V = \pi \left[H^2 \left(R - \frac{H}{3} \right) - h^2 \left(r - \frac{h}{3} \right) \right]$$

$$B = \frac{3}{4} \left[\frac{\left(R - \frac{H}{3} \right) \frac{H^2(2R - H)^2}{3R - H} - \left(r - \frac{h}{3} \right) \frac{h^2(2r - h)^2}{3r - h}}{H^2 \left(R - \frac{H}{3} \right) - h^2 \left(r - \frac{h}{3} \right)} \right]$$

7. Bevel ring

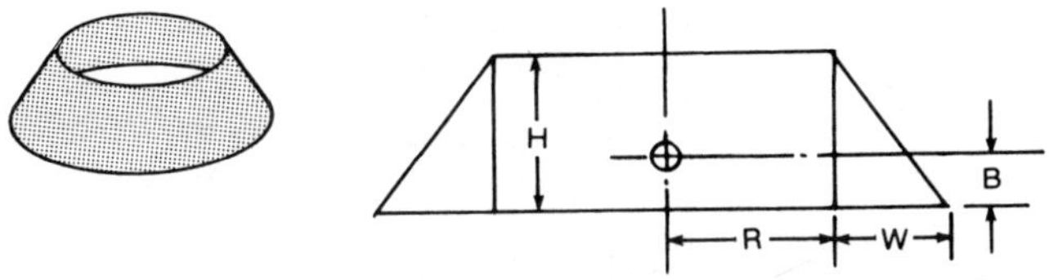

$$V = \pi \left(R + \frac{1}{3} W\right) WH \qquad B = H \left[\frac{\frac{R}{3} + \frac{W}{12}}{R + \frac{W}{3}} \right]$$

8. Hollow cylinder

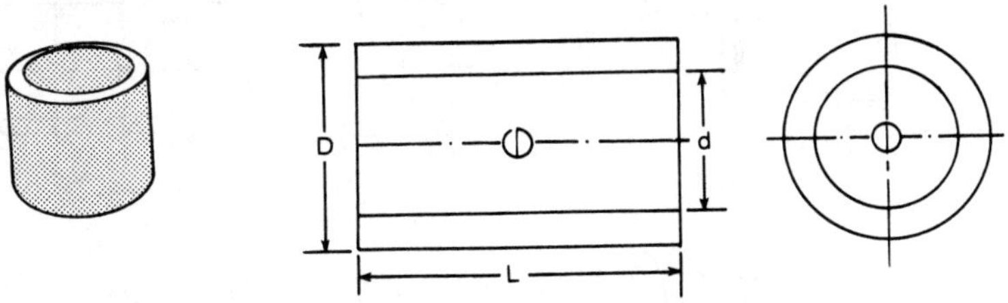

$$V = \frac{\pi L}{4} (D^2 - d^2)$$

CG at center of part

9. Half hollow cylinder

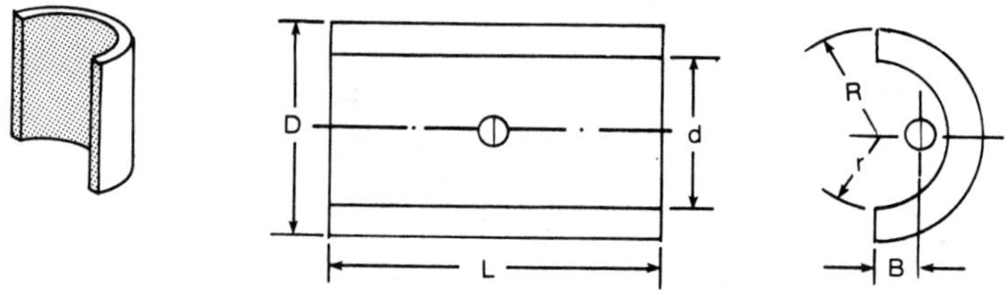

$$V = \frac{\pi L}{8}(D^2 - d^2) \qquad B = \frac{4}{3\pi}\left[\frac{R^3 - r^3}{R^2 - r^2}\right]$$

10. Hollow torus

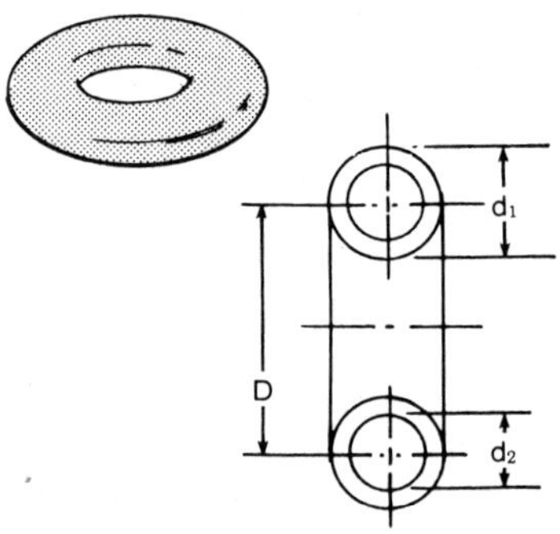

$$V = \frac{1}{4}\pi^2 D(d_1^2 - d_2^2)$$

11. Intersecting hollow cylinders (volume of junction box)

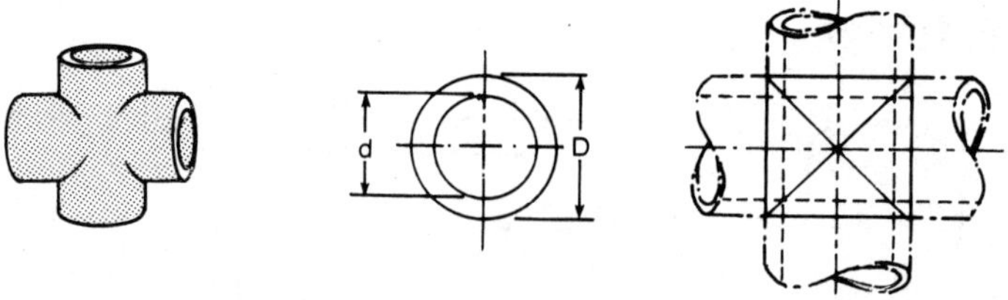

$$V = \left(\frac{\pi}{2} - \frac{2}{3}\right)(D^3 - d^3) - \frac{\pi}{2}d^2(D - d)$$

$$V = 0.9041\,(D^3 - d^3) - 1.5708\,d^2(D - d)$$

12. Frustum of hollow cone

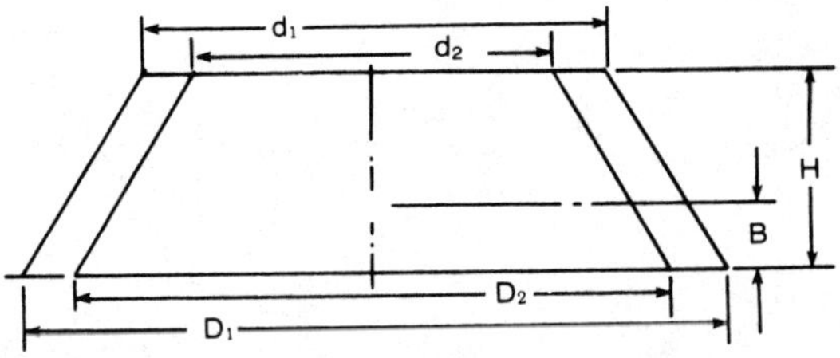

$$V = 0.2618H\,[(D_1^2 + D_1d_1 + d_1^2) - (D_2^2 + D_2d_2 + d_2^2)]$$

13. Hollow sphere

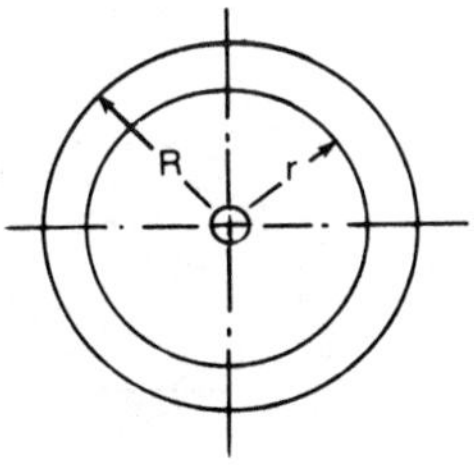

$$V = \frac{4\pi}{3}(R^3 - r^3)$$

14. Frustum of pyramid (with base of any shape)

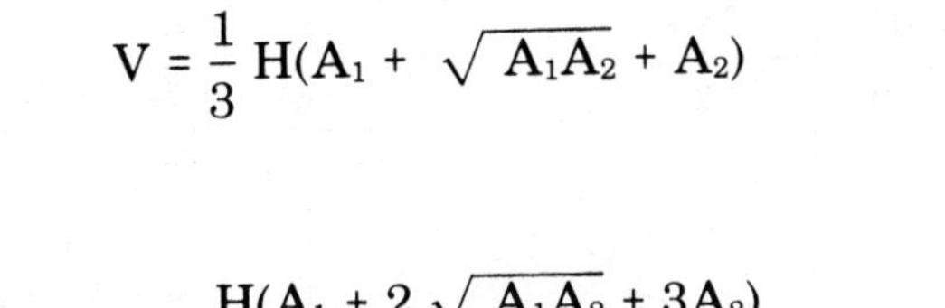

$$V = \frac{1}{3}H(A_1 + \sqrt{A_1A_2} + A_2)$$

$$B = \frac{H(A_1 + 2\sqrt{A_1A_2} + 3A_2)}{4(A_1 + \sqrt{A_1A_2} + A_2)}$$

15. Bevel ring

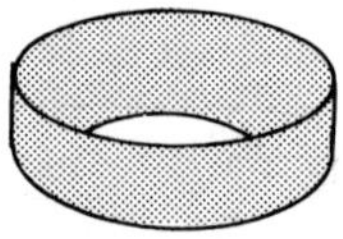

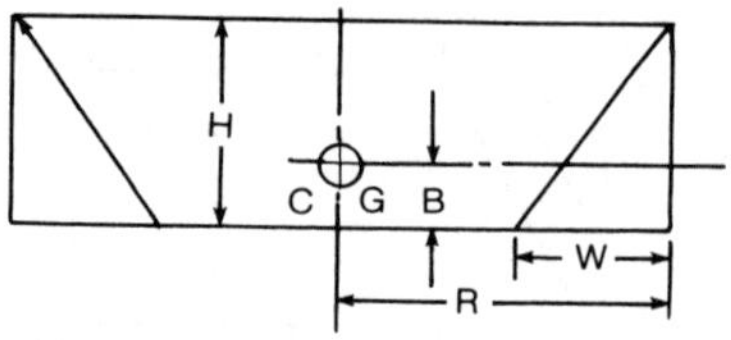

$$B > \frac{H}{3}$$

$$V = \pi \left(R - \frac{1}{3} W\right) WH \qquad B = H \left[\frac{\frac{R}{3} - \frac{W}{12}}{R - \frac{W}{3}}\right]$$

16. Quarter torus

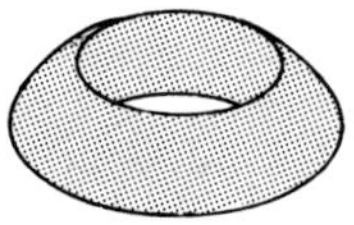

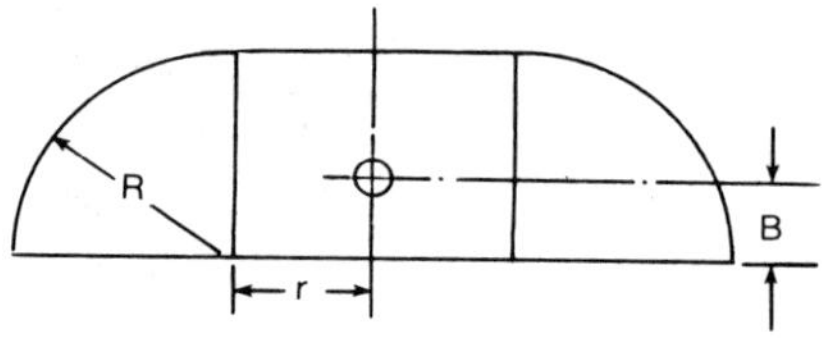

$$B < 0.4244R$$

$$V = \frac{\pi R^2}{2} \left(r + \frac{4R}{3\pi}\right) = 4.9348R^2(r + 0.4244R)$$

$$B = \frac{4R}{3\pi} \left[\frac{r + \frac{3R}{8}}{r + \frac{4R}{3\pi}}\right] = \frac{0.4244Rr + 0.1592R^2}{r + 0.4244R}$$

17. Quarter torus

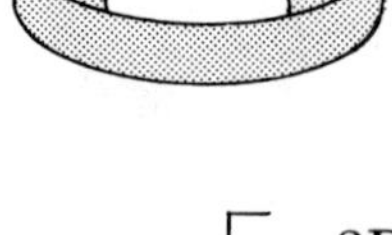

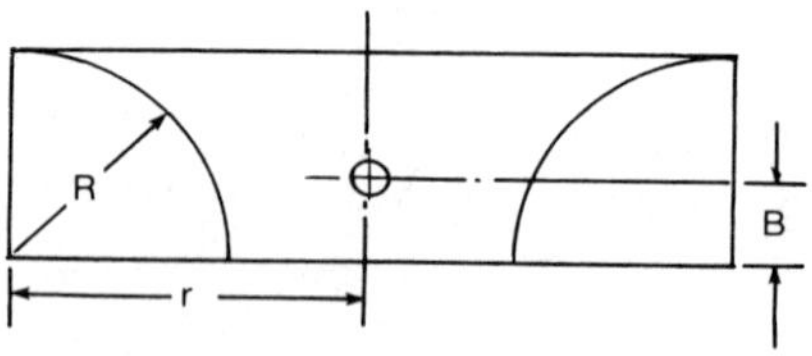

$$B = \frac{4R}{3\pi} \left[\frac{r - \frac{3R}{8}}{r - \frac{4R}{3\pi}}\right] \qquad V = \frac{\pi^2 R^2}{2} \left[r - \frac{4R}{3\pi}\right]$$

18. Curved shell ring

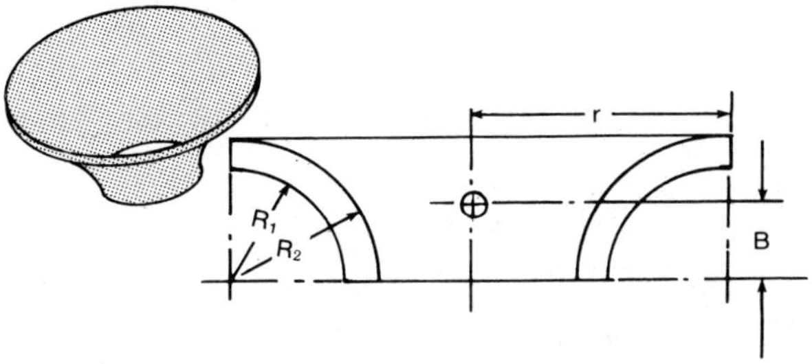

$$V = 2\pi \left\{ r - \frac{4}{3\pi} \left[\frac{R_2^3 - R_1^3}{R_2^2 - R_1^2} \right] \right\} \frac{\pi}{4} (R_2^2 - R_1^2)$$

$$B = \frac{4}{3\pi} \left[\frac{R_2^3 \left(r - \frac{3}{8} R_2 \right) - R_1^3 \left(r - \frac{3}{8} R_1 \right)}{(R_2^2 - R_1^2) \left\{ r - \frac{4}{3\pi} \left[\frac{R_2^3 - R_1^3}{R_2^2 - R_1^2} \right] \right\}} \right]$$

19. Curved shell ring

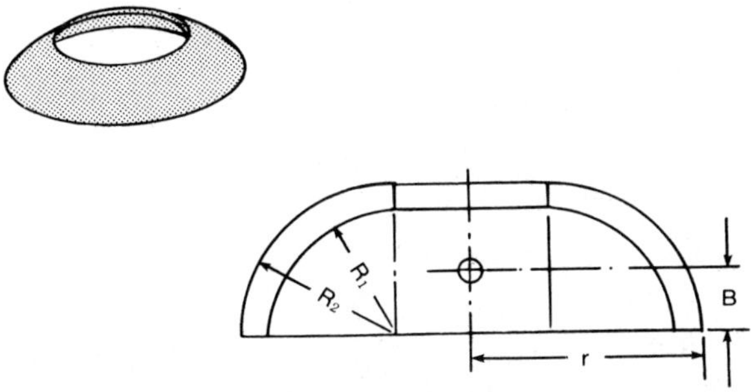

$$V = \frac{\pi^2}{2} \left[r(R_2^2 - R_1^2) + \frac{4}{3\pi} (R_2^3 - R_1^3) \right]$$

$$B = \frac{2}{\pi} \left[\frac{\frac{2r}{3} (R_2^3 - R_1^3) + \frac{1}{4} (R_2^4 - R_1^4)}{r(R_2^2 - R_1^2) + \frac{4}{3\pi} (R_2^3 - R_1^3)} \right]$$

APPENDIX

L

Riggers' Rules of Thumb

It is much easier to learn a simple formula than it is to memorize complex tables and graphs. These "rules of thumb" are reasonably accurate and most are on the "safe side" of recognized safety practices. But, *it must be remembered, these formulas are only close estimates and are not intended to replace manufacturer's specifications in critical situations.*

FIBER LINE

The *lay of rope* can be determined by running your thumb away from you along the strands in the line (if strands go to the right, it is a right lay, etc.).

Coil line in the direction of the lay (right lay, clockwise, etc.).

Uncoil a new bail counterclockwise to avoid kinks.

Natural fiber (Manila)

1. Breaking strength in tons = $\text{Diameter}^2 \times 4$
2. Safe working load in tons = $\dfrac{\text{Diameter}^2 \times 4}{10}$

   (safety factor = 10)

Synthetic Fiber (Nylon, Polypropylene)

1. Nylon = Approximately 2-1/2 times the strength of manila.
2. Polypropylene = Approximately 1-1/2 times the strength of manila.

TACKLE SYSTEMS

1. **Friction** in a tackle system = Approximately 8% per sheave @ 180° bend, 4% per sheave @ 90° bend. (For wire rope tackle, approximately 6% for bronze bushed sheaves and 3% for ball bearing sheaves.)
2. **A Simple Tackle System** = One line and one or more blocks.

Mechanical advantage = Number of lines on traveling block.

3. **A Compound Tackle System** = Two or more simple tackle systems compounded.
Mechanical advantage = The M.A. of each system multiplied in series.

WIRE ROPE

Determine lay, and coil like "Fiber Line"

1. **Lang Lay**—Wire laid up in same direction as strands of rope.
   Regular Lay—Wires laid up in opposite direction as strands of rope.
2. **Wire Core** (IWRC) is approximately 10% stronger than fiber core.
   Galvanizing reduces strength approximately 10%.
3. **Breaking Strength in Tons** = $(\text{Diameter})^2 \times 40$ (For 6 × 19 I.P.S. F.C.)
4. **Safe Working Load in Tons** = $(\text{Diameter})^2 \times 40 \div$ Safety Factor.
5. **Safety Factors** = Slings = 8
   Nonrotating lines = 8
   Cranes and Derricks = 6
   Derrick Guys = 5
   Hoisting Tackle = 5
   Guy Lines = 3.5
6. **Broken Wires** = Unsafe when 4% of total wires in the rope are broken within the length of one rope lay.
7. **Choker Stress** = $\frac{W}{N} \times \frac{L}{V} = T$

   N = Number of Chokers
   W = Weight of load in pounds
   V = Vertical distance in feet
   L = Choker leg length in feet
   T = Tension in pounds
8. **Sheaves**—The critical diameter of a sheave is 20 times the diameter of the rope.
9. **Drums**—For spooling a drum you face the lead block and place the right or left hand, whichever corresponds with the lay of the rope, palm down on the top of the drum for underwind. The thumb will point to the flange where spooling begins.
10. **Drum Capacity**—in feet = $\frac{.2618}{\text{Rope Diameter}^2} \times (A + B) \times A \times C$ (in inches)

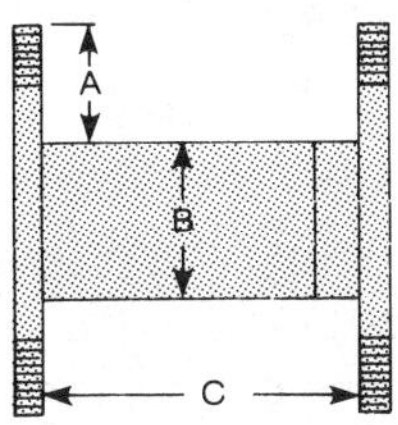

11. **Fleet Angle**—Maximum 4° on grooved drums and 2° on smooth drums. The distance to the lead block should never be less than 10 times the width of grooved drums or 20 times the width of smooth drums.

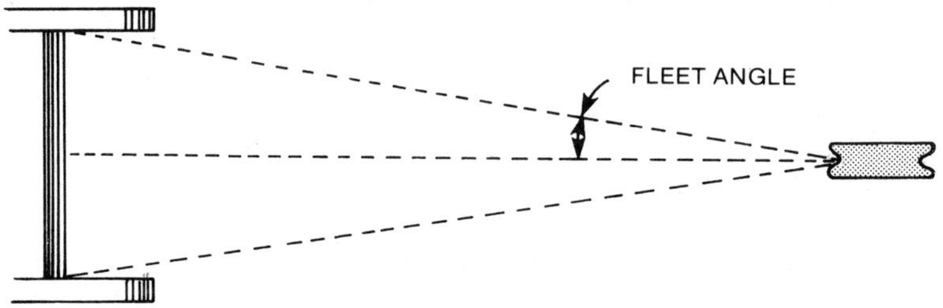

WIRE ROPE ATTACHMENTS

Safety factor of 5 included

1. **Crosby Clips**
   Number of clips required—4 × rope diameter + 1.
   Minimum spacing—6 × rope diameter.
2. **Hooks**—SWL in tons = $(\text{Diameter})^2$ where hook begins to curve or Diameter of eye in inches squared.
3. **Shackles**—SWL in tons = Diameter of pin in one fourth inches squared and divided by three (or select shackle pin one-eighth inch larger than choker size).
4. **Chains** (Alloy Steel)—SWL in tons = chain stock $(\text{Diameter})^2 \times 6$.
5. **Turn Buckles**—SWL in tons = bolt $(\text{Diameter})^2 \times 3$.

ESTIMATING WEIGHT OF STRUCTURAL STEEL

One cubic foot = approximately 490 pounds:

1. **Round Stock**—Approximate weight per lineal foot = $(\text{Diameter})^2 \times \frac{8}{3}$
2. **Plate, Flat Bar, or Square**—Approximate weight per lineal foot = thickness times width $\times \frac{10}{3}$ (All measurements in inches.)

(Angles, Tees, or Wide Flange Beams can be calculated in the same manner by measuring the legs, or web and flange individually as Flat Bar or Plate.)

SCAFFOLD PLANKS

Two inch planks (2 × 8, 2 × 10, or 2 × 12) of fir or spruce in first class condition can be considered safe to support the weight of an average man for a "span in feet" that is equal to, or less than the plank "width in inches."

Metric Equivalents of Standard Rope Sizes

Diameter		Circumference	
Inches	Millimeters	Inches	Millimeters
1/64	.40	3/64	1.2
1/32	.79	3/32	2.5
3/64	1.2	9/64	3.7
1/16	1.6	3/16	5.0
5/64	2.0	1/4	6.2
3/32	2.4	9/32	7.5
7/64	2.8	11/32	8.7
1/8	3.2	3/8	10.0
5/32	4.0	15/32	12.
3/16	4.8	9/16	15.
7/32	5.6	11/16	17.
1/4	6.3	3/4	20.
5/16	7.9	1	25.4
3/8	9.5	1-1/8	28.6
7/16	11.1	1-3/8	34.9
1/2	12.7	1-5/8	41.3
9/16	14.3	1-3/4	44.4
5/8	15.9	2	50.8
11/16	17.5	2-1/4	57.1
3/4	19.0	2-3/8	60.3
13/16	20.6	2-1/2	63.5
7/8	22.2	2-3/4	69.8
15/16	23.8	3	76.2
1	25.4	3-1/8	79.4
1-1/16	27.0	3-3/8	85.7
1-1/8	28.6	3-1/2	88.9
1-3/16	30.2	3-3/4	95.2
1-1/4	31.7	3-7/8	98.4
1-3/8	34.9	4-3/8	111.
1-7/16	36.5	4-1/2	114.
1-1/2	38.1	4-3/4	121.
1-5/8	41.3	5-1/8	130.
1-11/16	42.9	5-1/4	133.
1-3/4	44.4	5-1/2	140.
1-13/16	46.0	5-3/4	146.
1-7/8	47.6	5-7/8	149.
1-15/16	49.2	6-1/8	156.
2	50.8	6-1/4	159.
2-1/16	52.4	6-1/2	165.
2-1/8	54.0	6-5/8	168.
2-1/4	57.1	7-1/8	181.
2-1/2	63.5	7-7/8	200.

Metric Equivalents of Standard Rope Sizes (Cont.)

Diameter		Circumference	
Inches	Millimeters	Inches	Millimeters
2-3/4	69.9	8-5/8	219.
3	76.2	9-3/8	238.
3-1/4	82.5	10-1/4	260.
3-1/2	88.9	11	270.

Note: U. S. sizes designated by diameters. Circumferences shown here for comparison only.

Index